TRANSACTIONS

OF THE

AMERICAN PHILOSOPHICAL SOCIETY

HELD AT PHILADELPHIA

FOR PROMOTING USEFUL KNOWLEDGE

NEW SERIES—VOLUME XXV
APRIL, 1935

———

REPORT

OF THE

PENNSYLVANIA LOCAL GOVERNMENT SURVEY

THE COUNTY . . . . H. F. Alderfer
THE SECOND CLASS TOWNSHIP . . Edward W. Carter
THE FIRST CLASS TOWNSHIP . . Philip B. Willauer
THE BOROUGH . . . . Bradford W. West
THE THIRD CLASS CITY . . . . Blake E. Nicholson

———

PHILADELPHIA:
THE AMERICAN PHILOSOPHICAL SOCIETY
104 South Fifth Street
1935

PREFACE

The need for an intensive survey of local government in Pennsylvania has long been the subject of comment. An extraordinary number of proposals for readjustment or for complete reorganization of the local system have been made. The first step towards any informed judgment on these is the gathering of authentic data concerning the present system, its practical operation, its strength and weaknesses. With this end in view the American Philosophical Society has generously provided a grant of funds making possible the present Survey. The local units here described are the county, the second class township, the first class township, the borough and the third class city. These cover all the rural and minor urban local governments except those administering schools and welfare. A survey of school districts was made by the State Department of Public Instruction in 1930 and a report on welfare districts has been completed in 1934 by the State Department of Welfare with the aid of Federal funds. These two units are therefore omitted from the present report.

The method followed has been to prepare as to each unit:

a) A condensed summary of the applicable provisions of the State Constitution and statutes. Such brief historical data as space permitted have been woven into this summary.

b) A collection of material from official records in the State offices at Harrisburg, covering all examples of a particular type of unit throughout the State, e.g., reports on all second-class townships in the State Highways Department.

c) An analysis of those problems which are considered important by the various associations of local officers. These bodies are not merely protective but also devote their proceedings to the improvement of local government. Their views have been considered worthy of careful study.

d) An intensive field visitation by members of the Survey Staff lasting nearly three months, of many selected samples of each type of unit, during which all available data on important problems were sought at the source. While this part of the work was extremely arduous it is believed to have yielded compensatory results. Such novelty as the Survey may claim is to be found chiefly in the methods followed and the results secured in c) and d) above mentioned. These methods have disclosed a system of local government which in practice departs from its legal and theoretical basis.

As a result of these four general methods of approach there have been secured, on each unit, a mass of descriptive material, a series of conclusions or findings of fact and a set of recommendations.

Among the outstanding questions which here presented themselves were the following:

Is the SIZE of the unit, in area and population, suitable to the work which it performs?

Is its INTERNAL STRUCTURE or ORGANIZATION adapted to efficient results?

Is its RELATION TO OTHER LOCAL UNITS satisfactory?

Is its RELATION TO THE STATE satisfactory?

Is its PROCEDURE or OPERATION modern in method?

Is its FINANCIAL COST fairly proportionate to its results?

All these lead to the more comprehensive question - Can the system be so organized as to operate more effectively or at less cost?

The distribution of work among members of the Survey staff has been as follows:

The County - H. F. Alderfer, Pennsylvania State College
The Second Class Township - Edward W. Carter, University of Pennsylvania
The First Class Township - Philip B. Willauer, Ursinus College
The Borough - Bradford W. West, University of Pennsylvania
The Third Class City - Blake E. Nicholson, University of Pennsylvania.

The individual parts of the report are not uniform since the work of the various units is different in nature and scope. Each unit presents special unique problems which require an appropriate specialized treatment. A large amount of valuable and important data has been collected, particularly on the Borough and Third Class City, which could not be included without exceeding the appropriation for printing. It has been found possible, however, to give a fairly unified set of findings and recommendations from the viewpoint of the Survey as a whole. These have been summarized for convenient reference in the first Chapter, while more detailed conclusions are given at the close of each of the following Chapters.

The Survey staff members wish to express their deepest gratitude to the American Philosophical Society for its generous financial grants, making possible both the preparation and the publication of this report. The University of Pennsylvania by a liberal appropriation from its Research Fund has provided invaluable technical assistance in the preparation of statistical tables and has furnished office facilities.

Financial help has also been thankfully received from Mr. William B. Levis of Philadelphia. The Pennsylvania Economic Council and its Executive Secretary, Mr. Robert D. Dripps, together with members of its staff, especially Mr. David Kurtzman, Miss Elma Greenwood and Mr. Frank Wallas, have opened wide its invaluable facilities and stores of accumulated data and have given frequent and highly appreciated assistance and advice on many problems. The important work already done by the Council on costs and revenues of local units has been liberally offered and freely used and quoted. Special acknowledgment is also due to the following state and local officials and public men who have given of their time and help: Hon. Roland S. Morris, Hon. H. Edgar Barnes, Secretary of Revenue, Hon. Philip Sterling, Hon. Seabury C. Mastick, Prof. Luther Gulick of the New York State Commission for the Revision of the Tax Laws, Bernard G. Segal, former Deputy Attorney General, John H. Fertig, Director Legislative Reference Bureau, Henry W. Van Pelt, Director, Bureau of Statistics, Department of Internal Affairs, J. Herman Knisely, Director, Bureau of Municipalities, Department of Internal Affairs, Hon. Samuel S. Lewis, former Secretary of Highways, Harold Van Riper, Township Engineer, Department of Highways, H. A. Thomson, Secretary, and Garfield Bagshaw, President, State Association of Township Supervisors, Harry A. Fritschman, Secretary-Treasurer, and John J. Riordan, Vice President, State Association of Township Commissioners, Thomas Stephenson, Editor, "The Pennsylvania Road Builder", Dr. Leonard P. Fox, Research Bureau, Pennsylvania State Chamber of Commerce, Prof. Charles J. Rowland, Pennsylvania State College, Thomas F. Chrostwaite and L. W. Hagmaier of the State Association of Boroughs, Walter Greenwood, President of the League of Cities of the Third Class, Arthur Dunham, of the Public Charities Association of Pennsylvania. The staff members also express their cordial appreciation and thanks to the many local officials and employes who have patiently and cheerfully answered questions and given source information on the day-to-day operation of their local units.

H. F. Alderfer

Edward W. Carter

Blake E. Nicholson

Bradford W. West

P. B. Willauer

TABLE OF CONTENTS

		Page
Chapter 1	Summary of Major Conclusions	1
	Part I The County	
Chapter II	Internal Organization	11
Chapter III	Personnel	35
Chapter IV	Assessments	53
Chapter V	Election Administration	69
Chapter VI	Justices of the Peace	87
Chapter VII	Records	115
Chapter VIII	Costs and Revenues	127
	Part II The Second Class Township	
Chapter IX	Origins and Legal Basis	139
Chapter X	Diversity of Types and Population	145
Chapter XI	Administration	153
Chapter XII	Fiscal Problems	167
Chapter XIII	Fiscal Problems Continued	185
Chapter XIV	Roads and Bridges	211
Chapter XV	Debt, Consolidation, State Supervision	233
	Part III The First Class Township	
Chapter XVI	Creation and Corporate Powers	243
Chapter XVII	Internal Organization	255
Chapter XVIII	Administrative Methods and Procedure	265
Chapter XIX	Accounting Procedure	277
Chapter XX	Budgetary Procedure	289
Chapter XXI	Auditing Procedure	305
Chapter XXII	Revenues	315
Chapter XXIII	Functional Costs	335
	Part IV The Borough	
Chapter XXIV	The Borough	357
	Part V The Third Class City	
Chapter XXV	The Third Class City	403
Index of Tables and Charts		465
General Index		469

CHAPTER I

SUMMARY OF MAJOR CONCLUSIONS

Spirit and Scope of Survey Recommendations - In order to facilitate the use of this Report a summary of its major conclusions is here presented. Additional, more detailed recommendations are given at the end of each of the following chapters. At the outset it should be said that the Survey has not been undertaken in a spirit of criticism. After hundreds of personal interviews in all parts of the state it is the unanimous opinion of the Survey staff members that the local government officials of Pennsylvania, taken as a whole, are a devoted, loyal, actively interested body of men and that the same persons, if given a modernized local system with adequate facilities, supervision and technical advice, would produce results of incomparably greater value, and at less cost.

As to changes in organization, some of the chief proposals here voiced have been strongly urged by local officials themselves. A few others have already been found advantageous in other states or have been advanced by survey organizations therein. Having been undertaken in this spirit and conducted by direct observation in the field, the findings of the present Survey do not support types of proposals such as the total abolition of important local units and the centralization of their powers in the state government. This change might have been suggested from a purely library study of the problem and has indeed been so recommended and undertaken in other places. Nor does the Report contemplate the drastic, compulsory reorganization of the entire local system regardless of local wishes or the counsel of experienced local officials. Certain important changes of broad scope are here urged but it is recommended that they be undertaken after consultation with local officials and that their aim be to enlarge, strengthen and assist local units rather than to abolish them. The political wisdom of many centuries teaches that a vigorous, active local system is the foundation of strong state or national governments. Also that it is an indispensable means of training the citizenry in their public duties and rights. Approaching the problem from this viewpoint it is therefore proposed, not to solve all problems by state centralization of local functions, but rather to assure to each local unit a sufficient size, both in population and in taxable resources, to make its successful operation possible.

On the other hand strong central supervision has been unhesitatingly recommended in certain budgetary, accounting, recording and reporting matters, in order to assure reasonable standards of good business practice. Here the well established principle has been recognized that the substance and content of local financial administration should as far as possible be left to local decision but that the manner, form and general limits of such transactions should be prescribed by the state.

The major Survey recommendations are grouped under the following heads:

    I.   Internal organization of local units
   II. Their personnel
 III. Their Fiscal problems
  IV. Special recommendations on particular functions
   V. Relations of local units to the state and to each other
  VI. Special recommendations on each unit
 VII. Summary of constitutional changes

I. Internal Organization

Nature of Changes Recommended - In all the local units examined it was found that the adoption of three well established principles would greatly strengthen the system and enable it to operate at a lower cost. These principles are (a) reduction in the number of elective offices; (b) a simplified or functional organization of departments, i.e., the grouping of similar functions or duties in the same department; (c) the grouping of departments and offices under one executive head to assure concentration of responsibility. In all important organizations of either the business or governmental type these principles are now receiving general recognition.

A. Reduction of Elective, Increase of Appointive Offices - The qualifications and experience needed for local government work are not always best secured by election. This is especially true of officers whose duties are not policy determination but administration, that is, the carrying out of the details of policy. The essence of democracy is not in the popular vote on every minor office but rather in the popular choice of those responsible officials at the top, who determine general policy. This is especially applicable to counties, where the people elect from twelve to eighteen minor row officers every four years. In like manner, though to a less extent, it applies to all the other local units.

It is recommended that in the county, only the commissioners and the controller be elected and that they should in turn appoint the other county officers, or should choose a county manager who would have this responsibility. In the third class city the treasurer and controller should be appointed. In the borough and the township the tax assessor, the tax collector, the justice of the peace and the constable should be appointed by respective county authorities. This general change would enable the voters to concentrate attention upon a few important elective offices.

B. _Functional Organization_ - In carrying out the second principle it is recommended that wherever possible those local offices which have to do with the same general functions be grouped together in one department, also that this department be sufficiently broad in scope and organization to permit the employment of at least one full-time employe. In the county and in some of the other units this would involve a rearrangement of functions and a consolidation of offices. The following county organization, for example, would result, - department of records, department of treasury, department of the chief court officer, department of the district attorney, department of corrections and charities, and department of public works. In the smaller counties some of these could be further consolidated or, where not needed as separate independent departments, their functions could revert to the county commissioners. Similar changes could be made in the township, the borough and the third class city. This reorganization of administrative work upon a simplified, "functional" basis would undoubtedly remove many of the fundamental weaknesses of our present system, and would make local government more visible as well as more efficient.

C. _Responsibility_ - The third principle, responsibility, which it is proposed to incorporate, is probably the most urgently needed of all recommended changes. The reduction in elective offices would create a definite responsibility in the policy-determining authority in each local unit but the same step is also needed in the administrative offices which carry out, rather than determine, policy. The recommendation here is that the entire administrative branch in each unit be placed under the direction of one chief executive. This executive should not be a popularly elected officer but should be chosen by the policy-determining or legislative group. He should be a trained executive and should hold his position on the basis of efficiency.

Such a change, it is recognized, could not be made at once in all units. The smaller areas with less work and with few employes do not require such concentration nor could they afford it.

In the counties it is proposed that this strengthening of the organization be brought about by making the county manager, where one is established, the head of the proposed new Department of Records and vesting in him the present duties of the chief clerk, who is now in most counties the nearest approach to a general county executive. As head of the records department he would also have the duties and powers now delegated to the prothonotary, the clerk of the courts, the recorder of deeds, the register of wills and the chief clerk of the county commissioners. In the boroughs, townships and third class cities the electors should be permitted to vote on the question of adopting the manager plan. Experimentation with this plan in Pennsylvania has been confined largely to boroughs which have adopted it by act of council. Its successful operation in some thirty boroughs augurs well for its further adoption by popular vote.

II. Personnel

Training, Guidance and Supervision - The data gathered seem to leave no room for doubt that the present local personnel could be greatly strengthened, the standards of work raised and a substantial saving effected by some method of training and supervision. Three methods of doing this have been sufficiently tried out elsewhere to warrant their adoption in Pennsylvania: a) Handbooks of instruction on the duties and procedure of the respective offices, prepared by a competent state authority, would be of the utmost value, particularly in the fiscal services. b) Short courses of instruction for local officers should be given by a central state agency. This might well be extended to include correspondence work. Throughout the field-work of the Survey it was found that local officers were everywhere interested in the methods used by other units of the same type. A "clearing house" of the best practices would speedily lead to a substantial rise in standards of operation. c) The appointment of traveling field inspectors under the direction of a state agency, particularly in fiscal matters, would give a direct contact between the local units and the state, of a more personal, intimate and detailed nature. The cost would be slight as compared with the resulting savings and the gains in efficiency.

The Merit System - In every section of the Report emphasis has been placed upon the need for retention in office of capable local officials. The several authors of this Report separately reached the same conclusion, viz., that a merit system should be introduced into local government administration in Pennsylvania. Such a system does not mean a civil service in which public officers remain in office for life regardless of ability. A merit system could select administrative officers on a basis of examination, training and experience. The selection could be made applicable to the larger units at first and gradually extended if the method meets with approval. Its application should be confined to full time employes and entrance to administrative and clerical positions should be limited to appointees from the highest names on a list of eligibles created by examination. Where the merit system is recommended in the various parts of the Survey, the principles just described are intended rather than the system of permanent tenure now used by the federal government. It is believed that a modified system is more appropriate to local government conditions in Pennsylvania and that the power to discharge employes in the various offices should be retained. This power would not be subject to frequent abuse as long as the law required appointments to be made from a list of eligibles based on examination.

Similarly in the larger units where the amount of work to be done and the number of employes are sufficient to warrant it, the service of full-time employes should be used. This seems particularly desirable in the case of assessors, tax collectors, justices of the peace, constables, heads of county departments and other offices in the county. Even in the smaller local units where several part-time employes are engaged in the same general function, a smaller number of full time workers could give better results at less cost. If the unit is so small that the number of employes is not sufficient to make this change, there may be some question as to the advisability of maintaining a subdivision of such size.

III. Fiscal Problems

Nature of Fiscal Problems - The more important questions arising in this field are:

        A. Assessment
        B. Tax Levy
        C. Tax Collection
        D. The Budget
        E. Debt
        F. Accounting, Recording and Reporting
        G. Auditing.

A. Assessment - It is recommended that the county be made the unit of assessment for all local taxes. The county assessment should be utilized by all local subdivisions and the separate assessment by cities should be abandoned. In charge of this county assessment should be either the board of assessment and revision of taxes, which now exists in second and third class counties, or the chief clerk of the county (or county manager), or a subordinate of the chief clerk or manager. This would make possible the employment of a full time man experienced and trained in assessment methods. It would also lead to uniformity of assessments within the county and would make possible the adoption of a scientific method which is now so much needed throughout the state. Principles of property assessment are now well developed and are widely used in other states. The present organization of assessment methods in Pennsylvania does not permit the use of scientific assessment principles except in second and third class counties. The actual assessors, operating in the field, should be appointed by the county assessing authority and should be as far as possible full time officers paid on a salary basis. Assessment records should be kept on modern office forms. This reorganization would supplant the present elective method of choosing assessors and would transfer the assessment function from the township, the borough or the ward to the county. It would establish greater equity between local taxpayers and would remove the causes of much of the present complaint and sense of injustice.

B. Tax Levy - a) State statutes placing absolute, inflexible limits on the tax levy of the local subdivisions are not recommended. Such an inflexible limit would be either too high to offer any genuine safeguard for the taxpayer, or if low, would straitjacket those subdivisions whose particular circumstances of debt, etc., may temporarily require a higher levy. The underlying purpose of all proposals for tax limits is to control local extravagance. This is desirable. It is recommended that any statutory limitation enacted in the future shall be accompanied by provision for a state administrative agency authorized to permit local authorities to exceed the usual limits when exceptional conditions of debt, emergency or special need necessitate such action.

b) The field observations of the Survey disclosed that the tax levy in many units is now made without suitable provision for public hearings and that quite generally it is not based upon a careful examination of previous experience and present needs. Notably is this true in the all-too-frequent ignoring of large accumulations of tax delinquency. It is therefore recommended that the tax levy should be finally decided upon only after the preparation and examination of an adequate budget; also that full allowance be made for public hearings on the proposed levy before it is officially adopted. It should be the policy in both state statute and local practice not only to permit but to encourage local tax associations and other civic groups to examine all aspects of local financial problems and express their views thereon.

C. Tax Collection - Two important changes in the present system of collecting taxes are recommended. 1) All local taxes both current and delinquent should be collected by the county treasurer and his assistants. 2) The present method of selecting local collectors by popular election should be abandoned. The present system of collection is seriously defective and weak in that it entrusts this financial function of the highest importance to a disconnected organization of elected local officers. Definite responsibility and authority are thus divided. The concentration of this responsibility in the office of the county treasurer would immediately aid both the effectiveness of the work and the reduction of its cost.

There is no sound reason why this work should be performed by independent elected officers. There is or should be nothing of a political nature in the collection of the sums owing to local units. Discrimination, favoritism and partisan considerations should have no part in tax payments. Nor is the collection of taxes a function which requires the people to vote on party considerations. Likewise the fee system of paying collectors has amply proven its weakness and even its danger. The county treasurer and his assistants should be paid on a salary basis and should sit for limited periods in different parts of the county to facilitate the convenient collection in each local area.

Under the county treasurer it will also be possible to develop forms and procedure for efficient collection, upon standards uniform throughout the state and prescribed by a state agency.

D. Budgeting - A system of budgeting revenues and expenditures under a comprehensive plan should be adopted and enforced in every local subdivision. It is recommended that the statutes should establish a state agency with authority to prescribe budget forms for all the various local units, that these forms should be so drafted as to fit the different conditions existing in different subdivisions and that such state agency should be fully empowered to require that local budgets be filed within fixed time limits in the office of the state agency itself. The budget forms adopted, while allowing for local conditions, should scrupulously require such detailed statements of income and expenditure and other data as will fully inform the officers concerned, the state authorities and the people of the local units. Wherever possible the budget should be prepared under the direction of the chief executive officer of the local unit and adopted by the policy-determining council or authority.

E. Debt - Approximately one-third of the expenditures made for local government are for the payment of principal and interest on bonded debt. This is a major drain on the local treasury. The constitutional debt limits which now permit debts to be incurred up to twenty-eight per cent or more of the valuation of taxable property in a political subdivision of the state should be materially reduced. In addition, there should be statutory limits considerably lower than the limits now fixed by the Constitution. A state agency should be created with power to approve the issue of bonds in excess of the statutory limits (but within the constitutional limits) whenever the local authorities could justify such action. All long term bonds should be of the serial type and, as a rule, should be callable at fixed intervals at the option of the authorities issuing them.

F. Accounting, Recording and Reporting - The one central purpose of the accounting and reporting systems is to give the officials and the voters of a local unit the necessary accurate, up-to-date information on the fiscal operations of the unit in order that intelligent decisions may be made regarding not only its future but also its day-to-day transactions. Without such accurate information it is impossible to plan either administrative or financial policies. A second purpose is to enable the people and officers of the local unit to observe how other similar units are conducting their affairs and to profit from this knowledge. A third purpose is to keep the central authorities of the state informed in order that they may aid, supervise and legislate intelligently. Our present accounting and reporting systems fall far short of all three of these objectives. In doing so they make impossible any effective budgeting or auditing of local transactions since an adequate accounting system is the first essential to either budget planning or auditing.

It is therefore recommended that uniform methods of accounting and recording, adapted to the conditions of the different types of local units, be prescribed by a central state agency and that they be made mandatory upon local officers.

Another step needed in strengthening local accounting methods is the development of a better trained personnel. To this end it is recommended that the proposed central state agency in charge of local affairs should make provision for regular instruction of local officers and employes in the requisite accounting methods. A limited time, if only a few days each year, devoted to this purpose, would yield highly satisfactory results. The larger universities and colleges of the Commonwealth could readily develop brief courses of instruction either on their campuses or by extension work. The instruction force, the necessary knowledge of local accounts and the physical facilities are already available. All that is required is a small fund to cover the expenses of class work for limited periods.

In accounting, recording and reporting an immediate and noteworthy improvement could be secured by some adequate measure of state supervision. This has been shown in the report form now provided by the state highway department for second class townships. A broad extension of this practice to authorize a central state agency to prescribe not only local accounts but also the details of reports of local units to the state and to require the submission of such reports would make it possible for the officers and taxpayers of every local area to compare intelligently the results of their local government with those of other areas. It is difficult to imagine a more stimulating, tonic influence upon local government than would be afforded by this comparison.

The recommendations on improvement of county records, though specialized, are also believed to be of major importance. The county offices record both public and private transactions of intimate bearing upon all the people of the county. The forms used, the methods of keeping them, the fees charged and the administrative costs incurred are all so highly unsatisfactory as to call for a complete revision. It is recommended that a temporary state commission composed of members of the legislature, county recording officials, historians, archivists and administrative experts be appointed to redraft the entire plan of county recording and to submit to the legislature the detailed provisions for a new system.

G. **Auditing** - Upon completion of the improvements in method thus far proposed, one final step would be required to give the people of the local units that modernized, effective and reasonably economical system of financial administration to which they are entitled, viz., a real audit of local finances which would be worthy of the name. In order to do this, four steps are necessary:

1. The elective method for the choice of auditing personnel should be supplanted by appointment.
2. The part-time lay auditor should give way to the full-time trained comptroller. If the unit is not large enough to support such an officer, a periodic audit should be entrusted to competent persons selected from a list of qualified accountants to be maintained by a county or state agency.
3. Even where comptrollers exist there should be occasional independent audits of all county accounts.
4. The same benefits might be realized here from state supervision and instruction as in the tax assessment personnel already described. The state through a central agency should provide handbooks, field inspection of accounts and even short courses of correspondence instruction. That such methods are of the utmost value to all local units has been amply demonstrated by the experience of New York where a competent staff of field agents has immensely improved local accounting and auditing.

IV. Special Recommendations for Particular Functions

A. **Elections** - There are now nearly 8,000 election districts in the state. This is an unnecessary burden upon the taxpayers and is not required for the effective management of elections. It is recommended that the districts be enlarged and the number greatly reduced. The power to change election districts under present law is in the hands of the county court. It is recommended that the statutes be amended to provide for a maximum and minimum number of registered voters in each district and to authorize the county commissioners to reduce the number of districts within these limits. The commissioners should also be empowered and required to use modern business methods in keeping their election records and in purchasing election supplies. These changes would greatly reduce the cost of elections.

If the experiment in permanent registration which is now being conducted in second class cities proves successful in the elimination of frauds and the reduction of expense, it should be extended to other local areas wherever desired.

B. **Administration of Local Justice** - The present methods of dispensing local justice were found to be almost universally condemned. The causes lie chiefly in the method of selection of the local officials concerned, their payment, their lack of legal training and the sinister influences which have grown up under these conditions. It is recommended that the justices of the peace be appointed by the judges of the county court, that the number of justices be substantially reduced and be determined by the actual needs of the locality not by the mere accident of local boundaries, that the justices be learned in the law and be stationed at such points in the county as require the services of full time or at least half time officers, finally, that they be paid by salary rather than by fees. These changes would remove most of the widespread dissatisfaction with the present system.

There should be a similar revision of the constable's office. Constables should be appointed by the judges of the county court or by the chief court officer proposed in a previous section. They should be paid not by fees but by salaries. The chief court officer should replace the sheriff as the administrative agent of the county court. He should have charge of the details of court administration and should utilize county detectives for the service of writs and such other administrative duties as have heretofore devolved upon the sheriff and his deputies. The other duties of the sheriff should be assigned to other county officers such as the head of the proposed department of charities and corrections.

C. __Welfare__ - It is recommended that the various welfare activities of the county be consolidated in a department of charities and correction under a single appointed executive head. This department acting in close cooperation with the state department of welfare should be the center of all relief, correctional and other related activity. The administration of local public charitable institutions, of the jail and other correctional agencies should be under the direction of a consolidated welfare service and budgeted through this department. Where the county is too small to support such an administration, the union of two or more counties for this purpose should be permitted. It is vital that such a department should make use of trained personnel with a non-political tenure of employment during good service.

D. __Local Highways__ - In other states the small area, limited tax resources and crippled financial condition of local units have led to highly centralized plans of state road control. This is not recommended for Pennsylvania. Instead it is urged that the local road functions be maintained, particularly in the second class townships, with state aid and supervision and that such second class townships be reduced in number. Centralized state road control has been shown to possess grave political dangers in Pennsylvania. The present financial condition of the state also makes it doubtful whether the remaining local roads should be taken over by the Commonwealth. An enlarged and strengthened local administration under state guidance would perform this function satisfactorily and at a lower cost than at present.

V. Relation of Local Units to the State and to Each Other

State Supervision and State Survey - No one feature of local government has become so conspicuous during recent years as the dependence of local units upon outside assistance for the discharge of their functions. Not only has there been a marked change in public opinion as to which functions are local and which are state in character, but the local units have frankly admitted their financial inability to meet the burdens of ordinary local duties. The services expected of these units require cooperative action between all parts of the government, central and local. It is difficult to find a single government function which can be said to be exclusively within the purview of the one unit. This is especially noticeable in the administration of criminal justice, the assessment of property for tax purposes, highway maintenance, the administration of health, welfare and elections.

When this dependence upon outside aid and cooperation reaches a certain point, it is difficult to secure such cooperation by purely voluntary action. This raises the fundamental question - How far can our local system be preserved while at the same time giving the standards of service which our people have come to expect? Two answers are here proposed:

First, that a strong supervisory state agency be created, well equipped with expert service and a limited but efficient corps of trained administrators and field agents, to assist, guide and supervise certain operations of the local units.

Second, that a temporary state survey commission be created to report to the legislature on the reclassification and reorganization of the local system as a whole on lines of greater permanence, effectiveness and economy.

First. A state bureau of local government might well be founded upon the present Bureau of Municipal Affairs in the State Department of Internal Affairs. Regardless, however, of the exact form or location of such a bureau it should be furnished with supervisory powers and personnel sufficient to make it a strong and stimulating force in local administration. In such affairs as taxation, debt issues, budgeting, accounting, reporting, recording and auditing, the bureau could render invaluable assistance, advice and supervision. It would achieve a better integration of the efforts of various departments of the state government, such as health and welfare, in local areas. At present each state department acts as a separate unit locally and much duplication of effort and personnel is involved. The chief need for state supervision lies not in uniformity of actual service rendered but in a more general use of effective methods and procedure. Working under such a principle, local governments would retain their power to determine what they would do although they would accept the state's dictum as to how they would do it. The basic powers of local units would thus be preserved and they would benefit by the guidance of a central authority.

Second. A temporary state survey commission composed of members of the legislature, state and local officials, representatives of business, agriculture and labor, and qualified consultants could render a service of the highest value. Such a body acting as an arm of the legislature could collect important data which are otherwise inaccessible and could render authoritative recommendations on such difficult problems as the reclassification of local units and their equipment with the powers suited to their actual needs and financial resources, the annexation and consolidation of such units, their reduction in number far below the present unnecessary and highly extravagant figure (over 5000), the coordination of the different local services of health, schools, welfare, taxation, highways, etc., into a single or at most a duplicate organization and the eventual rearrangement of the relations of all local units to the state. By rearranging the local fiscal year and the periods of tax assessment and tax collection it could remove the need for short term loans in anticipation of tax collections and thereby save all the local units large sums in interest charges.

Such a commission sitting between the regular sessions of the legislature could draft a more comprehensive program and with greater weight of legislative authority than has hitherto been possible.

VI. Special Recommendations on Local Units

A. The County - Compulsory consolidation of existing counties is not recommended. Instead it is proposed that voluntary cooperative effort between counties be promoted by offering incentives, financial and otherwise. There is also a valuable opportunity for a state agency such as that proposed above, to examine and present to the people of particular counties the savings and the increases in effective service which such voluntary cooperation would make possible. Two noteworthy examples of this are seen in the possible union of small contiguous counties to employ the same district attorney or chief court officer, also in the possible joint employment of a chief assessor with trained assistants. While there are strong political reasons for dividing up the services and the personnel of the county into small jobs and using these to nourish party adherents, there is on the other hand an overwhelming need for better service and especially for financial savings in the unreasonably high cost of local government. The alternative which is here offered to the taxpayer is a choice between the disintegration of the public service in order to furnish political parties with patronage at a ruinous cost or the determined policy of higher standards and lower costs. It can hardly be doubted which policy the taxpayers will choose when they become conscious of the facts. It is fruitless to make the present complaints against a heavy tax burden while at the same time maintaining the system which aggravates this burden.

B. The Second Class Township - It is recommended 1) that the administration of the chief function of this unit, local highways, be placed in the care of a permanent qualified roadmaster; 2) that alternative forms of township government be provided and that the electors in each county be given the opportunity to decide whether they wish the township form to continue or a county unit to be substituted. 3) that the number of second class townships be reduced to not more than 500 in the entire state, with sufficient population and taxable resources to enable them to discharge their functions effectively. 4) that a new district type of government under supervision of a central state agency be provided for those areas, such as public and private forest land, in which the population and taxable resources have reached the vanishing point.

C. The First Class Township - Since many of these units are closely related to large municipalities and, although of small size, require a number of important municipal services, it is recommended that the legislative commission proposed in an earlier section draft a plan for the annexation or consolidation of the smaller townships with contiguous municipalities and that the procedure for voluntary consolidation be facilitated.

D. The Borough - 1. The creation of small boroughs should be made more difficult, the consolidation of existing boroughs with adjacent local units should be facilitated, as should also the annulment of existing borough charters, in the interest of econ-omy. The existence of many small boroughs with inadequate taxable resources and an excessive overhead caused by borough organization presents substantially the same problem as that already noted under the first class township.

2. For the protection of borough taxpayers the central state administrative agency al-ready proposed in other sections above should be authorized to exert some control over borough debt and budgets when the amount of delinquent borough taxes reaches a fixed point, or upon petition of a majority (in interest but not in numbers) of the borough taxpayers.

3. The practice of creating an independent school district for each borough, no matter how small, is a serious financial burden in the smaller units. This renders more important the annexation or consolidation of such units.

E. The Third Class City - 1. These units vary in size from 7,000 to 115,000 yet all are governed by the same laws. These laws were enacted primarily to meet the needs of urban communities with a population of at least 50,000. It is recommended either that the legislative commission proposed in the preceding section of this Report reclassify third class cities or that the present requirement of a minimum popula-tion of 10,000 be materially increased by law.

2. The commission form of government of third class cities combines both legislative and executive functions in each commissioner, contrary to well established princi-ples of administrative responsibility. The administrative authority should be con-centrated in a single executive head, either the mayor or the city manager accord-ing to the choice of the voters.

3. As recommended under the sections dealing with first class township and borough above, so also in the third class city it is recommended that small units adjacent to such cities should either be united with them to provide a single government for an entire urban area or, where this is not possible, that combined district govern-ments for fire, police, sewer and schools should be provided.

VII. Summary of Constitutional Changes

Local Debt and Local Officers - It is recommended that the State Constitution be amend-ed by (a) modifying the provisions limiting local debts and (b) repealing those which provide for the existence of certain local units and local officers and which prescribe their powers, duties and methods of election.

(a) Limits on local debt are clearly needed but the present limits are neither adequate nor elastic enough to satisfy present conditions. The legislature should be authorized to establish lower limits and to provide for a state administrative agency which would permit, for limited periods and under special conditions, temporary variations above the statutory but within the constitutional limits.

(b) All clauses of the Constitution fixing the title, the powers and duties, and the method of selection of local officers should be repealed. The legislature should have full freedom to offer to the people new and different forms of local organization cor-responding to their local needs. The modern trend in state constitutions is distinctly toward this greater freedom of action to permit greater elasticity in local forms. This recommendation would apply to Article 5, Sections 1, 5, 11 dealing with local judi-cial officers and justices of the peace, Article 8, Sections 14 and 16, prescribing the methods of choosing local election officers, and Articles 13 and 14 dealing with coun-ties and county officers.

Bibliographical Note

Special Studies in Pennsylvania Local Government - Among the more prominent recent in-quiries and reports on special problems in this field are "Preliminary Report", Penn-sylvania State Planning Board, 1934; "Cost of County Government in Pennsylvania", F. P. Weaver and H. F. Alderfer, 1933; "Collection of Local Taxes in Pennsylvania", B. E.

Nicholson, 1932; "Mandatory Expenditures of Local Governments", E. W. Carter, 1934. Much data of importance will be found in "Final Report", Poor Law Commission, 1925; "Study of Poor Relief in Pennsylvania", Pennsylvania Department of Welfare, 1934; "Final Report", Township Law Revision Commission, 1931; "Report", Pennsylvania Tax Commission, 1925; "Educational Surveys", Pennsylvania Department of Public Instruction, 1925, 1930.

Reports on Local Government in Other States - Among the more valuable recent inquiries pursued on local government in other states are the following: "Michigan Local Government Series", Michigan Commission of Inquiry into County, Township, and School District Government, 1933-34.

Reports, New York State Commission for the Revision of the Tax Laws, 1932, 1933, 1934, 1935.

"Township Government in Indiana", C. F. Snider; Vol.II, No 1, Indiana Studies in Business, 1932.

Report on State and Local Government, made to The Special Legislative Committee on Organization and Revenue, State of Wyoming, Vol. I and II, by Griffenhagen and Associates, 1933.

Report on a Survey of the Organization and Administration of State and County Government in Mississippi by the Institute for Government Research of The Brookings Institution, Washington, D. C., 1932.

A Survey of State and County Government in Alabama, Prepared by the Institute for Government Research of The Brookings Institution, 1932 (5 Volumes).

Report of the Commission on County Government, to the General Assembly of Virginia, December, 1931.

Report on a Survey of the Organization and Administration of the State, County, and Town Governments of New Hampshire, Prepared by the Institute for Government Research of the Brookings Institution, 1932.

Reports of Commission to Investigate County and Municipal Taxation and Expenditures in New Jersey, 1931.

CHAPTER II

INTERNAL ORGANIZATION

The county is the chief local government unit of the state. Its powers are broad and its services have always been of utmost importance to the people of the Commonwealth. Its general form is based upon English law of the 17th century. Many changes have been made in colonial times and during the 19th century, but its main outlines have altered so little that the county of today could easily be recognized as the direct descendant of the county of three hundred years ago. In fact, the sheriff, coroner, register of wills, clerk of courts, prothonotary, treasurer, the board of commissioners, recorder of deeds, the chief clerk, and others were functioning in the county before the Revolution. Of the important officers, only the district attorney, the controller and the auditors originated after 1800.

The earliest county government placed the county court in a dominant position in the administration of its affairs. The court appointed most of the officers, (many of them such as the beadle are now non-existent), and was an overseer of most of the functions carried on by them. Its power extended to assessing, levying, and collecting taxes, to roads and other public works, and to the care of the poor, indigent, and those convicted of crime. In 1715 the board of county commissioners was created to relieve the court of the work of assessments and taxes in general. Other powers were gradually taken away from the court and given to this board. New and additional functions as they arose were placed in the hands of the commissioners, but to this day the county court still exercises some power in county administration. After more than 220 years of development, moreover, the county commissioners have not yet been raised to a position of unquestioned responsibility in county affairs.

One of the most important changes in county government came with the Constitution of 1838. It made all the important officers of the county elective by the voters. This was in line with the theory of Jacksonian democracy that government was good only when the officers were close to the people. This theory has long been disproved by practical experience in American government, but to this day from twelve to eighteen county officers are elected every four years by the voters.

Although the development of business organization in the United States has influenced the course of national, state and municipal government, very little of this influence has reached the county. Its principal traces are to be seen in the establishment of the office of county controller and the gradual increase in power and responsibility of the county commissioners.

One of the characteristics of early development was the haphazard character of legislation concerning county government. Until the Constitution of 1874, the General Assembly could pass special laws for each county, and it generally exercised this power. This made for a lack of uniformity that defied even the most expert lawyers. Finally, as this power was abused to great extremes, a provision was adopted which forbade the General Assembly to pass special or local laws relating to counties. This was a move to the opposite extreme. It was unsatisfactory because there were wide differences in county needs. A law passed for Philadelphia could not be expected to apply to a unit with five thousand population such as Forest County. Thereupon, there developed classification of counties according to population for the purpose of recognizing these differences. In the beginning, such classification was almost as haphazard as special legislation, but at present the Constitution and the statutes recognize eight classes of counties according to population.

These are as follows:

1st class - those having 1,500,000 inhabitants and over - Philadelphia.
2nd class - those having between 800,000 and 1,500,000 inhabitants - Allegheny.
3rd class - those having between 250,000 and 800,000 inhabitants - five counties.
4th class - those having between 150,000 and 250,000 inhabitants - eleven counties.
5th class - those having between 100,000 and 150,000 inhabitants - four counties.

6th class - those having between 50,000 and 100,000 inhabitants - seventeen counties.
7th class - those having between 20,000 and 50,000 inhabitants - seventeen counties.
8th class - those having less than 20,000 inhabitants - eleven counties.

While classification was a significant step towards order and uniformity in county affairs, it did not go the whole way. Laws still in force have been passed and amended during the past two hundred years in a manner to suggest a constant patching up of the procedure and structure of the government. Even the Code of 1929, which codified and made uniform much of the law relating to counties, did not touch upon a host of important subjects. County law needs to be not only codified and clarified, but also modernized. An interesting side-light illustrating this point is the fact that the present county coroner derives his power of holding inquests over dead bodies from the Act of King Edward I, enacted in 1276. There is no statute of Pennsylvania that covers this subject.

One of the gradual developments in county government during the past several decades is the shifting of county functions. The Commonwealth has been gaining control bit by bit in what was once the county's field in police, health, roads, and welfare. This has been accomplished largely by the growth of state administrative authority over counties in these matters. This closer county-state relationship offers an effective means of solving problems that involve both state and county interests. It is helpful also in that it utilizes the services of experts in fields that have grown too complicated for the ordinary layman. On the other hand, the county has gained functions which once were performed by the smaller local units such as cities, boroughs, and townships. This is illustrated by its power to construct and maintain parks, playgrounds, recreation halls, airports, sewage disposal plants and libraries. The question might well be asked: Will the county develop into a governmental unit that could function as a regional government in a metropolitan area. While more pressing immediate problems confront us, this new question cannot long remain unanswered in view of the rapid changes in local needs.

Has the government of the county kept pace with the needs of a modern and complex society? Is it so organized as to render to the people efficient and economical services?

In this chapter, we are interested in the general features of the county as a whole: its form, finances, functions, and administrative procedure. We shall deal with the following subjects:

    I. Features of form and function.
        1. The elective officers
        2. Lack of responsibility in county finance
        3. Lack of modern methods in county finance
        4. The county courts as administrative agencies
        5. Functions of county officers

    II. Recommendations.
        1. Those which would not require a change in the Constitution
        2. Those which would require such change

I. FEATURES OF FORM AND FUNCTION

1. *The Elective Officers* - In keeping with the principle of Jacksonian democracy the important county officers were made elective under the Constitution of 1837. They remained so through the entire century, despite the general feeling that a long list of elective officers burdens the voter, develops the power of the political boss, creates friction, and causes waste in administration. Reorganization in state and municipal government since 1880 has been steadily in the direction of the elimination of most elective officers and the substitution of appointive officers under a chief executive, the latter either elected by the people, as is the mayor or governor, or elected by the council as is the city manager.

The following officers in Pennsylvania counties are elective: 3 county commissioners, 3 auditors or 1 controller, treasurer, sheriff, coroner, district attorney, 2 jury commissioners, prothontary, clerk of the courts, register of wills, recorder of deeds, clerk of orphans' court, and the surveyor. In addition, the judge of the district court (a state officer paid by the state) is elected by the people in the judicial district, which is usually the county. These officers are generally called "row officers," on account of their position in the county hierarchy of offices. Each of these elective officers is independent of the others. What goes on in one office has nothing to do with the others. One office may be busy for a month; the employees of the other offices may be idle, but there is nothing in the county system which would require one office to help out the other in the matter of seasonal clerical work or record keeping. This is true of virtually every possible relationship. When two offices have anything to do together, they retain in all respects their complete independence. As a result, there is duplication, waste, and inefficiency apparent to anyone who is interested in governmental structure. The heads of these elective offices are usually a law unto themselves alone, for the county government is headless. While they may be theoretically responsible to the people and to the law relating to their offices, this is more theoretical than actual. Since the people do not understand the complicated routine and functions of these offices, the law is brought to bear only when a case comes officially before the courts. The interpretation of the law must wait upon a final decision by the Supreme Court of the Commonwealth.

The following chart, taken from F. P. Weaver and H. F. Alderfer's County Government Costs in Pennsylvania, shows the structural arrangement of offices in Pennsylvania counties from second to eighth classes inclusive.

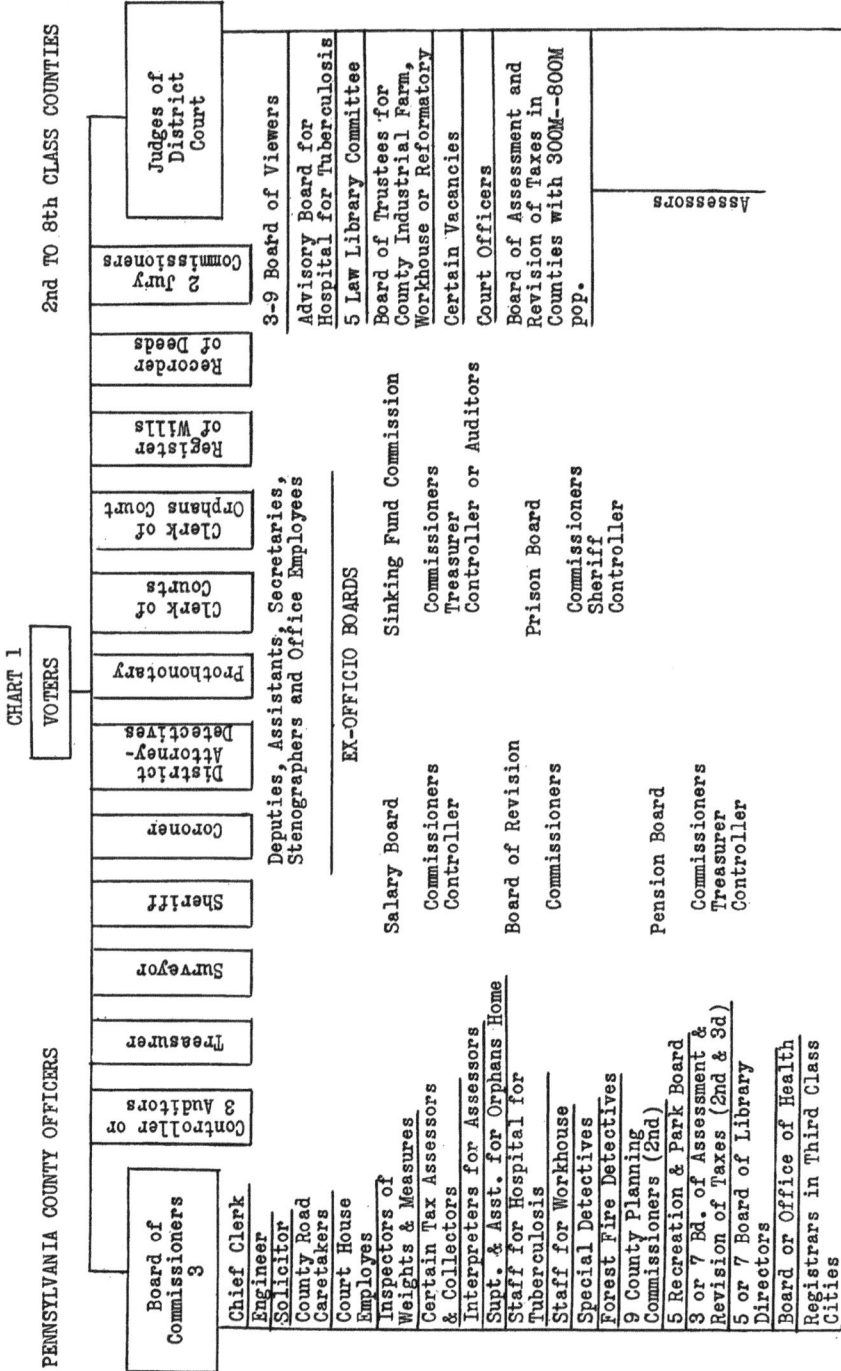

CHART 1

VOTERS

PENNSYLVANIA COUNTY OFFICERS

2nd TO 8th CLASS COUNTIES

County Officers (boxes):
Board of Commissioners 3 | Controller or 3 Auditors | Treasurer | Surveyor | Sheriff | Coroner | District Attorney-Detectives | Prothonotary | Clerk of Courts | Clerk of Orphans Court | Register of Wills | Recorder of Deeds | 2 Jury Commissioners | Judges of District Court

Under Board of Commissioners 3:
Chief Clerk
Engineer
Solicitor
County Road Caretakers
Court House Employes
Inspectors of Weights & Measures
Certain Tax Assessors & Collectors
Interpreters for Assessors
Supt. & Asst. for Orphans Home
Staff for Hospital for Tuberculosis
Staff for Workhouse
Special Detectives
Forest Fire Detectives
9 County Planning Commissioners (2nd)
5 Recreation & Park Board
3 or 7 Bd. of Assessment & Revision of Taxes (2nd & 3d)
5 or 7 Board of Library Directors
Board or Office of Health
Registrars in Third Class Cities

Deputies, Assistants, Secretaries, Stenographers and Office Employes

EX-OFFICIO BOARDS

Salary Board	Commissioners, Controller
Sinking Fund Commission	Commissioners, Treasurer, Controller or Auditors
Board of Revision	Commissioners
Prison Board	Commissioners, Sheriff, Controller
Pension Board	Commissioners, Treasurer, Controller

3-9 Board of Viewers
Advisory Board for Hospital for Tuberculosis
5 Law Library Committee
Board of Trustees for County Industrial Farm, Workhouse or Reformatory
Certain Vacancies
Court Officers
Board of Assessment and Revision of Taxes in Counties with 300M–800M pop.

Assessors

2. <u>Lack of Responsibility in County Finance</u> - While the county is in theory a unit of government, the responsibility for financing county offices and functions is dissipated among several authorities. This results from the decentralized structure of the county government itself.

a. <u>Mandatory Expenditures</u> - The state legislature passes laws making it mandatory for the <u>county to spend money</u> for certain items and in many cases specifies the amount to be spent. The following items, according to state law, <u>must</u> be paid for by the county.

> Salaries of the elected county officers
> Salaries of a number of appointed officers
> Compensation of officers paid per diem at a rate determined by state law
> Support of county inmates in state penal and correctional institutions
> Support of prisoners in the county jail
> Road damages for state and county highways
> Costs of the district court within the county
> Rooms for judge, women jurors, and juvenile offenders
> Premium on county officer's bonds
> Expenses of certain county officers to state conventions
> Expenses of commissioners in pursuit of duty
> Soldier and soldier's widow burials
> Markers for veterans' graves and compilation of war records
> G. A. R. Memorial Day observance
> Flags to decorate soldiers' graves on Memorial Day
> An appropriation for annual convention of school directors in county
> Office expenses, rent, and stenographer of county superintendent of
>   schools
> Registration of births and deaths
> Cost of general and municipal primaries and elections in county
> Cost of assessment of property for county purposes and registration of
>   voters
> Refunded and returned taxes on unseated and seated lands.

It has been calculated that in 1931 these 21 items, made mandatory by state law, ranged from 18 to 69 per cent of the total expenditures of the counties in the 67 counties of the state. †

b. <u>County Commissioners' Power of Appropriation</u> - The county commissioners, furthermore, <u>may exercise powers</u> of appropriation for over 40 enumerated purposes. The following are illustrations of such optional powers:

> Provide and maintain National Guard Armories
> Maintain a unit of the National Guard
> Appropriate for rifle-clubs in time of war
> Purchase burial lots for deceased service men
> Appropriate to veteran's organizations for Memorial Day.

c. <u>Joint Responsibility in County Expenditures</u> - The county commissioners are by no means wholly responsible for county expenditures. They divide that power with a number of county, as well as other local and state, agencies. This is illustrated by the joint responsibility in initiating, carrying through and finally approving public works of the county. There is a general lack of uniformity and standardization in practice. The following agencies share with the county commissioners control over the planning, execution and financing of public works:

> Board of Viewers
> Grand Jury
> Judge or Judges of Court of Common Pleas of Quarter Sessions
> Directors of the Poor
> Controller

† Weaver and Alderfer, <u>County Government Costs in Pennsylvania</u>, (1933). See also
  Edward W. Carter, <u>Mandatory Expenditures of Local Governments in Pennsylvania.</u> (1934).

Various State Administrative Departments, such as the
> Department of Health
> Department of Welfare
> Department of Highways

These following projects illustrate the authority required by law to carry out certain functions of county government:-

1. Altering and enlarging public buildings --
   Petition of citizens to court of quarter sessions, hearing, election, mandatory on commissioners.

2. Building county roads --
   In general - Commissioners with approval of grand jury and court of quarter sessions. Sometimes, initiation by petition of citizens is required. Sometimes, plans and surveys made or approved by Secretary of Highways.

There are more than a score of such combinations listed in the county law.

d. _The County Controller, County Auditors and County Treasurer._ While the county commissioners have some powers of appropriating county money, there are other financial officers of importance in the county. These are the county controller or three county auditors and the county treasurer. The controller audits and checks all county accounts and bills; he makes up a county budget as a guide for the county commissioners. He is the watch-dog of the county treasury. In counties in which there is no controller, three auditors make an annual check on county accounts. The county treasurer receives and pays out county money. He also collects delinquent taxes for all the local subdivisions. A detailed account of the functions of these officers is given later in this chapter. All of these officers, being elective, are independent of one another. The function of finance, however, is a unified process and cannot be handled adequately by separate agencies of government.

e. _The Fee Officers_ - The fee officers constitute another group who collect moneys due the county for services performed by their offices. Even when the officers and their employees are paid by salary, they collect these fees over the counter in their offices. This applies to the prothonotary, clerk of courts, sheriff, recorder of deeds, and register of wills. This necessitates the keeping of record books on sums received in these offices, a monthly audit by the controller, or an annual audit by the three auditors, the sending of reports to the controller, and the sending of money to the treasurer at least monthly. This means that in many cases each office has a separate account in a bank, is checked by the controller or auditors, and settles with the treasurer. There is an enormous duplication of records under such a system. In addition, the fees which the office earns for services performed by it for the county are paid either in cash or are credited by the treasurer to the office. In some cases, the treasurer writes monthly checks to the fee officers who place such sums in their monthly accounts and return them later to the treasurer along with the fees collected from individuals. This is made necessary by the law applying to such fee officers in third and fourth class counties, which provides that the aggregate amount of salaries of the office shall not exceed the amount of fees collected both from individuals and from the county in any one term of office.

The complicated and costly system of handling fees should be corrected by a centralization of collection machinery in the treasurer's office, a system of simple and uniform accounting, and the use of vouchers, which the controller's office should furnish individuals. Such vouchers would be taken to the treasurer's office with the money and in exchange for these receipts made out by the treasurer would be issued. All collections outside the county office should be made by deputy treasurers working as liaison officers with the offices rendering the services for which the money is to be collected. The use of the mails in collecting fees might well be utilized to a greater extent. There could be ample protection for the county in case of non-payment, the service either not being performed or not recorded officially until payment is made. No other office than the treasurer's should collect moneys due the county.

f. _The Collection of Taxes_ - The problem of tax collection is taken up in another section of this report. It is sufficient to state at this point that the commission-paid, elected tax collectors from boroughs, townships, and wards of cities should be abolished. Their place should be taken by the county treasurer as the collector of all local taxes. Sub-collectors could be stationed in various parts of the county when necessary. They should be paid either on a straight salary or on a per diem basis.

g. _Ex-officio Boards_ - Ex-officio boards have been established by law to integrate financial functions of two or more offices. The salary board, composed of the commissioners and controller, and in some cases of the commissioners, auditors, and treasurer, fixes the compensation and number of clerks and deputies in all the county offices in counties of the second to sixth class inclusive. This board is discussed in detail in the section of this report dealing with personnel. The sinking fund commission, composed of the commissioners, controller, or auditors, and the treasurer, administers county funds that are applied to the paying off of the bonded debt of the county. They select the banks in which such money shall be deposited and make investments in such bonds as are designated by law. Both the state Constitution and state law limit the amount of debt that the county and other local subdivisions may incur, and provide for the collection of an annual tax sufficient to pay the interest and principal within thirty years. Three-fifths of the electors voting at a public election may increase this limit from the ordinary seven per cent of the assessed valuation of taxable property of a county to ten per cent. The Bureau of Municipal Affairs in the Department of Internal Affairs of the Commonwealth is empowered to investigate, advise, and report on debt conditions of local subdivisions of the state.

h. _Summary_ - Thus we see the financial responsibility of the county divided among the following agencies:

1. The General Assembly of the Commonwealth
2. Various state departments
3. The county commissioners
4. The county controller or auditors
5. Judge of the district court
6. Grand jury
7. Citizens, as petitioners or voters
8. The county treasurer
9. The heads of the fee offices
10. The tax collectors

3. _Lack of Modern Methods in County Finance_ - The absence of modern financial methods is also a factor in the general situation. Methods and procedures long since adopted by municipal, state and national governments have not yet found their way into county government in Pennsylvania. This is not true in the same degree for all counties, since some units are progressive. As a rule, however, the county is characterized by loose and wasteful financial operations. This situation can not be blamed on the lack of opportunity for putting advanced methods into practice. Although the laws relating to counties usually do not require modern procedure they certainly do not prohibit such procedure. Even where the better practices are mentioned in the law, their application often is not in accord with the statutory letter or spirit. There is very little attempt to make county officers adhere to such practices. The underlying cause of this condition rests with the officers themselves. As long as they are elected on account of their political availability rather than because of their administrative qualities, progress towards a business-like administration of government must proceed slowly. Furthermore, when their appointed subordinates are also chosen on a political basis exclusively there is even more obstruction, because these subordinates conduct most of the technique of administration. An improved personnel system would automatically result in better financial administration, without necessarily changing one word of the law. Let us examine some of the more important financial functions as they are practised in twelve counties which were visited for purposes of observation. Names of the counties are not mentioned because it is not the purpose of this report to criticize particular county administrators, but to attempt to point out defects that inhere in the machinery of government in counties in general. As mentioned before, there are some bright spots in the picture. There are county administrations which are efficient, considering the

obstacles of organization; there are county officers who are earnest, industrious, and progressive and they achieve results that are commendable.

a. The Budget - The budget, a plan on which to run the county government for a particular year, appears to most observers as the sine qua non of good government. In counties where there is a controller, this officer is supposed to supply the commissioners with an estimate of expenditures and revenues for the coming fiscal year so that they may levy the tax millage with some certainty of providing enough money to run the government. In all the counties visited which have controllers, the controller did exactly as required by law. In only one out of six counties was such a budget published along with the controller's report of the same year. There was no indication that the budget is published in the newspapers or in any other way. The controller makes his budget from figures based upon what was spent during the last year and what differences can be expected during the current year. Estimates of revenue and expenditures are not usually gathered from the departments concerned, but are based largely on last year's figures. There are no budget estimate forms sent to individual departments.

In one county, estimates are furnished by the commissioners, who get them from the various offices. The estimate for roads is not accepted until it appears to be no higher than will be necessary to keep within the usual millage, which in this county is low. The implication is that as long as the aggregate of estimates will allow that millage, expenditures may be made without undue strictness. The expenditures for personnel indicate this situation since they appear to be much higher than necessary. In general, the budget items are not detailed, but include lump sums for departments with items for salaries, supplies, and incidentals. In another county it has been customary to put into the budget estimates for services not rendered or articles not bought, or expected to be bought. Thus, some leeway for incidentals was possible during the year. Another instance of abuse of the budget is seen in the comparatively large amounts provided for "incidentals" or "miscellaneous". In one county, the "incidental" item was used mainly for extra clerical help.

The most important fact about the budget is its flexibility. Offices may spend more than they are allowed without being cut off by the controller. As one controller said, "The budget is merely a working program and not a rigid instrument." In only one county of the six was there evidence that any expenditure not found in the budget could be incurred only after special action at a meeting of the commissioners. Here the unbudgeted expenditures received special treatment and consideration. Furthermore, there is much evidence that county commissioners alter the budget from time to time during the course of the year. Incredible as it may seem, the 43 counties without controllers have no budget. The commissioners estimate how much money is needed for the coming year, and figure the millage on that estimate. County offices do not work on restricted appropriations made in the beginning of the year. The commissioners may, and often do, refuse to incur expenditures in the interest of economy, but with no fixed plan. In some cases, other officers write vouchers without the approval of the commissioners, but these are mainly for court costs.

Before a final judgment is reached upon budget procedure it may be well to remember that sometimes more than half of the expenditures made in any year are mandatory by law. In all but a few counties of the state more than a quarter of the total expenditures are so fixed. Thus, it is impossible to make detailed plans for the amount to be spent on certain items. When the mandatory bills come in the county commissioners are forced to pay them whether they have planned for them or not. Sound budgetary practice will never be obtained as long as the commissioners are confronted by this situation.

b. Purchasing of Supplies - The law requires that all purchases of more than $300 be made by bids, with the exception of contracts for roads and bridges, where the limit is $500. The larger counties buy most of their material in bulk at the beginning of the year and bids are let to the lowest responsible bidder.

In one of the counties visited the following procedure is used in purchasing materials. At the end of the year, the county commissioners call for requisitions from the various offices for the coming year and collect these in a single list. The chief clerk prepares specifications, which may be obtained by any firm desiring to bid on all or any of the items which have been advertised according to law. Bids must be in by a certain date; they are opened by the commissioners and controller, after which the bids are awarded to the lowest responsible bidder on each item. The bids and the names of the companies bidding are placed in a purchase book in the office of the controller for future reference. In another county, when the materials are handed out to a particular office they are checked with the amount originally allocated to that office. If the office requires more than its allocated amount, a special reason must be given for this demand. In most counties, however, supplies are not allocated in this manner but are furnished as needed.

In smaller counties where purchases are comparatively few and small in amount, the chief clerk acts as purchasing agent. He buys all materials except those needed for bridges, roads and other public works. Supplies are bought when needed and in small quantities. Only rarely are contracts let through bidding. In some counties election ballots are let by bids but in most cases they are handed out locally. By custom, certain printers receive them by virtue of their political affiliation or their proximity to the county commissioners' office. Election supplies, assessment books, and other such supplies are often bought from a particular firm dealing in that specialized brand of supplies. Occasionally, a chief clerk begins "writing around" for estimates on his various articles and he is often surprised at the savings he can make by doing this, even from the companies with whom he has been dealing before. Sometimes bids are inaugurated for the purchase of supplies, with a large decrease in purchase prices. It is safe to say that chief clerks even in the smallest counties could save the county much money if they took pains to get estimates or bids and to buy in larger quantities. It is also probable that an examination of the cost of ballots would show surprising differences as between counties. Unfortunately, costs of election ballots are not segregated in most of the auditors' or controllers' reports. It was stated by the chief clerk of one county that the inauguration of bids for ballots brought the cost in one year down from $1,200 to $156.75. While this appears excessive, creditable savings could be made by using bids. "County printing," however, is too often deemed the prerogative of local and politically affiliated printing houses and newspapers. Another item which needs examination is the cost of county printing per column inch. Here, too, one finds great variations. The cost of printing is a fairly large item in the course of a year. It is common to charge the county prices which are far in excess of ordinary advertising rates.

    c. The County Auditors - As a rule, the persons elected as county auditors know little about the technique necessary to audit county accounts. Quite frequently they have no conception of accounts whatsoever. Sometimes one is elected who has some experience in business, and invariably he bears the brunt of the work. Sometimes a clerk is hired at the expense of the county and is required to do most of the real work. The auditors' reports show their general ineptness in this field; while there is much diversity all are indescribably bad, not only from the viewpoint of one who is trying to get a detailed picture of the financial condition of the county, but also from the viewpoint of the general public, which is interested in a simple, understandable account of county finances. For example, items are often marked John Jones, $4.98. For what? Neither this nor any other information about the expenditure is given. Frequently large sums are lumped together under an item such as "Commonwealth Costs." The average individual does not know what that means. Many of the auditors are anxious to do their best, and it goes without saying that most are honest. But they do not know how to audit. They spend a great amount of time on the work and do a very superficial job. As a rule they check the accounts of a certain office with the accounts of that office in the files of the treasurer. Sometimes they check over individual vouchers. Of the six counties where officers received fees as compensation, in only one case was there any attempt to find out how much was earned by these officers in fees. Even then, the auditors took the words of the officers and did not make a complete examination of the books of the office. In one county, the auditors had a dispute with the treasurer over his accounts so that he had to call in an independent auditor for his own protection. Because of this the auditor's report of that county for 1933 was not published until late in 1934. The other county officers generally have nothing good to say for the auditors, usually on the ground that the latter do not know what to do or how to do it. Each auditor usually has the chance to have his report printed in his favorite paper, usually at a price which he himself does not audit. One set of auditors, observed, at work for several afternoons, entertained visitors most of the afternoon and went to the movies occasionally. One very old man slept almost continually. One of the other auditors (of the opposite political party) greatly enjoyed waking him up suddenly by cracking a fly swatter down on a table. It was explained that the sleeping one always did things wrong and that they asked him not to do any tabulating because they had to do it all over again in any event. So he just sat around. All three were very obliging and two of them had a fairly good idea of county government. One admitted that auditing in this manner was excessively expensive and inefficient. Small counties, where a controller would be inadvisable, could hire the services of an independent auditing company or auditor. It would be far cheaper and better.

d. <u>Independent Audits.</u> - In the twelve counties surveyed, there was no instance of a thorough independent audit. Two such audits had been made in previous years when there was a discrepancy in one of the fee offices. In both cases, the county was found to be about $40,000 short. The officer in charge had not turned over the amount of fees due the county. Independent audits were called for after the bonding company had been brought into the case for adjustment.

This leads one to question whether the monthly audits of the fee officers by the controller are adequate. It would be very easy for an officer to keep aside a large amount of fees and to manipulate the books which the controller examines in his monthly audits. Each such office keeps its own books, and they are so complicated that it is indeed an active controller's office which makes a thorough audit even yearly. Instead of allowing the fee officer to handle the funds it would be wise to allow payments only to the treasurer by voucher from the fee office. This would be safer; it would eliminate much duplication and the need for monthly audits of each office.

Independent audits would be useful not only for keeping accounts in order but to aid the development of a sound system of efficient accounting. Those auditing would be able to suggest possible improvements.

e. <u>County Controllers</u> - The mere fact that a county has a controller instead of three auditors should not leave the impression that the auditing, scrutinizing and adjusting of county accounts are done efficiently. Much depends on the county controller - his training, experience, and interest in his work. Unfortunately controllers are elected and are political workers. As a result, some are without the prerequisites of a good controller, no matter how well they may wish to do their work, once they are in office. Most offices contain at least one good deputy or clerk who attends to the detail and routine of auditing. Most offices appear to be over-manned in comparison with the number of employees actually needed. Many offices simply adopt the policies of work developed in previous administrations.

Nevertheless, the office of the controller should be an important agency in the future county. It should be given more rather than less power, and should be allowed to exercise even stricter and more detailed control of the county expenditures. All court costs should be computed in this office instead of in the offices of the clerk of courts or the district attorney. In addition, the controller should be given power to audit receipts both from tax collectors of the county and from the various offices acting as state agents for the collection of Commonwealth fees and taxes.

f. <u>Controllers' and Auditors' Annual Reports</u> - No one who has had occasion to work with the controllers' and auditors' reports which are made annually to the county authorities and published for the information of the taxpaying public, needs to be told of the inadequacy of these reports. The controllers' reports, while more satisfactory than those made by the auditors, rarely give a clear, concise and accurate picture of county finance. The classification of expenditures is misleading, incongruous and frequently inaccurate. The word "incidentals" is used for an abundance of expenditures. There is no attempt to give the public a simple statement, but rather it appears that the more obscure the report the better it is considered. The auditors' reports, as mentioned before, are infinitely worse. They are almost useless as statements of the financial condition of the county. More interest in county government might be evinced if, in the few reports to the public that the county officials publish, greater clarity and sincerity of purpose were apparent. What is definitely needed is uniformity of accounting and reporting throughout the counties of the state. This would provide an opportunity to compare counties from year to year on general and detailed financial activities. The Pennsylvania Economic Council has set up a schedule of receipts and expenditures for county government which will enable comparison to be made. But the people of the Commonwealth should not be asked to rely from year to year on clarifications made by private organizations when this duty clearly rests on the public officials of the county.

g. <u>Mileage</u> - The cost of mileage for officials has long been an item of importance in county expenditures. The following officers or persons may receive mileage on certain occasions in performance of their duties: county commissioners, jury commissioners, sheriffs, coroners, desertion-parole-probation officers, juvenile officers, auditors, road viewers, detectives, deputy sheriffs, jurymen, witnesses, constables, election judges and inspectors, assessors, district attorneys, and sealers of weights and measures.

The salient points in the situation may be summarized as follows:

1. State laws are basically responsible for this system. The range allowed by law for different officers is from three to ten cents a mile. There is no reason why some officers should be given ten cents a mile while others receive three, four or six.

2. In many cases, such mileage is in addition to a salary; in other cases, it is part of the fees chargeable to counties for particular services.

3. While there are some of the lower rates above mentioned, the predominating rate is ten cents. There can be no question that this is more than the actual cost of the trip even when depreciation is figured very high. Automobile travel in this day does not cost that much.

4. In some cases, such as for auditors, mileage is allowed to and from destination only once a week. As auditors usually go home every night, such a regulation is without reason.

5. In the majority of cases, there is no actual check on the mileage handed in by these officers. Some merely submit accounts labelled "expenses" without reference to the number of miles covered or the places visited. Others hand in mileage with the number of miles but without the names of places visited. Many mileage bills are handed in monthly without any indication of the number of visits actually made. On a number of standard trips such as for jury, witnesses, or auditors from a particular town to the county seat, there is a chance for the chief clerk, controller, or other checking agency to hold to the standard rate. But in the case of constables, sheriffs, desertion officers, road viewers, and others there is no standard mileage possible because of the fact that such officers must not only travel to certain places, but often they must hunt around for the person sought, and they may carry out two or three errands on the same trip. Billing for duplicate mileage, when actually only one trip is made for a number of errands, is a common practice. Some counties allow certain officers only actual expenses, others pay a flat yearly rate for mileage. That there is a great possibility for overstatement of mileage is the consensus of opinion of chief clerks and controllers, who say that they do not know what to do about it. As a rule, they accept without much question the mileage statements. Some counties attempt to standardize trips more than others. Controllers and, in counties without controllers, chief clerks to the commissioners are responsible in most cases for checking mileage involving the payment of county moneys. But the clerk of the courts and the sheriff attend to mileage involving court costs and executions of court writs, especially when paid by individuals to the particular officers. All such computations should be centralized in the hands of the controllers or chief clerks.

6. In counties where officers are paid by fees plus mileage, part of the profit comes from mileage. For example, the sheriff is allowed ten cents a mile for hauling a prisoner from the county jail to a state institution. In the same car are the deputy sheriff and the prisoner. The deputy sheriff rates three and three-fifths cents a mile (one way). This all goes to the sheriff. At other times, the sheriff may haul three

prisoners at once and receive mileage for each.

The actual amount of fees paid out by the county is difficult to compute without making a thorough examination of the vouchers and day books (even then it is often impossible because mileage is not segregated from other payments). In one seventh class county the sheriff, it is estimated, receives over $5,000 a year in mileage alone. In another county, the sheriff stated that while he was not getting mileage fees but was using cars owned by the county, his predecessor had made the mileage business pay. He had figures to show that five trips made to five different state insitutions with prisoners actually cost $40.56. His predecessor, making the same five trips, had handed in a bill to the county for $505.36, making a profit of something like $464.80. Both included living expenses on the road. The predecessor had several entries for "taxi-cab -- $6.00" when he had, of course, the car. One trip which actually cost $ .75 was handed in at $59.20 by the predecessor. This was exceptional but it illustrates what abuses may result from the mileage situation.

Some counties actually make money on mileage. The law states that the sheriff may charge ten cents a mile circular in delivering writs and that this rate shall be charged to the parties on each separate writ. Thus, two or more writs may be taken on one trip. The deputies are paid only for one trip while the county pockets the rest. Thus the county, rather than the individual officer, profits. In one instance a county officer asked the county commissioners to make all mileage at the rate of six cents, and when they refused he too took his statutory rate of ten cents.

7. Counties vary as to rates of mileage allowed, even of those that are fixed by statute. There are instances where the law is either not known or not upheld on this point. Some officers who have the legal right to mileage fees are not granted them.

h. The Absence of Cost Analysis - Information as to unit costs of certain standard county services needs to be developed in a great many instances. If county officers could compare such unit costs with those of other counties, a real saving could readily be effected.

i. Finance and Personnel - Although the subject of personnel is treated in detail in the next chapter the relationships must be restated at this point to give a complete picture of the financial situation in counties. While salaries of the elective officers and certain others are made mandatory by state law, the salary board determines salaries of clerks and deputies in second-to sixth-class counties. The amount paid these subordinates, as well as the number in each office, varies from county to county. For example, the salary of the chief clerk in four fifth-class counties ranges from $2,160 to $3,000 per year. The list might be continued indefinitely. It might be said that what is paid in one county is the business of that county and not of the others. On the other hand, the nature and value of such services are often very similar in the same class of counties. There appeared instances where certain deputies knew and resented the fact that persons doing their kind of work in other counties were receiving a good deal more money than they were. Such instances are even found within a single county. A definite classification of salaries within the county is the least that can be expected.

The cost of extra help likewise deserves attention. Some counties and some offices within counties are notorious for the use of extra help at certain seasons. Much money could be saved if the work of the entire county were so coordinated and arranged that the full-time staff could be put on this seasonal work without unduly affecting their own duties. Some offices are crowded with work when their next-door neighbors have several idle or semi-idle workers reading newspapers and magazines and doing cross-word puzzles.

Tenure of office has also a direct connection with financial efficiency. Those who have more or less permanent positions with the county will attempt to master their work. Others will merely pass their time until a change of administration will inevitably throw them out of employment.

j. Accounting Systems and Financial Records - While an enumeration of the records kept in each county office will be found in Chapter VII it is obvious that meaningful records and a working accounting system are vital to good financial administration and must be mentioned at this point. The law does not specify that any particular accounting system shall be adopted; it does specify the form of keeping certain financial records. In the latter case, these are disconnected and desultory. Thus, each county has its own kind of accounting system, with its ledgers, day books, cost books, vouchers, warrants, postings, etc. Observations made in the various offices leave the impression that they are very complicated (more so than appears necessary), that there is a vast amount of duplication, and that a great deal of bookkeeping that should be done is not, and much that is done is worthless. It appears also that systems vary from county to county, and to some extent from year to year. As new officers come in, modifications are made. Sometimes the process is simplified, and at other times it is made more complex. Many officers freely concede the unnecessary duplication of records. Often a much-needed change in one office is prevented by its dependence on the form of records in another. Many officers in charge of records are unable to make changes because they have not the financial training which fits them for constructive criticism. Even to the uninitiated a uniform system of accounting and record-keeping throughout the offices within one county would be a boon as far as collecting information is concerned. There is a great deal of information, indispensable to the public and to students of the particular unit of government, which cannot be gleaned from the records without a great expenditure of time. Uniformity between counties is essential, likewise, for the comparison of counties as to operation. The training of accounting officers by some state agency would greatly benefit this service.

4. <u>The County Courts as Administrative Agencies</u> - During early colonial times, the judges of the courts of quarter sessions were the nucelus of county administration. They appointed many county officers, decided on public works, and administered assessments and taxation. Gradually other county officers sprang up to take some of the functions from the judges. But a great many odds and ends have survived two centuries of change. Today the judges of the court of quarter sessions still have (together with their functions as judges of the court of common pleas) a great many powers in county administration which make them an integral part of the county system. Indeed, some of their powers have increased. The development of probation, parole, and juvenile administration is one illustration. The officers engaged in such work are appointed by the judges and are subject to them. Many of the judge's administrative duties flow from his judicial position as an interpreter of the law. Many arise from the fact that in earlier days the judge was the only county officer learned in the law and therefore the final adviser, arbiter and dispenser of wisdom.

Without going into detailed statutory provisions let us indicate the general types of administrative power possessed by the judges. These are as follows:

Appointing --	Example: Appointment of court officers, probation, desertion officers, board of road viewers, etc.
Financial --	Example: Approve certain salaries of officers; determine other salaries are within statutory maximum.
Public Works --	Example: Approve most of the public works of the county such as roads, county buildings, and institutions.
Elections --	Example: Determine election districts, appoint overseers, etc.

These powers as exercised at present have the following results:

1. They add to the lack of responsibility in county government.
2. They take the time of the judge who is not interested in making decisions of a non-judicial character, and thus they are not always done with deliberation and study.
3. They bring the judge into politics and administration within the county in a way detrimental to his work as a judicial officer. Many judges are too much interested in the political aspects of the county. This is against the principle, long adopted in American practice, that the judicial officers

should be altogether judicial and separated from executive or political duties.

It would be comparatively easy to take non-judicial duties from the judges and transfer them to the county commissioners where they rightfully belong. This could be done in most cases by legislative action. In a few instances constitutional changes would be required. Under any circumstances it should be one of the points considered not only in county, but also in judicial reform.

It is true that this work is usually not very burdensome as far as time is concerned. With the decline of road building as a county function and the static nature of county public building projects, judges are not called upon as much as formerly to exercise their public works prerogatives. Nevertheless, there is still confusion of responsibility in these cases, and a host of other minor functions to perform. The power to determine the number of court officers and the amount of their salaries should be in the hands of the salary board of the county. Election administration should be lodged with the county commissioners, who are the election officers of the county.

5. Functions of County Officers - We shall now examine the functions of the various offices of the county government. These will be given in general outline without indicating the specific provisions of each law relating to function. Such provisions would in themselves make a voluminous hand-book. The object here is rather to give as simply as possible the general duties of each department and thereby to indicate the importance of each office in the general county set-up.

a. Functions of the County Commissioners' Office

1. Levying the annual county tax;

2. Supervising assessments of taxable property;

3. Appropriation of county money for purposes enumerated by state law;

4. Authorizing and arranging for temporary loans and bond issues;

5. Acting as a board of assessment and revision, except in counties where there is a separate board;

6. Issuing warrants for the payment of all bills and accounts against the county, except certain court costs;

7. Supervising and maintenance and repair of the court house and prison of the county;

8. Purchasing of materials to be used by county officers and offices;

9. Constructing and maintaining county roads and bridges (with certain limitations and restraints placed on them by other offices according to law);

10. Supervising and preparing for the conduct of primary and general elections;

11. Holding of commissioners' sales of tax delinquent property which the county has been obliged to buy;

12. Keeping records on financial, tax, assessment, criminal, election, and other activities and facts:

13. Checking bills for services rendered and materials bought in counties where there is no controller;

14. Appointment of clerks and deputies and determination of their salaries in all county offices as members of salary board, and appointment of subordinate

officers in their own office;

15. Certification to treasurer of tax returns sent in by local tax collectors;

16. Compilation of tax duplicates, assessment books and voters' lists;

17. The establishment and maintenance of certain charitable and welfare agencies, and membership in the prison board in certain counties;

18. Acting as directors of the poor in certain counties;

19. Acting as members of the sinking fund commission.

In order to assist them in the various functions of their offices, the county commissioners appoint a chief clerk, a solicitor, an engineer, a sealer of weights and measures, subordinate clerks, and other officers depending upon the county. The chief clerk organizes the work of the commissioners' office and conducts the administrative and clerical routine. It is he who acquires a knowledge of the detail, and on him the commissioners (who are only rarely full-time working officers, even though paid for full time) depend. He should therefore be an executive and office organizer of no mean ability in the larger counties. Even in the smaller counties he has a multitudinous variety of business to handle and, in all, he has a constant direct contact with nearly all branches of county administration. For this reason he might very well prove the nucelus of a more elaborate department of records which would conduct the general clerical and recording work of the county. The chief clerk as the head of such a department would supervise the records of the county commissioners, clerk of courts, prothonotary, register of wills and recorder of deeds.

The solicitor to the county commissioners might also occupy a more important position than he now does. He could supply all departments of the county with legal advice instead of having special solicitors for such offices as those of the sheriff, recorder of deeds and controller. These special attorneys are often highly paid and do but little work. A still more desirable alternative would be the elimination of all solicitors and the centralization of all legal matters in the district attorneys's office. The district attorney should be able to represent the county in civil as well as criminal actions.

The registered professional county engineer may be appointed by the county commissioners. He is in charge of the engineering work of the county, which includes roads, bridges, and other public works. In many counties the office of the elective surveyor has been allowed to go unfilled and his duties have been assigned to the engineer. With the taking over of the county roads by the state, this office as a permanent salaried office is becoming less important. Many counties have abolished the post. Other counties hire an engineer when occasion demands and pay him a stipulated fee. In several counties the engineer's chief duties involve the survey of damage done by state highways and bridges, to aid the commissioners in settling for such damages. Where the engineer has duties enough to keep him busy the office is properly placed under the county commissioners.

The sealer of weights and measures is required by law to test all instruments and machines used in weighing and measuring merchandise for sale. In one county none was appointed, the work being done by an agent of the state. It appears that this is a Commonwealth function and could be better undertaken by a state agent.

b. Functions of the County Controller's Office

1. Keeping of financial records of the county;
2. Countersigning all warrants made by the commissioners before payment is made by the treasurer;
3. Auditing at regular intervals all accounts of officers handling county money;
4. Scrutinizing bills and accounts owed by the county to see if they are

legally payable;

5. Preparing a budget for the county commissioners at the beginning of each financial year;

6. Seeing that bids and contracts are made in the proper manner;

7. Acting as member of the salary board, the sinking fund commission and, in certain counties, of the prison board;

8. Publishing annual controller's report.

c. Functions of the County Treasurer's Office

1. Paying out county money on warrants of the county commissioners, counter-signed by the controller (also on court costs not drawn and countersigned by these officers);

2. Receiving, recording, keeping, and depositing county revenues and unseated land taxes for all local governments which are returned to them;

3. Acting as agent of the commonwealth in the issuance of hunting, fishing, dog, and beer licenses; collecting mercantile taxes;

4. Keeping the county commissioners and the controller informed as to the condition of county finances;

5. Acting as a member of the sinking fund commission and, in certain counties, of the salary board;

6. Collecting delinquent taxes for all local units; conducting treasurer's sales on property delinquent in taxes;

7. Collecting county taxes in certain counties.

d. Functions of the County Auditors

1. Auditing accounts of all county officers handling county moneys;

2. Surcharging officers in case of deficiencies or illegality of expenditures.

e. Functions of the County Sheriff's Office

1. As officer of civil court:

a. Executing writs on real estate and personal property on parties interest-ed; executing other writs and summonses as directed by the court;
b. Conducting and advertising sales of property to satisfy writs of execu-tion for the satisfaction of mortgages and judgments;
c. Attending court;
d. Making collections of certain court costs and settlements;

2. As officer of the criminal courts:

a. Executing writs of the court;
b. Making arrests in certain cases;
c. Transporting prisoners to institutions;
d. Making commitments to jail or place of detention;
e. Making discharges of prisoners;
f. Collecting court costs in some counties;
g. Attending court; having charge of prisoners brought to trial;
h. Assisting in selection of jurors; summoning of jurors;

3. As officer of the peace:

a. Administering the county prison;
b. Making arrests on certain occasions;

      c. Suppressing riots and insurrections;
      d. Issuing fire-arm permits;

   4. As election officer:

      a. Proclaiming elections; listing election districts, polling places and candidates;

   5. Keeping records and financial dockets and accounts;

  f. <u>Functions of the District Attorney's Office</u>

   1. Preparing for the grand jury indictments of persons charged with crimes and violations of the law from returns of justices of the peace;

   2. Initiating proceedings for arrest and detention of persons suspected or accused of crimes;

   3. Prosecuting all persons against whom a true bill is found before the trial court. Sometimes, computing court costs of witnesses and juries;

   4. Taking charge of detectives to gather evidence against those to be prosecuted;

   5. Checking up on justices' and aldermen's courts to put a stop to cases which have not sufficient cause to go before the criminal court;

   6. Sometimes, administering desertion and non-support cases.

  g. <u>Functions of the Coroner</u>

   1. Investigating cases of sudden or suspicious deaths, or deaths while in prison or unattended by a physician; holding an inquest if necessary, having the power to compel the attendance of jury and witnesses;

   2. Acting as sheriff if vacancy exists.

  h. <u>Functions of the Prothonotary's Office</u>

   1. As clerk of the court of common pleas, has charge of all records, documents, papers, dockets, etc., arising out of civil cases.

  i. <u>Functions of the Clerk of the Courts' Office</u>

   1. Acting as clerk of the court of quarter sessions, and Oyer and Terminer and General Jail Delivery, has charge of all records, documents, papers, dockets, etc., arising out of criminal cases;

   2. Computing and collecting criminal court costs in most of the counties.

  j. <u>Functions of the Recorder of Deeds Office</u>

   1. Making records of all mortgages, deeds, letters of attorney, leases, assignments, charters, judgments, soldiers' discharges, searches, etc.;

   2. Making a monthly report to the commissioners, of changes in mortgages, judgments, etc., for the purpose of correcting assessment books;

   3. Keeping indexes of such records, and collecting fees for such services.

  k. <u>Functions of the Register of Wills and Clerk of Orphans' Court Office</u>

   1. Probating wills and granting letters testamentary;

2. Recording documents relating to the settlement of estates;

3. Issuing marriage licenses;

4. Attending sessions of Orphans' Court;

5. As an agent of the state, collecting inheritance taxes.

1. Functions of the Jury Commissioners

Drawing of the jurymen from list of persons selected and placed in the jury wheel, for all juries, grand, traverse, and petit.

In addition to these important officers, there are court officers, including messengers, tipstaves, criers, probation-desertion-parole officers, juvenile officers and road viewers. There are also subordinates in the various elective offices, as well as officers such as the assessors, boards of assessment and revision of taxes, tax collectors, election officers, constables, justices of the peace and aldermen, who are paid county money for services rendered. These, as well as many of the elective offices, are discussed more completely in other chapters of this report, especially those on personnel, justices of the peace, election administration and assessments. They are all mentioned at this point in order to present a complete picture of the county organization.

CONCLUSIONS

FINDING - Our general outline of the county as a working unit shows that its various activities and offices are disconnected. By reason of dissipated power and responsibility the county is weak in form, function and procedure.

Its most urgent needs are:
Integration of its internal structure by greater concentration of authority, i. e.
Functional organization of departments
Grouping of all operations connected with the same function in a single department
Freedom and elasticity of organization to fit needs of each area
Standards of performance and costs
Some degree of state administrative supervision arises as a necessity in order to satisfy these needs.

In presenting the following recommendations attention has been directed both to the urgency of these needs and to the bounds and limits of political possibility. It has been thought wise to offer alternative proposals wherever conditions permitted.

RECOMMENDATIONS - As the present State Constitution contains certain provisions relating to county government and enumerates certain elective county officers, we shall attempt to make recommendations (1) Those which could be adopted without changing the Constitution and (2) Those which should be adopted if the Constitution is revised. The suggestions are so arranged that both sets could be used if the Constitution were changed.

A. Recommendations Requiring No Constitutional Change

1. Relaxation of statutes making certain expenditures specifically mandatory on the county, viz.-

a. Salaries of all elected and appointed officers, excepting the members of the county salary board (county commissioners and controller); (see recommendations following chapter on Personnel).

b. Compensation of court officers;

c. Land damage claims on state highways within the county;

to the end that county financial officers would have greater freedom in making and enforcing a budget and in developing responsibility for county finance in the offices of the county commissioners, county controller, and county treasurer.

2. A thorough revision of statutes relating to compensation and fees received for services rendered by county officers, with a view to making them more flexible and more commensurate with services actually rendered. For example, salaries of elective heads of departments are, in most cases, too high (see chapter on Personnel).

3. A change in statutes on county financial offices, looking toward the centralization of financial power in the county commissioners, county controller and county treasurer

a. The county commissioners should be given more freedom to initiate or curtail county expenditures. They should be unrestricted by the administrative authority of the judges or by any other county agencies.

b. The controller should be given the power to audit all funds handled by county officers, and to determine the legality of all bills presented for payment to the county. At present his power is somewhat restricted in regard to court costs. His office should be in charge of the computation of all court costs and should have power to audit and scrutinize the accounts of all local officers paid by county funds.

c. The county treasurer should handle the collection of moneys due the county in:

1. Taxes
2. Fees for services of county officers:--
   Fee officers should collect no moneys due the county but should make out vouchers for payment to the county treasurer. This would eliminate duplication of bookkeeping and introduce modern methods of collection and computation.
3. Fees now collected on the road by the Sheriff or detectives should be collected by deputy treasurers.

These three offices, the county commissioners, controller and treasurer, are an excellent nucleus for the development of the centralization, consolidation and integration of financial power and responsibility so necessary in the county. If the powers of the salary board of which they are members were extended to include the appointment through their executive agent, the chief clerk, of all clerks and deputies in all county offices, as well as the determination of salary and number, these five officers could be regarded as the central administrative agency for a modern county government without the need of any constitutional changes. As members of the sinking fund commission, the salary board, the prison board, the board of revision of assessments (or as responsible for the appointment of the board of assessment and revision of taxes), these five officers (singly or together) are the financial powers that decide on tax levy, assessments, personnel, purchasing, auditing, collecting, appropriating and the rest of the powers of finance. This virtually makes them a commission of county government charged with policy determination and the financial functions of the county. All that is needed is an enlargement of their powers so that they might be less restricted by either state or county agencies.

4. The state should legislate less on content and more on form and procedure of county government. Thus it would be advisable to have state legislation making mandatory the adoption of modern financial practices. Improved practices in regard to budgeting, purchasing of supplies, annual financial reports, uniform accounting and classification of expenditures, independent audits, abolition of fees as compensation, a uniform rate and practice regarding mileage, the development of costs analyses, personnel methods, and other such modern governmental practices, should be adopted by state law and made mandatory for counties. On these points the Survey and other interested agencies have received a comprehensive series of suggestions as to accounting procedure

and other financial practices from Charles J. Rowland, Associate Professor of Economics, The Pennsylvania State College. Briefly these suggestions are:

1. Establish courses in colleges and universities for training financial officers and employees

2. Appoint officers on a merit basis

3. Establish a state agency to set up uniform accounting systems and practices in all local governments

4. Set up budgetary control

5. Require uniform financial reports at regular intervals

6. Have municipal auditing performed by certified public accountants

5. The judges of the district court, both as judges of the court of common pleas and of quarter sessions, should be relieved of county administrative duties as far as possible. The number and salaries of court employees should be fixed by the salary board. Duties regarding elections should be given to the county commissioners, as also those relating to public works. Such a change would aid in realizing a major objective- the concentration of administrative responsibility.

6. Another possibility of consolidating county functions in fewer departments without changing the Constitution lies in the clerical officers - the clerk of the courts, prothonotary, register of wills, and recorder of deeds. In eighth class counties, at present, all these offices are held by one person. In some of the higher classes, they are held by two or three officers. This consolidation of offices in the smaller counties is made possible by law. A complete consolidation of all four elective clerical offices in all counties is recommended. This would mean the election of one officer who would be prothonotary, clerk of courts, register of wills, and recorder of deeds, as is now done in the eighth class counties. Thus three large salaries would be abolished, the work would be allocated through deputies, there would be a flexibility of clerks to help out in the work, from one department to another, and the county would save money. All these offices should produce revenue for the county. New legislation should require that the fees earned by these offices as agents of the state, such as the register of wills as collector of inheritance taxes should revert back to the county treasury, and not to the pockets of the officers themselves. The salary board should be given the power to determine the number, the compensation, and the appointment of the subordinates in this consolidated office. The head of the office would, of course, be a member of the salary board, as at present.

7. Another consolidation could be made without constitutional change, in the functions and offices relating to public charities and welfare. The following suggestions are made by Mr. Arthur Dunham, Secretary of the Family Welfare and Relief Division of the Public Charities Association of Pennsylvania, who has completed an intensive examination of this problem.

    a. Establish a single public welfare board in each county to have charge of care of the poor, mothers' assistance, old age assistance, pensions for the blind, dependent children, insane, feeble minded, and epileptic persons, and unemployment relief.

    b. The board should consist of seven citizens of the county.

    c. The board should serve without pay, but be reimbursed for traveling expenses.

    d. The term of office should be six years with overlapping terms.

    e. The board should appoint an executive secretary and he, with the approval of the board, should select a staff of assistants. A state

civil service merit system should be used for the selection of the
executive secretary and staff.

> f. Salaries of director and staff should be fixed by the public welfare
> board in an annual budget and submitted to the county commissioners for
> approval.

8. Contrary to popular belief, the known facts do not warrant a recommendation for compulsory consolidation of counties at present. It is therefore proposed that
only permissive consolidation be contemplated. Although there are 28 counties of
50,000 or less population, and 11 counties of 20,000 or less, the problem of consolidation presents such formidable political difficulties, with such uncertain benefits,
that experiment over a period of years must determine the decision. There is also
serious doubt as to whether consolidation would save money for the taxpayers (see
Weaver and Alderfer, op. cit.).

B. Recommendations Requiring Constitutional Change - Other parts of this report
show that a basic weakness of county government is the existence of the disconnected
"row" elective officers. To remedy this weakness and to make county government simpler,
more effective, and less costly, the Constitution of the Commonwealth should be so
amended that the legislature could authorize the establishment in each county of such
officers as can most effectively administer its affairs.

Suggested County Organization - The following structure is suggested as being
especially adapted to Pennsylvania conditions. (For a graphic presentation, see the
Charts in the next chapter).

1. Three county commissioners should be elected in each county as at present.

2. In the larger counties there should be an elected controller; in the
smaller counties an annual audit should be made by auditors appointed by a central state
agency, and the other powers of the controller should be exercised by the chief clerk.

3. All other county administrative officers should be appointed. In the
larger counties the heads of departments should consist of a chief clerk, county treasurer, chief court officer, district attorney, chief of the Department of Corrections
and Charities, and chief of the Department of Public Works. In the smaller counties
the heads of the two last-named departments might do all the actual work of their
departments, or it might be performed by the commissioners themselves. In some cases
adjacent small counties might appoint the same man. Thus the district attorney might
have the same district as the judge of the court in the smaller counties. This district
frequently includes two or three counties.

4. As an alternative to 3 above, the commissioners should have power to create
the office of county manager. The manager should in such cases be the chief executive
officer of the county and should directly supervise the work of all administrative
departments. Subject to the approval of the commissioners, he should prepare the county
budget, appoint the heads of departments, and exercise the powers of the salary board.

5. All heads of departments should be appointed by the county commissioners,
but the appointment of the district attorney should be made from a list of eligibles
prepared by the county bar association, and should require the approval of the Attorney
General of the Commonwealth. The appointment of the chief court officer should be
subject to the approval of the judges of the county court. Appointed officers should
have four-year terms but should be removable by the commissioners at any time, after due
notice and hearing.

6. Allied functions should be grouped together and performed under the supervision of the department heads named in paragraph 3 above as follows:

(a) The chief clerk, as the head of the Department of Records, should have
the present duties of the chief clerk to the commissioners, the prothonotary, the clerk
of courts, the recorder of deeds, and the register of wills. His department should

also supervise the making of assessments.

(b) The county treasurer should perform the present duties of his office and, in addition, should collect all local taxes, both current and delinquent, and should directly collect all fees accruing to the county or state for services rendered by county officers.†

(c) The chief court officer should have the present duties of the sheriff to execute the writs of all courts, but should not have the present powers of the sheriff in making arrests, administering the county prison, transporting prisoners, etc. He should supervise court routine and should supervise such court officers as stenographers, tipstaves, criers, messengers, etc.

(d) The district attorney's office should have all the present duties of the office and, in addition, the duties of the sheriff with regard to the arrest of offenders, the transportation of prisoners, etc., the latter duties to be performed by the detectives attached to the office.

(e) The Department of Corrections and Charities should have charge of the county jail, probation, parole and desertion administration, and all county welfare institutions, such as hospitals, children's homes, etc. It should also control poor relief, which should be on a county-wide basis.

(f) The Department of Public Works should have the functions of the county surveyor and county engineer, and should be in charge of the construction and maintenance of all county public works. In the counties where there are not enough public works to make the department necessary, its functions should be performed by the county commissioners.

7. The office of coroner should be abolished.

8. The commissioners, controller, and the head of the department in which an appointment is to be made should constitute the salary board, with authority:

(a) to determine the number and compensation of subordinate employees:

(b) to develop modern and efficient methods of recruiting employees, classifying positions, providing for promotions and permanent tenure of office, and, with the treasurer, developing a pension system in the larger counties.

9. The local election of assessors, tax collectors, justices of the peace, registry assessors, constables and election officers in townships, boroughs, and wards should be discontinued. Registry assessors should be dispensed with entirely, while the tax assessors and election officers should be appointed by the Salary Board. Justices of the peace and constables should be appointed by the judges of the county court. In many cases the districts over which these minor officers have jurisdiction should be enlarged, either to give full-time work or to promote economy. Each of these officers should be under the control of a county officer; the assessors under the chief clerk, or the employee in the chief clerk's office who supervises assessment routine; justices of the peace under the judges of the county court; constables under the chief court officer; and election officers under the chief clerk. Deputies of the county treasurer should collect all county and local taxes, and other moneys due the county.

10. The present ex-officio boards and commissions should be retained, with their present or added functions. These are the salary board, board of revision of assessments, the sinking fund commission, pension board (if there is a pension system), and the prison board. The prison board should be changed to a board of county institutions, composed of the commissioners, controller, if any, and the chief of the Department of Correction and Charities.

The board of assessment and revision of taxes might be retained for second and

---

† The weaknesses of the present system of tax collection are fully discussed by Blake E. Nicholson, in "Collection of Local Taxes in Pennsylvania," 1932.

third class counties, but the practice of appointing three highly paid officers to supervise assessments is not feasible for the smaller counties. Furthermore, it is questionable whether such a board is best suited to supervise the making of assessments, except at the time when the new county-unit assessment plan is installed. There should, preferably, be an expert in assessment routine in the Department of Records. Assessment policy and the hearing of appeals should still be the duty of the county commissioners.

State Supervision - Elasticity and adaptability to varying conditions should be a fundamental feature of our local system. Yet certain aspects of county administration require not only supervision, but the collection of uniform data on matters of general importance, and helpful information and advice. To fill this need the Bureau of Municipalities in the Department of Internal Affairs should be given power to require uniform financial practices in the various counties; to enforce debt limitations; to collect all local government statistics; to instruct county officers in the best modern practices applicable to their work; to develop better administrative methods; and to advise on local problems. It should have similar powers over all local subdivisions and should advise the legislature on all proposed changes in the laws relating to local government.

CHAPTER III

PERSONNEL

INTRODUCTION - The problem of personnel is of major importance in the government of Pennsylvania counties. A large share of current county expenses, in one case over fifty per cent, is devoted to compensation for personal services. There is no civil service or merit system in operation, slight evidence of classification of positions, and no regulation designed to make county positions attractive as permanent employment. On the other hand, there is an increasing need for the trained official and the expert in county government. He is needed in assessments, personnel management, financial management, the new duties in social service, and in executive work generally. If, as the conclusions of this Survey advocate, the county is to develop as a main unit of local government, it must be made more economical and efficient. An essential step in this direction is the building up of a competent body of officers and employees.

The subject of personnel will be considered under the following main topics:

    I. Constitutional and statutory provisions

    II. Actual operation of personnel administration
        a. The number of county employees
        b. The salary board in operation
        c. Actual turn-over of county employees
        d. Solicitors in county government
        e. Cost of personal services in proportion to total county costs
        f. Are departmental heads necessary?

    III. Recommendations

I. CONSTITUTIONAL AND STATUTORY PROVISIONS

a. The Constitution on County Officers - The Constitution of Pennsylvania is the greatest obstacle to reform of county government. In Article XIV, section 1, we find:

> "County officers shall consist of sheriffs, coroners, prothonotaries, registers of wills, recorders of deeds, commissioners, treasurers, surveyors, auditors or controllers, clerks of the courts, district attorneys and such others as may from time to time be established by law; and no sheriff or treasurer shall be eligible for the term next succeeding the one for which he may be elected."

This means, in effect, that the legislature may not abolish these offices, although it may consolidate them as it did in the smaller counties. The Constitution likewise declares that county officers shall be elective, holding their offices for four years. Article XIV, section 2 declares:

> "County officers shall be elected at the municipal elections and shall hold their offices for a term of four years, beginning on the first Monday of January next after their election, and until their successors shall be duly qualified; all vacancies not otherwise provided for, shall be filled in such manner as may be provided by law."

Furthermore, one year's residence within the county and citizenship are necessary to qualify for appointment to a county office. Certain county elective officers: the prothonotaries, clerks of court, recorders of deeds, registers of wills, county surveyors and sheriffs, must locate their offices at the county seat (Article XIV, sections 3 and 4).

The compensation of county officers, according to the Constitution, shall be regulated by law. Article XIV, section 5 declares:

> "and all county officers who are or may be salaried shall pay all fees which they may be authorized to receive, into the treasury of the county or

State, as may be directed by law. In counties containing over one hundred
and fifty thousand inhabitants all county officers shall be paid by salary,
and the salary of any such officer and his clerks heretofore paid by fees,
shall not exceed the aggregate amount of fees earned during his term and
collected by or for him."

Counties containing one hundred and fifty thousand inhabitants and over include 1st,
2nd, 3rd and 4th class counties. In 1934, there were 18 such counties. Furthermore,
the Constitution provides that the General Assembly shall provide by law for the strict
accountability of all county, township, and borough officers, as well for the fees which
may be collected by them, as for all public or municipal moneys which may be paid to
them (Article XIV, section 6). Article XIV, section 7, provides for limited voting in
the case of the three county commissioners and three county auditors. Each elector
shall vote for two, the three highest winning. This is an attempt at minority representation. It sometimes does not operate as intended. This section likewise provides that
any vacancy in the office of county commissioner or county auditor shall be filled by
the court of common pleas.

    Thus the Constitution, as it now stands, does not permit the elimination of elec-
tive "row" officers. It also makes one year's residence a qualification of appointed
county officers. These provisions militate against that consolidation of county offices
which is so much needed to insure effective and economical administration.

    However, it is interesting to note that the constitutional amendment regarding the
consolidation in Allegheny County does not include these provisions and as a result the
new government of this area will not be restricted in the same manner as the other
counties. A similar amendment has been insistently urged for Philadelphia. The problems in these two counties of the first and second classes respectively can not be
discussed in this report.

b. _Statutory Enumeration of Elective Officers_ - The following officers shall be elected
by the qualified electors of the county: (1929, P. L. 1278, Article 3, section 51).

> Three county commissioners
> Three auditors or one controller
> One treasurer
> One surveyor
> One coroner
> One recorder of deeds
> One prothonotary
> One register of wills
> One clerk of the court of quarter sessions and court of oyer and terminer
> One clerk of the orphans' court
> One sheriff
> One district attorney
> Two jury commissioners

c. _Differences Among Counties As to Elected Officers_

(The Controller and Auditors)

    As indicated in the previous chapter, counties are classified for purposes of State
legislation into eight classes according to population. In counties of the 2nd to 5th
class, the office of controller is established and the office of auditor abolished.
However, the office of controller may also be established in the smaller counties
by decree of the court of common pleas on petition signed by 25 per cent of the
number of electors voting in the last election and by a number of county officers
(1929, P. L. 1278, Article 3, sections 131 and 136).

(The Clerical Offices)

    In 2nd, 3rd and 4th class counties, one person shall hold each of the offices of
prothonotary, clerk of courts, recorder of deeds, and register of wills. The clerk of

courts is clerk of both the courts of oyer and terminer and quarter sessions. The register of wills is also clerk of the orphans' courts.

In 5th class counties, one person shall hold the offices of prothonotary and clerk of courts, while one shall be recorder of deeds, and one register of wills and clerk of orphans' court.

In 6th and 7th class counties, one person shall hold the offices of prothonotary and clerk of the courts, and one person shall be register of wills, clerk of orphans' court and recorder of deeds.

In 8th class counties, one person shall hold all the offices of prothonotary, clerk of courts, recorder of deeds, register of wills and clerk of orphans' court.

There are some local laws still in effect which make special provisions for specific counties, differing from those which apply to other counties of the same class. Each county operating under special local laws may voluntarily abandon its special status by accepting the general law for its class. For example, Centre county (7th class) has one person for the office of recorder of deeds and one for the offices of register of wills and clerk of the orphans' court.

(The Surveyor and the Coroner)

Although these officers are enumerated in the State Constitution and laws as elected officers of the county, few counties elect surveyors. While most counties have a coroner, the duties of this office are in some places (e. g. Forest County) performed by one of the justices of the peace.

d. Salaries of County Officers and Employees - Four methods are employed to fix the compensation of county officers and employees. These are:

    1. By statute
    2. By salary board
    3. By head of department or county court
    4. By fees taken in by the office

1. Statute - There are a large number of county officers whose salaries are fixed definitely by statute. This group includes all elective officers in counties where payment is made by salaries instead of by fees, and a certain number of appointive officers in various classes of counties. Added to these are the officers who are paid a per diem rate, established by law.

2. Salary Board - The number and compensation of county employees not fixed by statute is determined by the county salary board in all counties from the 2nd to the 6th class, inclusive. In 2nd, 3rd and 4th class counties, it consists of the county commissioners and controller; in 5th class counties, of the county commissioners, the controller, and treasurer; in the 6th class, of the county commissioners and the treasurer. In 7th and 8th class counties, there is no body of this nature, since most of the county officers are paid by fees. Where salaries are paid, they are fixed by state law. The salary board plan was established by legislation in 1876. In 1915 and 1923, it was extended to include 2nd, 5th and 6th class counties. Philadelphia County does not operate under a salary board because fiscal affairs are in charge of the city council. When considering the number and compensation of employees in a certain office, the salary board includes the head of the particular office under consideration. Thus, when considering how many employees shall be hired for the sheriff's office, the sheriff is a member of the salary board. When employees of the prothonotary's office are being determined as to number and compensation, the prothonotary is a member of the salary board. A majority decision prevails, although appeal to the courts by anyone dissatisfied is made possible by the law. In sixth class counties, however, the salary board consists only of the county commissioners and treasurer, not including the head of the department under consideration. The acts for the various counties usually indicate that "deputies and clerks" shall be appointed by the salary board, although there are a few appointive officers who cannot be considered as either "deputies" or "clerks."

3. <u>Head of Department or County Court</u> - In some cases, the salaries of employees are fixed neither by the state law nor by the salary board. For example, some of the salaries of the court officers and employees of the district attorney's office are fixed or approved by the court. The county commissioners, as such, fix the compensation of certain officers appointed by them in counties not having a salary board. Such is even the case in the employment of stenographers for the office of the district attorney and for the probation office in 7th class counties, where such employees are not appointed by the commissioners, but where they must allow the compensation from the county fund. While there are not many such employees, the class illustrates the diversity of methods employed.

4. <u>Fees</u> - There are two main divisions of officers and employees affected by the amount of fees taken in by the office in which they are employed. The first includes those who are paid by salary in lieu of fees but whose salary is contingent upon the collection of enough fees to pay the total salary expenditure in the office. The second class contains those whose compensation is entirely contingent upon the fees collected by their offices.

The first class: Prior to the Constitution of 1874, most county officers were compensated by fees instead of salaries. In the more populous counties, the aggregate amount of these fees was often large even after the head of the office had paid his employees. Therefore, the Constitution provided that:

> "In counties containing over one hundred and fifty thousand inhabitants all county officers shall be paid by salary, and the salary of such officer and his clerks heretofore paid by fees, shall not exceed the aggregate amount of fees earned during his term and collected by or for him." (Article XIV, section 5)

This abolished the fee system for county purposes as far as compensation was concerned for counties of the first four classes and required that all salaries fixed by law must be paid from the amount of fees paid into the county treasury by the particular office. After ascertaining and deducting salaries due to employees and clerks, if there is not enough left to pay the salary of the head of the office, he will receive only such proportion of his salary as shall be equal to the aggregate of the net fees received and earned by him during such monthly period. However, if in subsequent months there shall be a sum of fees paid into the treasury which shall exceed the amount needed for current salaries, the deficit of the preceding month shall be made up. This act has been construed to mean that surpluses of subsequent or preceding years can be applied to deficits in any year of the term of the officer affected (1876, P. L. 13, section 6).

The law states, however, that certain specified officers shall be paid monthly or quarterly the full amount allowed by law. These include: the county solicitor, county surveyor or engineer, county detectives, county treasurer, interpreter of courts, district attorney and his assistants. The others shall be paid the amounts assigned by law only when the net receipts of their respective offices shall reach amounts fixed for them. These include the sheriff, recorder of deeds, clerk of courts, prothonotary, register of wills and coroner. This act applies only to 1st, to 4th class counties (1776, P. L. 13, section 16; 1887, P. L. 182, section 1; 1901, P. L. 261, section 1).

In all counties where salaries are fixed by law, these shall be in lieu of any fees, mileage, perquisites, or moneys received as compensation. All income to the office shall be turned over to the treasury. The fees received by officers such as the treasurer and register of wills, as agents of the state, still accrue to those officers personally. The same applies to any moneys collected by them as agents of the national government.

The second class includes employees and officers, mostly in the counties of the 7th and 8th classes, who are paid entirely by fees collected in their offices. Thus the officer hires as many employees as he deems advisable and pays them out of his own pocket at rates fixed by agreement. Under such conditions the compensation of deputies is much lower than if they were paid from county funds. There are few records of the income of such offices, but such as there are indicate in many cases an income far above the standard of ability and effort required in such work.

The salaries and fees of county officers as fixed by statute, together with the authority determining the rate of pay in cases where this is not specified by law, will be found in the "Report of the Salary Survey Commission to the Pennsylvania General Assembly", (1929), Page 48.†

e. Appointment of County Officers and Employees - The non-elective officers and employ-ees of the county are appointed by one of three major authorities:

    1. The County Commissioners
    2. The Judges of the District Court
    3. The Head of a Department of County Government

While the county commissioners are usually the chief appointing agency, the aggregate of appointments made by the heads of the other departments is greater. In many counties the number of places filled by judicial appointment is large, including probation offi-cers, tipstaves, court stenographers, messengers, juvenile officers and desertion offi-cers. In all counties this number is much greater than is popularly supposed. Judicial appointment of local officers is a survival of the time when the court was the real administrative authority of the county. It is also a high tribute to the respect and confidence in which the judge has been held. Under present conditions it has two un-fortunate consequences. By giving the judge important patronage (in Lackawanna county 40 employees) it confers on him a power which has far-reaching political implications. This in turn necessarily injects politics into the judicial office. Instances are not wanting in which a judge has become the recognized leader of his party in the county. These considerations suggest that the county salary board should have some authority in determining the number and compensation of the court officers except the tipstaves and criers, and that the board or an officer chosen by it should be consulted in the appointment of the other court employees.

It must be emphasized that salary boards do not now appoint the officers and em-ployees. This is done by the heads of the departments. The salary board merely de-cides the number and compensation of the deputies and clerks in the various departments.

Illustrations of the practical operation of appointments are presented in the following charts indicating the personnel in Montgomery, a third class county, Tioga a seventh class county and Forest an eighth class county. Montgomery is one of the large counties of the state, while the others are small. Forest is the smallest in population of the entire sixty-seven.

TIOGA COUNTY

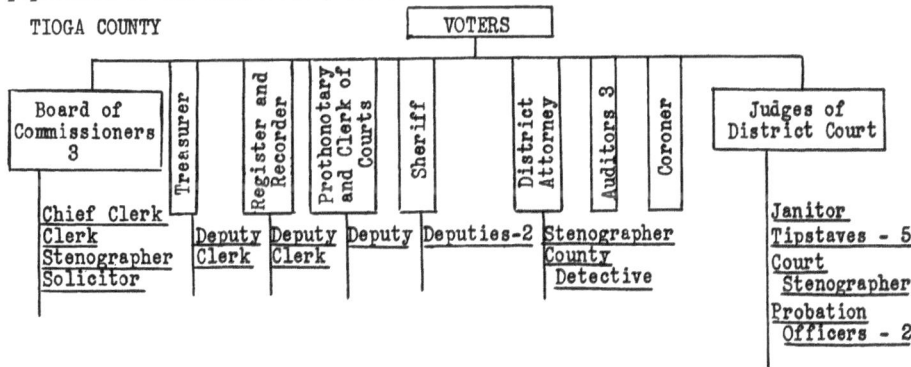

---

† The report of the Salary Survey Commission has been brought up to date in F. P. Weaver and H. F. Alderfer, "County Government Costs in Pennsylvania" (1933), Page 7.

MONTGOMERY COUNTY

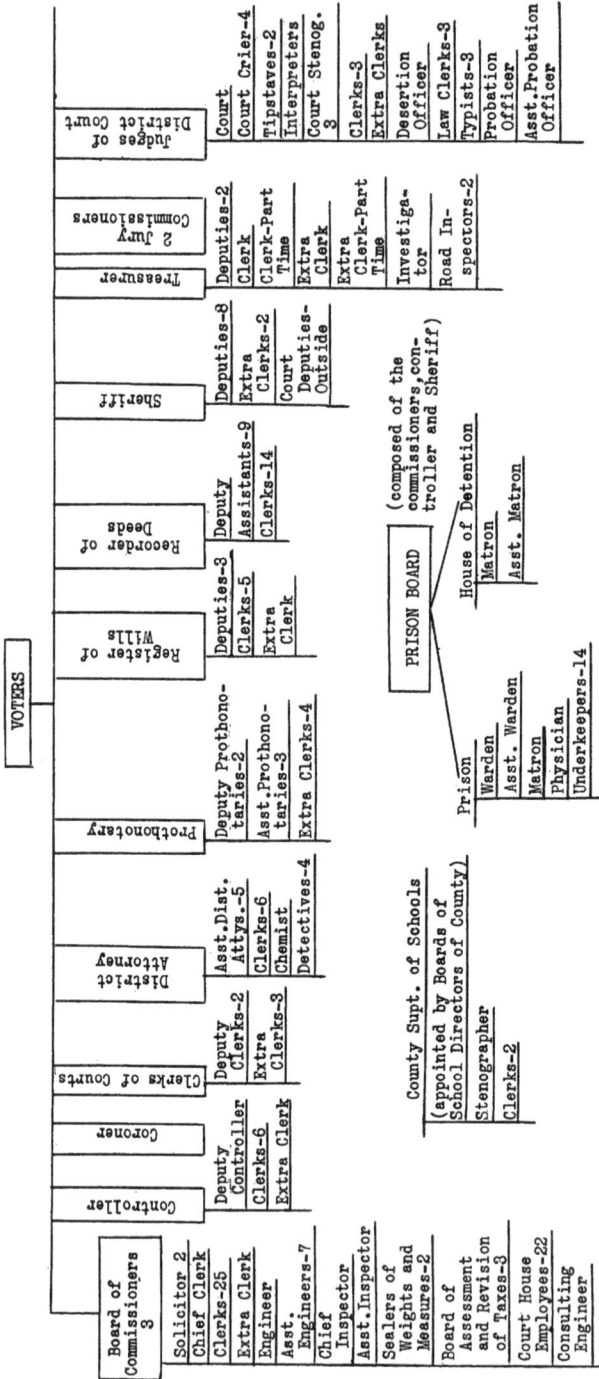

VOTERS

Board of Commissioners 3
- Solicitor 2
- Chief Clerk
- Clerks-25
- Extra Clerk
- Engineer
- Asst. Engineers-7
- Chief Inspector
- Asst.Inspector
- Sealers of Weights and Measures-2
- Board of Assessment and Revision of Taxes-3
- Court House Employees-22
- Consulting Engineer

Controller
- Deputy Controller
- Clerks-6
- Extra Clerk

Coroner

Clerks of Courts
- Deputy Clerks-2
- Extra Clerks-3

District Attorney
- Asst.Dist.Attys.-5
- Clerks-6
- Chemist
- Detectives-4

Prothonotary
- Deputy Prothonotaries-2
- Asst.Prothonotaries-3
- Extra Clerks-4

Register of Wills
- Deputies-3
- Clerks-5
- Extra Clerk

Recorder of Deeds
- Deputy Assistants-9
- Clerks-14

Sheriff
- Deputies-8
- Extra Clerks-2
- Court Deputies-Outside

Treasurer
2 Jury Commissioners
- Deputies-2
- Clerk
- Clerk-Part Time
- Extra Clerk
- Extra Clerk-Part Time
- Investigator
- Road Inspectors-2

Judges of District Court
- Court
- Court Crier-4
- Tipstaves-2
- Interpreters
- Court Stenog. 3
- Clerks-3
- Extra Clerks
- Desertion Officer
- Law Clerks-3
- Typists-3
- Probation Officer
- Asst.Probation Officer

County Supt. of Schools (appointed by Boards of School Directors of County)
- Stenographer
- Clerks-2

PRISON BOARD (composed of the commissioners, controller and Sheriff)

Prison
- Warden
- Asst. Warden
- Matron
- Physician
- Underkeepers-14

House of Detention
- Matron
- Asst. Matron

FOREST COUNTY

VOTERS

| Board of Commissioners 3 | Register Recorder Prothonotary & Clerk of Courts | Treasurer | Sheriff | Coroner | 3 Auditors | District Attorney | Judges of District Court | 2 Jury Commissioners |

Chief Clerk
Sealer

Clerk
County Sup. and Stenographer
Tipstaff
Court Crier
Court Stenographer
Probation Officer
Janitor
Matron

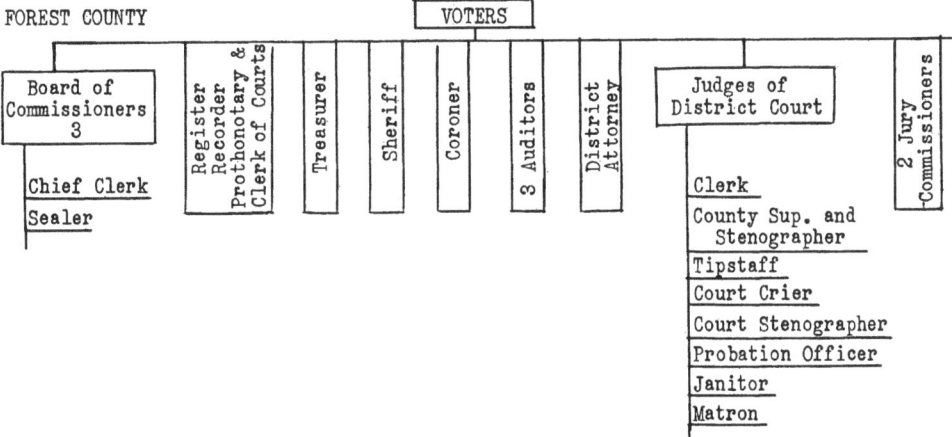

The differences in the system of appointment, if the suggestions made in Chapter 2 were adopted, are indicated by the following:

Proposed Forms of County Organization.

(Without a county manager)

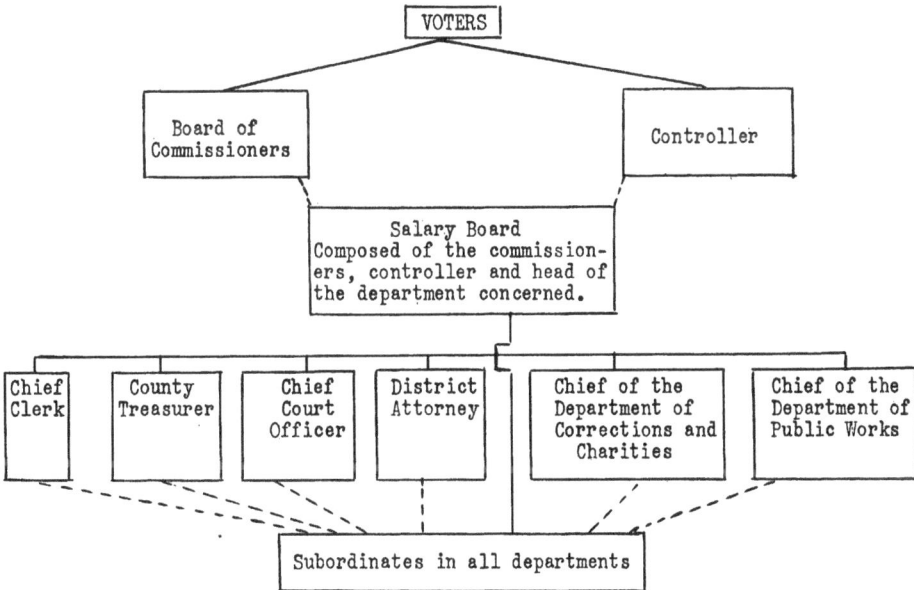

VOTERS

| Board of Commissioners | | Controller |

Salary Board
Composed of the commissioners, controller and head of the department concerned.

| Chief Clerk | County Treasurer | Chief Court Officer | District Attorney | Chief of the Department of Corrections and Charities | Chief of the Department of Public Works |

Subordinates in all departments

Proposed Forms of County Organization

(With a County Manager, Acting also as Chief Clerk)

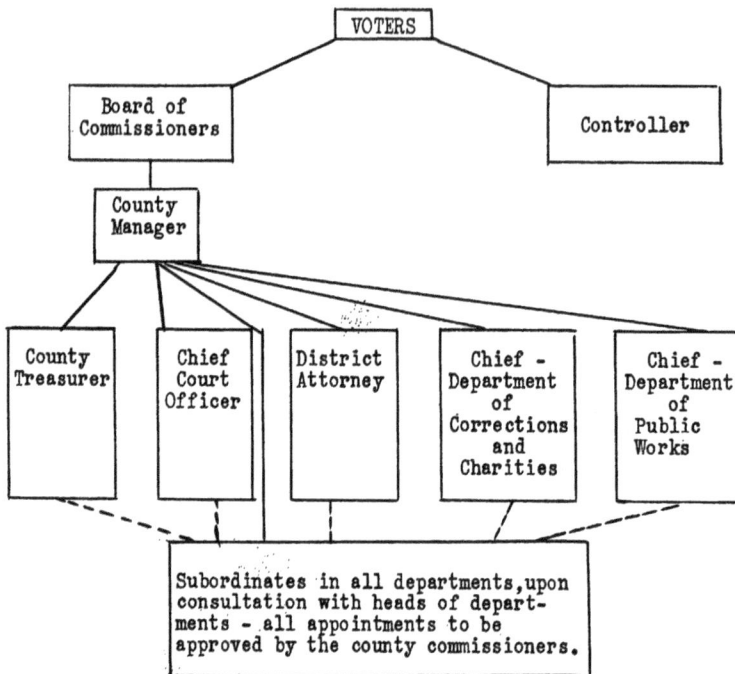

```
                              ┌────────┐
                              │ VOTERS │
                              └────────┘
         ┌──────────────┐                    ┌──────────────┐
         │ Board of     │                    │  Controller  │
         │ Commissioners│                    │              │
         └──────────────┘                    └──────────────┘
         ┌──────────┐
         │ County   │
         │ Manager  │
         └──────────┘
```

County Treasurer	Chief Court Officer	District Attorney	Chief - Department of Corrections and Charities	Chief - Department of Public Works

Subordinates in all departments, upon consultation with heads of departments - all appointments to be approved by the county commissioners.

f. **County Pension System** - Counties of the first and second class are authorized to establish pension systems for their employees. As these may develop into much future importance, their present legal basis in 2nd class counties is presented in some detail. An employee is defined as any person employed by such a county at a wage or salary payable at stated intervals. This does not include any elective officer. A pension fund shall be established under the supervision of a pension board, consisting of the county commissioners, the controller and the treasurer of the county. The treasurer shall receive $100 a year for his work as treasurer of the pension board.

The board shall keep a register of employees, containing the name, address, occupation, the time of entering county employment and other such information. The board shall adopt and alter, in its discretion, such regulations as it deems necessary. The head of each department shall file with the board a list of employees with certain items of information and shall report all changes in status of the employees to the board.

The county commissioners shall appropriate and pay into the pension fund not less than one-half of one per cent, and not more than two per cent, of all available moneys received by the county as taxes during the preceding year. Each employee shall pay into the pension fund three per cent of the amount received by him every month, not exceeding $8.00 per month. It is to be collected by the treasurer and paid into the pension fund. No employee shall be entitled to a pension unless such payment is made as required. The board invests this fund in securities that are considered safe. Persons over fifty years of age who have been with the county for twenty years, not necessarily continuous service, are entitled to a life pension. Persons with the county ten years or more who are permanently disabled are entitled to a pension. The pension paid shall equal 50 per cent of the average annual amount received by the employee as salary or wages during the two years immediately preceding the date of retirement. No pension shall exceed one hundred dollars a month, and pensions are payable in monthly installments.

This plan is described briefly as an example of an institution that can go far in bolstering up the morale of county employees. It must, however, go hand in hand with a plan for choosing personnel on a merit basis and a system of tenure of office during good service.

g. Other Provisions Concerning County Officers - There are several other general provisions relating to county officers such as provisions as to their bonds, the state associations to which they may belong as county officers, the qualifications for various offices, etc. It may be said that the qualifications for county office are so general as to be no qualification at all. Political appointments cannot help but be the rule once the system of elective "row" officers is understood.

ACTUAL OPERATION OF PERSONNEL ADMINISTRATION

Observations concerning the operation of personnel methods are summarized under the following headings:

(a) The number of county employees

(b) The salary board in operation

(c) Actual turn-over of county employees

(d) Solicitors in county government

(e) The percentage that cost of personal services is of total county expenditures for current purposes

(f) Are department heads necessary?

a. The Number of County Employees - Table I gives the total number of county full-time employees in 1933. These data were taken from the regular county pay-rolls, or from the county controllers' or auditors' reports when individual names were given. Full-time employees are those receiving $500 per year or over. Officers and employees who were paid from both the county fund and from fees are in this group. The fees were tabulated by an office-to-office canvass in counties employing the fee system. For purposes of comparison, the employees of county charitable institutions were omitted in all counties. Thus, employees of almshouses, children's homes, and county hospitals were omitted while employees of jails and detention homes for juveniles were included.

Other features of the table are the population of the county per employee and the total salary expenditure per capita. Employees engaged in personal services (assessments, tax collection, and charitable activities) and mentioned in other chapters of this report were excluded from the total of salary expenditures.

Table I.   County Personnel - All Offices
                                      1933

County	Class	Total Number of full time employees a	Population per employee	Total Salaries - full and part time employees	Salary Cost per capita
Lackawanna..................	3rd	H 223	1,392	P $441,715.56	$1.42
Westmoreland................	3rd	120	H 2,457	P 252,448.00	L .86
Montgomery..................	3rd	212	1,254	H P 507,389.84	H 1.92
Erie.......................	4th	108	1,623	P 222,223.00	1.27
Chester....................	5th	73	1,734	P 160,168.00	1.26
Northumberland.............	5th	64	2,008	P 153,759.00	1.22
Centre.....................	7th	b 35	1,323	P c 46,833.00	1.01
Adams......................	7th	b 27	1,377	41,825.76	1.13
Tioga......................	7th	b 31	1,028	P c 31,913.00	1.00
Monroe.....................	7th	b 29	975	P c 25,332.00	.90
Union......................	8th	b 19	919	c	
Forest.....................	8th	L b 13	L 399	L c 6,176.57	1.19
12 Counties................		1,054	1,393		

(a) Includes only employees receiving $500 or more; 3 auditors treated as one full time
employee if paid less than $500. Does not include employees of county almshouses and
hospitals, extra clerical and part time employees, assessors in the field, tax collectors, registrars and other election officers, poor directors and their employees, judges,
road viewers, employees of the mothers' assistance administration and of other charitable
enterprises.
(b) Both county and fee paid.
(c) Total does not include fees taken as compensation.
(P) From county schedules of the Pennsylvania Economic Council.
(H) High.   (L) Low.

     This table shows a wide range with respect to the population per employee in the
various counties. In the 12 counties there were 1,054 full-time employees, or an average of 1,393 inhabitants per county employee. The greatest population per employee was
found in Westmoreland County, with one employee to every 2,457 people. At the other
extreme was Forest, the smallest county in the State, with an employee for every 399
people. It must be understood that Forest County pays most of her officers small
salaries so that this comparatively high number is not reflected in salaries per capita.
It is interesting that Montgomery County is the only one of the six larger counties to
have a ratio of population to employees less than the average of the 12 counties, and
that the six smallest in population all have less than the average number of people per
employee. Again this is tempered, in the case of the smaller counties, by the low
salaries of many of the officers. Montgomery has the highest per capita salary cost with
$1.92, Lackawanna comes next with $1.42, while Westmoreland, another third class county,
is lowest of the twelve with $.86 per capita. This indicated that there is very little,
if any, relation between the size of the county and the amount of salary per capita.
The first two counties are prodigal in spending money for employees' salaries. Westmoreland County has about half the number of employees to perform the same services as
Montgomery and Lackawanna Counties. These latter counties have nearly the same population. Erie, Chester and Northumberland Counties are almost alike in salaries per capita,
although Northumberland has considerably less employees in proportion to its population
than either of the other two counties. The six smallest counties, with some of their
officers paid by fees, all pay salaries of approximately $1.00 per capita and, with the
exception of Forest County, have nearly the same population per employee.

     If any point is shown clearly by these figures, it is that there is little relation
between the number of employees and the actual services rendered the population. Westmoreland County has a government which appears to be as efficient as any observed, although it is by no means a perfect one. It operates with half the number of employees
of Lackawanna and Montgomery Counties, and renders comparable services.

     Table II shows the population per employee and the salary cost per capita of population for the 12 counties surveyed for the following county offices or groups of employees:

Commissioners' office
Controller
Treasurer
District Attorney
Court employees
Sheriff

Coroner
Court house employees
Register of wills
Recorder of deeds
Prothonotary
Clerk of courts

This is followed by Table III which shows the receipts of various county fee offices.

The figures in these tables are based upon tabulations of the number of employees found in county controllers' or auditors' reports, or county pay-rolls found in the controllers' offices, and from data secured in county schedules made by the Pennsylvania Economic Council.

Table II. Population per County Employee and Salary Costs, by County Offices, 1933

County	Class	Commissioners		Controller or auditors		Treasurer	
		Population per employee	Salary cost per capita	Population per employee	Salary cost per capita	Population per employee	Salary cost per capita
Lackawanna..........	3rd	6,898	.34	34,488	.07	34,488	.06
Westmoreland........	3rd	16,388	.14	49,165	.05	49,165	.04
Montgomery..........	3rd	5,537	.58	33,225	.08	26,580	.05
Erie................	4th	14,606	.17	35,055	.06	21,909	.05
Chester.............	5th	14,069	.26	63,315	.05	31,407	.11
Northumberland......	5th	12,850	.25	25,701	.11	64,252	.08
Centre..............	7th	6,613	.17	46,294	.03	15,431	.06
Adams...............	7th	6,188	.21	37,128	.03	18,564	.11
Tioga...............	7th	4,553	.24	31,871	.05	10,624	.08
Monroe..............	7th	4,041	.32	28,286	.04	14,143	.21
Union...............	8th	3,494	No data	17,468	no data	17,468	no data
Forest..............	8th	1,295	.41	5,180	.03	5,180	.15
		District Attorney		Court Employees		Sheriff	
Lackawanna..........	3rd	34,488	.11	7,759	.26	28,218	.07
Westmoreland........	3rd	22,691	.12	16,388	.12	49,166	.06
Montgomery..........	3rd	15,635	.18	12,082	.23	29,545	.11
Erie................	4th	29,213	.11	10,310	.18	21,909	.11
Chester.............	5th	25,326	.11	9,665	.12	42,210	.08
Northumberland......	5th	25,701	.10	25,701	.19	32,126	.07
Centre..............	7th	23,147	.08	5,787	.16	15,431	.17
Adams...............	7th	18,564	.06	7,426	.06	37,128	.21
Tioga...............	7th	10,624	.14	5,312	.18	10,624	.10
Monroe..............	7th	9,427	.11	4,041	.13	14,143	.03
Union...............	8th	17,468	no data	3,494	no data	17,468	no data
Forest..............	8th	5,180	.12	1,727	.04	5,180	.23
		Coroner		Courthouse Employees		Register of Wills	
Lackawanna..........	3rd	310,397	.01	8,389	.12	44,344	.06
Westmoreland........	3rd	98,332	.02	16,449	.09	73,749	.03
Montgomery..........	3rd	132,902	.01	9,165	.15	29,545	.10
Erie................	4th	175,277	.01	14,606	.09	25,039	.07
Chester.............	5th	126,629	.01	21,106	.04	42,210	.07
Northumberland......	5th	128,504	.01	42,835	.03	42,635	.06
Centre..............	7th	46,294	.01	15,431	.06	23,147	Fees
Adams...............	7th	37,128	.01	37,128	.02	24,752	.08

(Continued on next page)

Table II. Population per County Employee and Salary Costs, by County Offices, 1933
(Continued)

County	Class	Coroner		Courthouse Employees		Register of Wills	
		Population per employee	Salary cost per capita	Population per employee	Salary cost per capita	Population per employee	Salary cost per capita
Tioga................	7th	31,871	.01	31,871	.03	21,247	Fees
Monroe..............	7th		None	14,143	.06	18,857	Fees
Union................	8th	17,468	no data	17,468	no data	17,468	Fees
Forest..............	8th		.00	5,180	.12	20,720	Fees

County	Class	Recorder of Deeds		Prothonotary		Clerk of Courts	
Lackawanna...........	3rd	38,797	.05	23,878	.08	62,079	.04
Westmoreland.........	3rd	42,142	.05	42,142	.05	73,748	.03
Montgomery...........	3rd	10,632	.17	44,300	.08	88,600	.04
Erie.................	4th	17,528	.09	28,039	.08	87,638	.03
Chester..............	5th	18,079	.11	42,209	.06	42,209	.06
Northumberland.......	5th	42,635	.06	51,402	.06	51,402	.06
Centre...............	7th	23,147	Fees	46,294	Fees	46,294	.26
Adams................	7th	24,752	.08	18,564	.13	18,564	.10
Tioga................	7th	21,247	Fees	31,871	Fees	31,871	.12
Monroe..............	7th	18,857	Fees	28,286	Fees	28,286	.04
Union................	8th	17,468	Fees	34,936	Fees	34,936	Fees
Forest..............	8th	20,720	Fees	20,720	Fees	20,720	Fees

Table III. Receipts of County Fee Offices, 1933

Office	County	Amount of Fees					Amount Salaries are above or below fees
		From County	In cash	Total	Per employee	Per capita	
Sheriff	Lackawanna	15,615	15,728	31,343	2,849	.10	- 8,918
	Westmoreland	9,257	16,193	25,450	4,242	.09	- 5,545
	Montgomery	2,835	54,956	57,791	6,421	.22	-30,186
	Erie	2,216	25,370	27,586	3,573	.16	- 8,873
	Chester	---	13,820	13,820	4,606	.11	- 3,568
	Northumberland	4,134	6,048	10,182	2,545	.08	- 882
Register of wills	Lackawanna	---	15,705	15,705	2,244	.05	+ 1,616
	Westmoreland	---	14,982	14,982	3,745	.05	- 5,803
	Montgomery	---	38,574	38,574	1,542	.11	-11,384
	Erie	---	19,958	19,958	2,851	.11	- 6,858
	Chester	---	5,796	5,796	828	.05	+ 3,620
	Northumberland (a)	125	6,661	6,786	2,262	.05	+ 738
Recorder of deeds	Lackawanna	3,092	17,447	20,539	2,567	.07	- 3,481
	Westmoreland	1,672	13,199	14,871	2,124	.05	- 541
	Montgomery	2,029	39,647	41,676	1,667	.16	+ 2,778
	Erie	5,277	13,762	19,039	1,903	.12	- 3,925
	Chester	---	9,407	9,407	1,343	.07	+ 4,053
	Northumberland (a)	125	6,661	6,786	2,262	.05	+ 738
Prothon-otary	Lackawanna	11,291	25,564	36,855	2,835	.12	-10,994
	Westmoreland	2,296	18,873	21,169	3,024	.07	- 5,999
	Montgomery	6,343	28,973	35,316	5,886	.11	-13,276
	Erie	2,406	27,703	30,109	4,301	.17	-15,847
	Chester	1,404	9,549	10,953	3,651	.09	- 3,373
	Northumberland (b)	3,018	5,131	8,149	3,259	.06	+ 150
Clerk of - courts	Lackawanna	25,449	610	26,059	5,211	.08	-12,536
	Westmoreland	26,588	9,013	35,601	8,900	.12	-25,089
	Montgomery	10,705	917	11,622	3,874	.04	- 560
	Erie	5,389	4,642	10,031	5,015	.06	- 3,729
	Chester	5,177	2,620	7,797	2,599	.06	+ 233
	Northumberland (b)	3,018	5,131	8,149	3,259	.06	+ 150

(a) Fees of combined offices divided equally.
(b) Fees of combined offices divided equally.

b. The Salary Board in Operation - The salary board has the duty of determining the number and compensation of the clerks and deputies of the various offices of the county government. It does not do the actual appointing, this being left to the heads of the different departments. The salary board, adopted in 1876 for county government, illustrates a step in the direction of centralizing the administration of personnel in the major financial and policy-determining agencies of the county. This, it would appear, is commendable in theory but has brought very little actual improvement into the county personnel problem. There is little evidence of salary classifications to reward the same kind of work with the same pay; there is little or no attempt to provide standards for county positions; there is no development of the merit system either in recruiting or promoting county employees; there is no pension system outside of two counties of the sixty-seven. On the other hand, the salary board has often been used as a political club over the head of a county officer belonging to a different party or a different wing of the same party. This was mentioned several times by county officers who felt that sort of opposition. One officer declared that his deputies received far less than the average in the whole county government because he was a member of the opposite party. Likewise, he was allowed fewer employees than were necessary. In addition he was deprived of necessary office equipment by the county commissioners, who act as members of the salary board with a majority vote.

There is no evidence that the salary board in any county has undertaken a scientific study of the number of employees needed in each office. Factors influencing this number are (1) the number used during the previous year, (2) an increase in total amount of fees received in these offices, (3) the addition of a special function, (4) the desire to get as many people into the positions as possible. As a result, most county offices in the more populous counties are over-manned. This impression is confirmed by visits to many of these offices. There is much idling, visiting, reading newspapers and just plain sitting around. Of course, there are times when the whole office force moves with celerity. Even then, extra clerical help is used for a press of business.

In developing a classification of salaries, the salary board has not functioned efficiently. Generally the salaries of county deputies and first grade clerks are high when viewed in the light of the training, ability and experience needed. Most of the positions are of a routine, clerical nature. They can be held by anyone with a high school education and some business training. Some hold them successfully without these requisites. Nevertheless, the salaries of such employees are uniformly high. In one county, most of them are between $2,000 and $3,000 per annum, with quite a few over $3,000. In another county, most of these positions pay between $1,500 and $3,000. One county has a general salary rate of $2,080 for such officers. A tax payers' association in another county lamented that at $2,400 a year such clerical officers are receiving compensation at the rate of $1.40 an hour. In private industry, it pointed out, the same clerical help can be had for $65 to $100 a month. This association bulletin asks why deputies' salaries in their county are forty per cent higher than in other counties of the same class. The answer, they say, is in the salary board.

Erie County is typical in the range of salaries given to the more responsible and higher ranking deputies and clerks. For 1933 this was from $1,376 to $3,500. In the case of deputies and first assistants, there seems to be little or no relation between tenure of office and salary. Some who recently entered office have better salaries than those in office twenty years. There is no doubt but that deputies and first assistants should have compensation more nearly uniform as to type of work, experience and efficiency. Equalization of salaries for similar work, on a scale worked out either by statute or by the salary board of the county should be undertaken.

The Erie County salary rates for the more routine and less responsible clerks, stenographers and secretaries indicates that more uniformity exists in such offices. Ranges in 1933 were from $766 to $1,487 per year for the great majority. A few ran as high as $2,000 per year. Court stenographers, however, are in a class by themselves. They are uniformly well paid with an average salary in Erie County of $2,400 per year plus fees sometimes as high as $900. It is true that this work requires speed and accuracy but it is a question as to whether their compensation is comparable to that received by other county officers.

The case of Erie County is typical. Some counties are better and others are worse. No county has developed a system of equalization of salaries based upon a proper job analysis such as has been conducted in the state government and in private industry. Salary boards have the power to make such a classification and job analysis, yet it is never done. This is for the most part due to the political interests of county officers.

Another personnel problem is that of extra clerical help. This question could well be handled and investigated by the salary board. Much of it comes from seasonal rushes of business such as the preparation of the assessment books, or of the voting check and ballot check lists in the commissioners' office, or the handling of delinquent tax accounts in the treasurer's office. The total cost of extra clerical help is often enough to hire a good number of full time clerks. It would be well for a central agency such as the salary board to examine the possibility of reducing this cost by utilizing more completely the full-time clerical help through better schedules of work in order to eliminate, as much as possible, the seasonal rush. There are many months when the routine work in the office is light; in those months the clerks might be utilized on some of these special jobs wherever possible. Following is the amount spent for extra clerical help in 1933 in eight counties surveyed. In the other four counties no such expenditure was recorded.

Westmoreland	$16,805
Montgomery	18,965
Erie	3,053
Chester	10,753
Northumberland	6,603
Centre	602
Adams	7,451
Monroe	645

c. _The Actual Turn-Over of County Employees_ - Despite the fact that there is no permanent tenure of office in Pennsylvania counties there is a surprisingly high average length of service among employees. Some few have held their posts for more than twenty continuous years, many have worked for the county between ten and twenty years continuously. This can be partially explained by the dominance until recently of one party in the great majority of county seats. While different factions have wrested control from each other (and even this did not happen very often), the same party continued in power. This meant that many positions, originally political, took on a more permanent tenure. Different department heads have come in but have retained the several faithful employees who knew how to run the office. This was, of course, not altruism on the part of the department head. It relieved him of learning the work and of breaking in new employees who because they were politically appointed would find it difficult to learn the rudiments of a complicated routine position, especially when the department head himself needed instruction. The rather long tenure, despite the lack of protection, was found mostly in the offices of a routine, clerical nature, although some few belonged in the elective class.

From the knowledge gained from some of the older employees as to the years of service of the various employees in certain counties, it was estimated that the average length of service for employees on the 1934 payroll in Adams County was 4.4 years, in Tioga County 4.9 years, in Monroe 6.2 years, in Union 3.8 years and in Erie 7.7 years. This is considerably longer than the term of four years applicable to elective offices of the county. But it must be remembered that many elected officers hold their offices for more than one term. Some have held office for three and four continuous terms.

d. _Solicitors in County Government_ - The law allows a few offices in counties of certain classes to hire legal solicitors, whose compensation is generally fixed by the salary board. These offices include the county commissioners, the controller, the sheriff, and certain of the clerical offices. This is another example of the decentralization of such central services as legal advice. In fact, all solicitors could and should be abolished and the work given to the district attorney's office, except when legal disputes between different officers of the county arise. These solicitors are well paid, in view of the small amount of work required. The cost of solicitors for 1933 in Erie County is typical. It follows:

Solicitor	-- County Commissioner's Office	$2,650.00
Extra attorney	-- County Commissioner's Office	300.00
Solicitor	-- Controller's Office	800.00
Solicitor	-- Recorder of Deeds Office	826.20
Solicitor	-- Sheriff's Office	500.00
		$5,076.20
Solicitor	-- Poor Board	140.00
Solicitor	-- Poor Board	860.00
		$1,000.00
	Grand Total	$6,076.20

e. _Ratio of Expenditures for Personal Services to Total Current Expenditures_ - The following table shows what proportions of the total current expenditures of seven counties were spent for personal services. The percentages were computed from data in the

1933 financial schedules prepared by the Pennsylvania Economic Council. Because some of the officers and employees of the last three counties in the list are paid by fees, the amount of which cannot be ascertained, the computations are necessarily incomplete. The table shows that the comparative importance of expenditures for personal services varies greatly in different counties. Such expenditures took the greatest percentage of total expenditures in Lackawanna (53 per cent) and the lowest percentage in Northumberland (17 per cent). There seems to be no relation between the relative importance of personal services and the density or sparsity of population.

Table IV. Ratio of Cost of Personal Services to Total Current County Expenditures for 1933

County	Total Cost Personal Services	Per Cent of Total Exp.
Lackawanna	$ 692,963.00	52.5 %
Westmoreland	378,335.00	20. %
Montgomery	616,421.00	36.5 %
Erie	314,725.00	20. %
Chester	198,543.00	30. %
Northumberland	176,153.00	16.7 %
Centre	56,591.00	24.1 %

This includes personal services, full-time and part-time, under the following headings: commissioners' office, solicitor, weights and measures, courthouse maintenance, elections (voting machine maintenance only), tax assessments, treasurer, tax collector, controller, recorder of deeds, register of wills, sheriff, coroner, prothonotary, clerk of courts, district attorney, law library, courts, probation, county prison, juvenile detention home, almshouses, other hospitals and institutions, administration, superintendent of schools, county engineer and highway maintenance.

f. Are Departmental Heads Necessary? - To answer this question dogmatically would be impossible. There is clearly a need for a head in any office if it is to produce results. Nevertheless, the heads of county offices are very rarely the efficient executives which their high salaries indicate that they should be. Considering the problem only from the stand-point of actual worth as directors of departments, it must be conceded that most of the heads of the clerical offices (register of wills, recorder of deeds, prothonotary and clerk of courts) do not earn their present salaries. It is too well known that many are not greatly interested in learning their office duties. In fact, those particular offices do not need separate elected heads. They could function very well if all four were consolidated under one executive with a chief deputy in charge of each office. 'Advantages would be the elimination of useless overhead in the form of high salaries and better cooperation in the matter of work schedules to take up the slack from office to office. Better still, these clerical offices could be merged with the office of the chief clerk of the county commissioners and a consolidated Department of Records developed. The head of this department should be the chief clerk, appointed by the commissioners. The latter change would require a change in the Constitution.

CONCLUSIONS

1. FINDING: The Pennsylvania Constitution is the greatest obstacle to reform of county personnel.
   RECOMMENDATION: That the Constitution be changed by amendment or revision to permit the alteration, consolidation, and abolition of elective county offices by the General Assembly.

2. FINDING: The county salary boards now provided by law have not generally undertaken such simple requisites as classification of positions and provisions for equality of pay for similar work.
   RECOMMENDATION : Positions in county offices should be classified as to duties performed and experience required. Compensation should be graduated according to types of positions and the same pay should be given for the same type of work in different county offices.

FINDING: Only Allegheny and Philadelphia Counties are authorized by law to provide pensions for county employees.

3. FINDING: Appointments to county positions are based mainly on political considerations and tenure is governed by the "spoils system."
RECOMMENDATION: That a civil service system be established in counties of the first to fourth classes inclusive under the direction of the salary board. This would include provision for examinations, promotions, permanent tenure and pensions adapted to local conditions. Responsibility for this system should rest solely with the salary board.

FINDING: Salary expenditures vary greatly on a per capita basis among the different counties. The highest expenditure among the counties studied was in Montgomery County at the rate of $1.92 per capita. The lowest was in Westmoreland County which had a salary expenditure per capita of only $.86.

FINDING: There is little relation between the number of county employees and the actual service rendered to a given population. In the 12 counties visited there were 1,054 full-time county employees or an average of 1,393 inhabitants per county employee. The smallest county (Forest) in the state had the highest ratio - one county employee for every 399 people.

4. FINDING: Appointment of large numbers of employees by judges tends to make the judge an important political factor in the selection, retention and promotion of probation and parole officers, court clerk, court crier, tipstaff, interpreter, etc.
RECOMMENDATION: That the authority of the county salary board be extended at least to control the number and compensation of such officers.

5. FINDING: A large amount of extra clerical assistance is employed in county offices.
RECOMMENDATION: That the county salary board supervise the engaging of such extra help and seek to minimize it by more fully utilizing those regularly employed and by work schedules for periods of seasonal rush.

FINDING: Although there is no permanent tenure of office in county government, the length of service among employees has attained a relatively high average.

6. FINDING: Merging of county offices is urgently needed. There are too many independent department heads.
RECOMMENDATION: In at least four instances (register of wills, recorder of deeds, prothonotary, and clerk of court) the supervision could be performed by one officer. The most desirable change would be the merging of the above offices with that of the chief clerk of the county commissioners to make a consolidated Department of Records.

7. FINDING: No salary board is now provided for counties of the seventh and eighth classes.
RECOMMENDATION: That a salary board be established in every county as urged by the Salary Survey Commission in its report to the General Assembly in 1929. The salary boards should have power to fix the number and compensation of all appointed county officers receiving compensation from county funds.

8. FINDING: In counties of the seventh and eighth classes several county officers are compensated upon a fee basis.
RECOMMENDATION: That fees be abolished entirely as a method of compensation of county officers and that such offices be merged in order to permit the employment of an officer on a salary basis.

9. FINDING: The salaries of elective county officers are fixed by Act of Assembly.
RECOMMENDATION: If salaries are to be fixed by Act of Assembly the maximum should be accompanied by a sufficiently low minimum to furnish flexibility in the salary range and permit local control. An exception may be made in the case of the salaries of the county commissioners, controllers and treasurers who, as members of the salary board, should not have to decide upon their own salaries.

10. **FINDING:** No system of training or improvement of county personnel is in operation at the present time.
RECOMMENDATION: That some form of training for county personnel be developed. This might be done by the Bureau of Municipal Affairs, Department of Internal Affairs, of the Commonwealth, or by a department of government and administration in a college or university within the state. It could not be undertaken by any one county but it might succeed with the voluntary cooperation of the various counties. The plan could easily be extended to other local subdivisions. This training agency should issue pamphlets and bulletins; it might conduct intensive courses certain weeks in the year; it might advise and correspond with officials and employees directly concerned. Such training may be expected to occupy an important place in the future development of local government.

11. **FINDING:** The county personnel lacks a central administrative head. This lack of definite responsibility and authority is one of the outstanding sources of weakness in county government.
RECOMMENDATION: That an alternative form of county administration be provided wherein the county manager (as suggested in Chapter 2) becomes the personnel head. Under this system, the county manager acting as chief clerk and head of the Department of Records, would appoint the heads of the other five main departments: treasurer, chief court officer, district attorney, public works and welfare. He would also appoint all subordinates for each of the departments after consultation with department heads. All appointments would be subject to approval by the board of commissioners. In this manner, the county manager as chief clerk could develop a sound and permanent personnel policy with the merit system as a nucleus. He would be the day-to-day administrator to whom all personnel problems would automatically come. This would be possible and salutory in the larger counties, although a less elaborate system would be needed in the smaller counties.

CHAPTER IV

ASSESSMENTS

INTRODUCTION - The problem of assessments in Pennsylvania counties has long been recognized as one of great difficulty. The county assessment is used as the basis for taxation in all local subdivisions, except in second and third class cities. The day seems not far distant when this basis will be used by all local units. Furthermore, the great bulk of local revenue in ordinary times comes from the general property tax, and this fact has made the equitable and uniform assessment of property for tax purposes the foundation stone of good financial management. Even if general property is relieved of part of the cost of local government by the increase of grants-in-aid from the state (and perhaps national) government, and by the establishment of a state income tax, the development of a sound system of property assessment will still be essential in all local areas.

Since 1931, there has been in effect legislation establishing a board of revision and assessment of taxes for third class counties. In 1933 a codification, revision and clarification of assessment law was enacted. Important studies in assessment and its relation to local taxation have been made. The whole problem has received state-wide publicity and discussion, especially since the beginning of the depression. Many tax and citizens's leagues and associations have developed an interest in this question in their local areas.

The general subject of assessments will be discussed under the following topics:

    I.   Statutory provisions relating to assessments
   II.   The assessment system in operation
        a. In third class counties
        b. In fourth to eighth class counties
 III.   The cost of assessments in Pennsylvania
  IV. Recommendations

I. STATUTORY PROVISIONS RELATING TO ASSESSMENTS

Many years ago Pennsylvania achieved a complete separation of state and local tax systems as far as the general property tax was concerned. All proceeds from the general property tax are devoted to the upkeep of county, township, city, borough, school district and poor district governments and activities. Before 1913 the 4 mills tax on personal property was in part a state tax, but since that time the counties have used all of the proceeds from this tax, even though it is still popularly known as a "state tax." Thus there is no such need for state equalization as there is in states which use part of the general property tax for general commonwealth purposes. The system of assessments is a state problem only insofar as the state must act in the development of a sound local policy in this respect.

    a. Taxables - There are three kinds of taxables in the general property system in Pennsylvania. These are real estate; occupations, trades, or professions; and personal property.

All real estate is liable for taxation except that which is by law exempted. This latter category includes churches, burial grounds, charitable and educational institutions, court houses, poor houses, parks, posts of discharged soldiers' organizations, public property used for public purposes, and the like. Court interpretations have made this list impressive, as will be shown by statistics on tax exemptions. Taxable real estate therefore includes land, buildings, manufactories, and even unmined coal.

The second class of taxables includes all salaries, emoluments of office, professions, trades, and occupations of persons (except farmers) over 21 years of age. The county commissioners decide upon the schedule of assessments for different occupations and callings. These are not taxable in second and third class counties for county purposes but there is, instead, a poll tax of fifty cents on each resident over 21 years of age. In all counties, Federal employees are taxable by poll at fifty cents each.

Personal property, taxable at the 4 mill rate for county purposes, includes mortgages, money at interest, private and public loans, except those to Pennsylvania and the national government, shares of stock, except those taxable by the Commonwealth of Pennsylvania, annuities yielding more than $200 annually, buses, stages, hacks, etc. (actually not taxed or assessed) and property generally classed as "money at interest." (1933 P. L. 54, sec. 1.)

b. Assessing Authorities- The work of assessing this property is divided between the county commissioners or the county board of revision and assessment of taxes, and the assessors elected or appointed for the local areas. The combinations for the various classes of counties is as follows:

1st Class (Philadelphia) -- Board of revision (appointed by judges of the court of common pleas) appoints the assessors.

2nd and 3rd Class   -- Board of assessment and revision of taxes, appointed by the county commissioners, appoints sub-assessors.

4th to 8th Class   -- Assessors elected by townships, boroughs and wards. County commissioners act as board of revision.

The county may be said to have control over the assessment system inasmuch as the county commissioners pay the costs of assessments (except certain costs in first class townships), arrange and pay for the necessary clerical work, initiate the annual process, fill vacancies in the office of elected assessor, and act as, or appoint, the board of revision. As mentioned before, the assessment for county purposes is the base for taxation in other local subdivisions, except second and third class cities.

c. Compensation - The following schedule of compensation is provided in the 1933 codification of the assessment law (1933 P. L. 853).

Real Estate Assessors in 1st Class Cities - $5,000 per year.
Sub-assessors in 2nd and 3rd Class Counties -
    2nd Class - No statutory rate prescribed.
    3rd Class - Determined by salary board, not to exceed $2,000
            per annum.
Elected Assessors in 4th to 8th Class Counties -
    Cities of the 3rd Class - Determined by county commissioners.
    Boroughs and townships of the second class - $3.50 or $5.00 per
    day of work. Two statutes enacted in 1933 and approved on the
    same day fix two different rates. Prior to 1933 the rate was
    $5.00 per day.
    Townships of the 1st Class -
        Township Assessor - $10 per day of work and expenses.
        Assistant Assessor (both in triennial and intertriennial years) -
                $10 per day of work, and expenses.
        Assistant Triennial Assessor - $5 per day and expenses.

    (In first class townships, the law provides that the assessors
    make their returns to the county commissioners, who may examine
    the assessors as to the number of days handed in before allowing
    the claim.)

Mileage at three cents a mile circular is allowed assessors for necessary trips to the county seat or to places where appeals are held.

The board of assessment and revision of taxes in 3rd Class counties is compensated at a rate determined by the Salary Board of the County.

d. The Triennial Assessment - There are two kinds of assessment provided by law. One is the triennial assessment, held every three years (1933, 1936, etc.); the other is the inter-triennial assessment which is held in each of the two years following the triennial. The triennial assessment is supposed to be a complete house-to-house canvass and a new valuation of all taxable property and persons. The inter-triennial is an assessment only for the sake of recording changes in property and taxables which happen from year to year. The order in the triennial is as follows:

1. _Issuing of Precepts_ - Precepts are orders and instructions to the assessors to begin assessments in their districts. They are accompanied by the regular assessment books and other paraphernalia supplied by the county assessment authorities. In triennial years the assessor usually receives the previous year's assessment book for his district, a new one to be filled out, and a blotter on which to make his field investigation copy. Precepts are issued as follows:

> Counties of the 1st class -- by board of revision.
> Counties of the 2nd class -- by board of assessment and revision of taxes when they shall prescribe.
> Counties of the 3rd Class -- by board of assessment and revision of taxes when they shall prescribe.
> Counties of the 4th class -- by county commissioners on or before June 1.
> Counties of the 5th to 8th class -- by county commissioners on or before July 1 for 1st class townships, and September 1, for towns, boroughs, wards and 2nd class townships.

2. _Valuation of Real Estate_ - The law requires the assessor to assess real property on the basis of a fair selling price on the open market. He does this by making a house-to-house canvass. The law does not state by what method he shall arrive at the assessment, the implication being that he shall make his judgment based upon as much information as he can get from any source. The books which he fills out may be as detailed as the county assessment authority may provide, the law itself specifying only a few items that must be returned in the assessment. Timber lands, for example, must be returned in a separate column. As a rule, although not specified in the law, the following items are found in these books: name of taxable, address, occupation, real property, personal property. In the larger counties, a more detailed description of the property is usually required by the assessing authorities.

The law requires exempt property to be returned as to value. Assessors are required to make a list of all taxable persons and return it to the county commissioners or assessing authorities. Persons acquiring unseated lands are supposed to give facts relating to this land to the county commissioners. Coal lands are assessed even when they are severed in ownership from the surface land over the mines.

The recorder of deeds in the counties of the 3rd to 8th class are required to keep a daily record of conveyances of property and give them to the county assessing authorities, who in turn give them to the assessors. This is an aid in keeping up with changes in ownership.

3. _Valuation of Occupations_ - The assessors merely list the occupations of all persons over 21 years of age. The county commissioners determine the schedule on which such occupations shall be assessed for taxing purposes. A poll tax of fifty cents per resident over 21 takes the place of this tax in 2nd and 3rd class counties for county purposes.

4. _Valuation of Personal Property_ - Blanks for the assessment of personal property are furnished the assessors by the county commissioners. Assessors give these to the taxables, who return them to the assessor. Upon the refusal of a taxable to make return within ten days of notification, the assessor makes return for the taxable person from his best information. He examines records and lists of judgments, notes and mortgages returned by the prothonotary and recorder of deeds to the office of the county commissioners or board of assessment and revision of taxes, and adds any amounts found from other sources. This total return may be corrected by the county assessing authority, which sends for the taxable and swears him as to the proper amount of his personal property.

This amount may be increased by 50% as a fine for not returning, unless reasonable cause is shown. Information in the register of wills' office as to the estates of deceased persons may be examined. If taxable personal property is found that was not returned by the deceased, the tax on the difference for the 5 years preceding death is taken from the estate plus a penalty and six per cent interest charge from the date the tax was due. Certified records of personal property found in a county but taxable in another county are sent by the county in which such instruments are recorded to the county in which the owner is taxable.

5. Returns of Assessors - Return of assessors' books to the county commissioners or board of assessment and revision for the triennial assessment is provided as follows:

> Counties of 2nd class -- On or before 1st Monday in November
> Counties of 3rd class -- Determined by board of assessment and
>   revision of taxes
> Counties of 4th class -- Not later than September 1
> Counties of 5th to 8th class -- December 31 or before
>   (First class township assessors are given until
>   February 15 as the limit, if necessary.)

This return consists of one of the books given out by the county commissioners with the precepts. It is kept in the office of the county assessing authority. The other is kept by the assessor.

6. The Inter-Triennial Assessment - During the two years that follow the triennial assessment, county commissioners or the board of assessment and revision of taxes send transcripts of the triennial assessment to assessors for changes and alterations resulting from changes in property and persons taxable within the district.

Precepts are issued at times fixed by the board in 2nd and 3rd class counties; June 1, in 4th class counties; and before the 2nd Monday in September in 5th to 8th class counties.

Returns are due in most cases after ninety days.

Assessment of personal property is carried on in the same manner during the inter-triennial years as in the triennial.

7. Optional Reassessments - County commissioners may issue precepts March 1, returnable May 25, to assessors to record changes in assessment. This action is in addition to the fall assessments and may be taken each year.

8. Revision and Appeals - After all the returns are made to the county commissioners or board of assessment and revision, this office publishes a list of aggregate assessments made by each assessor under the heading of real estate, occupational and personal property. Dates for hearings are published and taxables may examine returns in the county offices. The county commissioners or the board of assessment and revision of taxes, as the case may be, may then make revisions. In counties of the 4th class, the county commissioners may hire competent persons to assist them in making revisions. This is done mainly for the purpose of equalizing assessments as between subdivisions. After the revisions of the triennial assessment have been made, it is the duty of the county commissioners, in counties where they are the board of revision, to cause accurate transcripts of assessments to be made, and to transmit these to the assessors of the districts before the 2nd Monday of April, the year following the actual assessment. The transcript is accompanied by a statement of the rate of the tax and the dates of appeal days. Assessors give written or printed statements of the assessment, rate of tax and date of appeal. Taxables are also notified in case the valuation of their property is increased. The notices in inter-triennial years need be made only for changes. Assessors attend the appeals in which cases in their district are concerned. Appeals may be held either in the county house or in the district for which they are being heard, at the discretion of the county assessing authorities. After the appeals are made and the changes accepted, the books are altered to that effect. However, changes may still be made after this date, and appeals from the board of revision may be taken to the courts.

9. Making the Tax Duplicates - After the appeals are over, the county commissioners order clerks to make a fair duplicate which includes the name of each taxable and
the value and description of his property. This includes only real estate and occupation, but not personal property. These duplicates are furnished to school districts of
the 3rd and 4th class and townships of the first class. These are used as bases for
the taxes in these districts. (The same is usually done for boroughs and townships of
the second class, but the law does not specify that this shall be done. In some counties these subdivisions make up their own duplicates from assessors' books.)

10. The Board of Assessment and Revision of Taxes in Third Class Counties - This
feature of the assessment system deserves special mention because of its place in
Pennsylvania assessment reform. Established by the law of 1931 (P. L. 1379), it
brought to third class counties a system already in use in second class counties
(Allegheny). It abolishes the elected assessors of the boroughs, townships and wards
in cities of the third class, and substitutes sub-assessors, appointed by a board of
assessment and revision of taxes, which board is in turn appointed by the county commissioners. There are three members, their salaries are determined by the salary board
of the county. They may hire a solicitor, engineers, and such clerks as they need,
whose salaries are decided upon by the salary board. They appoint sub-assessors, the
number of which they shall determine, with a salary determined upon by the salary board,
but not exceeding $2,000 per annum. The board may make rules and regulations regarding
the method of assessments, the time for precepts and returns, and other matters regarding assessments. Sub-assessors make house-to-house canvasses according to the board's
regulations, and make a list of inhabitants subject to the poll tax. The board has
the general power of the board of revision in other counties with regard to appeals
and revision. Assessments are used for the basis of all taxation except in second and
third class cities, and may be so used in the latter upon approval of the council and
mayor, and certification to the Governor. If this is done, the city and school taxes
are based on the county assessment.

In this step the legislature has established the county as the unit of assessment,
for experimentation at least. It has paved the way for a uniform and scientific assessment. It has made possible the development of a corps of expert assessors.

II. THE ASSESSMENT SYSTEM IN OPERATION

a. In Third Class Counties - Of the 12 counties surveyed, three (Westmoreland,
Montgomery and Lackawanna) were of the third class. It was here that the actual operation of the system allowed by the act of 1931 was observed. In all three counties
different methods were used in the application of the law. Only one triennial assessment (1933) has been undertaken under the provisions of the new law, although changes
have been made in assessments during 1934 and plans were being laid in 1932.
It might be well to compare the three counties on the following points:

1. Personnel and Salaries

	Westmoreland		Lackawanna		Montgomery	
	Personnel	Salaries	Personnel	Salaries	Personnel	Salaries
Board of Assessment	2	$9,720.	3	$10,800.	3	$10,500.
Chief Clerk	1	2,160.	1	2,280.	1	2,500.
Assistant Clerk	3	1,980.	1	2,040.	1	1,070.
Solicitor			1	3,000.		
Mining Engineer			1	2,400.		
Clerk			1	2,205.		
Personal Property or Tax Clerk	1	2,160.	1	2,016.	2	4,080.
Clerks	1		13	25,390.	15	30,600.
Sub-assessors, 1933			7	13,400.	(Same number as when elected 104)	
Extra Clerks, part time	4					

2. Application of the System - (Westmoreland County) - The system as applied in
Westmoreland county was outstanding as to the preparation and thought which were
devoted to the new plan. Three members of the board of assessment and revision of
taxes were chosen, not for their political activities, but for their previous experience in businesses that would be of practical benefit in administering the new system
of assessments. One was a real estate man, one a farmer and the third a contractor.
Emphasis was placed on the fact that the county wanted a uniform system of assessments.
The commissioners and the board sent a delegation to confer with Mr. Zangerle of
Cleveland on the preparation of a scientific system of assessments. This was completed
and is described fully in a large and profusely illustrated pamphlet entitled The
Principles of Land and Building Appraisals as Scientifically Applied in Westmoreland
County, Pennsylvania, 1932. It has been furnished to local officials in the county.

The purpose was to equalize assessments on a county-wide basis at 100% of the true
value at the time of assessment. Comparisons were formerly too difficult when appraisals were fixed on a percentage of true valuation. Discrimination usually crept in
through this apparently harmless method of assessment. It is emphatically stated as
an essential feature of the new plan "That PROPERTY be assessed--not individuals."
The pamphlet then sets forth the unit values as applied to different types of property:
land, lots, and buildings, and serves as a guide or instruction-book for assessors.
It is a sample of what is possible under the new legislation for third class counties.
A brochure explaining the new method was also published for the public in 1933.

The sub-assessors were appointed by the board of assessment and revision of taxes.
Fourteen were originally named in the twelve districts into which the county was
divided. At the highest peak of the work there were twenty-six. In June, 1934, there
were none. Six sub-assessors were needed for adjustments during the inter-triennial
year of 1934. Sub-assessors received $125 per month and expenses at the rate of six
cents per mile. Sub-assessors will again be employed for the triennial assessment to
be made in 1936.

After being instructed by the board of assessment and revision of taxes, the
sub-assessors go through their districts making a house-to-house canvass. They assess
according to the rules worked out by the board. They take with them property record
blanks on which they place all the items concerning the property that have any influence on the assessment according to the rules. These blanks show the amount of detail
sought for and illustrate the application of a scientific system of assessment. From
these work-sheets of the assessor (white), copies are made on the same kind of blank
(yellow) for deposit in the tax office, where they are bound in books according to
districts.

When assessors are in the field they send in semi-monthly reports stating the day-
by-day activities, including the number of properties assessed, the number of taxables
assessed for occupation, the mileage and other expense. This is checked over by the
tax office and compared with reports of others.

While it is early to hazard a final conclusion on the effectiveness of the new
plan, careful observation shows it to possess certain definite advantages. Taxpayers
calling at the tax office to complain that their assessments are too high are asked to
describe their properties. As each does so, the clerk checks up the statements made
with the facts shown in the property record. It then appears perhaps that the taxpayer has not counted all the rooms of his house while the assessor did. Many property holders leave the office much better satisfied after observing the exactness of
the assessment, the method used, and the assessments upon similar properties in the
neighborhood. Some must be told that they may file an appeal before the board of revision, but that the tax office itself cannot change an assessment unless an error in
copying has been made. The board of assessors likewise will not change an assessment
except upon a view of the property.

The objection is sometimes offered that the sub-assessor did not know much about
property. The answer to this is a three-fold one:- first, sub-assessors had been
chosen for their ability as well as their previous experience; second, the former
elected assessors possessed no special qualifications or experience; third, under the

new plan the sub-assessor was so carefully instructed in principles of accurate assessment that if he were a man of fair intelligence an extended experience was not required.

Little difficulty was experienced with the large industrial plants. Company representatives were usually sent to meet with and assist the board by furnishing plans, blue-prints and any other necessary information needed to appraise a large property. Assessments of some such properties were raised as much as half a million dollars, without complaint as long as the assessment was believed to be fair and uniform.

By this new assessment, 49 wards, boroughs and townships lost and 56 subdivisions gained in assessed valuation. The net loss for the county was about one million dollars. The losses were largely accounted for by readjustment of assessments on unproductive coal lands. The gains came from the adjustment of previous inequalities. The actual assessment was 50 per cent of the appraised value.

Lackawanna County - In this county, the new plan was established in 1932 and the regular triennial assessment of 1933, to take effect in 1934, was made under it. The new set-up was put into operation by the appointment of forty-four subordinate assessors, who worked about ten months in the field, making up their own blotters and turning them over to the clerks in the office. These sub-assessors received $160 per month. After the original assessment was made, only seven of them were retained to make changes in the annual assessment.

The work of the original new assessment was organized by teams and captains, the captains being fairly experienced in the work. The assessors in the field measured the lots and indicated the particular kinds of houses on them. In case of any doubt as to the application of the standards to any particular unit, the assessors conferred with their captain. This information was sent in to the office, and the office worked out the ultimate assessed valuation under the supervision of the board of assessment and revision. The valuation set by the board was largely determined from the area in which the property existed, as shown by maps, and by visiting localities in question. To get the value of a business property, the peak points in the business and suburban areas were located, and the value of the property was determined by its proximity to these peak points. There was no scientific valuation of large business places or factories, but reliance was placed on the experience and ability of the chief assessor.

Lackawanna County has a peculiar system of assessing its coal lands. It employs a mining engineer on full time, who makes a constant survey of the condition of coal lands which are owned only under surface, as well as those which are owned by the operating companies both as to the surface and sub-surface. A court order having fixed the assessed valuation for coal lands at $300 a foot-acre, all coal lands are assessed on that basis. Luzerne County, the adjoining anthracite region, assesses under a graduated scale which goes from $80 to $240 per foot-acre. The regular assessors therefore have nothing to do with any coal lands; and as the value of coal lands is a very large item in the assessed valuation of the county, this is an important peculiarity.

The city of Scranton has a separate assessment of its property for city purposes. It was last assessed in 1925 and has not taken into consideration the deflated values due to the depression. It is expected that in a few years the city will accept the county's assessment, partly because it was made in 1933 and took into consideration the effects of the depression, and partly because of the expense to the city of an entirely new assessment. There is no reason why this should not be done, as the county assessment appears to be generally approved throughout the county. The differences between the city and county valuations in the cities of Lackawanna County are as follows: (a) The total city valuation in 1934 exceeded the total county valuation by $12,957,334, or 11.8 per cent; (b) the city valuation of coal property exceeded the county valuation of the same property by $246,492, or 2.3 per cent; but the county assessors listed 19,103, or 32.6 per cent more persons liable for occupational taxes than did the city assessors.

After the new assessment was made, the regular appeal days were held. Because it was an entire revision, appeals took on an important aspect. In this way popular opinion could be ascertained. It was stated that the people of the city of Scranton and the inhabitants of the rural regions -- non-coal -- were fairly well pleased, with the usual exceptions. But when the board of revision sat as it did to hear appeals in the various coal boroughs, there was a great deal of trouble. In some places the police were used to stop impending riots, and the board needed protection from actual violence. Delegations from these boroughs made trips to the county seat in quest of "justice". This is an indication of the political habits of Lackawanna County. Most of these coal boroughs have local political leaders who made capital of the ignorance of their constituents. The new assessment provided them with such an opportunity. They aroused the citizenry, which as a rule knew nothing about it, to the injustice of the new assessment. Literally hundreds of appeals were made, most of which resulted in no change by the board of assessment. Many came to the offices and even sent their wives to seek adjustment. In either case it was very difficult for them to understand the English language as spoken by the office clerks. This caused much trouble and excitement. One local leader stated that he believed the assessment to be all right as far as he knew, but that it was to his advantage to champion the rights of his followers who protested.

The assessment of personal property is handled in the conventional way. Personal property return blanks are mailed to those who are on the books as having money at interest in any of its forms under previous assessments. They are then returned by mail to the office. There is no investigation, nor do the sub-assessors have anything to do with it. However, the offices of recorder and prothonotary send weekly reports to the assessor's office of changes in deeds, mortgages, notes and other papers relating to transfer of property.

The work of the assessment office includes the copying of assessment books and blotters, both alphabetically and by lot number and block, the making up of tax duplicates for the county, the making out of a school copy for taxing authorities using the county assessment, the handling of personal property returns and the hearing of appeals. The owners of all taxable properties get copies of the triennial assessment notice sent out on post-cards. Previously letters were used at a much greater cost.

Montgomery County - The new system as worked out in Montgomery county was more conservative in procedure than in either of the other two counties surveyed, and involved but slight change except the establishment of the board of assessment and revision of taxes. Instead of the elected assessors for boroughs, wards and townships there are appointive ones, but the formerly elected ones were appointed to carry on the new plan. The county is not re-districted, the same wards, boroughs and townships being used as assessment districts. The pay of the 104 sub-assessors is substantially unchanged. First class township assessors work full time and are paid partly by the county and partly by school districts.

Since the year 1933 was not considered a favorable time for a general revision of assessments or the establishment of a new plan for making them, and since raising assessments (and in order to equalize, much raising would be necessary) would cause agitation, the old system was maintained with certain modifications. One of the changes was an equalization of properties which fronted on the lines of wards, townships, and boroughs. Under the old plan each assessor developed his own scale of values, and as a result the properties lying adjacent to each other but in different assessing areas often were assessed very differently. This was generally eliminated by the action of the board of assessments in dove-tailing the assessed values more adequately. Better records of property were also worked out. A card system, not nearly as detailed as in Westmoreland, is used for office purposes and is filled out by the assessors. It contains information of location, kind of lot, construction of building, kind of street, characteristics of the inside of the house. etc.

The assessors are given new assessment books every year. In triennial years, there is also given a blotter on which to make a copy of the new assessment. In other years the assessors are given the old books on which they make changes. They then hand the books in to the board, and transcribers make new books. These books are kept in the office of the board until used again next year. Books are kept by wards, townships, and boroughs. About fifteen transcribers are used to keep these new assessment books up to date. They are employed all year round. Besides the three members of the board, there is a chief clerk, an assistant clerk and a stenographer.

If Montgomery county is outstanding in any one phase of the assessment process, it is in the assessment of the personal property tax. In 1932 some 44.5 per cent of the entire assessed valuation was in personal property. This is the result of two factors. First, about one-half of the county is a suburban area for Philadelphia. This is a comparatively wealthy residential district, and many of the resident taxables are rich in "money at interest." Second, the county makes an effort to ferret out the amount owned by each taxable. Compared with Lackawanna with 9 per cent of its assessed valuation in personal property and Westmoreland with 13 per cent, this is quite important. Even in Chester, where conditions are more or less comparable, only 25 per cent of the total assessed valuation is in personal property. This explains why Montgomery county can boast a 3 mill tax on real property and still spend heavily for county government.

Two "mortgage clerks" handle the office routine for the assessment of personal property. The assessors receive personal peoperty blanks and hand them to each taxable at the door of each house. This is done only on triennial years to get a check on every resident. Even when an inhabitant is indigent the blank must be returned to the assessor, and the county then has a list of all taxables, keeping a list of indigents separately. The assessor keeps a check on taxables every year, but not as completely as in triennial years.

It is estimated that $100,000 a year is received from non-returned mortgages alone. In the first eight months of 1934 it was stated that $106,000 was collected from personal property found in estates that had not been previously returned. Furthermore the clerks maintain constant watch over financial transactions in Philadelphia to bring up their personal property records to date. As to persons who do not return tax blanks and who are suspected of having "money at interest" in some form, a guess is made as to their financial position. On this is added a 50 per cent penalty. If there is no protest, the sum is raised the next year until the taxable comes in. The personal property records are sent to the assessors of the district in which the taxable lives and entered in their books.

The tax office also writes up the tax books (duplicates) for county, road, school, and borough taxes. These taxes are computed in this office and then sent to the tax collectors, upon filing of proper bond. The tax collectors write out the tax bills and send them out by mail.

b. In Fourth to Eighth Class Counties

1. Personnel- One of the chief obstacles to uniform and equitable assessments is the fact that tax assessors in these counties are elected from the various subdivisions of the county. The results, obvious to any student of assessments, are (1) an un- trained, irresponsible, and political body of assessors; and (2) a lack of uniform standards of assessment as between the various divisions of the county.

Although the county commissioners, as a board of revision, might legally pre- scribe rules of assessment which the assessors might follow, this, at least in the 9 counties surveyed, is never done. Consequently, the assessors use their own standards without regard even to the most elementary classifications of property, either land or buildings. Some assessors, interested in developing more accurate systems of judging property, may, in the course of their experience, arrive at some such methods, but this is the rare exception rather than the rule. Judgments are arrived at by a va- riety of influences, such as the owner's idea of the value of his property, the owner's economic and political status in the community, a general notion of property value on the part of the assessor, and the general consensus of opinion that low assessments reflect a triumph of the individual owner in his relations to the government.

In fourth class counties the county commissioners are allowed a supervisor of assessment. Erie county has such an employee, who was engaged until the 1933 triennial assessment was completed. While he called the assessors in for conference, he did not give them anything but general instructions as to how to assess. When interviewed he indicated that the prime difficulty in arriving at equitable assessments was the fact that elected assessors were the foundation stone of the system. Nothing could be done except if assessors were trained "to know property." He himself was an assessor of many years practical experience, although he had very little time for "scientific" assessment methods, which he felt might be badly managed in the hands of untrained assessors.

If elected assessors are to continue as a part of the system, it would be well to provide them with a manual of assessment which could be used with local modifications in all the counties of the state. This manual could include units of value for the assessors' use in arriving at valuations. At least, each county should have such a manual in which such unit bases of valuation would be set forth. This would be a great aid to newly elected assessors, who would welcome such assistance. There is no question but that assessors in the main try to do their work fairly but they are subject to so many influences that without a definite standard they have nothing tangible to support their judgments as to value. For the lack of such support, they are prone to yield to the more vociferous taxables.

Curious as it may seem, the small property-owner is oblivious to whether he is over- or under-assessed in relation to his neighbors. Persons with a great deal of property may often be under-assessed. They examine their assessment carefully and protest when it seems too high.

2. <u>Assessment Records</u> - The handling of assessment records varies in detail from county to county. The commissioners' clerk hands the assessment book of the previous year to the assessors. This may occur in September but is done in July in fourth class counties and in first class townships in all counties. Sometimes a duplicate of the old assessment book is given, the original book remaining in the court house. With the last year's book or duplicate goes a new book, and in triennial years a blotter (duplicate) for field work purposes. Thus new books for each assessing district are necessary every year, although the general practice is for the county commissioners' office to make up these new books from the changes handed in by the assessors in inter-triennial years. In this way the assessors make out their books completely every three years.

It appears that this method of keeping assessment books is cumbersome and expensive. A system of cards on which changes could be made in inter-triennial years from blotters of the assessors might be established. In some counties assessors make two complete copies every year, keeping one in their own possession and handing one to the county commissioners. Their own copy is used as a basis for the next year's assessment, and they receive only two new books every year. From the copy handed in, duplicates are made for the computation of taxes in the county and other local subdivisions. In one county there are three books given the assessors. The original is kept by the county commissioners, one is given to the secretary of the school board in each district, and the other is given to the tax collector. In the tax collector's book, the county tax is computed by the county commissioners' office, and this is in lieu of making out a separate duplicate for the tax collector. The local subdivisions do the same for their taxes, but use other books.

3. <u>The Occupational Assessment</u> - The occupational assessment is made by the simple process of having the assessors write into their books the occupation of each taxable. The standard of assessment is made by the county commissioners. These assessments are nominal and are not based upon any standard of income. Monroe is an example of many counties in this respect. This is not to say that all counties have the same occupational assessments, for they vary in amount and in designation from county to county. But they are much alike in the amount of assessment. The Monroe standard follows:

Housewife	$ 50.00	Lawyer	$400.00
Invalid	25.00	Banker	400.00
School Teacher	200.00	Retired	300.00

Laborer	$100.00	Merchants	$200.00
Office	· 100.00--200.00	Others	25.00
Physician	400.00	Aged	25.00

The occupational assessment is of little or no significance. It is not worth the effort of keeping the records. The straight poll tax would adequately meet the necessity, if there is such a necessity, of taxing all inhabitants whether or not they have assessable property.

4. The Personal Property Tax - The assessment of personal property ("money at interest") is a hit-and-miss affair, especially in counties from the fifth to eighth class. Few counties make much of an effort, outside of the work of the field assessor, to probe into this source of revenue. In a few of the fourth class counties a special clerk in the county commissioners' office, who is called the "personal property clerk," has charge of this branch of assessments. He sees that the personal property return blanks are given to the assessors, and sometimes supplements these by mailing blanks to prospective taxables. In Erie County, for example, the city of Erie is covered entirely by mail in this manner. The return is receivable after the taxable has the blank ten days. In counties where there is no special property clerk, the clerk to the county commissioners generally accepts the assessors' return without question. He has very little time to look into the matter, and for that matter, neither do the assessors, so that it becomes actually a system of self-assessment, except as the assessors utilize the records of the prothonotary and recorder of deeds as to judgments, mortgages and other such instruments recorded in county offices. In Erie County, it was stated that the personal property clerk actually did investigating in the field and through various sources of information. The office was created in 1932. The difference in assessed valuation of personal property in 1930 and 1933 might be some indication of the worth of this officer and his secretary, except for the fact that values declined so heavily in this period as to make impossible a fair comparison. The difference between 1930 and 1933 is $5,170,399, out of a total for 1933 amounting to $34,546,747. The tax on this difference, if collected, would amount to $20,681.00 and makes an adequate investment of the money spent on this office, which in 1933 was $4,110.48. This increase is also in the face of the depression which was especially hard on such properties.

The law relating to this assessment is not uniformly put into operation. In some counties there is a large return; in others it is small in comparison to the total assessed valuation. Naturally, such wealth is not evenly distributed throughout the counties and there is no measuring stick for determining what the return should be. Table 1 gives the entire assessed valuation of the 12 counties, the amount of personal property taxes, and the percentage that the personal property assessment formed of the total in 1932, according to the latest available figures compiled by the Bureau of Statistics in the Department of Internal Affairs in Harrisburg.

Table I

Personal Property Valuation in 12 Counties, 1932

Counties	Total Assessed Valuation	Personal Property Assessment	Percentage Personal Property Assessment of Total
Lackawanna	$212,213,148	$ 20,908,011	9%
Westmoreland	178,733,822	23,689,357	13.2%
Montgomery	498,337,993	221,860,598	44.5%
Erie	164,714,376	34,726,368	21%
Chester	155,238,159	39,570,299	25.4%
Northumberland	76,909,683	9,724,923	12.6%
Centre	21,029,659	4,534,871	21.5%
Adams	23,303,031	5,092,212	21.8%
Tioga	17,262,653	2,995,212	17.3%
Monroe	24,482,253	4,941,156	20.1%
Union	11,087,824	2,140,386	19.3%
Forest	3,071,429	1,184,782	38.5%

The percentage which the personal property assessment forms of the total assessment ranges from 9 per cent in Lackawanna to 44.5 per cent in Montgomery. While this may indicate different ratios of wealth in terms of "money at interest" to real estate values, it also reflects differences in assessing personal property. If the elected assessors are unable to make adequate assessments on real property, how much less able are they to make assessments of personal property, under the system now prevailing?

5. Proportion of Assessed Value of Real Estate to True Valuation - While differences in the assessed valuation and the true valuation make no difference in the taxing situation as between counties (for the property tax is entirely local in character), it has a great deal to do with inequalities that exist in tax burdens between the various divisions of the county in relation to county taxes. It is a well known fact that local assessors attempt to keep their assessments down to the very lowest level in the hope that it will make their taxables pay less county tax than the people of another district. In fact, assessors are often considered good or bad on this basis, and the community feeling that is aroused when it is shown that other districts pay less tax when conditions of wealth are approximately equal is often too intense for the assessor. He either cuts the assessments down or seeks no further reward at the hands of the electors. While the board of revision, the county commissioners, have the power to revise in order to equalize the load between districts, this is rarely done because of the lack of knowledge on their part or on anyone's part as to the correct proportion. The law calls for an assessment according to the selling price and actual value, but there are few districts where a 100 per cent actual value is used as the assessed value. The following table indicates the proportion of the assessed valuation to the true value as reported by the county commissioners and the board of assessment and revision of taxes to the Bureau of Statistics, Department of Internal Affairs. These figures, as reported, can not be considered accurate, but they offer some idea of the existing diversity.

County	Percent of True Valuation	County	Percent of True Valuation
Lackawanna	45% to 100%	Centre	33% to 100%
Westmoreland	75%	Adams	80%
Montgomery	33% to 100%	Tioga	80%
Erie	60%	Monroe	50%
Chester	55%	Union	50%
Northumberland	50%	Forest	50%

At this point it might be well to mention two valuable studies relating to assessments in Pennsylvania. These are:

F. P. Weaver, Rural Taxation in Pennsylvania, a part of a larger study: Some Phases of Taxation in Pennsylvania by Clyde L. King and F. P. Weaver, Bulletin, Pennsylvania Department of Agriculture, (1926,).

F. P. Weaver, The Rural Tax Problem in Pennsylvania, Bulletin, The Pennsylvania State College School of Agriculture and Experiment Station, (1931).

Dr. Weaver concludes that the wide variations and total lack of uniformity in making assessments which prevail throughout the Commonwealth, inevitably cause serious injustice. The weaknesses so obvious in the present system come largely from the incompetence of the assessors, the lack of concerted action of the county constituencies, and the influence of the property owners over local assessors. These wide variations occur not only between different tax districts, but also within the districts themselves. "Some property owners, both rural and urban," it is stated, "pay from two and one-half to eleven times as much tax in proportion to the value of their properties as do other property owners in the same taxing district, because of unequal assessments."

III. COST OF ASSESSMENTS IN PENNSYLVANIA

Table II indicates the number of taxables per county assessor, the amount of property per assessor and the cost of assessment per $1,000 of assessed valuation for all the counties in the state except Philadelphia, for 1932. These figures were compiled from official data taken from photostatic copies of "Assessment and Tax Statistics for Local Government by Tax Districts, 1932" arranged by the Division of Assessments and Taxes, Bureau of Statistics, Department of Internal Affairs. They were kindly furnished by Mr. Henry W. Van Pelt, Director of the Bureau. They indicate clearly the disparity in the work of the assessors and the cost of assessing property in Pennsylvania.

Table III shows the range in cost of assessment per $1,000 assessed valuation by districts within six counties for 1933. The figures were compiled from assessment cost transcripts in county commissioners' or controllers' offices. All costs are on the old rate of $5 a day for assessors' compensation.

Table II

Range in Volume and Cost of Assessments

County	Number of Taxables Per County Assessor	Amount of Property Per Assessor	Cost of Assessment Per $1,000 of Assessed Valuation
	1932	1932	1932
Adams	460	529,614.34	.469
Allegheny	1,942	14,854,927.72	.154
Armstrong	526	753,150.19	.178
Beaver	1,083	1,516,676.96	.154
Bedford	540	543,205.25	.105
Berks	1,567	3,267,429.98	.062
Blair	1,466	1,855,625.15	.198
Bradford	525	399,170.69	.286
Bucks	976	1,509,854.76	.120
Butler	856	1,555,577.16	.093
Cambria	1,009	1,955,449.58	.103
Cameron	359	320,670.67	.222
Carbon	912	1,068,460.46	.290
Centre	566	425,741.48	.250
Chester	898	1,805,094.87	.162
Clarion	528	474,401.63	.165
Clearfield	710	453,975.04	.284
Clinton	576	459,852.62	.247
Columbia	874	909,202.32	.163
Crawford	763	770,499.81	.191
Cumberland	998	1,440,363.43	.076
Dauphin	1,483	2,439,482.52	.114
Delaware	4,740	14,181,275.18	.120
Elk	1,268	795,485.38	.291
Erie	1,717	3,294,285.52	.146
Fayette	1,464	1,805,583.25	.160
Forest	321	341,269.89	.164
Franklin	1,428	1,795,804.96	.177
Fulton	460	242,300.67	.536
Greene	691	1,926,132.10	.063
Huntingdon	355	484,363.76	.116
Indiana	957	1,329,334.07	.117
Jefferson	913	976,261.26	.140
Juniata	569	343,746.06	.447
Lackawanna	4,088	4,421,107.25	.571
Lancaster	1,334	2,179,984.53	.071
Lawrence	1,590	2,406,415.97	.136
Lebanon	1,253	2,363,192.57	.074
Lehigh	1,680	4,222,481.11	.094
Luzerne	7,500	1,223,733.13	.177
Lycoming	771	776,025.08	.202

County	Number of Taxables Per County Assessor	Amount of Property Per Assessor	Cost of Assessment Per $1,000 of Assessed Valuation
	1932	1932	1932
McKean	934	1,319,891.83	.208
Mercer	978	1,122,134.17	.158
Mifflin	1,078	939,173.67	.136
Monroe	748	1,020,093.88	.253
Montgomery	1,159	4,018,854.78	.084
Montour	554	437,386.14	.287
Northampton	1,370	2,180,361.90	.108
Northumberland	1,076	1,183,225.89	.759
Perry	549	383,902.53	.205
Philadelphia		Does not include Philadelphia which is tabulated as a city	
Pike	558	855,473.00	.235
Potter	380	234,177.50	.359
Schuylkill	936	1,432,629.10	.097
Snyder	594	603,128.25	.123
Somerset	773	982,933.21	.188
Sullivan	371	280,594.77	.108
Susquehanna	470	505,190.54	.329
Tioga	517	431,566.33	.222
Union	628	652,224.94	.153
Venango	820	1,549,138.17	.142
Warren	637	734,466.63	.234
Washington	1,188	2,210,917.18	.050
Wayne	514	587,250.04	.323
Westmoreland	1,524	1,770,037.84	.094
Wyoming	399	264,378.32	.526
York	1,098	1,264,995.33	.111

Total 67 Counties 1,604

Table III. Range in Cost of Assessments within Six Counties, 1933.

County	Range in Cost per $1000 Assessed Valuation by District	County	Range in Cost per $1000 Assessed Valuation by District
Erie	Range $.07 to .70 Under .20 --- 14 Dist. .20 --.40 --- 21 " .40 --.60 --- 4 " Over .60 --- 1 "	Adams	Range .10 to .61 Under .20 --- 2 Dist. .20 --.40 --- 18 " .40 --.60 --- 10 " Over .60 --- 2 "
Northumberland	Range .07 to 1.11 Under .20 --- 3 Dist. .20 --.40 --- 11 " .40 --.60 --- 16 " Over .60 --- 7 "	Monroe	Range $.13 to .56 Under .20 --- 7 Dist. .40 --.40 --- 5 " .40 --.60 --- 8 " Over .60 --- 0 "
Centre	Range .03 to 1.62 Under .20 --- 14 Dist. .20 --.40 --- 15 " .40 --.60 --- 5 " Over .60 --- 6 "	Union	Range .09 to .43 Under .20 --- 11 Dist. .20 --.40 --- 3 " .40 --.60 --- 1 " Over .60 --- 0 "

CONCLUSIONS

A. Assessment of real property in third class counties -

1. FINDING: The act of 1931 affords a favorable opportunity for uniform and equitable assessments on a county-wide scale. Westmoreland County alone has realized
this opportunity fully and has developed highly satisfactory methods by setting up
standards for assessments, by division of the county into proper assessment districts and by the selection of assessors.
RECOMMENDATION: That all counties in this class should, in justice to their property holders, immediately avail themselves of the possibilities under the new act.

2. FINDING: Where a large number and variety of properties offer complexities
which an unskilled assessor finds difficulty in handling, the board of assessments
and revision is the only authority now provided to meet the difficulty. In practice
the routine of this work devolves upon subordinates.
RECOMMENDATION: The board should declare the general policy and make necessary revisions upon appeal, but it should also employ a chief assessor to direct the daily
work and supervise the assessors in the field.

3. FINDING: The sub-assessors are now employed on part time, except for a few who
are kept over for adjustments in the inter-triennial years.
RECOMMENDATION: The sub-assessors should be full time employes of the county and
should develop a continuous routine of assessment so that a smaller number would
function permanently. If the assessment process were continuously made, the work to
be done in the triennial assessment could be handled with slight if any increase of
force in the peak period.

4. FINDING: All counties require information and advice on assessment problems.
RECOMMENDATIONS: (a) The data now reported by the county to the Bureau of Statistics in the Department of Internal Affairs, should be given in more detailed form,
should be mandatory and should be reported either to the new bureau of local government proposed in other sections of this report or if such bureau is not created,
then to the present bureau of municipalities with enlarged power.
(b) The Bureau of local government or if it is not established the bureau of municipalities should be authorized and equipped to give both information and advice on
assessment problems.

5. FINDING: Modern assessment records have been satisfactorily developed in Westmoreland County.
RECOMMENDATION: A county assessment record using cards or other appropriate forms
should be installed in all counties of this class.

A. Continued - Assessment of real property in fourth to eighth class counties.

1. FINDING: Dissatisfaction with the present assessment system is wide-spread.
Such opposition as exists to a modern method is directed chiefly against employment
of assessors who do no not know the people and conditions of the locality. This
could be avoided.*
RECOMMENDATIONS: (a) Assessors should be appointed by the commissioners, not elected; the assessment process in all counties should be on a county-wide basis.* (b)
Assessors should be employed on full time especially in more populous districts.

2. FINDING: The study of assessment conditions and the development of a satisfactory system calls for expert service. Such service does not exist at present.
RECOMMENDATION: In the smaller counties the chief assessor, who is hereby recommended, might be the chief clerk to the commissioners, if some degree of permanence and
suitable qualifications were assured. In many places a single employe might act for
several counties.

---

* See note at end of chapter

3. FINDING: Equalization of assessments within the county has become a problem of importance. Satisfactory informal methods have been developed in Lancaster and Cumberland Counties.
RECOMMENDATION: The elected officers in the boroughs and townships should sit with the county commissioners to assist in the equalization of values in their respective districts.

4. FINDING: In counties of the fourth to eighth class inclusive the need for greater accuracy and uniformity of valuation is especially noticeable. A simple yet effective classification of farm property has been developed by Professor Weaver for Berks County, and is an example of what might be done in every county.
RECOMMENDATION: The county assessing authority, with the aid of interested groups of citizens and of agencies specializing in assessment technique, should prepare a manual for assessors.

5. FINDING: The old fashioned assessment books, with their meagre description of property, are relied on in the smaller counties. Their descriptions do not give necessary data on the more complicated types of property, nor do they classify tillable land.
RECOMMENDATION: A permanent, modern assessment record preferably of the card type should be established.

* 6. FINDING: The assessment of property by county assessors and by city and borough assessors now varies widely. There is also a double expense for assessment.
RECOMMENDATION: That the county assessment be made the basis of all local taxation.

B. Assessment of Personal Property.

1. FINDING: Assuming that the personal property tax is to continue in some form, the present lack of effort to find taxable property will result in greater injustice as other forms of property are forced to bear greater burdens. Montgomery County has shown that modern methods are effective in bringing personal property to bear its fair share.
RECOMMENDATION: All available sources of official information in the county offices should be used to trace property. The chief assessor, where such officer exists, should devise orderly means of securing this information within his own county and, by reciprocal arrangement, in other counties.

C. Assessment of occupation

1. FINDING: The tax on occupations has shown itself to be of little revenue value.
RECOMMENDATION: A poll tax on each resident would be sounder in principle and could be better administered. Persons who pay real property tax of more than a certain amount should be exempted from the poll tax.

---

* Note: Subsequent to the drafting of this Chapter the recommendations indicated were likewise made by former Budget Secretary Dr. Edward B. Logan in an intensive study, "Taxation on Real Property in Pennsylvania," 1934. Dr. Logan also offers the following additional recommendations:

Establishment of a State Tax Commission with broad powers over local taxation.
Elimination of the triennial assessment and the substitution for it of more frequent assessment.
Publication of all assessments, and uniformity in personal notices to taxpayers and in the times and places for revision and hearing of appeals.
Revision by sub-assessors, appeals on factual questions to county assessors, and appeals on legality to the court.
Broader tax base, including much property now exempt, and an orderly system of securing exemptions.
Assessment at full sales value.

CHAPTER V

ELECTION ADMINISTRATION

INTRODUCTION - In examining the election process in counties we may note the following steps:

1. The county commissioners advertise the election in newspapers of the county.

2. The commissioners receive petitions of local candidates to have their names placed on election ballots, and receive information as to candidates for state-wide offices from the Secretary of the Commonwealth.

3. Registration of voters by registrars and registry assessors under the supervision of the registration commission in first and second class cities, and the county commissioners in others.

4. Copying of the voting check list and ballot check list from the registry books of registrars and assessors, usually by clerical help in the county commissioners' office but sometimes by assessors and registrars.

5. Purchase of ballots and election supplies by county commissioners.

6. Delivery of election supplies, ballots, voting lists, etc., to judges of elections of election districts.

7. Holding the primary and general elections.

8. Making returns to county commissioners, recording the vote and recounting the vote when necessary.

9. Issuing warrants for payment of election officers, registrars and assessors, election supplies, etc., on the part of the county commissioners.

While the county commissioners are generally considered to be responsible for the conduct of elections, their power is at certain points quite limited. Sharing this authority with them are the judges of the court of quarter sessions, registration officers or registry assessors, and the election officers in each district. Furthermore the Secretary of the Commonwealth, as the state election officer, is clothed with some authority in the preparation of ballots and the purchase of voting machines. A number of local officers also have minor duties in regard to elections. For example, the prothonotary of the county receives and records the returns from the judges of election; the sheriff issues election proclamations; constables attend elections in their districts to preserve the peace; they also advertise local elections. Finally the election laws of the Commonwealth prescribe in detail the election duties of the county commissioners. The commissioners pay the election bills, including those for ballots and supplies, but they have very limited discretion, since the compensation of all election officers is fixed by law and the voters themselves decide whether they want voting machines.

County administration of elections will be considered under the following topics:

    I.   Election districts
   II.   The registration of voters
  III.   The cost of elections in election districts
   IV.   The total cost of elections

I. ELECTION DISTRICTS

Fundamental in the problem of election cost and administration is the election district. When such districts have few voters, they are expensive because the basic costs of election officers and rent are the same for all districts. When it is apparent that the same number of election officers in one district can handle two or three times the number of votes that are cast in other districts, there is an indication of waste.

It is true that some districts are of much larger area than others and that there must be some limit to the distance voters should be made to travel in order to exercise their suffrage, especially in the sparsely populated rural areas. On the other hand, many districts were laid out in the "horse and buggy" days, and better means of transportation have now made larger districts possible without much inconvenience to the voters.

Study of the 1,229 election districts in the counties surveyed reveals a wide range between election districts with regard to population and the number of registered voters. No standard exists as to the size on which election districts should be recast. The determination of such a standard, not only for the creation of new districts, but for the recasting of old ones, is essential if election costs are to be reduced.

The subject of election districts will be considered as follows:

    a. Constitutional and statutory provisions
    b. The number of election districts per local subdivision
    c. Population and registered voters per election district

Constitutional and Statutory Provisions - The State Constitution in Article 8, section 11, as amended in 1928, provides for the creation of election districts as follows:

> "Townships, and wards of cities or boroughs, shall form or be divided into election districts of compact and contiguous territory, in such manner as the court of quarter sessions of the city or county in which the same are located may direct; but the courts of quarter sessions, having jurisdication therein, shall have power to divide or change the boundaries of election districts whenever the court of the proper county shall be of opinion that the convenience of the electors and public interests will be promoted thereby."

The court of quarter sessions has plenary power to create election districts from townships and wards of boroughs and cities. This assumes that each township and borough and each ward of a borough or city becomes automatically a separate election district. For example, the division of a borough into wards automatically makes each ward an election district. One to three councilmen are elected from each ward, thereby necessitating distinct election districts. However, the consititutional provision gives the court of quarter sessions "power to divide or change the boundaries of election districts" whenever it deems this advisable. The Act of 1854 (P. L. 419) gives the court the power to "form an election district out of parts of two or more adjoining townships so as to suit the convenience of the inhabitants thereof," provided that no district so formed shall contain less than one hundred voters and that proceedings had in such case shall be the same as in the erection or alteration of the lines of townships. Furthermore, the Act of 1895 (P. L. 377) provides that upon petition of a majority of the qualified electors voting at the last general election in each of "two or more adjoining election districts in any township, borough or city, the same may be consolidated into one election district by the court of quarter sessions."

Briefly summarized the usual steps in the creation of a new election district are: the petition of electors; the appointment of commissioners of inquiry; the report of such commissioners to the court; the filing of exceptions; the review and confirmation of the report. The Supreme Court of Pennsylvania (in re Bern Twp., 115 Pa. 615) held that the court could divide a township under its constitutional power "without appointing commissioners or proceeding by view, review and exceptions."

It would seem, therefore, that the court of quarter sessions has power to divide or consolidate election districts in any manner and by any procedure it deems advisable for the convenience of the voter and the promotion of public interests.

Since the enactment of the voting machines law in 1929, statutory provision has been made for the court of quarter sessions to "redivide wards of cities or boroughs, where voting machines are adopted, into election districts of compact and contiguous territory," and to "consolidate election districts therein into new districts, each having five hundred registered voters as nearly as may be," except that new districts of less than five hundred voters may be created, if the court deems it advisable for the convenience of the public. Petitions may be presented to the court either from the county commissioners or from ten or more qualified voters in the districts involved. Even this provision does not appear to encroach unduly on the blanket power of the court to divide or change the boundaries of election districts, as given to it by the Constitution, although, as one court opinion put it, "it seems the better practice to conform to the statute." Furthermore, the statutory provisions are no practical hindrance to a court (or an interested body of citizens, or even a county board of commissioners), in making election districts follow more closely the dictates of economy. There are no statutory provisions requiring an election district to have a certain number of registered voters. The Act of 1854 states that no district shall be formed which contains less than one hundred voters. A constitutional provision, repealed in 1928, states that in cities of over 100,000 in population, districts which polled over 250 votes in the last election should be divided. The 1931 amendment to the voting machine law stated that new districts should be formed with as near to 500 registered voters as possible, although smaller districts could be formed (the original clause provided for 800 registered voters). County commissioners are empowered to abolish election districts in which less than ten qualified electors reside, although appeal may be made to the court of quarter sessions. As a result, there is a wide range in the number of registered voters in election districts.

b. **The Number of Election Districts per Local Subdivision** - In the 12 counties studied, there were 1,229 election districts in 447 townships, boroughs and cities. On the average there were 2.8 election districts per local subdivision. The range of county averages was from 1.1 in Tioga to 5.9 in Lackawanna. The number of districts is determined not only by population and area, but also by the customary lines of neighborhoods, political exigencies and other non-legal factors. Our inquiry must confine itself to a study of population and registered votes per election district because there is no available material relating to the area of these districts or to other peculiar local factors which have entered into the creation of those now existing.

Population and Registered Voters per Election District - As we have seen, there is no statutory requirement as to the number of inhabitants or registered voters per district. In the 1,229 districts here surveyed, there is an average population of 1,194 per district. This average ranges from 1,565 in Erie County to 288 in Forest. In general, the most populous and urban counties have a higher average per district although many individual districts within these counties have small populations.

The average number of registered voters per election district for the 1,229 units is 522, while the range by counties is from 789 for Montgomery to 164 for Forest. The latter, as its name implies, contains large woodland areas and is sparsely settled, while Montgomery is suburban to Philadelphia and is thickly populated. However, the factor of population per square mile is not always a deciding one. Union, Monroe, and Tioga are all sparsely settled counties, but their average number of registered voters per election district is approximately that of Chester and Erie, which are urban and populous.

The range in the number of registered voters per district within the counties is indicative of the haphazard size of these districts. The lowest number of registered voters in any district in the eleven counties surveyed is 31, the highest 2086. In all but two counties, Union and Montgomery, there are some districts with less than 100 registered voters; in all but two, Union and Forest, there are some districts with more than 1,000 registered voters.

Table I shows the average population per election district, the average number of registered voters per district, and the range in number of registered voters per district, by counties.

Table I.   Population and Registered Votes in Election Districts, 1933

County	Average Population per Election District	Average Number Registered Voters per District	Range in Number of Registered Voters per District
Lackawanna	1257	541	48 --- 1557
Westmoreland	1288	556	83 --- 1256
Montgomery	1460	789	204 --- 2083
Erie	1565	483	72 --- 1273
Chester	974	490	
Northumberland	1201	573	40 --- 1510
Centre	723	372	97 --- 1500
Adams	884	512	97 --- 1195
Tioga	708	441	
Monroe	912	541	70 --- 1048
Union	794	449	130 --- 934
Forest	288	164	31 --- 577
12 Counties	1,194	522	

These figures indicate definitely that the cost of elections, based as it is largely upon the arrangement of election districts, may be reduced by rearranging the size of these districts and eliminating those which are too small in population or in registered voters. Further evidence on the actual cost of elections and of registration of voters will support this position.

II. THE REGISTRATION OF VOTERS

There are three general methods by which voters are registered in Pennsylvania. The first is that in use in Philadelphia and in third class cities -- personal and periodic registration. The second method was inaugurated by the legislature in 1933 for the use of second class and second class A cities, Pittsburg and Scranton. It is a system of permanent registration, the effectiveness and economy of which will be interesting to all students of election administration and tax payers in general. The third is the system of registration which has been basic in Pennsylvania for almost a hundred years; voters in all areas of the Commonwealth at one time or another have been registered by this method. It is the system of registration by registry assessors, and is in use in all boroughs and townships of the state.

All three methods are employed in some part of the area covered in this survey of twelve counties. Therefore it is appropriate to sketch briefly not only the main provisions of the laws which govern the process, but also to give information concerning the actual operation of these law within the counties. The general subject of registration may be divided conveniently into the following topics:

     a.   The process of registration according to law
     b.   The actual operation and suggestions for improvement
     c.   The cost of registration for 1933 in 12 counties

The Process of Registration - (a) Personal and Periodic Registration - Personal and periodic registration is used in cities of the first and third classes. This includes Philadelphia (first class) and forty-four other cities ranging in population from seven thousand to one hundred fifteen thousand. While the laws relating to first and third class cities vary in some important details, the general process is much the same. Voters must register personally once a year before registrars sitting in the polling places of each district. In Philadelphia, these registrars are appointed by the registration commission of five persons which is in turn appointed by the Governor of the Commonwealth; the county commissioners appoint the registrars in cities of the third class. There are four registrars per district in Philadelphia and two per district in cities of the third class. The registrars sit at the polls in their districts three days in the fall and one day in the spring in the even-numbered years, and three days in the fall in odd-numbered years. The spring primary in even-numbered years accounts for the extra day. Two registers are made by the registrars in each election district and are turned over to the county commissioners who, in turn, give them to the election officers on election day for the sake of checking the eligible voters. One of these registers is known as the ballot check list and the other as the voting check list.

All registration costs are paid by the county. The per diem compensation of registrars is ten dollars.

(b) Personal and Permanent Registration - Cities of the second class, Pittsburgh and Scranton, have had a system of personal and permanent registration since 1933. The law for second class A cities (Scranton) provides that a voter need register only once while he continues to reside at the same address. One is stricken from the permanent list only if dead, or upon leaving the district, or upon failing to vote for four years. The original registration under this new law took place during 1934 on the regular registration days. Voters registered in the polling places and before the regularly appointed registrars. At this time, permanent records of each voter were made. These records, containing various items of information about the voter, were filed in the central office of the registration commission.

Once the new registration is completed, there will be no need for the regular registrars for each election district or for the regular registration procedure. Eligible voters not on the list may be placed there by applying to the registration commission at any time during the year or by registering with specially appointed registrars who will sit on the days and in the places indicated by the registration commission. The registration commission instructs the employees of the central office to make up the voting check list, the ballot check list, and the street lists of registered voters used by the election officers in the election districts. These lists are made up from the permanent records. The records are kept up to date by securing reports of deaths from the registers of vital statistics, reports on removals from municipal offices furnishing electricity, water or gas to householders, and reports from real estate brokers regarding removals from properties in their charge. Furthermore, the commission checks registers by mail every four years, and at frequent intervals sends out specially appointed inspectors of registration to compare addresses.

The registration commission, appointed by the Governor, is in complete charge of this function. It appoints all the registrars, inspectors of registration, and other employees, while the salary board of the county determines the number and compensation of such employees.

(c) Registration in Boroughs and Townships - The registration of voters in boroughs and townships is carried on by registry assessors. In boroughs, townships, and wards where there is only one election district, the regular tax assessor is the registry assessor; but in those subdivisions containing more than one election district, there is a special registry assessor. The duties of the regular assessor and the registry assessor in regard to the registration of voters are identical, and for the sake of simplicity they will be referred to hereafter as the registry assessors. They are elected by the voters in each district for terms of four years.

In the beginning of May of each year, registry assessors visit all homes and dwellings for the purpose of making a list of eligible voters for the year. This list is called the "original registry list," and is made in a book prepared for the registry assessor by the county commissioners' office. After completing the list in his district, the registry assessor makes a copy and hangs it on the door of the election house in his district. This must be done on or before the fourth Monday in May. The registry assessor retains the original list and makes necessary changes in it upon personal application. He sits at the polls in his election district one day in September before the November election to make such changes and additions. He next sends the changed list to the county commissioners' office, where a ballot check list and a voting check list are made. These lists are sent to the election officers for the election day. After the November election, the county commissioners return to the assessors the original list. On even-numbered years when there is a spring primary, the registry assessors sit one day, sixty-two days before the spring primary, to make changes in the list. The list is again sent back to the county commissioners, who make up their voting check list and ballot check list for the election districts to be used in the primary.

The compensation of assessors for the registration of voters in boroughs and townships, in counties other than those of the first and second class (Philadelphia and Allegheny), is ten cents for every name assessed and entered in the book, but the total amount paid to the assessors for each period of registration must not exceed an average of $3.50 a day. Besides this, the assessor receives $3.50 a day for the two sittings at the polls that are provided by law. The county pays all registration costs.

Previous to 1933, the assessors received $5.00 a day for the annual house to house canvass in May and $5.00 a day for sitting at the polls four days instead of two. The reduction was made on the demand of taxpayers groups that something be done to reduce tax burdens.

Registration in boroughs and townships is non-personal and periodic, but it is, in many cases at least, non-personal and permanent because of the actual failure of assessors to make a house to house canvass every May.

Actual Operation and Suggestions for Improvement-(a) Personal and Periodic Registration- Personal and periodic registration, while making possible a complete annual purging of the voters' lists of ineligible and indifferent voters, is troublesome to the voter and expensive to the taxpayer. It is interesting to note that the Registration Commission of Philadelphia approved the permanent registration bill for Philadelphia at the 1933 special session of the legislature. While this proposal passed the Senate, it was lost in the last minute rush in the House. The Philadelphia Commission in its 1933 Report indicates the safeguards which it considers essential in any system of permanent registration:

1. The signature of the elector at time of voting.
2. A compulsory annual check-up of registered voters by mail.
3. Keeping names on active registry cards for at least two years. If the elector moves from the division or fails to vote for two successive years, his name should be stricken off, after notice to him.
4. Proper protective locking devices for registers.
5. All-year opportunity to register at the office of the Commission.
6. All-year work by the commission in clearing lists.
7. Mandatory weekly report of deaths from Bureau of Vital Statistics.
8. The Commission to be free to appoint clerks and registrars with consideration only of their qualifications for the work, and without restrictions as to residence, other than residence within the county.
9. Sworn declaration by registrant at time of registration in case assistance will be required at time of voting.
10. Registration Commission should have power to order a complete re-registration whenever necessary.

(b) Personal and Permanent Registration - While it is too early to pass judgment upon the system of permanent registration in second class cities, some comments based on direct observation of the Registration Commission in Scranton may be appropriate. Voters come to the central office to report their removal to another election district, or to register upon coming of age. No sub-offices in other parts of the city are contemplated at present, nor are special days set aside for changes in registration. There is as yet no system of checking removals from districts unless they are reported to the central office. There is to be no regular check-up of registered voters during the year by a house to house canvass such as is conducted in many other systems of permanent registration, although a check-up by mail is provided in the statute for every four year period, at the discretion of the Commission. Furthermore the cards for registered voters in each election district are kept in a specially constructed book, which is locked securely and can be opened only by a key that is retained by the chief clerk of the Commission. Thus there can be no tampering with the records except with collusion of the chief clerk. The cards are serial-numbered, are made of durable material (8 x 5), and have spaces for a record of voting to last until 1965. Two cards have been made for each registered voter, most of them being typewritten. The Commission plans to make another copy of the cards which will give the office a record in triplicate. From these cards, lists of registered voters to be used as ballot check lists and voting check lists are made by the Commission. Street lists of voters will also be made.

It was impossible to obtain a sample registration card because they were numbered serially. However, the card contains spaces for the following items of information arranged approximately as follows:

Name. . . . .   Address . . . . . .
Occupation. . . . .   (Room, apt., flat or floor
Date residence in election district began . . ·.   occupied)
Place of last registration . . . . Designation of political
   (No. Street, City, State, year)   party for primary vote.
                                  Sex . . (year - affiliation)
Place of birth . . . . Color .   (12 spaces)

Personal Naturalization   Oath of qualification for
Date. . . . . .Court. . . . . voting.
Place . . . . .Certificate No. . . . . .
                                            Date . . . .

Derivative Naturalization
. .Father . .Mother . .Husband

Physical Disability   Inability to Write
State exact nature of disability to see or mark   If applicant is unable to write
ballot or operate voting machine or enter voting   his record, this information is
compartment or voting booth without assistance.   required: height, color of hair
                                                   and eyes.

                    (On opposite side)

Record of removals. . . . .(with spaces)
Record of voting at elections . . . .(with spaces for
     regular primary and general and special elections 1934 - 1965)

There are now four commissioners, each with the statutory salary of $3,000 per annum. The office is in charge of a chief clerk and an assistant clerk, who are assisted by additional part-time clerical help depending upon the amount of work to be done. The number of registrars, inspectors of registration and other employees who will be regularly needed is as yet not determined.

(c) Registration in Boroughs and Townships - The actual situation in regard to assessors in boroughs and townships varies as to detail from county to county. Assessors in some counties are allowed compensation for the days they return their list of voters to the county commissioners' office at the same rate which they receive for sitting at the polls; in other counties they are not allowed such compensation, and are asked to send their lists in by mail. In some counties they have been allowed mileage for distance covered in their work; in others, no mileage is allowed.

In some counties, the county commissioners give the assessors four registration books in which they are to copy the names of registered voters. These books are returned to the commissioners when completed, and are re-issued to the election officers; the clerks in the county commissioners office are not required to make up the voting check list or the ballot check list. In others, the assessors make only one copy to hang on the door of the election house, and return the original to the commissioners.

The new provisions relating to compensation have been applied with varying details. In rural counties, the problem of compensating assessors in sparsely settled townships gives rise to the practice of allowing the assessors to report more days than are actually worked in order to make up for mileage and time consumed in visiting the few scattered houses in their district. The assessors are supposed to keep a record of the number of hours worked, divide this total by eight (the number of hours agreed upon as constituting a working day), and add enough hours to cover the cost of mileage. The cost of registration is expected to be about the same as it was before the new law went into effect. In some counties, the rate of compensation per name registered is made lower than ten cents, and additional compensation is allowed for copying the names in the registry books. While no conclusions can be drawn as yet about the decreases in registry costs under the new law (no figures are available, since 1934 was the first year of its operation), it is doubtful whether much economy, outside of the decrease in the number of days spent at the polls, will be made. Even the decrease in per diem compensation from $5.00 to $3.50 will doubtless be neutralized in many places by an increase in the number of days found necessary to canvass.

While the actual registration of voters is accomplished with very little cause for complaint (it involves only the writing down of the eligible voter's name and a few facts relating to his occupation, address, and political affiliation), it is to be understood that non-registered voters may vote at the regular elections by proving their right and qualification to vote. Thus, any omission on the part of the assessors does not bring serious inconvenience. In effect, this means that one does not have to be registered in boroughs and townships in order to vote. Once having voted, however, the elector has his name added to the list of voters and is duly registered.

The most important consideration in the problem of registration in boroughs and townships is the lack of control by the county commissioners over the assessors. They give the latter the necessary books, receive these again when completed, and pay the bills as handed in. There is no way by which the clerk of the county commissioners, who handles this routine, can determine whether the bills handed in represent actual work done by the assessors in the May canvass of voters. There is no doubt that the canvass is not made in a great number of districts. In fact there is no reason why it should be made in most of the boroughs and townships of the state. The assessor knows who lives in almost every house and no person is barred from voting by reason of not having registered at the regular time. The names are copied from one year's registry book to another with very few changes. Nevertheless, the registry assessors usually hand in bills that represent the time that would have been spent had the canvass been made, even when they do not make the actual canvass or when they make it only partially. In one county, the assessors were called in by the county auditors and asked to explain the apparent discrepancies between the costs of registering in various districts. Various "explanations" were given, and the next year the auditors published the money spent on assessors. Though the assessors complained of the unfairness of publishing their individual accounts yet in the following year they reduced their bills by nearly one-half.

The clerk of the county commissioners rarely takes any initiative in checking up on these bills. The assessors are elected officers, politically important in their districts, and the county commissioners usually allow their claims. Only rarely do the auditors act. Even in counties where there are controllers, a check-up on the amount of time which the registry assessors spend seems next to impossible. Furthermore, in townships, wards and boroughs with only one election district the offices of registry assessor and property assessor are held by the same person. He may send his bill in unitemized, and it often contains the number of days for both the registry and property assessment. Again, some assessors hand in their bills when they "come to town", perhaps once a month or once a year. Thus in some counties it is impossible to estimate how much was spent for registry assessment apart from assessment of property. This explains why the county commissioners accept the bills of assessors as handed in.

Two possibilities of improvement suggest themselves. First, how can the registry of voters be put on a more business-like basis. Second, can a form of permanent registration be adopted in boroughs and townships under the supervision of the chief clerk of the county commissioners.

To answer the first, several possibilities occur. The elected registry assessors might be retained and the clerk of the county commissioners might be given power to work out a standard scale of allowance for time, based upon a study of past performance in that particular district or similar districts, and power to hold the registry assessors to this scale in recommending payment to the county commissioners. This would be possible under existing statutes. While assessors could bring action against the commissioners in the courts, this would be a rare occurrence because of the fact that a study of the subject had been made, a definite and fair plan had been worked out by the chief clerk after discussion with the registry assessors themselves, and because of the pos-. sibility of adverse publicity. If the commissioners were really interested in keeping down such costs to a fair basis for the work done, very little hindrance would be placed in their way. The present obstacle is that as yet no standards of performance have been worked out for the purpose. Again, the county clerk could devise a plan whereby the registry assessors would hand in itemized bills showing the days when work was done, the number of places visited, and the area covered during a certain day. Then by a few judicious telephone calls he could check up the work of the assessor and refuse recommendation for payment if he found out that the assessor had not covered the territory in question. Such bills are called for in some counties, but there is no checking up. Bills should be submitted at the end of each month in which work was done and should not be allowed to accumulate.

The problem of applying permanent registration to boroughs and townships should not be more difficult than applying the principle to urban areas. In fact, it should be more simple. The county commissioners could act as the registration commission in each county, and the chief clerk could act as the general administrator. First, a thorough registration of voters could be made by the elected registry assessors on forms indicated in the new law, which would cover the new permanent registration. Then the office of elected registry assessor would automatically cease to exist. The cards, made in duplicate, would be sent to the county commissioners office, where they would be kept permanently in modern files. From these cards, yearly ballot check lists, voting check lists, and street lists could be typed and sent to the election officers. Changes in registration could be made throughout the year by appearing in the commissioners' office or by appearing on certain dates in the various townships and boroughs before inspectors of registration, to be appointed by the county commissioners. Once on the registration list, no change would be necessary unless the elector had not voted for four years. In fact the principles of permanent registration as worked out in many states and recognized by experience as good policy could be modified to meet the conditions in Pennsylvania counties. The details of such a plan should be developed by a temporary commission composed of experts in election administration, including representation from the registry assessors, county clerks, election reform associations, the Department of State, the General Assembly, and other interested and helpful agencies. Such a law should be drafted, however, only after a detailed examination of present registration methods in boroughs and townships.

The Cost of Registration in 12 Counties - Included in the figures for the cost of registration of voters in the 12 counties surveyed is the compensation of the registration commission and its employees and the cost of supplies in second class cities (only Scranton in Lackawanna county), the compensation of registrars in third class cities and of registry assessors in boroughs and townships, and some minor incidental expenses of such officers. Not included are the cost of registry books or the cost of clerical help to copy them when this is done in the office of the county commissioners. While the costs in the various counties are fairly comparable, there are variations depending on the application of the registration laws customary within each county. These variations have been indicated for the most part in a previous section of this report.

The range in cost of registration per registered voter for 1933 is from 39.6 cents in Forest to 9.4 cents in Union. In all except these two counties the average cost per vote is between 10 and 20 cents. Five counties have an average of fourteen cents per registered vote. In only two rural counties is the cost per registered vote near the ten cent standard which is provided in the new legislation. It is interesting to note that the highest costs are in Forest, the county with the least population, and in Lackawanna, the county with the highest population of the counties surveyed.

The range in cost per registered voter, by election districts within the counties is even more varied. One district in Northumberland has a cost of 3 cents per registered voter, another has a cost of $1.23. The highest one is in Monroe with a cost of $1.50. Union and Centre have the lowest "high" cost, 24 and 26 cents respectively. They also have the lowest average cost per county.

The reason for the wide range in average cost is not to be found in the number of registered voters in the district, although this factor may have some bearing on individual cases. It is due to the work of the assessors themselves, or rather to their supposed activity as represented in their bills. When this activity is as unstandard-ized as it now is, there can be very little relation between cost and actual work. If the county commissioners made a study of the actual amount of time required to register voters in particular districts and attemped to establish a norm, there could be a great reduction in the cost of assessment. The size of the district is also an impor-tant factor in cost.

Table II gives the detailed figures for registration and costs per registered voter by counties:

TABLE II

Total Number of Registered Voters; Average Number per Election District; and Average Cost per Registered Voter, in Specified Counties, 1933

County	Total Registered Voters 1933	Average Number Registered Voters in Election District	Average Cost per Registered Voter	Range in Cost per Registered Voter by Districts
Lackawanna	133,568	541	$.359	Part range in some boroughs and townships $.07 --- .13 Scranton excluded
Westmoreland	127,435	556	.148	Range .06 --- .82 Under .20 ---174 Dist. .20 ---.40--- 50 " .40 ---.60--- 3 " Over .60 --- 1 "
Montgomery	143,535	789	.147	Range .05 --- .50 Under .20 ---125 Dist. .20 ---.40--- 54 " .40 ---.60--- 3 " Over .60 --- 0 "
Erie	54,121	483	.146	Range .08 --- .83 Under .20 --- 85 Dist. .20 ---.40--- 26 " .40 ---.60--- 0 " Over .60 --- 1 "
Chester	63,723	490	.147	No data
Northumberland	61,288	573	.201	Range .03 ---1.23 Under .20 --- 61 Dist. .20 ---.40--- 35 " .40 ---.60--- 6 " Over .60 --- 5 "
Centre	23,805	372	.105	Range .07 --- .26 Under .20 --- 45 Dist. .20 ---.40--- 7 " .40 ---.60--- 0 " Over .60 --- 0 "
Adams	21,492	512	.190	Range .08 --- .48 Under .20 --- 18 Dist. .20 ---.40--- 22 " .40 ---.60--- 2 " Over .60 --- 0 "
Tioga	19,850	441	No data	No data

TABLE II Con't.

County	Total Registered Voters 1933	Average Number Registered Voters in Election District	Average Cost per Registered Voter	Range in Cost per Registered Voter by Districts
Monroe	16,777	541	$.141	Range $.08 ---1.50 Under .20 --- 15 Dist. .20 ---.40--- 10 " .40 ---.60--- 2 " Over .60 --- 1 "
Union	9,879	449	.094	Range .05 --- .24 Under .20 --- 21 Dist. .20 ---.40--- 1 " .40 ---.60--- 0 " Over .60 --- 0 "
Forest	2,956	164	.396	Range .13 --- .97 Under .20 --- 2 Dist. .20 ---.40--- 8 " .40 ---.60--- 6 " Over .60 --- 2 "
12 Counties	678,429	522	**.189	*Range .03 ---1.50 Under .20 ---546 Dist. .20 ---.40---213 " .40 ---.60--- 22 " Over .60 --- 10 "

* 10 Counties.
** 11 Counties.

III COSTS IN ELECTION DISTRICTS

The following items are included as costs in election districts: the compensation of a judge of elections, two inspectors of elections, two clerks of election, and a constable; the rent and cost of heat and light of polling places, and some minor costs that vary in different counties, such as the cost of taking down and putting up voting booths, and mileage allowed the judge of elections and minority inspector for making returns. Previously, assessors of the districts had been directed to sit at the polls on election day and receive compensation, but this provision was recently repealed. Judges and inspectors are elected by the voters of the district for two year terms. Each inspector appoints one clerk.

The statutory provisions regarding compensation of election officers are as follows:

In Philadelphia, judges receive $20 a day; inspectors and clerks, $10 a day; return judges, $2 in addition.

In counties of the second class (Allegheny), judges receive $15 a day; inspectors and clerks, $10 a day; return judges, $2.50 a day in lieu of mileage and expenses.

In other counties, minimum pay for judges, inspectors, and clerks is $5 per day. In districts where more than one hundred votes are cast in any election, each officer receives $1 for each 100 ballots, or fractional part thereof cast after the first hundred ballots, in addition to the minimum pay, provided the maximum pay shall be $10 a day. Return judges receive $2 in addition and mileage. (Clearfield, Susquehanna and Bradford Counties operate under old special acts relating to this point.)

Constables in all counties receive $5 a day for attending elections, and this fee includes compensation for notifying persons elected at local elections.

The cost of rent, heat and light for polling places and other costs incidental to the furnishing of them are determined by the county commissioners. In many counties ten dollars is paid for each polling place per election; in others there was some variation, the highest noted being $50 dollars a year in the central area of Scranton.

All these costs are paid by the county treasurer on warrants from the county commissioners. In all counties surveyed these costs were handed in to the commissioners by the judge of elections on blanks specially prepared. Each receipient of fees or expenses signs his name opposite the amount to which he is entitled. On the same blank is a space indicating the actual number of votes cast. It was from these transcripts that the basic figures on the cost of the primary and election of 1933 in election districts were obtained.

Since the greater part of these costs is for compensation of the election officers as provided by statutes, it follows that the size of election districts is the determining factor in such costs. Based on the actual vote cast, the cost per vote will be high when few votes are cast and low when many are cast. This of course is limited by the fact that there is a graduated scale of compensation, but this graduation is insufficient to neutralize the effect of extreme ranges in the number of votes cast. The following figures for the 1933 primary and election (municipal) indicate the wide range in cost per vote cast in approximately 1,200 election districts, about one-sixth of the number in the entire state.

In the 12 counties surveyed the average cost of the 1933 primary for each of the 408,106 votes cast was 16 cents per vote. The cost per vote cast ranged from 10 cents in Lackawanna to 49.2 cents in Forest. But the range within the counties by election districts is more startling. The lowest cost per vote cast, by election districts, is 4 cents; the highest is $3.00, or twenty-five times as much as the lowest. In each county there are great extremes. While the majority of districts show a cost of less than 20 cents per vote cast, many have a higher rate.

A similar story is told in the cost per vote in the 1933 election. Here the average for all votes cast in the 12 counties is 14.9 cents per vote. It is lower than the average in the primary because more voters came to the polls. The cost per vote cast by counties ranges from 9.8 cents in Lackawanna to 38.9 cents in Forest, while the cost per vote by election districts ranges from 5 cents to $3.30. Only 751 of the 1,229 districts have costs under 20 cents per vote cast.

TABLE III gives in complete form the results indicated above for the general election of 1933.

TABLE III

Cost of Election in Election Districts -- 1933 Election

County	Cost in Election District of 1933 Election	Number of Votes Cast in 1933 Election	Cost per Vote in County for 1933 Election	Range in Cost per Vote in 1933 Election by Election Districts
Lackawanna	$12,265.66	124,459	$.098	Range $.05----1.13 Under .20----228 Dist. .20 ---.40---- 14 " .40 ---.60---- 3 " Over .60 ---- 2 "
Westmoreland	11,983.17	80,307	.149	Range .05---- .98 Under .20----140 Dist. .20 ---.40---- 67 " .40 ---.60---- 17 " Over .60 ----- 3 "

TABLE III Con't.

County	Cost in Election District of 1933 Election	Number of Votes Cast in 1933 Election	Cost per Vote in County for 1933 Election	Range in Cost per Vote in 1933 Election by Election Districts
Montgomery	13,448.52	87,752	$.153	Range$.07 --- .83 Under .20 --- 122 Dist. .20 ---.40--- 54 " .40 ---.60--- 4 " Over .60 --- 2 "
Erie	4,491.10	34,043	.132	Range .07 --- .69 Under .20 --- 86 Dist. .20 ---.40--- 17 " .40 ---.60--- 7 " Over .60 --- 2 "
Chester	6,864.56	34,787	.197	Range .09 ---1.15 Under .20 --- 47 Dist. .20 ---.40--- 71 " .40 ---.60--- 6 " Over .60 --- 3 "
Northumberland	7,769.71	46,550	.167	Range .06 ---1.40 Under .20 --- 60 Dist. .20 ---.40--- 33 " .40 ---.60--- 11 " Over .60 --- 4 "
Centre	3,504.24	14,616	.239	Range .08 ---1.45 Under .20 --- 14 Dist. .20 ---.40--- 23 " .40 ---.60--- 15 " Over .60 --- 9 "
Adams	2,545.24	13,263	.192	Range .08 --- .81 Under .20 --- 18 Dist. .20 ---.40--- 20 " .40 ---.60--- 3 " Over .60 --- .1 "
Tioga	2,289.38	12,157	.188	Range .05 ---3.14 Under .20 --- 15 Dist. .20 ---.40--- 25 " .40 ---.60--- 1 " Over .60 --- 4 "
Monroe	2,018.71	9,239	.218	Range .09 ---1.23 Under .20 --- 9 Dist. .20 ---.40--- 14 " .40 ---.60--- 5 " Over .60 --- 2 "
Union	1,480.12	5,356	.276	Range .13 --- .76 Under .20 --- 6 Dist. .20 ---.40--- 7 " .40 ---.60--- 7 " Over .60 --- 2 "

TABLE III Con't.

County	Cost in Election District of 1933 Election	Number of Votes Cast in 1933 Election	Cost per Vote in County for 1933 Election	Range in Cost per Vote in 1933 Election by Election Districts
Forest	$663.67	1,723	.389	Range$.13 ---3.30 Under .20 --- 3 Dist. .20 ---.40--- 2 " .40 ---.60--- 3 " Over .60 --- 10 "
12 Counties	69,324.08	464,252	.149	Range .05 ---3.30 Under .20 ---751 Dist. .20 ---.40---348 " .40 ---.60--- 82 " Over .60 --- 44 "

Table IV shows, for twelve counties, the number of boroughs and townships in which election costs in election districts amounted to $.20 or more for each vote cast. These high cost election districts total 472 out of an aggregate of 1,229 election districts in these counties.

TABLE IV
High Cost Election Districts, 1933 Election

Range in Cost per Vote in Election District in 12 Counties	Borough	Wards in Boroughs	Townships	Districts or Wards in Townships	Total
Total $.20 --- .40	114	43	104	86	347
Total .40 --- .60	18	6	40	16	80
Total Over .60	13	2	14	16	45
Total Over .20	145	51	158	118	472

PERCENTAGES

	Borough	Wards in Boroughs	Townships	Districts or Wards in Townships	Total
$.20 --- .40	32.7%	12.4%	29.9%	25%	100%
.40 --- .60	22.5%	7.5%	50%	20.1%	100%
Over .60	28.8%	4.4%	31.1%	35.6%	100%
Total Over .20	30.7%	10.8%	33.4%	25.2%	100%

IV. TOTAL COST OF ELECTIONS

The following objects of expenditure are included in county election costs:--

1. Printing of registry books
2. Pay of registration officers and registry assessors
3. Advertising elections, etc.
4. Printing of ballots
5. Providing election supplies
6. Providing and repairing election places, booths, guard-rails
7. Delivery of election supplies and scoring ballot boxes
8. Voting machine supplies and upkeep
9. Pay of election officers in election districts

As mentioned before, the greater part of this cost is mandatory on the county commissioners, who have little chance to cut costs by their own management. The pay of election officers and assessors is statutory. There are some possibilities for the exercise of good business practice in the purchase of supplies. For example, ballots are bought on bids in surprisingly few cases. In all counties surveyed the ballots were bought from a concern doing business in the county, usually at the county seat. There is good reason for this. Last minute changes in ballots may be made with a minimum of inconvenience, and the commissioners can check up the kind of work done. Sometimes each commissioner takes one-third of the ballot order and gives it to his favorite printer. The county pays for three separate printings of these ballots and, as could be expected, the cost is correspondingly higher. Where competitive bidding on ballot printing and supplies contracts has been established, savings as high as 66 per cent have been effected. In the smaller counties, bids are not usually asked on ballots. The fact that there are only a few printers and that they have maintained an important place in county politics makes it almost impossible to deny them the work when the party which they support has control of the county commissioners' office. In some of the larger counties bids for ballots are asked from printers within the county. In some counties the rotating system is used.

Election supplies such as instruction sheets, penalty notices, record books and other items used in the polling places are usually bought from a company which makes up packages for each election district according to law and sells them to the county commissioners. These are delivered to the election officers without any trouble on the part of the county commissioners. They are called "election boxes." The W. G. Johnson Company of Pittsburgh fills most of these orders. The same holds true with registry books. They are made up according to law and often bought directly in the quantities needed.

In minor expenditures for elections the county commissioners might effect further savings. Examples are seen in cost of delivery of election supplies to the officers, the storage of ballot boxes, the costs of "advertising" in newspapers of the county, and similar items. A serious effort to reduce county expenses for elections would begin with the setting up of standards of cost based upon the best practices observed in well managed counties and the adoption of these for all counties.

The current costs for the maintenance of voting machines in counties which have adopted them include the cost of custody, repair, demonstration, and hauling. Districts which have voting machines dispense with the services of the two clerks to the inspectors.

Table V contains figures for the entire cost of the 1933 election in the counties surveyed, as to total cost, number of votes cast in the primary and in the election, cost of elections per vote cast, per registered voter, and per capita of population. The number of voting machines in the county is also shown.

TABLE V

Total Cost of Elections for 1933, in 12 Counties

Counties	Total of Election Costs for 1933 in County		Number of Votes Cast in Primary and Election for 1933 in County	Cost per Vote Cast 1933	Cost per Registered Voter 1933	Cost per Capita Population 1933	Number of Voting Machines Used 1933
Lackawanna	$96,389.89	C.	246,628	$.39	$.72	$.31	330
Westmoreland	56,712.59	P.	152,823	.37	.45	.19	None
Montgomery	59,859.00	P.	153,566	.39	.42	.23	None
Erie	29,630.95	C.	64,043	.46	.55	.17	131
Chester	28,820.00	P.	60,818	.47	.45	.22	None
Northumberland	35,488.18	C.	92,210	.38	.58	.28	70
Centre	11,629.00	P.	23,172	.50	.49	.25	5
Adams	11,675.62	C.	22,089	.53	.54	.31	None
Tioga	9,272.01	C. (#1)	25,259	.37	.47	.29	None
Monroe	8,952.10	C. (#2)	18,924	.47	.53	.32	None
Union			9,834				None
Forest	2,837.16	C. (#2)	2,992	.95	.96	.55	None
12 Counties			872,358				

C.   Controllers Report, 1933
P.   Pennsylvania Economic Council Schedule, 1933
(#1) Registration of Voters estimated at $50 per election district. (45 districts)
(#2) "Election expenses" in Auditors' Report.

CONCLUSIONS

1. FINDING: The size of election districts in the counties surveyed varies widely, the
smallest having 31 registered voters, the largest having 2,086. Many districts have
less than 100 and some have more than 1000.
RECOMMENDATIONS: (a) The state Constitution should be amended giving the legislature
power to determine the size of election districts throughout the state. Rearrangement
of existing districts should be based on the findings of an inquiry made by a state commission or agency acting with the cooperation of the courts and local officials. To
prevent any injustice or great inconvenience to the voters in a particular district, provision should be made for appeal to the courts.
   (b) The size of the districts should be fixed so as to avoid either excessively
large or small districts but any rearrangement should take into consideration the characteristics of certain types of districts such as those using voting machines, rural districts, urban districts and forest districts. Consolidation or reallocation of some of
the present districts would materially reduce the cost of elections.

2. FINDING: There are three general methods of registration:

   a. Personal and Periodic
   b. Personal and Permanent
   c. Non-personal and Periodic

   The first method exists in cities of the first and third-class, the second in cities
of the second-class and the third in boroughs and townships.
RECOMMENDATION: Personal and permanent registration should be considered for all local
units of government and should be accompanied by such safeguards as the signature of the
voter at time of voting; annual check-up of registered voters; protective locking devices
for the register; all-year work by the registration authorities in checking lists; control by registration authorities of clerks and registrars under a merit system; in cases
where assistance will be required at time of voting, a sworn declaration to be made by
voter at time of registration; and finally, the registration authorities to have power to

order a complete new registration whenever necessary. This recommendation is based on data secured and conclusions reached by the Philadelphia Registration Commission. In third-class cities, boroughs and townships, permanent registration should be under the supervision of the county commissioners in charge of the chief clerk. This would abolish elective and appointive registry assessors and registrars.

3. FINDING: In boroughs and townships little or no control is exercised by the county commissioners over the elected assessors, yet the latter incur expenses which in the aggregate are heavy.
RECOMMENDATION: If permanent registration is not adopted for boroughs and townships then:

    a. Registry assessors should be appointed by the county commissioners.
    b. The commissioners' clerk should establish a standard scale of pay for assessors based on the time actually devoted to the work involved.
    c. Assessors should be required to present itemized statements of expenditures.

4. FINDING: Election costs vary widely in the counties surveyed. While some costs are mandatory on the county commissioners, the purchase of supplies, newspaper and advertising contracts, the printing of ballots and other material offer a field for economy.
RECOMMENDATION: (a) That newspaper advertising be contracted for on some standards or basis of circulation and population.
(b) That printing of ballots and other materials and purchase of supplies be made on the basis of estimates in all cases.
(c) That the purchase of all election supplies and materials be made by one officer and not distributed between the commissioners as is the practice in some counties at the present time. The above changes would produce material savings.

5. FINDING: The state has a vital interest in local as well as state elections. Irregularities and negligence will continue in the absence of authoritative observation and intervention.
RECOMMENDATION: There should be more supervision by the state over local elections and the conduct of local election officials. This supervision should be exercised by the Elections Bureau of the Department of State under the Secretary of the Commonwealth. Among other things the bureau should collect, record and publish municipal election returns, and keep a record of the amount of money reported as spent by the different candidates for office.

CHAPTER VI

JUSTICES OF THE PEACE

INTRODUCTION - Justices of the peace, aldermen, magistrates, district court judges, constables, and the sheriff constitute the executive officers of the district court. This entire local judicial structure is part of the county government, and county funds are used to pay a large share of the expense of such services. It is from the viewpoint of costs and procedure, and only incidentally from that of functional efficiency, that this problem has been approached. The following subjects have been considered:

    I. Constitutional and statutory provisions.
        a. Election of justices and constables.
        b. Statutes governing court costs.
        c. Fees and compensation of justices, aldermen, and constables.
    II. Cost of justices, aldermen, and constables in the county.
        a. The number of justices in counties.
        b. The methods of paying justices, aldermen, and constables by the county.
        c. The cost of justices, aldermen, and constables in twelve counties.
        d. Notes on justices, aldermen, and constables in various counties.
    III. Court costs.
        a. Analysis of court costs.
        b. The assessment of court costs.
        c. Computation and collection of court costs.
          In general
          Notes on eleven counties.

I. CONSTITUTIONAL AND STATUTORY PROVISIONS

    a. Election of Justices, Aldermen and Constables - The Constitution at present provides for a term of six years for justices and aldermen. Justices of the peace are elected by the voters of boroughs, wards of boroughs, and townships. Aldermen are elected by the voters of wards in third class cities. The Constitution further provides that no township, ward, district or borough shall elect more than two justices of the peace or aldermen without the consent of the majority of electors in such district, and that no person shall be elected to such office unless he shall have resided in the district for one year preceding the election. Not more than one alderman may be elected to one ward or district in cities of more than fifty thousand inhabitants. The Governor fills vacancies in these offices. Justices and aldermen cannot be removed except by the Governor on the address of two-thirds majority of the Senate. The Governor commissions them.

    The statute in force (1839, P. L. 376 as amended) provides for the following number of justices and aldermen:

        Township, two justices.
        Borough, not divided into wards, two justices
        Ward, in borough divided into wards, two justices
        Ward, in third class city, one alderman
        Ward, in second class city, one alderman

    Constables are elected by the people for a term of four years. They may be removed for certain causes by the court of quarter sessions. The Constitution has no provisions relating to the constables, so they may be called statutory officers.

The number of constables provided for by law is as follows:

        Wards in cities of second and third class, one constable
        Boroughs, divided into wards, one constable in each ward

Boroughs, undivided into wards, one constable
Townships, one constable

b. **Statutes Governing Court Costs** - Justices, aldermen and constables are compensated for their services by fees which are fixed by state law. These fee payments are prescribed for all the many official acts performed in administering justice, criminal and civil, within the jurisdiction of the minor courts. In civil cases, those which involve disputes between two parties, costs are paid by the plaintiff or defendant, as the court decides. In criminal cases, costs may be placed on the prosecutor, defendant, or county, according to the type of case. When a person is acquitted, costs are paid from the county funds. In the event of conviction, the payment of costs falls upon the defendant. If the county pays the costs in such a case, it has the power to attempt collection from the person convicted.

The following may be regarded as typical of the manner in which statutes fix costs incidental to trials conducted by justices of the peace and the court of quarter sessions:

> Assault and battery cases. Where the prosecution is not well founded, justice of peace has power to assess costs in whole or in part, upon the defendant, the prosecutor, or the county. In default of payment, the person defaulting may be committed to the jail one day for each dollar of costs (1919 P.L. 306).

> Where an innocent person has been discharged by a justice of peace or a court, after having been held for a crime, costs must be paid from county funds (1791, Sm. L. 37). This act applies to preliminary examinations prior to presentment of a bill to the grand jury. It does not apply to summary proceedings before a justice.

> In cases of acquittal by petit jury (felonies excepted), the jury may determine whether the defendant, the prosecutor or the county shall pay the costs. In default of payment, the prosecutor or defendant may be committed to the county jail unless he furnishes security within ten days (1860 P. L. 427).

> Where a misdemeanor or felony is charged, the costs of prosecution are borne by the county, on bills ignored by the grand jury; and where a bill of indictment is returned ignoramus by the grand jury. Such costs are paid by the county in any event, if the party convicted defaults (1874 P. L. 132; 1887 P. L. 138).

> In cases of conviction of any crime, costs shall be paid by the party convicted, but where such party is discharged without payment of costs, the costs of prosecution are paid by the county (1860 P. L. 427).

It will be noted that these costs usually fall upon the county. There are approximately 20 statutes governing the payment of costs in such cases, the oldest dating back to 1791, and the most recent having been enacted in 1933.

As far as justices, aldermen, and constables are concerned, there are three types of cases with respect to costs paid by the county. These are dismissed or discharged cases, in which the justice finds no cause to hold the defendant, and charges the costs to the county; summary convictions, in which the justice finds the defendant guilty and sends him to jail or fines him, but, in default of fines and costs, charges the county with the cost of the case; and court cases, in which the justice binds the defendant over for trial in the quarter sessions, which is preceded by indictment by grand jury. In all such cases the county may be called on to pay costs, although this does not necessarily follow.

c. **Fees and Compensations of Justices, Aldermen and Constables** - There is no relation between the fees of justices, aldermen and constables and the various duties they are called upon to perform.

The comment of the Governor of the Commonwealth in approving a bill establishing·
new fees for justices and aldermen in 1929 is indicative of the executive viewpoint.
The bill provided for great increases in amount of fees for services rendered. The
Governor stated: "This bill authorizes the increase of fees chargeable by justices of
the peace and aldermen. A similar bill was passed in 1927 and vetoed by me. The
present bill passed the house by 196 yeas to 7 nays and the Senate by 47 yeas to no
nays. This indicates popular approval of the proposed increase and I, therefore, give
it Executive approval."

The average citizen has little knowledge of the many intricate details whereby
the minor judiciary is enabled to measure justice on the piece rate basis. Almost
every motion the justice makes during trial has its appropriate statutory fee. For
filing a complaint on behalf of the Commonwealth against the defendant, the justice
receives 75¢. Upon issuance of the warrant against the defendant, the justice is entitled to another 75¢. If additional defendants are added to the warrant, 10¢ per
person is allowed. Next the justice enters the action in behalf of the Commonwealth
in his docket, and for this he receives 75¢. A hearing may be the next development,
entitling the justice to $1.00 for each day so occupied. Upon opening the hearing,
the justice receives a dime for administering the oath in a criminal case, but if the
defendant wishes to confess or plead guilty, this entry upon the docket entitles the
justice to 75¢. A few of the many steps through which the justice conducts the defendant from entry on the docket to ultimate commitment or discharge are: the hearing,
the taking of testimony, the receiving of bail, entering judgment, recording conviction
or acquittal and recording sentence. Each step has its statutory fee, and the total
of all these items constitutes the costs which will be paid by the defendant or the
county. The following may be regarded as typical of the manner in which fees of
justices, constables and aldermen are fixed by law:

Information or complaint on behalf of the Commonwealth, one defendant -- $ .75
Warrant or capias on behalf of the Commonwealth, one defendant --------- .75
Each additional person or defendant named on information or warrant ---- .10
Docket entry of action in behalf of the Commonwealth -------------------- .75
Hearing in each criminal case, each day occupied or fraction of day ---- 1.00
Administering oaths in criminal cases ---------------------------------- .10
Docket entry of confession or plea of guilty in criminal cases --------- .75
Making docket entry of testimony in cases of summary conviction,
                                                each witness ------ .50
Taking bail for a hearing or for appearance at quarter sessions, each
                                                defendant --------- .75
Entering judgment on conviction for fine ------------------------------- .75
Recording conviction --- .50
Recording sentence --- .50
Commitment of each defendant --- .75
Discharge to jailor -- .75
Entering discontinuance in criminal cases ------------------------------ .50
Transcript and certificate for any purpose to any one in criminal cases 1.00
Issuing summons -- .50
Every continuance of suit -- .30
Trial and judgment in civil case --------------------------------------- 1.00
Administering oath in civil case --------------------------------------- .10
Entering satisfaction in civil cases ----------------------------------- .30
Entering discontinuance of civil cases --------------------------------- .30
Entering amicable suit or confession of judgment ----------------------- .75
In all cases of summary convictions, in which the evidence is not suf-
ficient to convict and the defendant is dismissed, the costs of the suit
shall be paid by the county.
In all cases of summary convictions, in which the defendant is convicted
and sentenced to jail in default of payment of fine and costs imposed, the
costs of prosecution shall be paid by the county.

There are more than 126 separate fees similar to those listed above prescribed by
numerous acts of Assembly for various purposes. The enumeration is intended to show
(1) the peculiar statutory detail by which the costs of justice may be measured and

(2) the costs in numerous cases are placed upon the county by statute.

II. THE COST TO THE COUNTY OF JUSTICES, ALDERMEN AND CONSTABLES

a. The Number of Justices in Counties - It is interesting to note that while each township, borough and ward of a borough is allowed two justices, and each ward of second and third class cities is allowed one alderman, the number of actually commissioned justices and aldermen is much smaller than the number allowed. This is true particularly in boroughs and townships rather than in wards of cities. Furthermore, the number of active justices and aldermen is very much smaller than those actually commissioned, according to studies of transcripts handed in by justices to county financial authorities. This leads to the conclusion that there is no real need for the large number of justices (and this is true of constables as well) in counties. The automobile has made possible the work of a smaller number of active officials. In fact, justices now hear a great many cases originating outside of their districts, the more active of them getting the most of the business at hand. The constables for the justices may roam outside of their district to bring in cases. The more popular, influential or industrious constables bring in the most business to their justice. Many possible justiceships go begging, due to the lack of any candidates to run for the offices in elections.

Table I gives, for the twelve counties surveyed, as far as available materials allow, the number of justices and aldermen actually commissioned, the number it is possible by law to have, and the number actually handing in transcripts for payment to the county.

Table I
NUMBER OF JUSTICES BY COUNTIES - 1933

County	Class	Number of Justices Commissioned in County	Number of Justices Allowed by Law	Number of Justices Actually Receiving fees from County 1933
Lackawanna	3rd	94	182	b
Westmoreland	3rd	133	171	b
Montgomery	3rd	134	200	21
Erie	4th	52	88	28 c
Chester	5th	35	149	b
Northumberland	5th	51	119	21
Centre	7th	40	82 a	16 c
Adams	7th	35	72	6
Tioga	7th	40	80	b c
Monroe	7th	24	40	b
Union	8th	18	34 a	8
Forest	8th	5	18 a	b
12 Counties		661		

a. Boroughs tabulated as without wards; two justices for each borough.
b. Transcripts not available in form to make such tabulations.
c. Dismissed and summary cases only. Justices without dismissed or summary cases are usually also without court cases.

b. <u>Payment of Justices, Aldermen and Constables by the County</u> - The steps in the process of paying the fee bills of justices, aldermen and constables are as follows:

1. Justice hands bill in to county authority --
    In counties in which there is no controller --
    In discharge cases -- to county commissioners
    In summary cases -- to clerk of courts (for certification and docketing)
        who turns it over to county commissioners.
    In court cases -- to clerk of court, who turns it over to county
        commissioners, after certification and docketing.
    All bills remain with the chief clerk of commissioners.
    In counties in which there is a controller --
    All bills eventually reach his office, either directly from the justices,
        or indirectly from the office of the clerk of courts, who certifies the
        bill.

2. Checking the bill --
    In counties where there is no controller, it is the work of the solicitor
        of the commissioners to pass on the legality of these bills. This he
        does according to his own interest and initiative. Some solicitors go
        over each bill very carefully in order to discover any flaws in the
        transcript as to duplicating returns, over-charges, addresses of witness-
        es, etc. Some solicitors do not even examine them. Even the best of
        checking is merely as to legality; no solicitors in the smaller counties
        actually make a check-up as to the actuality of cases. Practices in
        different counties vary, but it may be said that what the solicitor does
        not do is not done, as the chief clerk is not engaged in the work of check-
        ing up these cases. Mileage bills of constables are usually accepted
        without check because it is difficult to ascertain whether the right
        route was taken or whether the constable had to hunt around a town for
        a person. However, when a sheriff or constable goes in search of a man
        and does not bring him back, some counties do not award mileage. Some
        constables make long trips in search of persons who could be found more
        easily by inquiring of the police in the vicinity of the person wanted.
    In counties where there is a controller:
    In these counties all checking on bills of justices, aldermen, and
        constables is done by the controller's office. In the larger counties
        there is usually a cost clerk aided by the solicitor who is responsible
        for checking as to legality. In some counties a great deal of attention
        is given to this phase of the work, as the justices are apt to hand in
        inflated bills. Some county controllers consider that their local jus-
        tices are pretty well "trained," but an examination of the transcripts
        leads one to believe that they are trained in form only and not in
        spirit. Lackawanna county has a cost clerk on the road checking up on
        the reality of summary cases. The cost clerk in Westmoreland county
        has saved to the county his entire salary in cases disallowed. In
        Northumberland county a taxpayers' association representative takes
        every dismissed case brought in by the justice and checks up as to legal-
        ity and actuality by going out on the road. This is an exceptional prac-
        tice, but it seems to operate effectively since from January to September
        of 1934 only four discharged cases were allowed by the solicitor. In
        Montgomery County the controller's office is extremely cautious in allow-
        ing claims on any discharge or summary cases, and very little money is
        paid out for this purpose in comparison with other counties. The con-
        troller there takes the position that the law allows him to refuse pay-
        ment on most of these cases, and he does that. There has been, as yet,
        no court case on the question in Montgomery. In some counties payment
        is refused on summary convictions.

The importance of discharged cases may be seen in that aldermen, justices, and magistrates are unrestricted in assault and battery cases so far as the Act of 1919 is concerned in determining who should pay the cost when the defendant is discharged. The justice may assess the cost upon the prosecutor, the defendant, or the county. Cases of assault and battery are numerous, and, for the most part, petty. To place the cost of such cases upon the county is to require the public to pay for large numbers of trifling offenses. Where the defaulting prosecutor or defendant has been committed to jail it has been proposed that the controller or clerk of county commissioners should verify the transcript by consulting prison records. At the present time an expensive investigation is usually necessary to determine whether the case for which the justice requests payment is bona fide. The purpose of the proposed law is to prevent suits in justices' courts by persons lacking financial resources to pay the costs but anxious to annoy another at public expense. The proposed bill would add to the constables and justices fees already paid, but the number of cases which would be brought would be greatly lessened. At the annual convention of the County Controllers' Association of Pennsylvania at Pottsville, September 28, 1933, it was unanimously decided that the convention was opposed to any justice anywhere in the state accepting trivial cases in which the county might be required to pay the costs.

c. The Cost of Justices, Aldermen, and Constables in 12 Counties - Table II gives the costs of justices, aldermen and constables in the counties surveyed:

Table II. Costs of Justices of the Peace and Constables Paid by Counties.

| | Justices of the Peace | | | | Constables | | | |
| | Disposition of Case | | | | Disposition of Case | | | |
	Dis- missed	Sum- mary	Court	Total	Dis- missed	Sum- mary	Court	Total
Lackawanna	$10,838	---	$9,845	$20,683	$5,009	$2,058	$3,804	$10,871
Westmoreland	7,722	---	7,348	15,070	7,974	---	22,154	30,128
Montgomery	70	282	9,434	9,786	---	1,016	4,786	5,802
Erie	4,796	1,876	3,035	9,707	4,325	2,649	3,009	9,983
Chester	512	---	3,210	3,722	---	---	6,893	6,893
Northumberland	5,757	2,981	2,002	10,750	3,192	4,192	3,735	11,076
Centre	---	---	---	3,192	---	---	---	2,889

d. Notes on Justices, Aldermen and Constables in Various Counties

In the following notes detailed information is given as to the number of cases, cost per case, and routine connected with each county. It indicates a difference of attitude in relation to enforcement of those provisions of law that apply to all counties. The detail is here included to show the need for more intensive examination of this interesting and obscure field of county government. In some counties it is almost impossible to search out the transcripts of justices because they are placed in the controller's files according to date rather than subject matter. It would require days to hunt them out. In other counties it is relatively easy to examine some of the transcripts. In some counties they were counted and classified. While the material, therefore, is not complete, it brings to light some matters seldom considered.

1. Lackawanna County - a. The controller allows payment only on discharged and court cases; summary cases are not recognized for payment under his interpretation of the law.

b. An interesting case of summary conviction by an alderman is Groezinger vs. Lackawanna County, in the court of common pleas, September 1933. The case and the decision are stated in the following opinion by Judge Newcomb:

"The proceeding is had to test the liability of the county for certain costs incurred by plaintiff in the exercise of his official functions as alderman of this city. It appears that one Michael Sopke was charged with disorderly conduct on the complaint of Joseph Herko, Jr. The locus in quo was at the corner of Lackawanna and Washington Avenues, one of the busiest places in the city of Scranton, at which there is a police officer at all hours of the day and night. The occasion, therefore, for complaint at the instance of a private prosecutor, who, apparently, is quite as obscure as the defendant, gives rise to the suspicion that the proceeding was what is commonly termed a "racket." At all events, the case proceeded to judgment at the hands of the plaintiff in his official capacity, a fine was imposed, and, in default of payment thereof for a period of ten days, the defendant was committed to the county jail, at the expiration of which he was discharged by the alderman."

"Whatever the motive of the county controller for refusing the alderman's bill, he made no mistake in that particular....."

The court indicated that a statute allowing costs to be born by the county in such cases is applicable only when the cases are free from jurisdictional defects. The court quoted the act as follows, "if any person or persons shall willfully make or cause to be made any loud, boisterous, or unseemly noise, etc."....the court continued, "It will be observed that the complaint here does not aver that the alleged disturbance was willfully made." The claim was refused on these technical grounds.

    c. So important is this matter of discharged cases in Lackawanna County that the controller's office has an investigator on the road, devoting most of his time to the investigation of such cases. He inquires as to the merits of each case determining what happened, whether or not money was collected from the defendant, whether the witnesses were real, whether the prosecutor was real, etc. This keeps down the amount of money that is spent by the county for discharged cases. The controller has a summary of the costs paid to justices and constables on discharged court cases since 1902. The amounts paid in various years fluctuate widely:

Discharged Cases

Justices	Constables
1902--$ 7,998.85	$ 7,304.80
1919-- 25,471.44	14,481.55
1930-- 9,017.48	5,052.47
1933-- 10,837.74	5,008.91

Court Cases

1902--$11,672.80	$15,168.86
1919-- 5,142.01	2,945.29
1925-- 6,951.73	4,240.65
1933-- 9,844.68	3,803.97

In 1920 the controller really began to investigate the cases on the road. Prior to that there had been an investigator, but, apparently, an inactive one. The cost immediately dropped. One can note the comparatively low totals during the last few years, in spite of the fact that justices' fees were increased 50 per cent by the statute of 1929.

    d. In discharged cases, the justice hands his transcript to the controller's office; in court cases, it goes to the clerk of the courts' office.

    e. The following is the form of investigation sheet sent to the defendant and used by the controller's office in checking upon justices' cases.

INVESTIGATION SHEET

CONTROLLER'S OFFICE

LACKAWANNA COUNTY

SCRANTON, PA.,

M..

....

Dear Sir:

Kindly fill in answers to the following questions and mail at once to this office in the enclosed stamped envelope, and oblige.

Yours truly,

W. G. Watkins, County Controller

Alderman

1. Were you arrested before Justice of the Peace
Answer..

2. Who had you arrested? Answer..

3. What did he have you arrested for? Answer...........................
....

4. Was there a hearing in the case? Answer............................

5. What day was the case heard before the Alderman-Justice of the Peace?
Answer..

6. Did you pay any money to the Alderman, Justice of the Peace or Constable
in this case? Answer...

7. If you did pay any money in the case, how much did you pay? Answer $....
What did the Alderman--Justice of the Peace say he did with the case?
Answer..
....(Sign)

2. <u>Westmoreland County</u> - a. This county spent more than $45,000 for justices
and constables in dismissed, summary and court cases. The 1638 dismissed and summary
cases alone required $15,811 in 1933, averaging a cost to the county of $9.04 each.

b. The controller's office (the cost clerk) computes the witness,
justice, and constable fees and the subpoena costs in all magistrates and court cases.
The office is a very active one in this county and actually saved the county the
following items in such cases in 1933:

Ninety dismissed cases, costs not allowed by controller's office:

Justices.................... $533.23
Constables................. 719.45.....................$1,252.68

Reasons for not granting:
Fees asked exceed statutory fees.
No address of prosecutors or witnesses.
Several informations on one case.
Insignificant cases, example, very petty larceny.
Cases as to which there is no statute.

In court cases, payment was refused on justices, constables, witness
fees... $5,612.59
In dismissed cases partially refused by controller's office......... 353.17

Total.. $7,218.44

These savings more than pay the salary of several cost clerks in the controller's
office.

(c) The county received in fines from summary cases of thirty aldermen
and justices of the peace the sum of $911.48.

3. <u>Montgomery County</u> - This county is very strict as to claims for fees in
dismissed and summary cases. In 1933, there were only forty-four cases allowed to
twenty-one justices of the peace. Of these, thirty-four were summary convictions and
ten were dismissed cases.

4. <u>Erie County</u> - (a) Transcripts of magistrates' cases in the court of
quarter sessions for the year 1933 revealed 493 court cases as having originated in
justices' courts.
(b) The county controller's office had transcripts of both summary and
discharged cases for each justice and alderman in the county. The following table
indicates the number of such cases handled by each justice and alderman in the county:

Justice or Alderman		No. Discharged Cases	No. Summary Cases
Alberstadt	-Erie	163	15
Barber	-Erie	121	31
Boland	-Borough	3	3
Ceeder	-Township	4	18
Conyngham	-Erie	205	15
Dick	-Township	1	0
Fisher	-Township	8	3
Gazkowski	-Erie	372	52
Goodwin	-Borough	70	70
Gordon	-Corry City	0	2
Henry	-Township	0	1
Hicks	-Borough	4	8
Hugger	-Township	0	2
Knoll	-Township	0	3
Noles	-Township	22	53
Morrison	-Borough	4	0
Nuss	-Township	5	11
Pease	-Borough	11	23
Porter	-Corry City	6	7
Randall	-Borough	2	17
Rhoads	-Township	2	0
Robbins	-Borough	2	14
St. John	-Borough	5	12
Schabacker	-Erie	9	3
Shaw	-Township	41	24
Shugart	-Erie	259	16
Stker	-Township	10	28
Wilcox	-Borough	39	38
Wimnersberger	-Township	2	11
Total -- 29		1,370	480

The preceding tabulation indicates that in 1933 five Erie City aldermen out of twenty-nine justices and aldermen in Erie County handled 82% of the discharged cases in which the costs fell upon the county. The same five justices conducted 27% of the summary cases of this type.

In one year more than one thousand discharges of arrested persons were made by these five Erie aldermen. Either the complaints and arrests were on very trivial grounds or the magistrates were very lenient. In any event the latter profited from the county on each case by $4.00 to $10.00, while the constables did likewise. The same aldermen in Erie County had only 132 out of 480 summary cases, indicating that they did not convict with the same latitude with which they gave discharges. This is the more remarkable when it is considered that throughout the county other than in the City of Erie justices returned more summary cases than discharge cases. This is the reverse of the situation in Erie County and may be shown as follows:

	Justices Returning More Summary Cases Than Discharge Cases.	Number of Justices Making Returns
Township	8	12
Borough	7	9
City of Corry	2	2

Despite the trend in other parts of the county, all aldermen in the City of Erie with but one exception had a large number of discharged cases in comparison to the number of summary convictions for which the county was required to pay costs at rates established by law.

There are 16 boroughs, 22 townships and 2 third-class cities in Erie County. The law provides for 2 justices of the peace in each borough, borough ward, and township, making a potential total (cities excepted) of 78 borough and township justices of the peace. In most jurisdictions, the full number of justices are not elected, and it is frequently difficult to find persons to fill such vacancies. The county controller's deputy in charge of returns from justices, has only 21 township justices and 19 borough justices on his list. Therefore, only 40 such justices are active in office, although there are 78 possible places. Not included on his list are those who may have taken office but who never make returns to the county.

In addition to township and borough justices, Erie City has 6 aldermen and Corry City has 4, or one for each ward. This makes a total of 88 aldermen, and justices in Erie County including cities, only 50 of which are listed as active by the county controller, for the purpose of payment of costs. In 1933, only 29 of these aldermen and justices made returns for discharged or summary cases.

Among the 29 justices who made returns to the county in 1933, 12 were from townships, 9 were from boroughs, 6 were from Erie, and 2 from Corry. In other words, 12 out of a possible 44 township justices, 9 out of a possible 32 borough justices, 6 out of a possible 6 Erie aldermen, and 2 out of a possible 4 Corry aldermen made returns to the county in 1933.

The concentration of the magistrates' business in the hands of a few individuals has been a matter of frequent comment; and studies of particular counties including the present one of Erie confirm this view. The discharged case is of special significance, because of the large number in which costs fall upon the county. In such cases, the magistrate is sure of payment of his costs. Other important points are (1) the amazing diversity of practice among justices of the peace, and (2) the absence of any records or adequate control of costs on the part of either the county or defendant. Upon compliance with formalities in the submission of bills, the county justices are entitled to costs in many cases. In other cases costs are paid to the justice by the defendant and unless the latter takes an appeal there is no record as to whether costs were properly imposed nor the amount of income to justices from such public sources. Since the facts suggest that justices of the peace are inactive, as far as returns to the county are concerned, and since the county is the most lucrative source of revenue, it may be fairly concluded that the whole system might be improved by having larger areas for such officers, especially in rural districts.

d. The cost book of the controller, reveals the following amounts paid by the county to various justices of the peace in court cases, discharged cases, and summary cases in 1933:

Summary -- Justice of the Peace Payments from County, 1933

Justice of Peace		Summary Convictions	Discharged Cases	Court Cases	Total
Wilcox	Girard Borough	$ 123.60	$ 173.00	$ 135.65	$ 432.25
Shaw	Waterford Borough	73.10		31.65	104.75
Randall	Albian Borough	53.80		33.10	86.90
Alberstadt	6th Ward - Erie	29.25	1,097.50	613.59	1,740.34
Shugart	4th Ward - Erie	63.90	604.40	296.25	964.55
Barber	1st Ward - Erie	90.75	419.60	323.60	833.95
Hicks	Girard Borough	23.45	22.70	21.25	67.40
Stoker	Mill Creek Township	69.75	29.20	49.70	148.65
Ceeder	Springfield Township	44.10	3.35	12.75	60.20
Pease	North East Borough	65.30	13.50	43.75	122.55
Conyingham	5th Ward - Erie	94.85	868.05	253.95	1,216.85
Gazkowski	2nd Ward - Erie	265.50	935.40	355.36	1,556.26
Robbins	Union City Borough	33.85	6.10	96.52	136.47
Moles	Mill Creek Township	156.75	37.85	45.40	240.00
Gordon	Corry	7.70		12.80	20.50
Fisher	Lawrence Park Twp.	12.00	12.75	16.25	41.00
Goodwin	Wesleyville Borough	444.02	304.80	382.90	1,131.72

Justice of Peace		Summary Convictions	Discharged Cases	Court Cases	Total
St. John	Edinboro Borough	$ 37.50	$ 13.95	$ 4.10	$ 55.55
Dick	Harborcreek Township	2.25		4.00	6.25
Nuss	Lawrence Park Twp.	43.45	4.05	7.80	55.30
Wimnersberger	Elk Creek Township	20.50		25.10	45.60
Porter	Corry	29.50	4.00	68.70	102.20
Bolard	Union City Borough	16.15	13.55	21.60	51.30
Schabacker	3rd Ward - Erie	12.45	17.50	40.55	70.50
Hugger	Summit Township	4.50			4.50
Henry	Venango Township	3.75			3.75
Knoll	Harborcreek Township	2.25			2.25
Morrison	Wesleyville Borough			17.00	17.00
	Total	$ 1,823.97	$ 4,581.25	$ 2,913.32	$ 9,318.54

Some aldermen appear to get most of the cases. As explained by the assistant district attorney in charge of such cases, this is largely due to:
1. Type of population in ward or township
2. Popularity and ability of alderman or justice
3. Popularity of his constable
4. Industry of his constable

Constables do not have to function in their jurisdiction, but may work outside the district from which they are elected. As a result, especially in Erie, constables who are popular, energetic, and influential get many more calls from complainants than do constables who are less active or popular. These constables bring their cases before their own justices and the entire procedure assumes the appearance of an established business. It is estimated that only about 30 per cent of the cases are started by police officials; the other 70 per cent originate from private complainants to constables. This estimate includes only cases that ultimately reach the criminal courts.

In contrast to the thousands of dollars paid out to justices and aldermen by the County of Erie in 1933, the following amounts were paid over to the county by justices in the form of fines:

City of Erie	Justice Moles	$127.00
"	Shaw	125.00
"	Wimnersberger	22.00
"	Rhodes	20.00
"	Wilcox	12.25
"	Nuss	10.00
"	Gordon	10.00
		$336.25

One county officer indicated that while aldermen justices and their constables could find numerous cases in which the county was liable to pay costs these officers never could find many cases in which fines could be charged.

5. Northumberland County - In this county, as previously mentioned, a representative of a taxpayers' association takes all dismissed cases transcripts and goes out into the field to check them up as to legality and reality. This service is an exceptional one, bringing about substantial savings in 1934. The practice is made possible by the new law in effect in 1934 requiring each transcript to have the names of the defendant, prosecutor, and witnesses. In 1933, 21 justices and aldermen participated in county payments on dismissed and summary cases, which totalled almost $9,000. Thirty-four constables were paid about $7,000 in summary and dismissed cases.

The total number of cases, dismissed and summary, in which the county paid fees to justices was 1728. The average for 265 cases was $5.24 for justices' costs and $5.63 for constables costs.

6. Centre County - a. The financial routine of cases within the jurisdiction of the justices of the peace is as follows:

1. Summary cases in which the defendant is adjudged guilty and pays fine and costs of justice, constables, and witnesses.
    Defendant pays to justice, who pays others.
    No report is made to county.
    No records as to whether state or county receives fines.

2. Summary cases in which defendant is adjudged not guilty and the county pays the costs.
    Justices hand in fee bill to county commissioners.
    Constables or sheriff hand in fee bill to justice.
    Witnesses hand in fee bill to justice.
    County solicitor checks these bills.
    If justice does not include names and addresses of prosecutors and witnesses, the county does not pay. (A list of those bills not approved in 1933 is included. They were in almost all cases not approved for this reason.)

    The solicitor does not as a rule check these bills on any other point because it would require too much time and expense to make personal investigation on each case. Mileage bills of constables are usually accepted without much check-up. It is felt by most people that too little rather than too much mileage is reported. When the sheriff or constable goes in search of a man and does not bring him back, mileage is not awarded. On certain occasions constables have gone to Pittsburgh to hunt some one when it would have been easier to call for help on the part of the Pittsburgh police.

    Chief clerk approves and sends order to treasurer, who makes payment from the General Fund.

    County auditors check county commissioners' books and vouchers with treasury disbursements. A list of cases dismissed by justices with the county paying the costs is also included.

3. Summary cases in which the defendant is adjudged guilty and sentenced to jail or sentenced to pay a fine which he cannot pay, going to county prison in default.
    Justice submits bill to clerk of courts. Constables and witnesses submit bills to clerk of courts through justices.

    This officer certifies it to the county commissioners, after entering it on quarter session docket. County commissioners approve it and certify it to treasurer for payment.

4. Cases in which justices bind over defendants for quarter sessions or oyer and terminer.
    Justices, constables, and witnesses send bill to clerk of courts who certifies it to county commissioners, who in turn certify it to treasurer for payment.

5. Several grand jury reports in 1932 and 1933 indicated that justices should be careful in sending in transcripts to court.

6. Several county officers indicated that these transcripts often appeared to be excessive in number and amount.

7. Some fee bills are handed in on the old schedule of rates in force previous to the new one passed in 1929. The new law fixed the justices fee at $2.25 in motor cases. State highway patrolmen, and borough police do not receive fees in such cases. Borough police as constables receive regular fees of constable. Bills found by solicitor to be excessive in amount are not approved by county commissioners.

8. No one checks those fee bills which are handed in to clerk of courts.
           (b) Justice and constable fees not allowed by solicitor of
county commissioners - 1933. The following tabulation shows the justices' and constables' fees which were rejected in Center County. There is indicated, as to each case,
(1) the total costs; (2) the share of costs going to the justice of the peace; (3) the
share of costs going to the sheriff and constable; (4) the purpose for which the claim
was made, that is, assault and battery, issuance of fraudulent checks, neglect of duty,
etc.

Classification By Type Of Case -- Each Is A Separate Case

Type	1.	2.	3.	4.	Total
Assault and Battery					
Justice of the Peace	$ 4.85	$ 6.20	$ 6.20	$ 5.95	-$ 23.20
Sheriff and Constables	3.50	6.70	8.60	2.00	20.80
Total	$ 8.35	$12.90	$14.80	$ 7.95	$ 44.00
Fradulent Checks					
Justice of the Peace	$ 3.25	$ 3.25			$ 6.50
Sheriff and Constables	8.80	10.80			19.60
Total	$12.05	$14.05			$ 26.10
Neglect of Duty					
Justice of the Peace	$ 6.80	$ 6.00	$ 5.00		$ 17.80
Sheriff and Constables	14.40	25.50	19.00		58.50
Witnesses		10.00			10.00
Total	$20.80	$41.50	$24.00		$ 86.30
All other Cases					
Justice of the Peace	$ 5.50	$ 4.75	$ 1.50		$ 11.75
Sheriff and Constable	2.50	2.50			5.00
Total	$ 8.00	$ 7.25	$ 1.50		$ 16.75

Grand Total $173.15

   Although these were cases rejected by the solicitor as being improperly presented
for payment, they nevertheless indicate typical costs, both in amount and in the manner
in which they are assessed upon defendants. In default of payment by the defendant in
trivial cases, payment of costs becomes mandatory upon the county. Below is indicated
the distribution of fees for the same cases.

Classification by Distribution of Fees

Type	Justice of Peace	Constables & Sheriffs	Witnesses
Assault and Battery	$ 23.20	$ 20.80	
Fraudulent Checks	6.50	19.60	
Neglect of Duty	17.80	58.50	$ 10.00
All Other Cases	11.75	5.00	
	$ 59.25	$103.90	$ 10.00

GRAND TOTAL . . . $173.15

   The classification indicates the distribution of fees among justices of the peace,
constables and sheriffs, and witnesses. The constables and sheriffs received substantially more than either justices or witnesses, but the number of cases involved is
too small to permit conclusions to be drawn.

   According to the county solicitor, these fees were not allowed by him because
they did not comply with the new law requiring justices to include the names and
addresses of the prosecutors and witnesses in each case. Only rarely was there any
other reason for disallowance. All rejections were made solely from the standpoint of

whether the law was fulfilled in making these returns both as to form and as to fees fixed by law. No check-up was made on either mileage or on the cases themselves.

(c) The costs paid by Centre county in 1933 upon transcripts of cases dismissed by justices of the peace totalled $492.00. Of this amount, $357.00 went to justices, $110.00 to sheriffs and constables. The remainder was divided among costs of subpoenas and witnesses. There were 92 dismissed cases in that year, making an average cost of $5.35 for each dismissed case.

7. Adams County - There were only 31 cases for which transcripts were filed in Adams County in 1933 as dismissed and summary trials by justices of the peace. These cases were divided among 6 justices, 3 of whom had only one case each. The costs were almost equally divided between justices of the peace and constables, the total being nearly $200.00. The average cost per case in Adams County was $6.41, for those cases the cost of which fell upon the county.

8. Monroe County - The justices' and constables' transcripts of discharged and summary cases are turned in for payment to the county commissioners' office. Here they are checked by the clerk and solicitor for legality, etc., and then recommended for payment to the county commissioners. There is little evidence of unauthorized transcripts. This is perhaps due to the fact that there are actually few cases per year. A search through the files of the transcripts for 1933 showed that payments for 9 summary cases to justices totalled $42.00, or approximately $4.71 per case. The fees of justices in 11 discharged cases in that year amounted to $92.00, or an average justice's fee per case of $4.86. In 28 cases in which constables' fees were filed, the amount of fees was $54.00, or an average fee per case of $1.87. A comparison of the above fees with those described in other counties indicated little variation in average cost per case.

9. Union County - In 1933 Union County paid a total of 8 dismissed cases, at an average of $4.88 for justices' costs and an average of $5.12 for constables' costs. There were 11 summary cases, with an average of $4.12 for justices and $.84 for constables. Variations as to constable costs are due to mileage turned in and the fact that when state highway officers or borough or city police turn in cases they cannot charge for it. There were 43 court cases in which payments were made to justices and constables. The average for justices was $6.19 and for constables $4.02, based on the cost for 24 cases.

Of all cases, summary, dismissed and court, one justice received 29 cases, one 13, one 9, one 2, while four received 1 case each. Only eight justices received pay from the county in 1933. The law allows 28 elected justices.

The solicitor of Union County disapproved 8 overcharges by justices of the peace in 1933. The total costs in this county are estimated at $819.00, divided as follows: court cases, $684.00; dismissed cases, $80.00; summary cases, $55.00.

10. Forest County - Total justices' costs in Forest County in 1933 amounted to $230.40. No other information was available at the county seat.

III. COURT COSTS

a. Analysis of Court Costs - The items included ordinarily in criminal court costs are:

        Justices' fees
        Constables' fees, including mileage
        District attorneys' fees
        Clerk of Courts' fees
        Witness fees
        Sheriffs' fees
        Jury fees
        County fees (for "use of county"; often in lieu of district attorney fees.)
    In addition, where fines are imposed, these are added to the total bill.

Not all cases have all these fees attached, but the typical criminal case here considered is one in which the preliminary hearing is held in the justice's court, the arrest having been made by the constable; and the defendant is bound over for quarter sessions court, indicted by grand jury, and tried by petit jury before the judge of quarter sessions. Such a case has the above items of cost.

b . The Assessment of Court Costs - The statutes give authority to assess the costs of the case on either the defendant, prosecutor or the county, or on a combination of all three, according to the nature of the case and the findings of the jury and the court. The authority who assesses such costs is the justice of the peace or alderman, the grand jury, the petit jury or the court of quarter sessions, as the case may be. In all cases state statutes control, except when the court has common law supervisory authority over costs that transcends even the statutes. By refusing payment for the county, it may be said that the county commissioners and the controller have some power in the matter of costs.

c . Statutory Regulation of Court Costs - Statutes supply in some detail the rates chargeable by officers in the administration of criminal and civil justice. Each service rendered by a county officer has its official rate; no more nor less can be paid. These rates have been adopted over a long period of years and have very little, if any, relation to actual costs. Some are far too low for the service rendered; others are much too high. Even when all the fees which are due are collected, they do not cover the cost of holding court and other allied services incident to the handling of a case. In practice very little of the money is actually collected.

In counties where many of the officers are paid by fee the payment of these costs is the direct compensation of such officers. Their income depends upon the amount of work they have to do. But it must be made clear that there is little relation between the payment of these officers and the fact that individuals owe the fees. In almost every case the county ultimately pays the officer and the individual pays the county. Thus the officer does not lose his costs in criminal cases.

In counties where most of the officers involved in the administration of criminal justice are paid by salary, their services are still computed by these statutory fees. In other words, when the clerk of court performs certain services in docketing such cases, he is credited in the controller's office with the amount of money earned by his services according to statute . These credits are labelled "fees earned from county", and are added to "fees taken in cash" in computing the gross earnings of the office on which his salary and that of his office help are based. (The aggregate amount of salaries in the office cannot be higher than the income of the office.) Even in the larger counties therefore, the statutory fees are important to the majority of the "row" offices.

d . Computation and Collection of Court Costs - In General. While the clerk of the court of quarter sessions, in his capacity as clerical officer of the criminal courts, computes costs in most of the counties, there is a diversity of practice in this process. Statutes do not specify the details. A similar situation exists as to the collection of court costs. It is impossible to formulate general principles about the process for the twelve counties surveyed. The routine is best exemplified by taking a number of individual counties.

Before doing this, the main problems related to the process should be stated. It may be assumed that the computation of court costs, important though it is, is a routine duty best handled by a cost clerk who has specialized in this work. What usually happens is that the computation of costs is entrusted to a number of clerks scattered generally through several offices, including the clerk of court's, the district attorney's, the commissioners' and the controller's. If this work could be centralized in one office, or under a liaison officer working in the several offices at this one task, much improvement could be secured.

The next problem is the collection of court costs and fines after they are assessed. An appallingly low percentage of such costs are actually collected. This, it is claimed, is due to the depression, but the difficulty lies deeper than that. In the first place, there is very little attempt to collect such moneys. There is little or no collection machinery. The elected officers have nothing to gain by attempting to collect. They actually would lose favor. So they generally wait for collections to come in over the counter. Some costs are obtained in this manner but very few. Westmoreland County keeps a detective of the district attorney's office out on the road doing nothing but collecting court costs and fines. He takes installments from those assessed, investigate their position to see whether they are earning money, and keeps after them. The plan has brought surprising results, and the detective has more than earned his salary. It is doubtful whether all counties could keep a man on the road full or part time. Centre County seems to get results by having the probation and desertion officer clothed with that duty.

The next difficulty is the absence of records relating to the problem of court costs. Very few records are kept with regard to the amount of outstanding costs and fines. No one seems to know about them. It is averred that they might be computed, but that it would be a tremendous task. Such business practices are an indication of the underlying reason for meagre collections.

There are a great number of fines and costs which neither the court nor any county officers expect to collect. If a man is sentenced to jail and to fines and costs no collections are actually expected unless there is property that is convertible. The sentence is merely a bookkeeping affair. Then, too, many persons are placed in jail in default of fines or costs. The jail sentence cannot run very long on account of the insolvency law. Persons are usually held not longer than thirty days. Such cases are doubly expensive to the county because of the additional jail maintenance.

Computation And Collection of Court Costs By Counties.

1. Lackawanna County - a. The clerk of courts is responsible for the computation and collection of court costs, with the exception of costs of witnesses, justices, and constables in court cases (done in witness fee office), and in justices' and constables' discharged cases which are handled by the controller's office. In other words, the clerk of courts' office computes court costs as to the following items: fees of the clerk of courts, district attorney, jury, sheriff, and the fines; it collects all court costs except those which are handled by the probation department, or by the sheriff (in cases where there are costs against persons whose property is being sold for liquidation of debt or delinquent taxes). It is interesting to note that in the year 1933 the sheriff's office collected only $157.14. Its part in the collection of court costs is, therefore, negligible.

b. A typical analysis of court costs is as follows: clerk, $5; district attorney, $7; jury, $4; and sheriff, $1.50, making a total of $17.50. A typical non-support order has the following fees: clerk, $5.50; district attorney, $7; sheriff, $1.50, making a total of $14.00 without justices', constables', or witness' fees. The cash earned in this way by the various offices is credited to their account in the controller's office as fees earned by them and chargeable to the county. This is not, of course, a cash transaction. The fees are specified by statute for each particular service. The statute relating to these offices requires them to earn at least as much in cash fees or fees chargeable to the county as the total of their salary schedule. If they do not earn that much, salaries are curtailed accordingly.

c. The fees earned from the county on court cases by these offices and by the jury for the year 1933 were as follows:

Clerk of Courts Reports, 1933

Fees Chargeable to County on Court Costs - Credit Earnings

Clerk	District Attorney	Jury	Sheriff	Total Fees Chargeable to County
$19,474.60	$7,119.00	$1,752.00	$1,525.50	$29,871.10

The total cash earnings of the offices from court costs or, in other words, the amount of court costs collected by the clerk of courts and allocated to various offices were as follows:

Clerk of Courts Reports, 1933

Total Cash Earnings -- Chiefly Collections of Court Cost Allocated

Clerk	District Attorney	Jury Fee	Sheriff	Fines	Cash Total
$457.15	$179.00	$80.00	$48.00	$879.23	$1,643.38

d. It is interesting to see that in 1933 the court costs in the form of fees chargeable to the county as credit earnings for the offices of the clerk, district attorney, sheriff, and the jury, total almost $30,000; this does not include the cost of justice, constable, and witness fees in court cases, nor justice and constable fees for discharged cases, which total approximately $42,500. Nor does it include other court expenses listed in the controller's report at about $128,000. A total of the entire court cases would be close to $200,000. Compare this with the almost futile showing of $1,643.38 collected in 1933 in fines and costs. It must be understood, however, that not nearly all this cost is by court order to be paid by defendants in lieu of jail sentences. There are no figures, without going over each case in the docket, to ascertain either how much in costs and fines is due the county from one year's cases, or how much in costs and fines is outstanding from previous years. This would be very valuable information, but it is not ascertained by the county officers because it might reflect discredit. A common excuse is that is very difficult to collect costs and fines in times of depression and that the great majority of defendants are without property and employment. There is verification of this but it is certain that much more could be collected if proper measures were taken. In Lackawanna County there is no one on the road investigating or collecting costs and fines. This practice might pay in dollars and cents, as it does elsewhere.

e. The witness fee office has a clerk from each of the controller's and county commissioner's offices. Its function is to check witnesses and witness fees. It is a peculiar office, not being found commonly in county organizations. The witness registers here at the beginning of every day that he is called in for court. He brings with him a card sent to him from the district attorney's office. The district attorney makes out the card which he stamps in his office. The witness brings this card to the witness fee clerk, and has it stamped. A record is made of these witnesses, and when the case is completed, it is copied on the regular docket with the names of witnesses and the time served. This docket is called "Controller's Criminal Docket," and it is a cost book for criminal cases.

This office computes the cost of witnesses, justices, and constables in court cases only. The controller's office, as has already been mentioned, handles all discharged cases. The copies of the justices' transcripts are made from the original, found in the clerk of the court's Office. After computation of these costs, this office gives out vouchers for court costs to justices, constables, and witnesses. The controller approves them, and they are paid by the treasurer. Both the clerks in this office, from the controller's and commissioner's offices, approve the costs as to legality and mileage.

2. Westmoreland County - a. One of the most important functions of the office of clerk of courts is the computation of court costs and allocation of credit to the various offices participating in the court procedure.

While the controller's office computes the witness, justice, and constable fees (all fees that require actual money to be paid out to persons outside the county government in criminal court cases), the clerk of courts computes the costs of jury, sheriff, district attorney, clerk of courts and defendant's bill.

The method of handling witnesses fees is as follows: Witness brings subpoena card to clerk on day of trial before trial. Clerk punches card. Witness collects at the end of trial if defendant is not guilty, or sentenced to jail, or committed, or if defendant pays fines and costs. Witness does not collect if defendant is granted time to pay fine or costs, or both, until they are fully paid; this also applies to constables and squires. In cases involving desertion and surety to keep the peace, officers are paid immediately; witnesses are not paid in these cases.

b. Computation of Costs for Services Rendered by County Offices in Criminal Court Procedure - The district attorney, the sheriff and the clerk of courts are the officers who render services in criminal procedure. The cost of such services is computed by the clerk of courts according to the statutory fees allowed. Although these officers and their employees are paid by salary (it being a third class county), nevertheless the law requires each office to have earned enough fees to pay for its salary schedule. If it does not, the amount paid out cannot exceed the amount taken in for fees. In most cases the actual money collected by these offices in cash and from outside sources does not nearly equal the amount of salaries paid out, but every service that these offices render, in cases for which the county receives no compensation, is computed on a fee basis, and credit is given these offices by the controller every month. The clerk of courts computes these amounts for his office, for the district attorney, and, in criminal cases, for the sheriff.

In 1933, the following amounts were credited to these offices as "fees chargeable to county." The amounts represent the total fees earned by the offices. Presumably all or most of the amount is for services in criminal cases and is therefore computed by the clerk of courts.

Sheriff	$ 9,257.07
District Attorney	15,374.00
Clerk of Courts	11,184.32
	$ 35,815.39

Monthly computations are made by the clerk and submitted to the controller's office, which credits them to the various offices as if they were actual cash earned. However, no checks are issued to the various offices for these services to the county as is done in some other counties.

At the same time that these credits are sent to the controller's office, the actual cash received in costs and fines is allocated to the various offices by the clerk of courts, and is then sent to the controller's office, where it is checked. The money is then sent to the treasurer. The clerk of courts collects court costs and fines in his office.

c. The collection of court costs is centered in the office of the clerk of courts. The procedure is as follows: The judge, on "sentence day" proceeds to sentence the number convicted in the last session of court. Sometimes as high as seventy are sentenced on one day. A sentence without jail penalty is usually for fines and costs, the latter including the various fees for services rendered by county offices, justices and constables. Many of the costs and fines are paid the same day, and the defendants or their representatives come to the counter of the clerk of courts office to find the amount of the costs. Some pay immediately, and they are then crossed off the books. On the morning of "sentence day" more than $1,200 was taken in by the clerk of court.

Several offenses, notably those under the new liquor law and the highway code, result in fines which are fixed by statute. These fines are supposed to go to the state treasury but may be deflected by the judge. If the sentence is a "fine," the money must go the state, and it must be of the statutory sum. Instead of a fine the judge may impose "suspended sentence on payment of costs and $25 for the use of the county." No money then goes to the state, the county gets $25 and costs, and all parties are satisfied, - even the defendant because he probably escaped a $200 statutory fine. Frequently the defendant is willing to pay or can pay the smaller amount, and thus imprisonment at the expense of the county is avoided. The state may not like this any too well, but little can be done since the Commonwealth cannot appeal, especially not in the name of the county, nor will the defendant appeal. This is the usage in several counties and is one reason why the state receives little revenue from county courts. In liquor cases, however, a tax on each unit of liquor confiscated does go to the state treasury.

When the costs are paid to the clerk of courts, he allocates them according to the services represented and the fines paid. He then transmits them to the treasurer and forwards an itemized report to the controller at the end of each month. The controller credits the various offices.

Before March, 1934, costs were collected only by waiting for them to come in at the clerk's office. Naturally, business was not so brisk as it might have been with some outside pressure. In March, Westmoreland County inaugurated a system whereby one of the detectives from the district attorney's office was put in charge of cost collection. He secures from the clerk of courts, after a reasonable time has elapsed since "sentence day," a statement of the amount owed. He then tries to collect and investigates the financial condition of debtors. He makes installment collections, if necessary, and uses his judgment as to how much to push them. While no "bench-warrants" for non-payment of costs are reported, they are a future possibility. So far, the detective has earned his way. The following collection of costs before and after the detective was installed shows the difference:

Income from Costs and Fines, Westmoreland County

Before	January	$ 715.16
	February	1,256.64
After	March	2,457.81
	April	1,827.07
	May	1,086.90

Naturally, the first month had the most collections because the detective was able to get many of those who were only waiting for a little persuasion to pay. The detective is paid by salary and not on the amount he collects.

There is yet a great deal to be collected and some that never will be. Especially during the depression have court costs been difficult to secure. It is estimated that at the beginning of 1934 there was $42,000 outstanding due the county. No estimate can be made of what will or can be collected.

When persons are jailed with fines to pay in addition, such fines are rarely collected. Many are confined in jail in lieu of costs and fines, but the county does not keep them there longer than a month.

In June, 1934, about 400 accounts were partially paid. This shows the extent of the installment business in court costs.

One of the important records of this office is the Account Book Collection Docket. In this record are found the cash collections by day and month as collected and turned over to the treasurer. It indicates for what service each amount has been paid, whether for fees of district attorney, sheriff, or clerk, and also includes what fees are chargeable to the county by the clerk of courts' office. The money collected by this office in 1933 was as follows:

Clerk of Courts -- Account Book -- Collection Docket -- 1933

"A" Clerk	"C" District Attorney	"C" Sheriff	Jury Fees	Fines	Collected Costs Due County	"B" Fees Chargeable to County
$3,990.35	$4,440.00	$483.00	$400.00	$2,835.73	$7,414.79	$12,680.57

"C" Fees paid by defendants but not the total of those earned in criminal cases.
"B" Fees earned by clerk of court and paid by county.
"A" Fees paid for clerk of court's services by defendant.
     "A" plus "B" Total earning of office of clerk of courts.

The Account Book is now made in duplicate loose-leafed form. A carbon copy of fees and costs handed in is sent to the controller with a receipt from the treasurer. This is a good step, as it eliminates copying the entries a second time.

d. The following is the fee bill used by clerk of courts in computing costs -- (third class counties):

Clerk of Courts Fee Bill

	Clerk	District Attorney
Pleas:		
Entering information	$ 3.00	
* Entering submission and Jud.	.75	
Recording Bill and Verdict	1.00	
Discharge	1.00	
Taxing Costs	.50	
	$ 6.25	$ 15.00
# Commitment	1.00	
Com. Subpoenas	.50 each	
Trial:		
Entering information	$ 3.00	
* Swearing Jurors	1.50	
Recording Bill and Verdict	1.00	
Discharge	1.00	
Taxing Costs	.50	
	$ 7.00	15.00
# Commitment	1.00	
Com. Subpoenas	.50 each	
Nolle Pros.:		
Entering information	$ 3.00	
Filing Petition and Order	2.50	
Taxing Costs	.50	
	$ 6.00	10.00
Nolle Pros. After Bill:		
Entering information	$ 3.00	
* Filing Petition and Order	2.50	
Recording Bill	1.00	
Taxing Costs	.50	
	$ 7.00	10.00
# Commonwealth Subpoenas	.50 each	
Ignoring Bill:		
Entering information	$ 3.00	
Recording ignoring bill	1.50	
	$ 4.50	5.00
Commonwealth Subpoenas	.50 each	
Detention and Non-support Surety:		
Entering information	$ 3.00	
All services	1.50	
Taxing Costs	.50	
	$ 5.00	10.00
* Set Costs		
# Extra Costs		

	Clerk	District Attorney
Bond	$1.00	
Discharge	1.00	
Committment	1.00	
Commonwealth Subpoenas	.50 each	

Montgomery County - a. Justices hand transcripts to the clerk of courts in court cases and to the controller in dismissed and summary cases. The cost clerk in the controller's office looks over the court transcripts of justices and checks them for legality of fees, duplication, and witnesses. He also checks the officers, the constables and local police as to the mileage they turn in. Fees can be collected only once for one day, no matter how many cases are involved. The clerk of courts checks the jury by calling roll in court room each morning. The deputy of the clerk of courts taxes or computes the cost of each case.

The grand jury may declare that no fees are to be paid; they may ignore the bill and even place costs on the magistrates or on the officers. There have been several such cases.

The witness gets a subpoena card from the district attorney's office, served by officers. The witness brings this card to the clerk on the morning of the trial. The clerk punches the card, and after the case is over he collects the cards from the witnesses to use as a check with the witness sheet.

The routine in court cases is as follows: The transcript comes in to the clerk of courts and is recorded in the transcript docket. It is then sent to the district attorney's office, where the bill of indictment is drawn for the grand jury. Then comes the trial, with a record of the case listed in the criminal docket. Next comes the computing of tax costs and the rendering of the bill to the sheriff, who collects the costs. If commitments are rendered, the sentence is passed and the prisoner committed to custody.

b. The sheriff collects court costs in his office. An itemized bill is sent to the sheriff by the clerk of courts. A deputy sheriff is sent out on the road to collect costs and fines, as his full-time job. This system of collection was started in Montgomery County one and one-half years ago.

Costs Collected, July 1933 - 1934

Total costs	$10,600.01
County Fines	2,221.00
State fines	700.00

There are no figures on outstanding fines and costs and it is impossible even to estimate the sum of money due the county from this source. The installment plan may be used in collecting some of these costs but in many cases it is impossible to collect anything.

4. Erie County - a. The clerk of courts' office has the function of collecting court costs but does not appear to do a very good job of it. There was no one on the outside collecting costs and an experiment in having a detective do this did not appear to work well. This was strange because in other counties, for example, Westmoreland, the detective was very successful. It appears that no effort is made to collect except when those who owe costs come into the office to pay. Sometimes the probation officer collected costs from probationers, but he made no practice of it.

The extent of cost collection may be seen from the following figures for 1933:

Fines collected by the Clerk of Courts	$ 950.87
Costs collected by the Clerk of Courts	3,247.50
Total collected	$4,198.37

Costs collected include:

Transcript fees	$ 946.12
District Attorney fees	1,368.35
Commonwealth bills	645.07
Miscellaneous	287.96
	$3,247.50

There were no figures available in this office on costs outstanding and unpaid. The clerk indicated that very few of them were collected. It appeared that in this important function the office was not active and much more could be done, to save the county money by better collections. A very small proportion of the costs and fines due the county were actually paid.

5. Chester County - a. The bills for court costs and fines are made out by the clerk of the court and sent to the sheriff, who enters the bill in the docket. After the sentence has been pronounced, the sheriff takes the debtor into custody to take him to the clerk of courts for his bill. He pays the bill at once or goes to jail, unless the court has given him thirty or sixty days to pay costs and fines. Some of these postponed bills are outstanding, and there is no one on the road collecting them. In some cases, where a fine is imposed with a jail sentence, nothing is paid.

Collected in Fines and Costs, 1932	$ 10,000.00
Collected in Fines and Costs, 1933	5,600.00

b. Costs in Criminal Cases

Felony --	District Attorney	$ 10.00
	Clerk	10.00
	Sheriff	1.25
Misdemeanor --	District Attorney	5.00
	Clerk	4.50 to $5.50
	Sheriff	.25

6. Northumberland County - a. At present the county commissioners' office, through the docket clerk, computes and collects all court costs. The clerk of courts checks the time of jurors. The witnesses must register with the controller each day in order to collect their fees. The probation officers collect these court costs from the outside.

Persons sentenced make settlement with the county commissioners. At the discretion of the court, the defendants either go to jail in lieu of costs and fines or they may be discharged and try to pay as best they can in installments.

b. The following is a record of court costs in one session of criminal court as given by the docket clerk in the county commissioners' office. This was for the February session of 1932, and was computed as of June, 1932. There were 71 cases in all, but the average session usually has from 150 to 170 cases.

Costs for Criminal Court, February 1932

Paid and unpaid costs due county	$ 2,912.61
Costs received by June, 1932	365.95

Unpaid balance of costs	$ 996.75
Paid witness fees	918.30
Unpaid witness fees, June, 1932	168.14
Paid justice fees	279.45
Unpaid constable costs, June, 1932	617.97
Court cost received and handed to treasurer	120.00

7. <u>Centre County</u> - The desertion, parole and probation officer under the direction of the court collects all the Commonwealth costs, fines, costs for support in desertion and non-support cases, lying-in and other costs in bastardy cases and restitution funds. The income through his office during the years 1932 and 1933 was as follows:

	1932	1932
Fines	$ 4,104.50 to county	$ 1,553.23 to county
Costs	2,988.33 to county	2,113.58 to county
Support	9,960.49 to individuals	10,927.21 to individuals
Restitution	1,489.25 to individuals	4,572.68 to individuals
Lying-in	360.25 to individuals	209.00 to individuals
	$ 19,002.82	$19,375.70

According to the secretary of the office, there was in 1933 about $9,651 outstanding in uncollected costs and fines due the county for this and previous years. It is very difficult to collect costs and fines especially during the depression. From some, nothing is collected; for others, the installment plan is used, according to the ability to pay. Facts such as the delinquent's job, his income, and the needs of his family are taken into consideration. According to the secretary, these costs and fines remain uncollected for "a year or so" and, if nothing is paid, a bench warrant (non-bailable) is issued and the defendant is brought in. He either pays, makes some arrangements to pay, or goes to jail.

The money due for these purposes is paid to the desertion, parole, and probation officer. This officer gets certification of costs from the quarter sessions docket in the clerk of courts office. He keeps his own books, and sends in monthly accounts to the county commissioners. The function of collecting costs and fines for Commonwealth cases is not by law the function of the parole officer. The court directs him in this case.

The desertion, parole and probation officer stated that the law required all fines and costs to be paid within one year, but that this was not done in Centre County due to conditions and individuals. The court, he said, was lenient. It placed a $100 condition upon the defendant "for the use of the county," not a fine, and added costs and mileage.

8. <u>Adams County</u> - a. The court costs are computed and collected by the clerk of courts. There are no figures available on costs outstanding and unpaid. Fees "for use of county" are $5.00 per case in lieu of the district attorney and jury fees. The fees for witnesses, clerk, justice, constable, and sheriff are added to the $5.00 "for use of county".

There were 102 cases in the session of criminal court in 1933. The fines and costs collected for 1933 were as follows:

Fines collected	$ 1,120.00
Costs collected	2,051.23
	$ 3,171.23

9. <u>Monroe County</u> - a. Costs of criminal cases under the following headings were taken from the quarter sessions docket for 1933, (witness fees and fines were not tabulated):

Cases listed in the docket

Justices of the Peace	$ 473.34
Constables	287.30
District Attorney	410.00
Clerk of Courts	410.00
Sheriff	145.10
Total	$1,725.74

The clerk of the court is charged with the responsibility of computing these costs, although he stated that the district attorney was responsible. They are set down in the quarter sessions docket and in a cost book.

The clerk of courts and the sheriff divide the work of collecting court costs. The sheriff turns over to the clerk of courts what he collects. They have an understanding that each gets 1½ per cent of all collected by both the sheriff and at the desk by the clerk.

When questioned about how much of the costs and fines was outstanding, neither the treasurer, clerk of courts, nor the sheriff had any idea, except that it was considerable. When a person is committed to jail in lieu of costs and fines, the law fixes a maximum sentence of three months. Many are not incarcerated for non-payment of costs and fines because they are sometimes only too glad to have a home in jail. The sheriff appeared to have no records to show how much of the costs and fines he collected. He stated that in his previous term, 1924-28, he collected from $15,000 to $18,000 in the four years. These sums were locally attributed to "boom times" and also "prohibition times", when a great many "New York bootleggers" were brought into the toils of the law. The clerk appeared to have no idea of how much was collected in one year. One could only ascertain this by going through the docket and checking the receipted items, or the cost book, or perhaps, both. This is another sign of the lack of records in the county. The procedure takes care of the interests of the property owner, the creditor, and the lawyer -- not those of a citizen interested in better and cheaper government.

The law states that fines from prohibition cases are to be placed in a fund to be used for the law library. At one time this fund had $56,000 in it. It was not, however, spent for law books. By order of the court, it was used both for emergency relief and for part of the construction cost of the new addition to the county building.

10. <u>Union County</u> - a. The judge imposes sentences of costs and fines, which are computed by the clerk of courts. The defendant or prosecutor, whichever the judge designates, pays the clerk of courts. The county officers and justices are not paid their fees until the defendant pays his bill. If, however, after a lapse of two, three, or four months the defendant has not paid, the county officers and justices may then get an order from the court requiring the county to pay the costs. When the defendant pays the county is reimbursed. In about 15 per cent of the cases the county pays the costs immediately, and is reimbursed when the defendant pays. The clerk of courts states that when the court still has power over the costs, the defendants pay better, and that when the court hands costs over to the county commissioners, it is very hard to collect. The clerk of courts estimates that 50 per cent of costs are paid by those who are not jailed.

The judge often gives time in which to pay the costs, but if he feels that the defendant can pay immediately, he issues a bench warrant. Sometimes the defendant is committed and then friends help him out. The court here is very much interested in getting costs.

b. Typical Cases in Court Costs in Criminal Cases

1. County $ 4.50
    Clerk   8.50
    Justice of Peace   7.05
    Sheriff   11.50

                             $ 31.55

2. County $ 4.50
    Clerk   7.50
    Justice of Peace   6.95
    Sheriff   9.00

                             $ 27.95

3. County $ 4.50
    Clerk   7.00
    Justice of Peace   7.40
    Sheriff   5.00
    Witness   4.92
    Witness   3.60
    Witness   4.92

                             $ 37.34

The clerk of courts fees for 56 cases in criminal court in 1933 averaged in costs $8.14 per case.

11. Forest County - a. The clerk of courts computes and collects the court costs and fines. One out of 10 criminal cases pays for its costs. When the defendants do not pay, the costs are certified to the county commissioners, who pay the witnesses, justices, and other county officers. There are no figures on the unpaid court costs and fines, and there seems to be no effort made to collect them.

CONCLUSIONS

A. Justices of the Peace and Constables.

FINDING: The constitutional requirement for the election of justices and aldermen has rigidly imposed an inefficient, irresponsible and costly system of minor judiciary upon the Commonwealth. About one-half of the justiceships provided by law are actually filled by persons commissioned as justices of the peace. Less than half of the justices commissioned actually perform substantial services relative to their office. The result is that the actual business of the minor judiciary is conducted by a relatively few active individuals.
RECOMMENDATION: That the Constitution be amended to eliminate the elected justice of peace as a constitutional officer. The General Assembly should provide for the appointment of justices, magistrates, aldermen and constables by the judges of common pleas of the county, or by the judges and county commissioners. Only persons learned in the law should be eligible.

FINDING: Justices of peace serve the same governmental area under the fourth Constitution of the Commonwealth that they did under the first, despite the rapid development of transportation and communication in recent years.
RECOMMENDATION: Districts in which justices of peace and aldermen serve should be greatly enlarged to serve modern needs and to conform to present day methods of transportation. In addition, the number of justices, aldermen and constables should be reduced. A circuit should be utilized to put the office on a half time or full time basis. These measures can be effected without constitutional change.

FINDING: The rates which may be charged by justices and constables in civil and criminal suits are fixed by statute. Practically every act has its corresponding payment fixed by law. The county is obliged to pay the amount prescribed. Amounts paid were in many instances fixed years ago and have been increased from time to time. The rates have no relation to the service rendered or the actual cost. In some instances the authorized rate is too low and in others it is excessive.

RECOMMENDATION: Statutory fees of justices, aldermen and constables should be revised and readjusted by the General Assembly according to schedules commensurate with the service rendered.

FINDING: The investigation of cases in which fee bills are submitted, is expensive and the amount of the bill in each case is small. The number of cases, however, is large. In one county the fee average for justices' costs in 265 cases was $5.24 and for constables' costs $5.63. $9,000 in fees went to 21 justices and aldermen while $7,000 went to 34 constables. In a single county the controller has disallowed over $1,2000 of fees in a single year. This situation is so general that the County Controllers' Association of Pennsylvania at its Annual Convention in Pottsville, September 28, 1933, unanimously recommended that no justice should accept cases where the county must pay the costs. Frequent grand jury reports have cautioned justices as to their practices on this point. Many county officers state that the bills submitted often appear to be excessive in both number and amount.

RECOMMENDATION: County officers authorized to receive transcripts for fees of justices, aldermen and constables should refuse acceptance of any bill not itemized, or which fails to include names and addresses of all parties.

FINDING: In the more populous counties, where an investigator has been employed by the county controller for the purpose of verifying the various bills and travelling expenses submitted by these officers, the savings resulting to the county were so large that one is forced to the conclusion that the procedure of some justices of the peace in minor cases is nothing more nor less than a raid on county funds.

RECOMMENDATION: That in large counties a county detective be assigned at regular intervals to check fee bills and expenses.

FINDING: In 43 counties having no controller there is little supervision, although fee bills are sometimes disallowed by the solicitor for lack of addresses, failure to itemize and other legal deficiencies.

RECOMMENDATION: That the clerk of the county commissioners in these counties be authorized by law to examine and reject fees.

FINDING: There is uncertainty as to whether the county is liable for fees in cases dismissed by justices of the peace. In Montgomery County payments are withheld for dismissed cases; in Westmoreland County, payments are made in such cases.

RECOMMENDATION: That the General Assembly by law prohibit the payment of county money in dismissed cases. These costs should be paid by the plaintiff or by the defendant.

FINDING: A large number of cases are settled by justices of the peace without recourse upon the county for costs or fees. The exact number is not known because the necessary data are not available.

RECOMMENDATION: That reports be required from justices of the peace in all cases, and that these reports be filed with the county records. Reports are now filed only on those cases in which costs are paid by the county.

B. Costs of Criminal Court Procedure.

FINDING: The collection of costs in criminal court trials is undertaken in different offices and under different authorities, including the clerk of court, sheriff and district attorney. Duplication results in the type of work performed, and frequently several receipts must be issued in the same case.

RECOMMENDATION: That the costs of criminal court trials be collected by the county treasurer, who should be present in court in person or by deputy on sentence day. In order to facilitate payment of costs in each case they should be computed in the controller's office and a bill issued which should be taken to the county treasurer's office for payment.

FINDING: Statutory fees relating to criminal court costs bear no relation to the work involved. It is not reasonable to expect that the costs collected will equal the cost of conducting the trial. Present costs were based upon a schedule fixed several years ago and changed from time to time.
RECOMMENDATION: The General Assembly should revise the statutory costs and statutory charges in criminal cases to sums commensurate with the work performed.

FINDING: A low percentage of court costs are actually collected. In many counties no effort is made to collect and no officer is assigned to this duty. One county assigned a county detective to the task with staisfactory results.
RECOMMENDATION: A county employe should be designated to collect court costs and fines.

FINDING: The records of court costs in criminal cases are frequently confused, illegible and difficult to secure.
RECOMMENDATION: That adequate records be maintained in the county office in order that receipts may be computed to indicate the amount of collections. Such records would permit audits where necessary.

CHAPTER VII

RECORDS

The public services rendered by the maintenance of an accurate and adequate system of records are of much importance to the people of the county. The authoritative determination of property and legal relationships often depends upon these records. Close examination shows that this is one of the county functions which are in great need of readjustment if reasonable standards of service and cost are to be established. Records are kept in the manner required by law, but the law itself is haphazard, confused, and incomplete. In some cases it declares in great detail the form in which records are to be kept. In others it provides merely that they are to be kept. In still others, it makes no mention of records even when there is an apparent need for them.

Furthermore, the law is often of ancient origin and has not been adapted to the possibilities of mechanical systems of record-keeping. Many property records are still entered in long-hand at great expense of time, money, and space when photostatic apparatus would be more satisfactory and economical. Most of the officers in charge have little knowledge of record-keeping. Direct observation reveals that important records are inexcusably ill-kept. For example, some officers were found to receive a high fee for entering records in a particular docket, yet they merely scratched the titles in the docket and charged the regular fee in the margin. Other records were found to be so inadequate or confused as to be unintelligible; some were so needlessly complicated as to require an expert for their deciphering. It is small cause for wonder that lawyers must be employed to ascertain some simple fact which any man of ordinary intelligence should be able to find for himself without additional cost.

It has already been suggested in another chapter that a Department of Records be established. This office would unite the records of the county commissioners, prothonotary, clerk of the courts, register of wills and recorder of deeds under one head. Modern methods could be installed at a great saving to the taxpapers and yet enable a more complete service to be rendered.

Before enumerating in full the records in the county, we may first divide them into two general classes. The first includes the official and permanent records and documents relating to important facts of property, inheritance, tax payment, debt, marital relations, settlement of disputes, and other such legal relationships as exist between individuals who have connections within the county. The county is the depository of these documents, which go to the very vitals of community life.

The second class contains a continuous chronicle of the activity of county and other local governments. In general, the first class is better kept than the second, chiefly because there is constant need for the former; the second class has never been felt to be of the same importance. Lawyers and property-owners insist that records which they need and use shall have at least some semblance of order. The second class is usually found to be in a deplorable condition. There is duplication, omission of important items, and a general lack of arrangement. It has often been necessary during this Survey to look in cubby-holes for material which is officially recorded and only a year old. It may be said that the use of records for the purpose of improving county government is indeed rare, even on the part of the officials themselves. Exceptions are seen in the controller's and auditor's reports which seem to be the Alpha and Omega of county financial chronicles as far as the general public is concerned.

The records of the first class have a great significance to students of history, economics, and the social studies generally. In them can be found primary material relating to the economic and social behavior of the county inhabitants. These records, moreover, have been in general preserved from the date of the original establishment of the county. They are a veritable gold-mine, but have not yet been explored to one per cent of their possibilities. The reason for this was their general complexity and inaccessibility. A vitally important function in this respect was performed by the Civil Works Administration during the winter of 1933-34. Under the direction of Dr. Curtis W. Garrison, Archivist of the State Library of Pennsylvania, in the Department of Public Instruction, a complete survey of county records was made. Two or three skilled workers made a list of records in all but a few of the Pennsylvania counties. The direction and the carrying out of this project was a comprehensive and highly technical piece of work.

Copies of these records are now in the State Archives. Thanks are due to Dr. Garrison, as well as to Dr. J. Paul Selsam, of the Department of History, Pennsylvania State College, for the use of the list of records of eleven counties which were intensively surveyed for this report.

In view both of the highly unfavorable conclusions here reached and of the importance of the county's record service to the people, it has seemed unjustifiable to deal in mere general statements. The results of the field survey in this matter are therefore given in full detail. A list of those records in present use in the eleven typical counties surveyed is presented in the following pages, the arrangement being in tabular form. The records are tabulated under the offices of the county. Where they are kept under special or peculiar conditions, those conditions are indicated. Where no records are kept, names of counties are not inserted. The following typical counties are thus tabulated as to all their collections: Lackawanna, Westmoreland, Montgomery, Erie, Northumberland, Adams, Centre, Monroe, Tioga, Union and Forest.

RECORDS IN COUNTY OFFICES (in present use)

Board of County Commissioner's Office

1. Minutes of Proceedings - All counties have these records.

2. Tax Records - Tax lists; all counties, except Montgomery and Adams, have these lists.

3. Tax Records - Assessment lists; all counties have these lists.

4. Tax Records - Record of Exoneration, etc; Erie, Northumberland, Centre, Adams and Union counties do not have this record. Montgomery County - in treasurer's office.

5. Tax Records - Record of Deeds for Taxation; Montgomery, Erie, Lackawanna, Northumberland, Centre, Adams, Tioga and Forest counties do not have this record.

6. Tax Records - Record of Mortgages for Taxation; Westmoreland, Montgomery and Centre counties do not have this record.

7. Accounts Records - Westmoreland, Montgomery, Northumberland and Centre counties are without these records. Lackawanna and Erie counties have them in the controller's office.

8. Records of Unseated Lands - Lackawanna, Montgomery, Centre, Adams and Tioga counties do not have these records. Northumberland has them recorded in Tax Reports to the State. Monroe county has these records in Property Assessors Book and Unseated Land Books, the latter being with the county treasurer.

9. Record Book of Unseated Lands (purchased) - Montgomery, Northumberland, Adams and Tioga counties are without this record.

10. Accounts of Poor House - Lackawanna, Westmoreland, Northumberland, Adams and Union counties do not have these records. Montgomery county has this record with the controller's or treasurer's office. while Centre county has no Poor House.

11. County Board Records .- Montgomery, Erie, Centre, Tioga, Monroe and Union counties do not have these records. In Lackawanna they are kept in the Quarter Sessions Docket, while in Adams County they are recorded in "Communications of County Commission."

12. County Budget Records - Erie, Centre, Tioga and Union counties are without these records. For Adams County, see 1 and 7.

13. Veterans' Records - Centre, Adams and Tioga counties are without these records. Lackawanna county has these records in the Recorder of Deeds' Survey and the

Soldier's Discharge Docket.

14. <u>Miscellaneous</u> - Lackawanna: board of county commissioners' office also has the
        Jurors' Time Book, commissioners' Inquest Docket, Soldiers' Burial Record,
        Widows' Burial Record, Polling Place Dockets, Primary Election Dockets, Re-
        cords of Prisoners in Penitentiary, Reformatories, County Jails; in the
        Witness Fee Office the same county has the Oyer and Terminer Docket, Criminal
        Court Docket, Juvenile Court Docket, Quarter Sessions Docket and the Witness
        Fee Receipts.
            Westmoreland: records the discharge of prisoners in the board of county
        commissioners' office.
            Montgomery: keeps the Personal Property Estates Docket in this office.
            Erie: keeps the voting machines and election returns in the commission-
        er's office.
            In the same office Northumberland County keeps the Commissioners' Crim-
        inal Docket, Inquests (oaths and bonds), Subpoenas, the Commissioners' Enroll-
        ment Book, Grand Jury Reports, Prison Discharge Docket, Commitments and Dis-
        missed Cases, Births and Deaths.
            Centre: records all foreign born unnaturalized persons in the county
        in this office.
            Tioga: records Land Returned in this office.
            Monroe: board of county commissioners' office keeps the registry lists
        for boroughs and townships and the records from the prothonotary's office of
        enlargement and satisfaction of judgments.
            Adams and Union: no miscellaneous records in commissioners' office.
            Forest: keeps in the commissioners' office the Auditor's Book, the
        Commonwealth Docket, County Disbursement records, records of the poor, odd
        file boxes, treasurer's transcript of taxes, the Assessment Record and the
        records of mortgages and judgment subjected to taxation.

<u>COUNTY CONTROLLER'S OFFICE</u>

    Among the counties surveyed only the following have this office:
Lackawanna, Westmoreland, Montgomery, Erie and Northumberland.

1. <u>Daily Reports of Tax Receipts</u> by Treasurer - All of the counties having this office
        have these daily reports.

2. <u>Register of Treasurer's Daily Receipts</u> - All of the counties have this register.

3. <u>Monthly Reports from Fee Offices</u> - Westmoreland County is without these reports.

4. <u>Ledger and Account Books</u> - All of these counties are thus equipped.

5. <u>Warrants: Files Containing Paid Warrants</u> - All of these counties are thus equipped.

6. <u>Warrants: Register of Warrants</u> - All counties have this register.

7. <u>Tax Duplicates by Districts</u> - Westmoreland County is without these duplicates.
        Northumberland County has these recorded in the commissioners' and treasurer's
        reports.

8. <u>Properties Returned to Treasurer for Sale</u> - For Lackawanna County see #5 under
        board of county commissioners above. Montgomery County does not have these
        recorded.

9. <u>Oaths of Witnesses as to Days in Court</u> - Erie and Northumberland counties do not
        have these recorded.

10. <u>Court Cases Discharged by Alderman</u> - Northumberland has these recorded in the
        office of the commissioners. Montgomery has these recorded in the County
        Appropriation Book.

11. **Miscellaneous** - Lackawanna: the following items are recorded in this office:
daily record of receipts, daily record of delinquent taxes, coroner's report,
wardens' reports, index to vouchers, the monthly statement to the commissioners, contracts and bids, and Lackawanna Jurist (publication).
    Westmoreland: records the following in this office: outdoor relief,
County Home papers, election records, the transfer of Controller's Cost Books,
telephone calls, printing and coal bids.
    Montgomery: cancelled warrants, prison records, unemployed disburse-
ments, monthly statements, county cash receipts, contracts and supplies,
County Almhouse order and account books.
    Erie: records witness fees, subpoenas, Poor Record, Poor Bills, Tax
Ledger, coroner's inquests.
    Northumberland: eleven other items are included under this office: Dock-
et of Court Costs and Fines, Held-up Transcripts of Dismissed Cases, National
Guard, County Farm Boardage, Mothers' Assistance Fund, Tax Collection and
Exoneration List, Bond Coupons, Jurors, Election Officers' pay, contracts and
bonds, Insurance Policies.

COUNTY TREASURER'S OFFICE IN 12 COUNTIES

1. **Day Books and Ledgers** - All of the counties surveyed are thus supplied, except
Monroe County, which uses only ledgers.

2. **Tax Records and Receipts from Taxes Paid** - Montgomery, Erie, and Adams counties
have no such record. In Lackawanna County these are carried in Disbursement
Register, and the receipt is filed with the controller. Northumberland County
has these in the commissioner's office.

3. **Tax Returns** - Northumberland and Centre counties do not record these. Tioga County
records returns for unseated lands, but not for seated lands. Monroe County
records these in Tax Lists, while Adams County does the same in the Tax Register. Lackawanna County records returns in the Delinquent Tax Register.

4. **Tax Sales** - Lackawanna, Erie, and Adams counties do not record tax sales. Northumberland transferred these records to the office of the commissioners. Union
County does not record Tax Sales for seated lands.

5. **Tax Liens** - Tioga and Monroe counties have no such records. Lackawanna, Erie, and
Adams counties have these recorded in prothonotary's office, while Montgomery
County records them in the commissioners' office.

6. **Mercantile Assessments** - Union County does not have these records.

7. **Tax Records Receipts** - Northumberland, Centre, Tioga, and Forest counties do not
have these recorded. Adams County has these filed with the assessments, while
Monroe County records them in the Commonwealth Tax Book.

8. **Licenses: Dog** - Westmoreland and Union counties do not record dog licenses. Monroe County records these in Commonwealth Tax Book.

9. **Licenses: Fishing** - No information of such a record was obtainable in Union county.
Monroe County records them in the Commonwealth Tax Book.

10. **Licenses: Hunting** - In Union County no information of such records was obtainable.
Monroe County keeps records of them in the Commonwealth Tax Book.

11. **Licenses: Beverage** - Erie county keeps no such record. Adams County keeps this
record in a Day Book from which it is transferred to the State Book. Monroe
County keeps this record in a special Cash Book.

12. **Fees Paid Jurors and Witnesses** - Centre and Erie counties do not have these
recorded. Northumberland County has these fees taken care of by the controller and prothonotary. Lackawanna has these fees carried in a Disbursement

Register Docket. Adams County has no special records but enters them in a day book. Monroe County records them in the Commonwealth Tax Book. Union County records them in an Order Book.

13. Miscellaneous - Westmoreland: includes in this office the Register of Lands Bought by the county commissioners.
    Montgomery: includes records of money paid to the treasury, cancelled checks, Department of Revenue, exoneration sheets, and dance hall licenses.
    Erie: Treasurer's office also includes the Treasurer's Receipt Book, and the Collector's Account.
    Northumberland: Receipt Book, Bicycle Duplicates, County Bond Book, controller's warrants, alderman's fines, Daily Cash Slips, General Fund Cancelled Checks, and Dealer's Cancelled Checks.
    Adams: keeps the Surveyor's Book in this office.
    Tioga: records the Tax Collector's Monthly Account in this office.
    Monroe: vouchers for poor relief.
    Forest: Collector's Accounts and Treasurer's Unseated Land Book.
    The following counties have no miscellaneous records in this office: Union, Lackawanna, and Centre.

CLERK OF THE COURT OF QUARTER SESSIONS AND OYER AND TERMINER

1. Minutes of Court - Forest County does not record the minutes. Centre County uses Quarter Sessions Docket for this purpose. For Union County see prothonotary - non-consolidated.

2. Quarter Sessions Court: Dockets - All counties surveyed use these.

3. Quarter Sessions Court: Papers - Used in all counties surveyed.

4. Oyer and Terminer Court: Dockets - Forest County includes these in the Quarter Sessions Court; all others use them.

5. Oyer and Terminer Court: Papers - Tioga and Forest counties file these with the Quarter Sessions Court. Union County makes no use of these.

6. Coroner's Inquest - Northumberland County has some of these records in the controller's Office and some in commissioners' office. Tioga County apparently has such records filed in the Quarter Sessions File. In Union County such records are found scattered in the Quarter Sessions Dockets. Forest County enters these records in a Miscellaneous Docket.

7. Road Papers - Erie County files these in the Engineer's Office.

8. Road Docket - Erie County files these with the Quarter Sessions Docket.

9. Annual Reports of Borough and Township Auditors - Montgomery County files and records these in Miscellaneous - see section 12. Monroe County has had incomplete reports from several townships during the past few years.

10. Oaths of Commissioners - In Lackawanna County the oaths of the constables are filed here, while those of township and borough commissioners are filed with the clerk of borough and township officers. Westmoreland county has such filed in the prothonotary's office. Montgomery county has these recorded in the Quarter Sessions Docket and filed in Miscellaneous. Adams County has these oaths of commissioners filed in the prothonotary's office.

11. Municipal Indebtedness Records - Montgomery County does not have such records, and Forest County has had none since 1932. Adams County files these in the commissioner's office.

12. Miscellaneous - Lackawanna: Juvenile Court Cases - Surety, Non-support and Appeal cases.

Westmoreland: Fee Dockets; Quarter Sessions Cost Book, Tax Collector's Book, Constable's Bonds, Expense Account, Candidate's Accounts, Constable's Return, Welfare Reports.

Montgomery: Constable's Returns - Tipstaves and Jurors.

Erie: Juvenile Court Docket, Commissioners' Book, Surety Book, Appointment of Deputy Constables, Constable's returns.

Northumberland: Miscellaneous Books and Papers.

Centre: Juvenile Docket.

Adams: Election Returns, Tax Collector's Bonds, and "Miscellaneous."

Tioga: Constable's Bond.

Monroe: Commonwealth Cases, Commonwealth Cost, and Dockets.

Union: Constable's Returns, miscellaneous files of little value.

Forest: Bonds of Township Officers.

RECORDER OF DEEDS OFFICE

1. Record Book of Deeds - All counties have this.

2. Record Book of Mortgages - All counties have this.

3. Bond Books - Westmoreland and Centre counties have no such books. For this record in Lackawanna County see the Prothonotary Survey. This record in Northumberland County is recorded in Deed Books. Adams County includes it with the Commissioners' Books, as does Forest County.

4. Commissioners' Books - Centre and Union counties have no such books. Lackawanna County records its bonds in the prothonotary's office. Montgomery County interchanges these books with Bond Books. Erie County combines these with Bond Books. Northumberland County generally records these records in Deed Books. Adams and Forest counties join these with Bond Books.

5. Miscellaneous Books - Westmoreland County uses a Grantor index. Montgomery County has these miscellaneous records but they are scattered. Northumberland County puts the miscellany of this office in Deed Books. Centre County has a miscellaneous record and a miscellaneous index book. Adams County uses a release book, a miscellaneous book, and a miscellaneous index. Tioga County uses a Grantor index. Lackawanna, Erie, and Monroe counties have no such records in this office.

6. Charter Books - Centre, Adams, Tioga, Monroe, and Union counties are not thus equipped.

7. Plats and Surveys - Northumberland and Union counties record these in Deed Books. Westmoreland County includes these in Charter Books. Erie, Adams, and Forest counties have no such records.

8. Miscellaneous - Lackawanna: power of attorney, miscellaneous, limited partnership, co-op associations, commissions.

Westmoreland: powers of attorney, soldiers discharge book.

Montgomery: powers of attorney, discharge of soldiers.

Erie: miscellaneous docket, soldiers discharge, satisfaction papers.

Northumberland: soldiers discharge index, letters of attorney, legal journals, scire facias certificates.

Centre: discharges of "World War Soldiers," deed entry book.

Adams: indices of deeds and mortgages, etc., original papers, unclaimed deeds and mortgages.

Tioga: soldiers discharge.

Monroe: letters of attorney, soldiers discharge.

Union: soldiers discharge, Federal Land Bank Commission.

Forest: No such records at present.

REGISTER OF WILLS OFFICE

1. Estates of Original Wills - All counties have these records.

2. Record Book of Wills - All counties are thus equipped.

3. Petition Book Containing Applications for Probate - Montgomery, Adams, and Monroe counties have no records of this sort.

4. Accounts of Fiduciaries - All counties have these.

5. Auditors' Reports - Westmoreland, Montgomery, and Tioga counties, do not have these. Union County has these docketed in the office of the prothonotary.

6. Inventory of Personal Estate of Descendants - In Montgomery County the originals are filed with the will. Adams County records these with Vendue Lists. Monroe County: see #7. Lackawanna County: under 10. This is not used in Centre County.

7. Inventory and Appraisement Dockets - All counties use these.

8. Inheritance Tax Dockets - Monroe County uses the direct tax only.

9. Vendue Lists - Montgomery, Erie, Centre, Union, and Forest counties do not make use of these. For Adams County, see #6. In Tioga County these lists are filed in the registrar's files.

10. Miscellaneous - Lackawanna: miscellaneous docket and return of death docket. Montgomery: Day-book of Wills.
   Erie: Proceedings index.
   Northumberland: disputed letters, register's work, Orphan's Court and register's work, Administrators' and Executors' certificates from other counties.
   Adams: soldiers discharge, Register and Recorder Counter Book, and Bonds-Administration.

CLERK OF ORPHAN'S COURT

1. Orphan's Court Docket - All counties use this.

2. Estates of Decedents - Lackawanna County: see under register of wills; all other counties have these records.

3. Guardians - Records of Guardians in Erie County will be found in Proceedings index #10 reg. of wills office. Centre County does not have this record in this office.

4. Trustees - All counties have this record.

5. Partition in Orphan's Court - All counties except Forest have this record.

6. Bond Books - None in Westmoreland County. For Adams County, see #10 reg. of wills office.

7. Argument Lists - Lackawanna, Centre, Tioga, and Forest counties do not have these records.

8. Adoption Record - All counties have this except Adams County.

9. Miscellaneous - Lackawanna: Specific Performance Contract Docket, Motion Docket, Order Docket, Orphan's Court Minutes.
   Westmoreland: Appeal Docket, Distribution Docket, Widow's Appraisement.
   Montgomery: Minutes, General Index, Marriage Licenses.
   Erie: All matters pertaining to Orphan's Court are filed in the Proceed-

ings Index.
  Northumberland: General index, index of minors and of Orphan's Court, widows' election in Deed Book, disbarment certificates, births and deaths, and marriage license docket.
  Centre: Escheat Docket and Marriage Books.
  Adams: 12 kinds of papers in file boxes. 1 docket.
  Tioga: Marriage Licenses.
  Monroe: Widows' Appraisement; index.
  Union: lunacy docket; medical and dental regulations; fictitious names; peddlers licenses.

THE PROTHONOTARY'S OFFICE

1. Minute Book - Centre and Forest counties are without these.

2. Appearance judgment Dockets: Judgment Docket - Montgomery County is the only one without this record.

3. Appearance and judgment Dockets: Plaintiff's Index - Adams County does not have this record.

4. Appearance and judgment Dockets: Ad Sectum Index - Lackawanna, Tioga, Monroe, and Union counties do not have this record.

5. Equity Docket - All counties supplied.

6. Partition Docket - Montgomery, Centre and Tioga counties do not have this record. Northumberland County has this in the Orphan's Court Record. Adams County has this record in Equity of Miscellaneous Dockets.

7. Mechanic's Lien Docket - All except Forest county have this record.

8. Municipal and Tax Lien Dockets - Erie County does not have this record. Adams and Forest counties enter this record in the Mechanic's Lien Docket.

9. Papers, etc. in Regard to Cases in Common Pleas - Northumberland and Union counties do not have such records.

10. Record Book of Sheriff's Writs - Erie and Westmoreland counties keep these writs in the Execution Dockets. Adams County's Sheriff's writs are filed in the Appearance Docket.

11. Record of writs of Execution - For Lackawanna County these writs are apparently in the Continuance Docket. Westmoreland County has these records in the Execution Docket. Erie and Union counties are without these records.

12. Dockets of writs; fi-fa - Lackawanna County has these entered with writs of execution. Westmoreland County includes these in Execution Dockets; Montgomery County, in judgment Dockets. Northumberland County has the outgoing writs in the Execution Docket, while the incoming are in the Appearance Docket. Centre County has no such docket, but the matter is entered in the Execution Docket. Adams County has none likewise, but the matter is recorded in the Appearance Docket. Tioga County has this material entered in the Execution Dockets. Erie County has no such dockets.

13. Records of Actions of Adjustment - Lackawanna County enters these in the Continuance Docket. In Northumberland County they are found in the Appearance Docket with a separate index. Adams County has these records filed in the Appearance Docket.

14. Narrs - Lackawanna enters this in the Continuance Docket, as do Westmoreland, Montgomery, and Forest Counties. Erie County records this in the Judgment Dockets. Centre and Adams counties file this in the Appearance Dockets. Tioga County files it with the Common Pleas. Northumberland and Union counties are without

this record.

15. Bills, Mortgages, and Bonds Satisfied and Entered - Lackawanna enters bonds in the Continuance Docket. Montgomery files these in the judgment Dockets. Northumberland records these in the Conditional Sales Docket and also enters them into the office of recorder of deeds. Tioga enters this matter with the Common Pleas, as does Forest County.

16. Transcripts from Accounts in Orphan's Court - Lackawanna County enters these on the Equity and Continuance Docket. Erie County records them in the Appearance Docket. In Northumberland, Montgomery, Centre and Adams counties these records are found in the prothonotary's office. Tioga County handles them through the Orphan's Court, whereas Forest County enters them in the Orphans' Court Docket.

17. Record Book of Accounts of Assignees, etc. - Lackawanna County records this in the Continuance and Equity Docket. Adams County accomplishes the same purpose by entries in the Account Book of the Sheriff's Distribution Docket. Forest County enters this matter in the Common Pleas Docket.

18. Sheriff's Deed Books - Lackawanna County records this with the Recorder of Deeds, as do Westmoreland, Erie, Northumberland, Tioga, and Monroe counties. Adams County refers this matter to the Register of Wills. Montgomery, Centre, Union and Forest counties are without such records.

19. Tax Records: Tax Collectors, Amount Due and Unpaid - Lackawanna, Westmoreland, Northumberland, Tioga counties leave this matter to the Commissioners' Office. Montgomery County records it in a Miscellaneous Docket. Erie County puts this in the Appearance Dockets, while Monroe County records it in the Treasurer's Office. Forest County enters it in a Continuance Docket. Union County has no such record.

20. Appeals from Assessment to Common Pleas - Westmoreland and Forest counties enter these appeals in a Continuance Docket. Erie, Northumberland, Centre, and Tioga counties enter the matter in Appearance Dockets; Montgomery County in a Miscellaneous Docket. Lackawanna and Adams counties keep no such records.

21. Docket of Treasurer's Receipts of Unseated Land - Lackawanna County keeps these records in County Treasurer's Survey. Westmoreland, Northumberland, Tioga and Monroe counties record this in the Treasurer's Office. Montgomery, Adams, and Forest counties have no records of this.

22. Reports of County Auditors - Lackawanna, Montgomery, and Erie counties have these records from the County Controller. Westmoreland has recorded this matter in a Continuance Docket. In Northumberland County this matter is found in the Office of the Commissioner, and the Office of the Clerk of Court. Centre County has no such reports.

23. Election Returns - Lackawanna and Erie counties hold these records for two years, as does Forest County. Adams County, however, does not have these records in this office.

24. Jury Lists - Montgomery, Erie, and Adams counties keep no such records.

25. Naturalization - Lackawanna, Montgomery, Erie, Adams, and Forest counties do not keep a record of naturalization in this office.

26. Miscellaneous - Lackawanna:

      1 Locality Indices-- 2 Satisfaction Docket--3 Assignment Docket--
      4 Arbitration Docket--5 Receipt for Files Docket--6 Conditional Sales--
      7 Fictitious Name Docket--8 Optometry Register Docket--9 Dental Register--
      10 Judgment Index Docket

26. <u>Miscellaneous</u> - Westmoreland:

      1 Medical Register--2 Dental Register--3 Optometry Register--
      4 Stallion Register--5 Lunacy, Divorce, and Drunkenness--6 Issue Dockets--
      7 Conditional Sales Docket--8 Conditional Sales Contract--
      9 Federal Tax Lien Docket--10 Argument Lists--11 Cost Book--12 Cash Book--
      13 Index to Assigned Judgment

Montgomery:

      1 Tax Records, Tax Collectors Amounts Due and Unpaid--2 Appeals from
      Assessment to Common Pleas--3 Bar--4 Fictitious Names--5 Conditional Sales

Erie:

      1 Treasurer's Deed Book--2 Law Student Docket--3 Medical License--
      4 Commission Books

Northumberland:

      Miscellaneous Books and Papers

Centre:

      1 Conditional Sales Docket--2 Osteopathic Register--3 Dental Register--
      4 Memorial Register--5 Fictitious Names--6 Stallion Register--7 Degrees and
      Divorces--8 Causes for Trial--9 Appointments of Borough Officers--10 Report
      of Township Auditors--11 Petition for Peddler's License--12 Miscellaneous
      Papers and Commission Bonds--13 Conditional Sales--14 Township Lines

Adams:

      1 Miscellaneous Docket--2 Bond Books--3 Assignee Bonds--4 Fictitious Names--
      5 Conditional Sales--6 Oaths of Office of Justice--7 Precipate--8 Acceptance
      of Justice of Peace--9 Oaths and Bonds Justice of Peace Commissioners--
      10 Treasurer's Bonds and Oaths of County Officers--11 Assignees Accounts
      and Discharges--12 Inventory and Sales Lists--13 Budge Contracts--14 Lunacy
      Proceedings--15 Partial and Final Writs--16 Deeds of Assignments--
      17 Insolvent Debtors--18 Equity--19 Conditional Sales Contracts--20 Certifi-
      cates from Land Office--21 Power of Attorney--22 Peddler's License for
      Soldiers

Tioga:

      1 Business Index of Fictitious Names--2 Divorce and Lunatics--3 Medical
      Register--4 Dental Register

Monroe:

      1 Oaths of Constables, Justices, etc.--2 Medical Register--3 Dental
      Register--4 Records to Practice Optometry--5 Lunacy--6 Conditional Sales
      Docket--7 Constable's and Collector's Bonds--8 Surplus Bond Book--
      9 Charter Books--10 Divorce Index--11 Marriage License Dockets

Union:

      1 List of Rules etc.--2 Court Trial Lists

Forest:

      1 Registry of Automobiles--2 Treasurer's Deeds--3 Conditional Sales--
      4 Soldier's Discharge--5 Fictitious Names--6 Execution Docket--6 Admission
      of Attorneys--7 Medical Register--8 Dental Register--9 Deposition Docket--

10 Arbitration Docket--11 Argument Docket

In the above description it will be observed that certain important features stand out. The prominent place occupied by "miscellaneous," in all offices, the varying contents of this file in the different counties, the lack of indices of a clear comprehensive nature, the different contents of the similar files in different counties and the absence of important records in many others, all point to the need of some comprehensive arrangement based on scientific principles. To this we must add the laxity of recording methods, the general continuance in manual copying and the inevitable waste of time and funds involved.

CONCLUSIONS

FINDING - The present system of county records is disconnected and confused, inefficient, inconvenient, and financially wasteful.

RECOMMENDATIONS -

(1) All county records should be integrated by concentrating them in a Department of Records under the Chief Clerk of the County.

(2) The personnel engaged in record-keeping should be trained in this work.

(3) The technical methods of record-keeping should be modernized and full use should be made of mechanical devices to increase accuracy, speed and economy.

(4) The general plan or arrangement of the new integrated system should be drafted by a temporary state commission composed of archivists, historians, librarians, county officials, accountants and administrative experts.

CHAPTER VIII

COSTS AND REVENUES

Most of the material which would logically appear in this chapter has necessarily been presented in preceding chapters devoted to particular aspects of county government. This is especially true of the chapters dealing with administrative organization (Chapter 2), personnel (Chapter 3), assessments (Chapter 4), elections (Chapter 5), and justices of the peace, constables and the courts (Chapter 6). An intensive examination of the subject has also been made in recent local government literature including F. P. Weaver and H. F. Alderfer, County Government Costs in Pennsylvania, 1933, and Edward W. Carter, Mandatory Expenditures of Local Government in Pennsylvania, 1934. The Pennsylvania Economic Council is engaged in making a detailed study of costs and revenues of local subdivisions, including counties. That organization is performing an important public service by preparing schedules of expenditures and revenues which present items in auditors' and controllers' reports in a uniform and orderly manner.

There remain, however, certain important questions which have not been answered in the data thus far presented. They require analysis here in order to give a more complete picture of the relative importance of different sources of county revenue and expenditures respectively, and the chief trends both in totals and in individual items. Among these questions are the following:

1. What are the chief present sources of county revenue?
2. What is the present trend in revenue receipts?
3. Has the burden of county taxation on property been substantially eased by the gasoline and liquor taxes?
4. Does the failure to collect court costs and fines, noted in Chapter VI, show any trend towards correction or improvement?
5. In how far have departmental earnings and the cash receipts of county institutions respectively been developed and what is their present trend?
6. What is the trend in county expenditures and in their relation to revenues?
7. What is the proportionate cost of each major county function?
8. Where have economies been chiefly effected?

Answers to these and certain other questions have been attempted in this chapter, in tabular form. The basic data for nine selected typical counties were generously supplied by the Pennsylvania Economic Council from its county schedules. These counties which were intensively surveyed, were so chosen as to assure typical representation of such factors as density of population, assessed valuation per capita, taxes levied per capita, industrial characteristics, etc. It is believed that they offer a fair cross section of county conditions.

Revenue Receipts - Table 1 shows the per capita revenue receipts of nine selected typical counties from all sources in 1930 and 1933. The data contained in this table are subdivided in tables 2 to 7, inclusive, to show the per capita receipts from property taxes, liquor licenses, gasoline taxes, court costs and fines, departmental earnings, and institutional earnings, respectively. Tables 2 to 7 show also what proportion of total county receipts was derived from each of these sources. State relief payments are omitted in all cases because of their extraordinary character.

Table 1. Total Revenue of
County per capita

County	1930	1933
Lackawanna	6.58	3.77
Westmoreland	6.18	4.56
Montgomery	11.11	7.11
Erie	6.90	7.57
Chester	6.78	5.25
Northumberland	5.08	4.82
Centre	4.96	3.56
Tioga	7.44	5.68
Monroe	12.66	8.05

The actual and per capita receipts of 8 of the 9 counties dealt with in Table 1 decreased substantially from 1930 to 1933, in three cases the decrease being more than one-third. Only Erie county increased its revenues. The range in per capita receipts was:

In 1930: $4.96 (Centre) to $12.66 (Monroe)
In 1933: $3.56 (Centre) to $8.05 (Monroe)

Table 2 shows that the decrease in county receipts is due almost entirely to the greatly shrunken returns from property taxes. In 4 of the 9 counties the decrease in returns from this source was even greater than the net decrease in receipts from all sources. In Lackawanna county the decrease was almost 50 per cent.

The range in per capita receipts from property taxes was:

In 1930: $3.83 (Centre) to $11.19 (Montgomery)
In 1933: $2.70 (Centre) to $7.11 (Montgomery)

Table 2. Property, Poll and Occupational Taxes

County	Per cent of total County Revenue		Property etc. Taxes per capita	
	1930	1933	1930	1933
Lackawanna	86.17%	78.61%	5.67	2.96
Westmoreland	83.10%	80.36%	5.14	3.67
Montgomery	83.2 %	77.2%	11.19	7.11
Erie	77.5%	88.8%	5.35	6.13
Chester	75%	70.9%	5.07	3.80
Northumberland	81.5%	79.21%	4.14	3.82
Centre	85.8%	76.1%	3.83	2.70
Tioga	87.2%	84.5%	6.56	4.80
Monroe	88.7%	76.9%	11.23	6.23

These taxes are by far the most important source of county revenue. In 1930 they constituted from 75 per cent (Chester) to 88 per cent (Monroe) of total county receipts; in 1933, from 70 per cent (Chester) to 88 per cent (Erie). However, in all counties except Erie their relative importance decreased between 1930 and 1933.

Table 3. County Liquor Revenue

County	Per cent of total County Revenue		Liquor Revenue per capita	
	1930	1933	1930	1933
Lackawanna	----	1.97%	----	.07
Westmoreland	----	1.97%	----	.09
Montgomery	----	1.2%	----	.08
Erie	----	.8%	----	.06
Chester	----	1.7%	----	.09
Northumberland	----	2.32%	----	.11
Centre	----	2.4%	----	.09
Tioga	----	.8%	----	.05
Monroe	----	4.2%	----	.37

Table 3 shows the per capita revenue received by the same counties from liquor licenses in 1933 and the proportion that these formed of total county receipts. These receipts were, in all counties except Monroe, less than 3 per cent of all receipts. In Erie and Tioga counties they were only .8 per cent of all receipts, while in Monroe county they amounted to 4.2 per cent of the total. Only in Monroe county, where they yielded $.37 per capita, were they a particularly important source of revenue. In the other 8 counties their yield was from $.05 to $.11 per capita.

Table 4. Gasoline Tax

County	Per cent of total County Revenue		Gasoline Tax per capita	
	1930	1933	1930	1933
Lackawanna	6.41%	6.43%	.42	.42
Westmoreland	7.76%	10.63%	.48	.49
Montgomery	6.9%	10.8%	.76	.77
Erie	10.2%	10.3%	.70	.71
Chester	12.3%	15.1%	.84	.83
Northumberland	9.72%	10.39%	.49	.50
Centre	6.0%	19.8%	.27	.69
Tioga	8.1%	11.7%	.61	.67
Monroe	8.9%	14.7%	1.13	1.19

The gasoline tax returned to the counties by the State is, as shown in Table 4, the second most important source of county revenue. In 1930 the proportion of all county revenues derived from this source ranged from 6 per cent in Centre county to 12.3 per cent in Chester county, with only two of the nine counties receiving more than 10 per cent of all their revenue from this source; in 1933 the range was from 6.4 per cent (Lackawanna) to 19.8 per cent (Centre). Except in Centre county this relative increase was accounted for chiefly by the shrinkage in receipts from other sources. The per capita receipts from the gasoline tax remained almost constant in the other eight counties during the period from 1930 to 1933, but there was a great degree of variation as between different counties, from $.27 (Centre) to $1.13 (Monroe) in 1930, and from $.42 (Lackawanna) to $1.19 (Monroe) in 1933.

Table 5. Revenue from Court Costs and Fines

County	Per cent of total County Revenue		Court costs and fines per capita	
	1930	1933	1930	1933
Lackawanna	1.62%	.09%	.11	.003
Westmoreland	2.51%	.75%	.16	.03
Montgomery	1.3%	.7%	.01	.005
Erie	.2%	.08%	.01	.005
Chester	----	----	----	----
Northumberland	.67%	.80%	.03	.04
Centre	3.4%	1.9%	.15	.07
Tioga	2.4%	1.4%	.02	.08
Monroe	.9%	----	.13	----

Table 5 deals with court costs and fines. The amount of these actually collected formed a negligible part of county receipts in either year, but especially in 1933. In that year three of the seven counties shown derived less than one-half cent per capita from this source, and the highest per capita amount collected in any of the other counties was $.08. In comparison with total receipts, court costs and fines yielded from .2 per cent (Erie) to 3.4 per cent (Centre) of the total in 1930 and, in 1933, from .08 per cent (Erie) to 1.9 per cent (Centre).

Table 6. Total Departmental Earnings

County	Per cent of total County Revenue		Total Departmental earnings per capita	
	1930	1933	1930	1933
Lackawanna	4.93%	6.90%	.32	.26
Westmoreland	6.27%	5.82%	.39	.27
Montgomery	6.6%	9.2%	.73	.66
Erie	8.1%	7.7%	.56	.53
Chester	8.3%	6.7%	.56	.35
Northumberland	7.34%	7.15%	.37	.34
Centre	----	----	----	----
Tioga	----	----	----	----
Monroe	----	----	----	----

The earnings of the offices of the treasurer, recorder of deeds, register of wills, prothonotary, clerk of courts, sheriff, etc., are shown in Table 6. The per capita and percentage figures presented in this table are based upon the total earnings of these offices, exclusive of fees paid to them by the counties. Data are presented for six counties where the heads of these departments are paid fixed salaries. Centre, Tioga and Monroe counties, included in the other tables, are omitted from Table 6 because in those counties officers are paid with the fees collected and therefore the fees are not accounted for as county revenue.

The range in per capita earnings of all county departments was:

In 1930: $.32 (Lackawanna) to $.73 (Montgomery)
In 1933: $.26 (Lackawanna) to $.66 (Montgomery)

Such earnings decreased from 1930 to 1933 in each of the six counties. Their relative importance did not vary greatly as between the counties shown. In all cases they constituted between 6.3 per cent (Westmoreland) and 8.3 per cent (Chester) of total county receipts in 1930, and between 5.8 per cent (Westmoreland) and 9.2 per cent (Montgomery) in 1933.

Table 7. Institutional Earnings

County	Per cent of total County Revenue		Institutional Earnings per capita	
	1930	1933	1930	1933
Lackawanna	.64%	1.00%	.04	.04
Westmoreland	.17%	.15%	.01	.008
Montgomery	1.3%	.5%	.15	.04
Erie	2.1%	1.6%	.14	.11
Chester	4.1%	2.4%	.28	.13
Northumberland	.20%	.06%	.01	.003
Centre	----	----	----	----
Tioga	1.1%	1.4%	.09	.08
Monroe	.9%	.4%	.12	.06

As is shown in Table 7 there is a wide range of variation from county to county in the per capita cash receipts of county homes, prisons, and similar institutions. The range was:

In 1930: $.01 (Northumberland and Westmoreland) to $.28 (Chester)
In 1933: $.003 (Northumberland) to $.13 (Chester)

This variation is accounted for largely by differences in the number and kinds of institutions operated by different counties. Therefore comparisons of total institutional earn-

ings of different counties are meaningless. It is noteworthy, however, that in nearly
every county whose institutions earned comparatively high amounts in 1930, such earnings
dropped very sharply in 1933. This is accompanied by a general increase in expenditures
for charities, as shown in Table 13 below.

County Expenditures - Tables 8 - 18 deal with county expenditures. Table 8 shows the
total per capita expenditures of nine counties in 1930 and 1933, exclusive of expenditures
for permanent improvements and those made from state relief funds. The range of per capita expenditures was:

In 1930: $3.89 (Northumberland) to $14.60 (Monroe)
In 1933: $4.25 (Lackawanna) to $9.80 (Tioga)

Three counties increased their per capita expenditures in the latter year while six showed
decreases. The highest per capita cost each year was incurred by one of the two smallest
counties but other small counties had average or less than average per capita costs. The
per capita expenditures of the three largest counties shown were consistently although
very slightly below average.

Table 8. Total Expenditures
          per capita

County	1930	1933
Lackawanna	5.25	4.25
Westmoreland	6.79	6.41
Montgomery	6.85	6.34
Erie	7.11	8.95
Chester	6.48	5.22
Northumberland	3.89	4.93
Centre	5.14	5.07
Tioga	8.13	9.80
Monroe	14.60	6.54

Administrative Costs - Table 9 shows the total per capita expenditures of the same
counties for the operation and maintenance of their general administrative departments.
Included are expenditures for the commissioners, solicitor, sealers of weights and measures, surveyor, court house maintenance, elections, assessments, treasurer, tax collector,
controller or auditors, recorder of deeds and register of wills. Important operating
costs excluded from this table are those of the courts, corrections, charities, and highways, which are dealt with in later tables.

The range of expenditures for administrative purposes was:

In 1930: $1.01 (Westmoreland) to $1.98 (Monroe)
In 1933: $.98 (Centre) to $1.80 (Montgomery)

Six of the nine counties decreased such expenditures from 1930 to 1933 and one showed
no change, while Montgomery and Northumberland counties increased their expenditures
slightly.

Table 9. Total Administrative Expense
          of County per capita

County	1930	1933
Lackawanna	1.36	1.36
Westmoreland	1.01	1.00
Montgomery	1.70	1.80
Erie	1.40	1.20
Chester	1.35	1.23
Northumberland	1.10	1.17
Centre	1.11	.98
Tioga	1.32	1.22
Monroe	1.98	1.60

Court Expenses - Judicial expenses, including the cost of the sheriff, coroner, prothonotary, clerk of courts, district attorney, law library, courts, justices of the peace and constables, are shown in Table 10. The range of per capita expenditures was:

In 1930: $.67 (Monroe) to $1.30(Centre)
In 1933: $.54 (Monroe) to $1.06(Centre)

All but two counties showed a large percentage of decrease in such expenditures and the increases in those two counties were accounted for by their low costs in 1930.

Table 10.   Total Judicial Expense of
County per capita

County	1930	1933
Lackawanna	1.00	.86
Westmoreland	1.09	.72
Montgomery	.91	.94
Erie	1.06	.85
Chester	.88	.81
Northumberland	.86	.87
Centre	1.30	1.06
Tioga	1.05	.79
Monroe	.67	.54

Law Enforcement Offices - The importance of law enforcement as a county function is shown in Tables 11 and 12. In Table 11 are given the 1933 total and per capita expenditures for the following law enforcement agencies: sheriff, coroner, courts, district attorney, justices of the peace and aldermen, constables, probation officers, and correctional institutions. The range in per capita expenditures for these offices was from $.80 (Monroe) to $1.46 (Erie). Their cost represented from 12.4 per cent (Monroe) to 30.7 per cent (Lackawanna) of total county expenditures for all purposes.

Expenditures for the prothonotary, clerk of courts, register of wills, and recorder of deeds are grouped together in Table 12. Although the work of these officers is closely related to that of the other law enforcement agencies, their cost is shown separately, because all four offices are engaged in work of a clerical nature. As has been suggested in earlier chapters these offices might well be combined.

Their per capita cost to the county ranged from $.01 in Monroe County to $.43 in Montgomery County. In the former, however, these officers are paid by fees of which no record is available, so that the cost shown is no indication of what they actually cost.

Excluding the fee counties from consideration these clerical offices took from 3.0 per cent (Westmoreland) to 7.85 per cent (Montgomery) of all county expenditures. As has been pointed out before, there is no relationship between the population of a county and the cost of its fee offices, the greatest variation occurring in counties of approximately the same size.

Table 11.   Law Enforcement Offices

County	Total Law Enforcement Expenses - 1933	Cost per capita	Percent of Total Expenditures of County - 1933
Lackawanna	404,759.00	1.30	30.69%
Westmoreland	311,195.00	1.05	16.46%
Montgomery	359,202.00	1.35	21.32%
Erie	256,682.00	1.46	16.35%
Northumberland	131,858.00	1.03	20.80%
Chester	178,589.00	1.41	27.03%
Centre	66,984.00	1.45	28.54%
Tioga	40,404.00	1.27	12.86%
Monroe	22,962.00	.80	12.41%

Table 12. Clerical Offices - 1933

County	Total Expense of Clerical Offices - 1933	Cost of Clerical Offices per capita	Per cent of Total County Expenditures - 1933
Lackawanna	73,763.00	.24	5.59%
Westmoreland	57,374.00	.19	3.04%
Montgomery	115,382.00	.43	7.85%
Erie	54,671.00	.31	3.48%
Northumberland	35,407.00	.28	5.58%
Chester	42,116.00	.33	6.38%
Centre	11,911.00	.26	5.08%
Tioga	3,950.00	.12	1.26%
Monroe	1,032.00	.01	.56%

Cost of Charities and Corrections - Table 13 shows the per capita expenditures of the same counties for charities and corrections in 1930 and 1933. The expenditures shown for charities are exclusive of state relief funds, nor do they include the expenses of poor districts in Centre and Union Counties. The range in per capita expenditures for charities was:

In 1930: $.18 (Centre) to $1.93 (Erie)
In 1933: $.36 (Centre and Northumberland) to $3.10 (Erie)

There was an increase in such expenditures in every county except Tioga, and in all except Lackawanna such increases were very substantial.

Per capita expenditures for corrections are based upon the combined cost of probation offices, county prisons, juvenile detention homes and other county correctional institutions, and the cost of maintaining children and convicts in state institutions. The range in per capita expenditures for these purposes was:

In 1930: $.33 (Northumberland) to $.92 (Erie)
In 1933: $.31 (Monroe) to $.76 (Erie)

Five counties decreased, two counties increased, and two did not change their per capita expenditures from 1930 to 1933.

Table 13. Per Capita Expenditure for Charities and
Corrections, 1930 and 1933.

County	Charities		Corrections	
	1930	1933	1930	1933
Lackawanna	.52	.53	.62	.57
Westmoreland	1.24	1.38	.47	.43
Montgomery	.74	1.39	.54	.54
Erie	1.93	3.10	.92	.76
Chester	1.19	1.37	.84	.74
Northumberland	.21	.36	.33	.33
Centre	.18	.36	.65	.69
Tioga	2.37	1.90	.43	.61
Monroe	1.90	2.61	.49	.31

Highway Expenditures - Table 14 shows the total per capita expenditures of the counties for highways, and the proportion of such expenditures to total county expenditures. Included are the costs of highway maintenance, road and bridge construction, road damages, viewers' fees and the salary of the county engineer. Per capita expenditures for these purposes were within the following range:

In 1930: $.37 (Montgomery) to $7.50 (Monroe)
In 1933: $.20 (Westmoreland) to $.83 (Monroe)

In every county there was a sharp decrease in such expenditures, the greatest being in Monroe (from $7.50 to $.83) and in Westmoreland (from $2.70 to $.20). The retrenchment in expenditures for this function was generally greater than in any other case.

Table 14. Total Highways Cost

County	Total Highways Cost per capita		Per cent of total County Expenditures	
	1930	1933	1930	1933
Lackawanna	1.09	.31	20.7%	8.9%
Westmoreland	2.70	.20	40.1%	3.1%
Montgomery	.37	.27	4.9%	4.2%
Erie	.99	.23	13.9%	2.6%
Chester	.69	.67	10.5%	12.8%
Northumberland	.94	.26	24.2%	11.9%
Centre	1.40	.36	27.8%	7.1%
Tioga	1.59	.31	19.7%	3.1%
Monroe	7.50	.83	39.2%	12.6%

Maintenance of County Buildings - The cost of maintaining county buildings was also curtailed materially between 1930 and 1933, as is shown in Table 15.

Table 15. County Buildings and Court House Maintenance Costs

County	Total cost per capita		Total Cost per cent of Total County Expenditures	
	1930	1933	1930	1933
Lackawanna	.56	.24	10.6%	5.7%
Westmoreland	.27	.14	3.9%	2.4%
Montgomery	1.97	.34	23.3%	5.4%
Erie	2.60	.17	35.3%	1.9%
Chester	.16	.19	2.4%	3.6%
Northumberland	.09	.22	2.2%	4.5%
Centre	.19	.14	3.7%	2.8%
Tioga	.79	.23	9.7%	2.4%
Monroe	----	.24	----	3.7%

In 1930 this cost was between $.09 per capita (Northumberland) and $2.60 per capita (Erie); and in 1933, from $.14 (Centre and Westmoreland) to $.34 (Montgomery.

Relative Cost of Administrative Offices - Table 16 groups together expenditures for the general administrative offices (commissioners, solicitors, treasurer, controller or auditors, and sealers of weights and measures) and shows their relation to the total cost of county government. Similarly, Table 17 presents data as to the relative cost of eleven items of expenditure. These two tables supplement similar data in Tables 11, 12, 14 and 15.

Table 16. Financial and Administrative Costs

County	Total cost per capita		Total Cost per cent of Total County Expenditures	
	1930	1933	1930	1933
Lackawanna	.49	.43	11.6%	9.6%
Westmoreland	.32	.23	4.9%	3.6%
Montgomery	.51	.55	7.4%	8.7%
Erie	.47	.40	6.6%	4.5%
Chester	.49	.44	9.5%	8.4%
Northumberland	.44	.47	11.3%	9.5%
Centre	.37	.38	7.3%	7.5%
Tioga	.51	.51	5.4%	5.2%
Monroe	.78	.71	5.3%	10.8%

Table 17. Percentage of expenditures of departments of Total Expenditure of County for 1933

County	Coroner	Clerk of Courts	District Attorney	Courts
Lackawanna	.27%	1.03%	2.76%	9.97%
Westmoreland	.32%	.56%	2.07%	3.85%
Montgomery	.28%	.66%	3.29%	5.73%
Erie	.23%	.42%	1.44%	3.27%
Chester	.46%	1.28%	2.75%	5.49%
Northumberland	.42%	1.53%	2.23%	6.35%
Centre	.21%	5.04%	2.97%	5.83%
Tioga	.13%	1.21%	1.99%	3.10%
Monroe	----	----	2.19%	4.93%

County	Justices and Aldermen	Constables	Probation	Correction
Lackawanna	1.57%	.82%	.83%	12.61%
Westmoreland	.81%	1.59%	.27%	6.4%
Montgomery	.58%	.34%	.52%	8%
Erie	.63%	.64%	.35%	8.17%
Chester	.56%	1.04%	.69%	13.55%
Northumberland	1.70%	1.74%	.43%	6.18%
Centre	1.36%	1.23%	2.29%	11.25%
Tioga	----	----	.43%	5.72%
Monroe	----	----	.58%	4.16%

County	Charities	Debt Service minus Temporary Loans	Personal Services[†]
Lackawanna	12.58%	12.44%	52.54%
Westmoreland	22.00%	22.74%	20%
Montgomery	21.91%	.72%	36.5%
Erie	34.64%	22.95%	20%
Chester	26.21%	3.28%	30%
Northumberland	7.21%	15.67%	16.7%
Centre	7.19%	6.80%	24.1%
Tioga	19.38%	14.33%	
Monroe	39.87%	13.40%	

[†] Personal services are also included in the other items in this table.

We have already seen that law enforcement accounts for a very large part of county expenditures (Tables 11 and 12). In 1930 highways took from 4.9 per cent (Montgomery) to 40.1 per cent (Westmoreland) of total county expenditures, but in 1933 they were much less important, ranging from 3.1 per cent (Westmoreland and Tioga) to 12.8 per cent (Chester) of the total (Table 14). Similarly, maintenance of county buildings decreased in relative importance, dropping from a range of from 2.4 per cent to 35.3 per cent in 1930 to one of from 1.9 per cent to 5.7 per cent in 1933 (Table 15).

The group of offices responsible for the general administration of the county (Table 16) decreased in relative cost between 1930 and 1933 in some instances and increased in others. In 1930 these offices took from 4.9 per cent (Westmoreland) to 11.6 per cent (Lackawanna) of the total; in 1933 the range was from 3.6 per cent (Westmoreland) to 10.8 per cent (Monroe).

Though unnecessary, the office of coroner is a minor expense in all counties, in no county taking more than 0.46 per cent of total expenditures in 1933 (Table 17). Other relative costs shown in Table 17 ranged as follows:

> Clerk of Courts: .42 per cent (Erie) to 5.04 per cent (Centre)
>
> District Attorney: 1.44 per cent (Erie) to 3.29 per cent (Montgomery)
>
> Courts: 3.10 per cent (Tioga) to 9.97 per cent (Lackawanna)
>
> Justices and Aldermen: .56 per cent (Chester) to 1.70 per cent (Northumberland)
>
> Constables: .34 per cent (Montgomery) to 1.74 per cent (Northumberland)
>
> Probation: .27 per cent (Westmoreland) to 2.29 per cent (Centre)
>
> Correction: 4.16 per cent (Monroe) to 13.55 per cent (Chester)
>
> Charities: 7.19 per cent (Centre) to 39.87 per cent (Monroe)
>
> Debt Service minus Temporary Loans: .72 per cent (Montgomery) to 22.95 per cent (Erie)
>
> Personal Services: 16.7 per cent (Northumberland) to 52.54 per cent (Lackawanna)

Of these the expenditures for the courts, charities and corrections were invariably important. Although the percentage of county expenditures allocated to debt service was only 0.72 in one county, 3.28 in another, and 6.80 in another, this was a major expense of the other six counties.

The relative amounts spent for personal services in all departments varied greatly as between counties, from one-sixth of the total expenditures of Centre County to more than one-half the expenditures of Lackawanna County.

There remains a further question of general interest - What is the order of county expenses in proportion of the total county outlay? The following list answers this question in part by giving for each item the proportion which it forms of the total. In order to avoid misapprehension this proportion is not given as an average but as a range of percentage, high and low:

.
> Law Enforcement Offices: (Sheriff, coroner, courts, district attorney, justices, constables, probation, corrections) - from 12 to 30 per cent of total. (Mainly cost of courts and corrections).
>
> Debt Service: (minus temporary loans) - from 1 per cent to 23 per cent of the total.

Highways: from 3 to 42 per cent of the total.

Administrative and Financial: (including county commissioners; controllers' or auditors' and county treasurers' offices) - from 4 to 11 per cent of the total.

Clerical Offices: (Prothonotary, clerk of the courts, register of wills, and recorder of deeds) - from 1 to 8 per cent of the total).

County Buildings and Courthouse Maintenance: from 2 to 6 per cent of the total.

Debt Service - Table 17 showed the proportion of the total expenditures of these counties spent for debt service (exclusive of temporary loan repayments) in 1933. Table 18 shows the total per capita expenditures for debt service and Table 19 per capita interest payments. The range for total debt service was:

In 1930: $0.10 (Montgomery) to $4.89 (Monroe)
In 1933: $0.80 (Montgomery) to $5.12 (Tioga)

and the range for interest payments was:

In 1930: $0.00 (Westmoreland) to $1.13 (Monroe)
In 1933: $0.02 (Montgomery) to $1.21 (Erie)

	Table 18. Total Per Capita Debt Service		Table 19. Total Per Capita Interest Paid	
County	1930	1933	1930	1933
Lackawanna	.93	.53	.35	.21
Westmoreland	1.66	2.71	0.00	.70
Montgomery	.10	.80	.02	.02
Erie	1.46	2.85	1.11	1.21
Chester	1.01	.61	.20	.17
Northumberland	.84	1.36	.30	.32
Centre	1.04	1.63	.20	.21
Tioga	2.14	5.12	.55	.71
Monroe	4.89	.88	1.13	.86

Net bonded Debt - Table 20 shows the per capita net bonded debt of the nine counties, which ranged:

In 1930: from $0.00 (Montgomery) to $19.59 (Erie)
In 1933: from $0.03 (Montgomery) to $20.98 (Erie)

Table 20. Total Net Bonded Debt per capita

County	1930	1933
Lackawanna	6.35	5.05
Westmoreland	9.23	10.86
Montgomery	00.00	.03
Erie	19.59	20.98
Chester	3.33	3.09
Northumberland	6.21	5.54
Centre	.83	.56
Tioga	9.59	12.30
Monroe	10.28	12.40

CONCLUSIONS

FINDING - Despite many years of discussion of various proposals to reduce the burden on general property, the property tax still remains the chief reliance for county revenue.

FINDING - From 1930 to 1933 county revenues have dropped at an alarming rate, especially those from collected taxes. In one case this decrease has been 50 per cent. (In Chapter IV it was also shown that delinquent tax payments have grown to grave proportions). The revenue from the gasoline tax has remained stationary and a slight additional income has been realized from the new liquor tax.

FINDING - Expenditures in general have not decreased in proportion to revenue. While less has been spent upon highways, the outlay for charities and debt service has risen sharply. Such items as general administration and financial officers' expense have decreased only slightly or have remained stationary.

For further findings and for recommendations reference is made to Chapters II, III, IV and VI.

ORIGINS AND LEGAL BASIS

<u>Methods Employed in Township Survey-</u> In its growth from colonial times, the township has been less subject to change than either the state or federal structure. With few substantial alterations since William Penn's day, it has taken firm root in the life and traditions of the Pennsylvania community. The diversity of conditions surrounding the township, the differences of size, geography, economic resources, race, traditions, and industry, all increase the complexity of its problems. The present study is a cooperative effort to analyze the experience, needs, services, and operations of communities under these diverse conditions.

More than 110 townships were visited, and over 200 township officials, residents, and business men were called on for assistance in presenting a picture of the township in operation. Although the lack of any previous study of this kind in Pennsylvania has added to the difficulty of the work, the township Survey has proceeded with cooperation from local officials of communities representing every section of the state. It contains their suggestions, as well as those of the author, aimed at lessening the burden of local taxation and raising the standards of operation without increase of cost.

Through the kindness of the officers in the Bureau of Statistics, Department of Internal Affairs, and the Township Engineer's Office, Highway Department, unpublished data were secured which are shown on various Tables. Without this help an effective analysis of the township could not have been presented. The careful methods employed in securing this information by the two departments named, render it an important primary source for all who are seeking authentic data.

<u>Early Township Planning</u> - Although planning and direction are lacking in the recent development of local government in Pennsylvania, this was not always the case. A great deal of thought was given by the Proprietor and Governor, William Penn, to the form of local government most suitable for his Province. The plans formulated in England had as their first objective "a large Town or City, in the most convenient Place upon the River for Health and Navigation." ("The Conditions or Concessions to the Province of Pennsylvania," July 11, 1681, see Pennsylvania "Acts of Assembly of the Province of Pennsylvania," p. viii, Philadelphia 1775.)

Originally among the principal factors in our community life were convenience, health, and accessibility. While these were secondary to tolerance for religious worship and for expression of political opinion, they were nevertheless an express part of the "Conditions" signed in England in 1681 by Penn and some of his prospective Quaker settlers.

<u>Origin of Township in Pennsylvania</u> - Planning and direction were given to the development of types of government in Pennsylvania. A Surveyor-General, Thomas Holme, located city boundaries with wide, connecting streets and provision for public parks at Philadelphia in 1682. He next surveyed the three upper counties (Philadelphia, Bucks, and Chester) and then three lower counties (Newcastle, Kent and Sussex). The counties were subdivided into townships. At Uplands (Chester, Pa.) William Penn convened the first General Assembly of Pennsylvania in 1682. He attended this Assembly in person and submitted to it a "Frame of Government." This was the first Constitution of the Province of Pennsylvania. It provided that the counties should be divided into "hundreds" or townships following the custom which had prevailed in England. The "Frame of Government" provided:

> "XVl. That.....the general assembly.....shall be enlarged as the
> country shall increase in people.....proportioning.....the choice
> .....most equally to .....the hundreds and counties, which the
> country shall hereafter be divided into....." ("Frame of Government" drafted by William Penn. (April 25, 1682) See Samuel Hazard,
> "Annals of Pennsylvania 1609-1682" p. 561, Philadelphia, 1850.)

When William Penn returned to England in 1684 there were 22 townships settled or in the process of settlement. They were divided among the six counties already mentioned.

Early legal provisions relative to the township seemed obscure. The act of 1705 informs us that the township officers were Justices of the Peace, and Overseers of the Poor. It indicates that the Overseers had authority to levy general property and poll taxes for poor relief.

Constitution of 1873 - While the township was mentioned in previous constitutions, more is to be found affecting townships in the Constitution of 1873 than in any of its predecessors. Of the four important provisions applicable to the second class township, two restrictions relate to indebtedness (one limits the amount, the other requires an annual sinking fund tax for its retirement). The third clause prohibits the General Assembly from enacting special legislation affecting townships, while the fourth permits the same authority to classify townships for the purpose of general legislation.

A summary of the provisions of the present Constitution (1873) which expressly affect the second class township is presented below:

Limitation of amount of indebtedness, (Article 9, Section 8).
Legislature authorized to classify townships, (Article 3, Section 34).
Legislature prohibited from enacting local or special laws.
Regulating the affairs of townships,
Authorizing opening, altering or maintaining roads,
Vacating roads,
Erecting new townships, or changing township lines,
Creating offices, or limiting powers and duties of officers,
Regulating practice, jurisdiction, and fees of justices
of peace,
Delegating power to a special commission to make municipal
improvements, levy taxes, or perform any municipal
function, (Article 3, Section 7 and 20.)
Duty to levy an annual sinking fund tax for indebtedness sufficient
to pay interest and principal in 30 years, (Article 9,
Section 10).
General Assembly not to authorize township to become a stockholder
in any company, or to lend its credit to any corporation or
individual, (Article 9, Section 7).
General Assembly to provide for strict accountability for fees and
all public moneys paid to township officers, (Article 14,
Section 6).
No street passenger railway to be constructed within the limits of
any township without consent of its local authorities,
(Article 17, Section 9).
Commonwealth not to assume debt of township unless contracted in
war or insurrection, (Article 9, Section 9).
No township to elect more than 2 justices of the peace without the
consent of a majority of the qualified electors, (Article
5, Section 11).
All elections for regular terms of township officers to be held at
municipal election, (Article 8, Section 3).
Laws regulating elections and registration to be uniform, but
General Assembly shall permit voting machines at option of
electors of township, (Article 8, Section 7).
Townships to be divided into compact election districts as the
court of quarter sessions may direct, (Article 8, Section 11)
No township shall be divided in the formation of a senatorial district, (Article 2, Section 16).

Reasons for Constitutional Revision - During the past five years emergency conditions have arisen. Economies are now necessary which 10 or 15 years ago would have been regarded as impossible. If the legislature is to deal effectively with numerous local situations requiring special treatment, it must have greater freedom. Lower taxes can be secured through strengthening and enlarging our scheme of local government to enable efficient work to be done by larger units of the same type. A proper reorganization to attain effective administration over larger areas would save substantial sums annually.

The Commission on Constitutional Amendment and Revision in 1920 advised a complete revision of our Constitution because it no longer fitted existing conditions. The Commission proposed 132 changes of substance, which were not adopted. Since 1921, 31 amendments to the Constitution have been proposed and 19 adopted. These adoptions have not eliminated the necessity for further revision.

So far as the provisions affecting townships are concerned, it is apparent that substantial changes in the Constitution of 1873 should be made. While no extended treatment of the subject can be undertaken here, some suggestions can be briefly noted.

1. Debt Limitation and Retirement. The constitutional provision limiting the amount of township indebtedness at 2%, or 7% by a vote of the electors is too generous to be construed as a limitation. This provision of Article 9, Section 8, is discussed in the chapter dealing with indebtedness.

The requirement for an annual sinking fund tax sufficient to pay principal and interest within 30 years could be strengthened to advantage by requiring payment of principal and interest within the estimated life of the improvement - in no case to exceed 30 years. (Article 9, Section 10). There appears to be small likelihood of such limitations fulfilling their proper functions under present conditions. Much could be accomplished by a helpful state supervising agency acting co-operatively with township and other local officers in financial matters.

2. Special Township Legislation. The provisions of the present constitution which prohibit the Legislature from enacting local or special laws should be amended as soon as possible. (Article 3, Sections 7 and 20). The Legislature should be free to deal with township problems for certain enumerated and approved purposes.

Since the care of public roads is the principal cost of township government, it appears short sighted to prohibit local laws vacating or altering roads. It is entirely possible that if the Legislature appointed a Commission of township officials to cooperate in a tax-saving study of their problems that many miles of roads would be proposed to be vacated. This criticism applies with equal force to that section of the Article which prohibits the Legislature from enacting local legislation, erecting new townships, changing township lines, or regulating the affairs of townships. This provision is a constitutional barrier to prevent the strengthening and enlargement of many townships that are now too small or without sufficient resources to carry a reasonable tax burden. The provision was adopted in a period of popular disapproval of local and special legislation. Its purpose was to prevent particular bills intended to give special political treatment to one locality.

3. Voting Machines. Another constitutional provision which has harmful consequences and should be removed is the requirement that the Legislature permit voting machines at the option of the electors of the township. (Article 8, Section 7). Most of the second class townships in Pennsylvania are regarded as entirely too rural to permit of effective use of such machines. The Legislature or the court of quarter sessions or both should be authorized to restrict the adoption of voting machines by the electors in districts in which the number of registered voters does not warrant their use. Dissatisfaction has appeared in Westmoreland, Schuylkill and other counties due to the adoption or threatened adoption of the voting machines by electors of rural communities.

Constitutional Status Undesirable. The township does not now have and should not have a constitutional status in the sense that the county so unfortunately has in the Constitution. The township is the creature of the legislature. The Legislature created it and can modify or abolish it. If enlarged to meet present day needs, the township should continue to render its important local services as directed by statutory enactment.

While the best view would seem to be to leave the township out of the Constitution so far as possible, and make statutory provision for it, this method has not been followed in other states. In Ohio, Missouri, Illinois and Nebraska constitutional pro-

visions permitting referendums on the continuance of the township form of government
have been adopted. In these states a dual system of local government may develop since
some counties continue the township organization and others discard it. (See "Recommendations on Township Government" Report No. 3 of the Committee on County Government
of the National Municipal League, by Arthur W. Bromage. Supplement to the National Mun-
cipal Review, February, 1934.)

Modern conditions require lower real estate taxes, to be made possible by more efficient local governments. The township must be enlarged and its revenue supplemented
from state funds. The grant of funds from the state should be conditioned upon local
reorganization to secure more effective administration. No constitutional restrictions
should hamper the legislature in enacting laws for such purposes.

Early Statutory Basis - Some of the early development of township law in Pennsylvania
has already been traced. The first step was the "Frame of Government" of 1682, which
was submitted to the first General Assembly of the Province of Pennsylvania by William
Penn. The Poor Relief Act of 1705 indicated statutory duties of township officers, including Overseers of the Poor and Justices of the Peace.

The first Constitution of the Commonwealth (1776) contained a provision (Section
30) which conferred on the General Assembly the power to make laws affecting townships.
These laws were largely confined to the duties of township officers.

Creation of Townships - There exists in the older records of township law a clear indication of the relationship maintained between the township and the county. Petitions
for change in township area apparently have always been under the County Court's jurisdiction. Townships were created by County Courts under the Act of 1834 (P.L. 537,
Sec. 13-14), now repealed. Various other acts in 1836, 1849, 1857, 1885, 1893, 1909,
1913 and 1917 were in the nature of additions or re-enactments of existing law which
have also been repealed.

A study of the statutory sources of township law was made by the Legislative Reference Bureau as a result of a measure adopted by the 1929 Session of the General
Assembly. This measure authorized the revision and restatement of township law by a
commission of ten members (1929 P.L. 842). The Report of the Commission shows that the
most important statutes and greatest activity in township legislation from the point of
view of number of enactments occurred in the years indicated above.

The final Report of the "Township Law Revision Commission" was made to the General
Assembly in 1931. It proposed a Code for first class townships which was adopted with
changes in the same year. It further reported to the legislature the draft of a proposed code for second class townships.

Subjects Excluded from Second Class Township Code. The Commission did not undertake to
revise general laws affecting all classes of municipalities. It noted that there had
been eliminated from the various municipal codes those provisions on which uniformity
in practice should be preserved. Among the provisions excluded from the proposed code,
either because uniformity was desirable or for some other reason, were the following:
>   Municipal and tax liens
>   Election officers
>   Municipal indebtedness
>   Assessments
>   Collection of taxes
>   Poor districts and school districts
>   Justices of the peace and constables
>   Township reward
>   State Roads, State-aid roads, and private roads
>   Validations of elections, bonds, ordinances and acts of
>   corporate officers.
The Commission undertook to revise and bring up-to-date the law contained in the Township Code of 1917. Bills were also proposed by the Commission on the subject of Township Reward. This was not regarded as proper subject matter for inclusion in the Code,
although it is a subject which relates particularly to townships.

Present Statutory Basis. The second class township code submitted by the Commission in 1931 was adopted by the General Assembly but was vetoed by the Governor. In the 1933 session of the General Assembly it was reintroduced and passed. A number of changes were inserted and the Governor signed the new bill. It became effective July 1, 1933. This code forms the fundamental legal basis upon which second class townships in Pennsylvania conduct their affairs (1933 P.L. 103).

Second class townships include Losch Act townships, both having as their legal basis the "Second Class Township Code" just described. Both types of units are governed by a board of three locally elected residents known as Township Supervisors. The code provides that all townships with a population of less than 300 per square mile are comprehended within its provisions (1933 P.L. 103, Section 201). The principal function of second class townships and Losch Act townships is the maintenance of the roads and bridges in rural communities.

Losch Act Township Statutory Basis. Certain sections of the Code are especially applicable to Losch Act townships (1933 P.L. 103, Sec. 1190-1196). These sections re-enact the so-called "Losch Road Law", authorizing taxpayers of a township (coal mining corporations, for example) to petition the court of quarter sessions to direct the township supervisors to make a contract with a corporation for the maintenance of roads and bridges. Due notice must be given the township supervisors and a bond of $500 per mile for faithful performance [given by sureties approved by the court] must be posted with the township. The company agrees to make and repair highways and bridges, to pay all expenses and officers' salaries, and to protect the township from damage suits. The township supervisors agree not to assess or collect a road tax during the fiscal year for which the contract is made. This contract for road purposes prevails in 26 Pennsylvania second class townships, 16 of which are in Schuylkill County, and 4 are in Luzerne County. Of the other coal-producing counties, Northumberland and Carbon contain two Losch Act townships each, while Columbia contains only one. There is one such township in Lebanon County. The largest Losch Act township in population is Hazel Township, Luzerne County (10,636); the smallest is Cold Spring Township, Lebanon County (31).

CONCLUSIONS

FINDING - The rural township as introduced by William Penn early became a constitutional unit of government. In the Constitutions of 1776, 1790, 1838 and 1873 it is expressly mentioned and provisions governing it are made in the fundamental law. The Constitution of 1873 limits the General Assembly in the enactment of laws governing townships, particularly laws designed to provide for special local needs. These restrictions have become burdensome and greater legislative freedom is urgently needed. The same is true of the recently adopted constitutional provision restricting voting machine legislation. RECOMMENDATION - These restrictions should be modified substantially or repealed.
If such repeal is impossible, amendments should be adopted to (1) authorize local legislation affecting townships on limited subjects (2) put voting machine adoptions under control of the General Assembly; and (3) further restrict township indebtedness.

FINDING - Constitutional status for the township is undesirable.
RECOMMENDATION - In constitutional revision, clauses referring to the structure, organization, functions and officers of the township and its finances should be eliminated.

CHAPTER X

DIVERSITY OF TYPES AND POPULATION

Before considering the administrative organization, functions and finances of the township we shall examine its types, its economic diversity, its population and similar factors. These influences play an important role in the development and functioning of the township government. Subsequent chapters are largely determined by these diversities. Laws must be framed by the General Assembly for the regulation of township affairs in a manner suited to the differences indicated. A clearer picture is obtained of the complexity of local government and of the many obstacles to state regulation if these factors are kept in mind. Differences of race, religion, language, and ancestral tradition also play an important part in community and township activity. Many townships in Berks, Northampton, and Lebanon Counties contain the same sturdy independent German farm folk who have contributed so much to Pennsylvania's development. Similarly, other counties bear evidence of other strains, including Italian, Polish, French, Irish, Scandinavian, Dutch, German, and English. While it is not possible to analyze the exact part played by these various racial elements, their influence upon government is often felt.

Types Of Townships. In population the second class township is the smallest unit of local government prevailing generally throughout the State. Since there are several types of townships, it is necessary to distinguish them as follows:

Type of Township ①	Population 1930	Number of Townships ②
Second Class Township	2,186,877	1,488
Losch Act Township	90,687	26
Special Road District	33 ③	2
First Class Township	472,120	61
Total	2,749,717 ④	1,577

The manner in which Pennsylvania is divided into 1,577 townships is illustrated on Chart 1. The light lines indicate township boundaries; the heavy lines show county boundaries. The 1,488 second class townships (with which this section of the report is chiefly concerned) had a population of 2,186,910 in 1930. This was nearly 23 per cent of the population of the State.

The manner in which township lines have divided the county is further depicted on Chart II. Here Chester County is shown, since it has more second-class townships than any county in the State. The 57 second-class townships in this county require 57 different sets of officers, 57 tax collectors, 57 road maintenance units, etc. The possibility of reducing the cost of government by reducing the number of second class townships to eight or ten per county has been clearly stated by the Secretary

---

①A second class township is an unincorporated rural sub-division of the county containing less than 300 persons per square mile. A first class township is a similar unit containing more than 300 persons per square mile. A popular referendum is necessary to effect a change of status.
A Losch Act Township contracts with a private corporation which undertakes to perform all township functions. A special road district is a road maintenance unit remotely situated with almost no population.

②As of July 1, 1934

③Pine, Clearfield County, Population 33. Population figures were not available for East Fork, Potter County.

④For population distributed by types of local government see Chap. XIII Page I.)

CHART 1.   PENNSYLVANIA TOWNSHIP AND COUNTY LINES - 1982

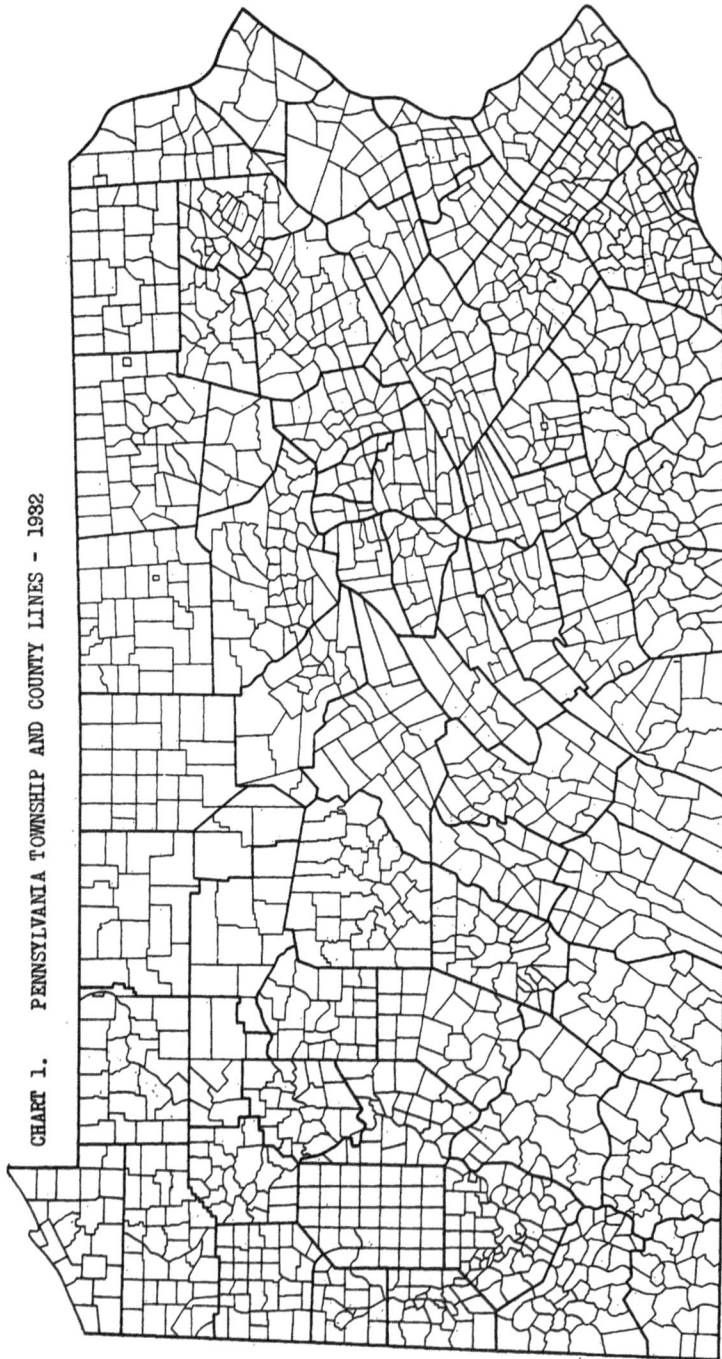

CHART 2

57 TOWNSHIPS IN CHESTER COUNTY

57 ▢ Townships
15 ● Boroughs
1 ◩ Third Class City - Coatesville

of the Township Supervisors' Association, Mr. H. A. Thomson who has strongly advocated a total of not exceeding 500 second class townships. No single change would so greatly strengthen township government.

Pennsylvania has an average of 23 second class townships for each of its 67 counties. The 10 counties having the largest number of townships, and the 10 having the least number of townships are as follows;

County	Number of Townships	County	Number of Townships
Chester	57	Cameron	5
Berks	44	Forest	8
Lycoming	42	Montour	9
Lancaster	41	Sullivan	9
Bradford	38	Union	10
Schuylkill	37	Elk	10
Crawford	35	Mifflin	10
York	34	Pike	11
Luzerne	32	Delaware	11
Butler	32	Fulton	11

Average 23

Economic Diversity. Realizing the significance of the part played by economic diversification in township government, an effort has been made to include every possible type for purposes of study. It is believed that the Survey has included a fairly representative cross-section of township economic life. A majority of these townships were visited during the summer of 1934. The outstanding economic characteristics of the 161 townships which were examined from the viewpoint of economic diversity were as follows:

Economic Characteristic	Number of Townships ⑤
Farming	146
Oil & Gas	12
Coal	35
Wooded	75

To be classified as farming, 50% of the township acreage must be devoted to that purpose. Farming includes dairying. Wooded includes densely forested regions, except where otherwise noted. There is a suburban type of second-class township which should be listed among those designated as exceptional. Examples of this type may be found adjacent to large centers of population in Delaware, Montgomery and Allegheny Counties.

Farming is sometimes conducted on soil which has little or no value and which should never have been used for the purpose. There is a great diversity of taxable value among townships; and Chart 3 shows the township per capita taxable valuation as being less than $400 in 17 counties; between $400 and $700 in 36 counties; and more than $700 in 13 counties.

---

⑤ Many townships are classified three times, for example; coal, wooded, and farm characteristics are frequently important in the same township. A number of exceptional townships might also be noted. Among these were some which were formerly boroughs, such as Frankstown Township, Blair County; Pulaski Township, Beaver County; Putnam, Osceola, and Nelson Townships, Tioga County; Columbus Township, Warren County. Other exceptional townships were Aleppo Township, Allegheny County, having only .7 of a mile of road, and Buck Township, Luzerne County, having no miles of road. Putnam Township, Tioga County consists of .6 of a square mile; and Pulaski Township, Beaver County contains 1 square mile. Derry Township, Westmoreland County, contains 98.8 square miles; while Hempfield Township, Westmoreland County, contains 88.8 square miles. These are probably the smallest and largest townships in the state.

CHART 3.   PENNSYLVANIA SECOND CLASS TOWNSHIPS - TOWNSHIP PER CAPITA TAXABLE VALUATION 1932

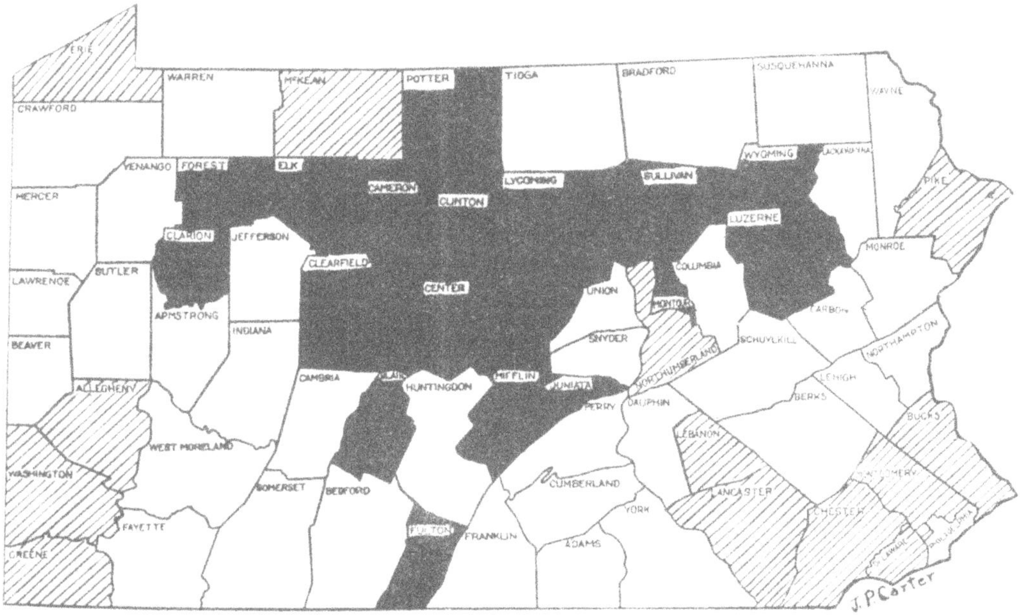

Per Capita Taxable Valuation

    Less than $400
    $400 to $700
    More than $700

Number of Counties

17	Solid	
36	White	
13	Lined	
66		

Population Trends and Distribution. - The population in some parts of Pennsylvania has declined due to the failure of lumbering, mining, and agriculture. In other sections, modern industrial development has brought a growth of population which has sharply changed the economic and social status of the local governments. There are now 9,631,350 persons in Pennsylvania, the population having doubled since 1880. The redistribution of population, with sharp increases in some sections and equally marked declines in others, has followed redistribution of opportunity for employment in industrial, mineral, agricultural, transportation, trade, professional, and other occupations.

The percentage of population in second class townships has steadily declined as the following partial tabulation indicates:

Year	No. of Second Class Townships	Percent of State Population
1930	1,512	23.6
1920	1,508	25.6
1910	1,512	29.0

The number of inhabitants in second-class townships in 1930 was 86,000 in excess of the number of inhabitants in 1910. This is relatively a considerable loss, since other local governments have advanced faster proportionately. Constant encroachment upon the taxable value and population of the second-class townships is encountered from the expansion of adjacent boroughs, cities and first-class townships. Suburban development frequently results in the poorest parts of the rural areas affected being left with insufficient taxable value to support a township government. In numerous instances a section of a second-class township has secured first fire protection, then street lighting, sometimes followed by water and sewer service. As these functions are added, with special taxes for their support, the city or borough nearby is expanding into suburban developments. In time, the city or borough, acceding to the demands of such sections for paved streets and better police protection, annexes the developed part of the second-class township, leaving the remnant of the township to shift for itself. Examples of this are seen in the following changes, listed by the Pennsylvania Department of Internal Affairs and included in the Pennsylvania Manual, 1933. All but a few of those listed were changes of status occurring during the period 1920-1930. They were as follows:

Changes of Status Affecting Second Class Townships, 1920-1930 (Incomplete).

Second class townships losing territory to borough city, and 1st-class township	99
Second class townships subdivided into two second class townships	8
Boroughs reverting to second class townships	8
Part of first class township reverting to second class township	1
Total	116

At least 55 new boroughs have been formed in recent years from townships or parts of townships.

Density of Population. There is considerable diversity in density of population in second-class townships in Pennsylvania. This diversity is shown by counties on Chart 4. There are 24 counties with a township population per square mile of less than 30 persons; 24 counties have from 30 to 75 persons; while 18 counties have more than 75 persons per square mile. This computation excludes the inhabitants resident in first-class townships and the square miles of area of first-class township (Ch. XIII Table I). Philadelphia County has no townships and is also excluded.

Population Range. The range of population among 1,514 second-class townships (including 26 Losch Act Townships) was as follows:

CHART 4 PENNSYLVANIA SECOND CLASS TOWNSHIPS – DENSITY OF TOWNSHIP POPULATION PER SQUARE MILE 1930

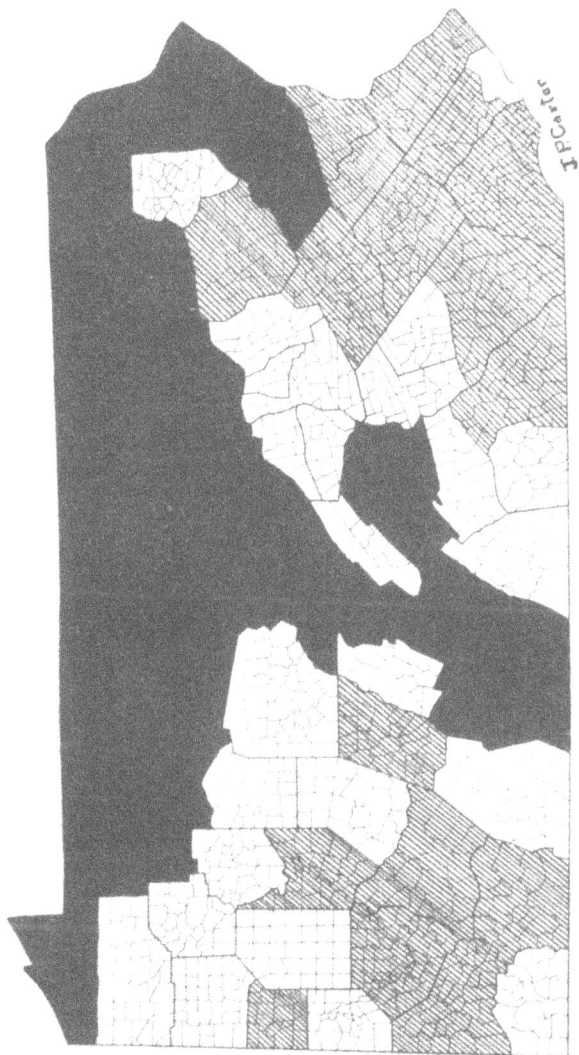

J P Carter

Population Per Square Mile

Number of Counties

Less than 30   24
30 to 75   24
More than 75   18
                      66

Range of Population 1930	No. of Rural Twps.	No. of Urban Twps.	No. of Rural & Urban Twps.
Less than 100	10	-	10
100 to 200	38	-	38
200 to 300	41	-	41
300 to 400	66	-	66
400 to 500	101	-	101
500 to 1,000	491	-	491
1,000 to 1,500	300	-	300
1,500 to 2,500	248	-	248
2,500 to 5,000	-	157	157
5,000 to 10,000	-	53	53
Over 10,000	-	9	9
Total	1,295	219	1,514

A different view of the diversity of population in second-class townships may be had from the following list showing the ten smallest, together with the population for each.

Township	County	POPULATION 1930	1920	1910	1900
Cold Spring	Lebanon	31	22	29	29
Pine	Clearfield	33	35	32	....
Cooke	Cumberland	33	69	162	242
Porter	Pike	55	51	50	53
Elkland⑦	Tioga	60	100	59	66
Portage	Cameron	73	69	143	246
Elk	Tioga	81	562	453	630
Stewardson	Potter	84	102	717	2,299
Pleasant Valley	Potter	99	125	255	276
East Keating	Clinton	106	128	171	225

The population in these ten townships has either remained fixed at a low number, or has been subject to rapid decline during the period 1900-1930. Assuming that one-third of the inhabitants are under age and that an additional third do not vote, one can only conjecture as to how the requisite number of township elective and appointive officers can be secured.

Small Townships in State Forests - Since all of the ten smallest townships in population are found in densely wooded areas of the state, the question arises as to the possibility of special treatment for such townships in governmental matters. All but two of these townships are in counties which include extensive acreage of state forest. One half of Pennsylvania is wooded, including parts of nearly all counties. Thirty-three counties contain state forest lands and nine have state forest parks. Cameron County is occupied to almost one-half its area by state reserves and like other counties its remaining space also contains much private wooded land.

Potter County contains 24 townships and 1 road district (East Fork). The state average as we have seen is 23 townships. Nearly 40% of the county is owned by the state and is covered with state forest. Nearly all of the remainder is wooded, although not state owned. Nevertheless the county is divided into townships irrespective of whether the land is forested or not.

⑦Old Nelson township taken to form Elkland township. Change made since 1910. Cornplanter Indian Reservation is in Elk township, Warren County, but is independent thereof. Its population was as follows:

Year	1930	1920	1910	1900
Population	34	35	77	81

The significance of such facts as these and the palpable unwisdom and waste of conducting full-fledged local government units for trees instead of people may be seen from the following figures.

State Forest in Counties. - The ten counties with the highest per cent of area in State forest land (as of January 1, 1934) were as follows:

County	Per Cent of Area State Forest
Cameron	47.4
Clinton	41.3
Potter	39.2
Union	28.7
Mifflin	20.9
Lycoming	19.4
Pike	17.1
Centre	16.8
Snyder	13.9
Tioga	13.8

State Forest in Townships. - Many townships are almost entirely covered by State forest in addition to those occupied by private wooded areas. With the State owning the land over large districts, why maintain supervisors, auditors, tax collectors, and secretary-treasurers for the few individuals situated therein?

The ten townships with the highest per cent of area in State Forest land as of January 1, 1934, were as follows:

Township	County	Per Cent of Area State Forest
Stewardson	Potter	92.4
Lewis	Union	91.7
Noyes	Clinton	89.2
Cooke	Cumberland	88.7
Elk	Tioga	81.2
Portage	Potter	77.8
Grove	Cameron	77.2
Wharton	Potter	76.7
Abbott	Potter	76.0
Lumber	Cameron	71.8

The two road districts, Pine, Clearfield County, and East Fork, Potter County, have respectively 77.9% and 86.0% of their area in State Forest.

Largest Townships in Population. - The ten largest townships are listed below, together with population for the period 1900-1930. The largest second class township in Pennsylvania is Hempfield, Westmoreland County, with a population of 19,947. Although its population has doubled since 1900, Hempfield is rapidly losing territory to the city of Greensburg and the boroughs of Jeannette, Hunkers, and Youngwood. It is located in the bituminous coal region of western Pennsylvania. In addition to mining, there is considerable farming within its limits.

Township	County	POPULATION			
		1930	1920	1910	1900
Hempfield ⑧	Westmoreland	19,947	18,598	16,926	9,256
Redstone	Fayette	17,211	13,396	9,525	1,187
German	Fayette	16,341	14,582	11,844	5,154
North Union ⑨	Fayette	14,013	12,762	11,968	9,617
Derry ⑩	Westmoreland	12,857	13,419	11,002	9,495
Mt. Pleasant	Westmoreland	10,918	12,583	12,997	10,228
Unity	Westmoreland	10,717	12,269	11,343	9,855
Luzerne	Fayette	10,662	8,790	4,332	1,155
Hazel ⑪	Luzerne	11,336	10,932	11,014	15,143
North Huntingdon	Westmoreland	9,384	8,360	7,800	7,438

Classification of Townships, Based on Density of Population and Referendum. - Townships are divided into two classes based upon population. This classification exists by virtue of Article 3, Section 34, of the Constitution of Pennsylvania which provides:

"The legislature shall have the power to classify. . .townships according to population, and all laws passed relating to each class, and all laws passed relating to, and regulating procedure and proceedings in court with reference to, any class, shall be deemed general legislation within the meaning of this Constitution:" (Article 3, Section 34, Amendment of November 6, 1923). This provision allows classification on a basis of population without violation of Article 3, Section 7, of the Constitution, which limits the legislature in the passage of special or local legislation. The classification adopted by the legislature designates townships having a density of population of more than 300 persons per square miles as first class townships, and those having a density of. less than 300 per square mile as second class. Formerly a township upon reaching a density of 300 automatically assumed first class status. Under the new first class township code, (1931 P. L. 1206, Sec. 205-208), a vote of the qualified electors of a second class township is necessary to effect a change of status. The second class code (1933 P. L. 103, Sec. 225-226 provides that a township of the first class may, irrespective of population, be reestablished as a second class township by a vote of the qualified electors. The classification of townships differs from that of counties, the latter resting total population regardless of density. A result of using density as a basis is that some townships of the first class have a population of less than 1,000 while 62 townships of the second class have a population of more than 5,000 inhabitants!

CONCLUSIONS

FINDING: There is wide variation in the number of second class townships per county, ranging from 57 townships in Chester County to 5 in Cameron County. The average for the entire state is 23 per county. This variation has no relation whatsoever to local government needs and should be re-examined and revised to meet local conditions.
RECOMMENDATION: The number of townships per county should be greatly reduced.

FINDING: Economic diversity plays an important part in township government, but no adequate adjustment of the governmental unit to economic differences has been attempted.
RECOMMENDATION: This diversity should play a larger part in the reallocation of township lines.

⑧ Parts of Hempfield township annexed to Greensburg and Jeanette boroughs since 1910. Hunkers borough was incorporated from part of Hempfield in 1929. Parts of Hempfield annexed to Greensburg City and to Youngwood borough.
⑨ Changed since 1910 - Parts of North Union township annexed to Uniontown City, 1923, 1924, 1925, 1926, and 1927.
⑩ Totals include population of Cokeville borough (409 in 1919; 674 in 1900), annexed to Derry township. Parts of Derry township annexed to Derry and Latrobe boroughs.
⑪ Part of Hazel township annexed to West Hazelton in 1926.

FINDING: Relative to other governments, the per centage of the state population in second class townships has declined substantially during the past 20 years, although the actual number of township inhabitants has slightly increased. The second class townships show wide diversity in range of population, 10 having less than 100 inhabitants each and 9 having more than 10,000 inhabitants each.

FINDING: The smallest townships show a rapid decline in number of inhabitants during the past 30 years; prospects of a continuing decline during the immediate future make township and even county government for these areas inappropriate and wasteful. This is particularly true in townships in which the land is 70% or more state forest.
RECOMMENDATION: The special, peculiar needs of depopulated areas should be met by a new district form of government which in area, functions, administrative structure and financial cost would fit the conditions. Such an area might be drafted, organized and supervised by the central state agency (bureau of local government) which is proposed or discussed at numerous other points in this report.

FINDING: Annexations and incorporations of parts of second class townships continue to occur without regard for the remnants of townships which are left. This is especially noticeable in the larger second class townships contiguous to more densely settled units (First class townships. boroughs and cities).
RECOMMENDATION: The tendency to annexation and incorporation into larger units should be encouraged but should be accompanied by some measure of state planning and supervision. Such supervision should assure the maintenance of a satisfactory minimum of taxable resources in any remnants left from annexation or incorporation. The minimum might be fixed by the central state agency above proposed.

FINDING: Elasticity in the township form as it now exists under the code, allows the electors of each township to determine substantially its form of government. The classification as second class townships, however, of areas with respective populations of 31 and over 19,000 and the fact that some first class townships with a more elaborate government system have a population of less than 1,000 while 62 townships of the second class have each more than 5,000 seems to show that the separate classification of the two types has lost its significance.
RECOMMENDATION: The state bureau of local government, when created, should be authorized to undertake, with the cooperation of township officers of both classes, a reclassification, based on a single township code with elastic features providing for variations to be adopted by the electors. The new classification should be subject to limitations of population-density and total property valuation.

CHAPTER XI

ADMINISTRATION

The general administrative powers of second class townships are set forth in the township code (1933 P.L. 103, Sec. 701). They include the power to sue and be sued in the township name; and to purchase, hold, lease and convey real and personal estate for township purposes.

The code also authorizes the performance of numerous municipal functions. This was intended to accommodate the needs of the more populous townships as well as to provide for future growth and development by existing units.

<u>The Corporate Powers of Supervisors</u> - The corporate powers of second class townships are exercised by the township supervisors. Although the scope of these powers as fixed in the Code is very broad, in actual administration it is quite limited. As a rule road and bridge construction, maintenance and repair are the only powers exercised. Some townships of this class have also undertaken municipal functions such as street lighting and fire protection, but these are exceptions, and even here the outlay for such purposes is slight as compared with that for roads and bridges. Municipal function townships spend large amounts for the surfacing of their macadam roads.

Approximately 2% of the second-class township road mileage is composed of bituminous macadam surface. This is a fair indication of municipal functions as distinguished from road and bridge functions.

The second-class township code provides (702) ‡ that the supervisors may exercise the following corporate powers:
1. To light and illuminate the streets, highways, and other public places.
2. To contract and levy taxes for lighting.
3. To contract and levy taxes for fire purposes.
4. To erect watering troughs.
5. To appropriate money for the expenses of Memorial Day services.
6. To subscribe for publications on roadbuilding.
7. To provide for the removal of ashes, garbage, and other refuse, and to collect reasonable fees for this purpose.
8. To erect and operate traffic lights and signals.
9. To purchase materials, equipment, machinery, tools, and other necessary implements for the construction and repair of roads and bridges.
10. To appropriate money to any forest protection association.
11. To prohibit accumulation of garbage and rubbish on private and public property.
12. To appropriate money or convey land for armories for the National Guard.
13. To exercise eminent domain to secure public lands, easements, and property for the use of the National Guard.
14. To purchase plots of ground in any cemetery for the interment of deceased service men.
15. To maintain and repair memorials.

<u>Outlay for Roads and for Municipal Functions, Illustrated</u> - One of the largest second class townships in township taxable valuation in Pennsylvania is Tredyffrin Township, Chester County ($7,901, 890 - see Chapter 13, Table 4). Since services of a municipal nature so frequently coincide with high taxable value, it may be helpful to examine the extent of such services where the taxable property is most concentrated. The municipal and road outlays respectively as determined from its audited expenditures were as follows:

---

‡ In certain instances in this and succeeding chapters code sections of the second class township code (1933 P.L. 103)are designated by section number only. For example, (702).

	1932		1933	
Road and bridge	$46,496	78%	$38,387	76%
Fire	3,732	6	2,355	5
Street Lighting	3,673	6	3,866	8
Police	5,700	10	5,300	11
Total	$59,601	100%	$49,908	100%

In Tredyffrin Township the item for police is regarded as a road and bridge expenditure. The policeman patrols the Lincoln Highway which traverses the township and although the protection of private property is one of his duties, it is a matter of opinion whether the regulation of traffic in the interest of safety is not his principal function. The municipal services in Tredyffrin Township are from a financial viewpoint over-shadowed by the road and bridge item, which make up three fourths of the expenditures for township purposes, excluding schools. Tredyffrin may be regarded as a township requiring a high quality of road and bridge service. Nearly 27 of the 38 miles of road in the township are of macadam surface. More than 22 miles were classified as "Good" by State Highway Department engineers in 1933. Only 5 miles of unimproved earth road may be found in the township. This township is typical of a small group of relatively densely populated, almost suburban, second class units.

Township Officers - The management of second class township business is entrusted to locally elected officers and to persons appointed by them as authorized by the township code. No special qualifications are required for any office, but only electors of the township are eligible. The township officer is an individual of high calibre. There is little money attached to his office and such as there is, is spent within the township. Although township officers as a group may lack formal education and special training, they have a wide practical experience coupled with shrewd native ability. Electors in second class townships know the family history and practical background of their officers. A few administrative provisions of the second class township code apply to all township officers. These include requirements such as the filing of an oath of office and the petition to declare an office vacant for failure to perform duties.

Bond Requirements - Tax collectors (571 and 427) and township treasurers (530) are the only officers specifically required to file bonds. However, the supervisors may require bonds from roadmasters and superintendents, (514).

Although the township code (535), authorizes banking institutions to act as depositories of township funds, it does not expressly require a bond from such institutions. Such a bond is required under the first class Code and was proposed for second class townships by the drafting commission but was not included in the Code as finally enacted. It is important that this ommission be supplied, so that second class township funds may have the same protections accorded first class townships (1931 P.L. 1206, Sec. 808), boroughs (1933 P.L. 818, Sec. 1006) and other local governments.

Elective and Appointive Officers - The officers of the township may be classified as elective or appointive (1933 P.L. 103, Sec. 402-590) as follows: †

Elective	Term	Appointed by Supervisors
3 Supervisors	6 Yrs.(one every 2 Yrs.)	Secretary-Treasurer
3 Auditors	6 Yrs.(" " " ")	Roadmasters
Tax Collector	4 Yrs.	Engineer
Assessor	4 Yrs.	Solicitor
		Police

In order to distinguish the officers directly responsible to the electorate from those appointed to office, Chart 5 has been prepared.

† Constables and justices of the peace are important elective officers but, having already been discussed in Chapter VI, they are omitted here. An interesting account of the township and of local government in Pennsylvania may be found in "Pennsylvania Government, State and Local", 1933 by Professors J.Tanger and H.F. Alderfer of Pennsylvania State College.

CHART 5. SECOND CLASS TOWNSHIP ADMINISTRATION

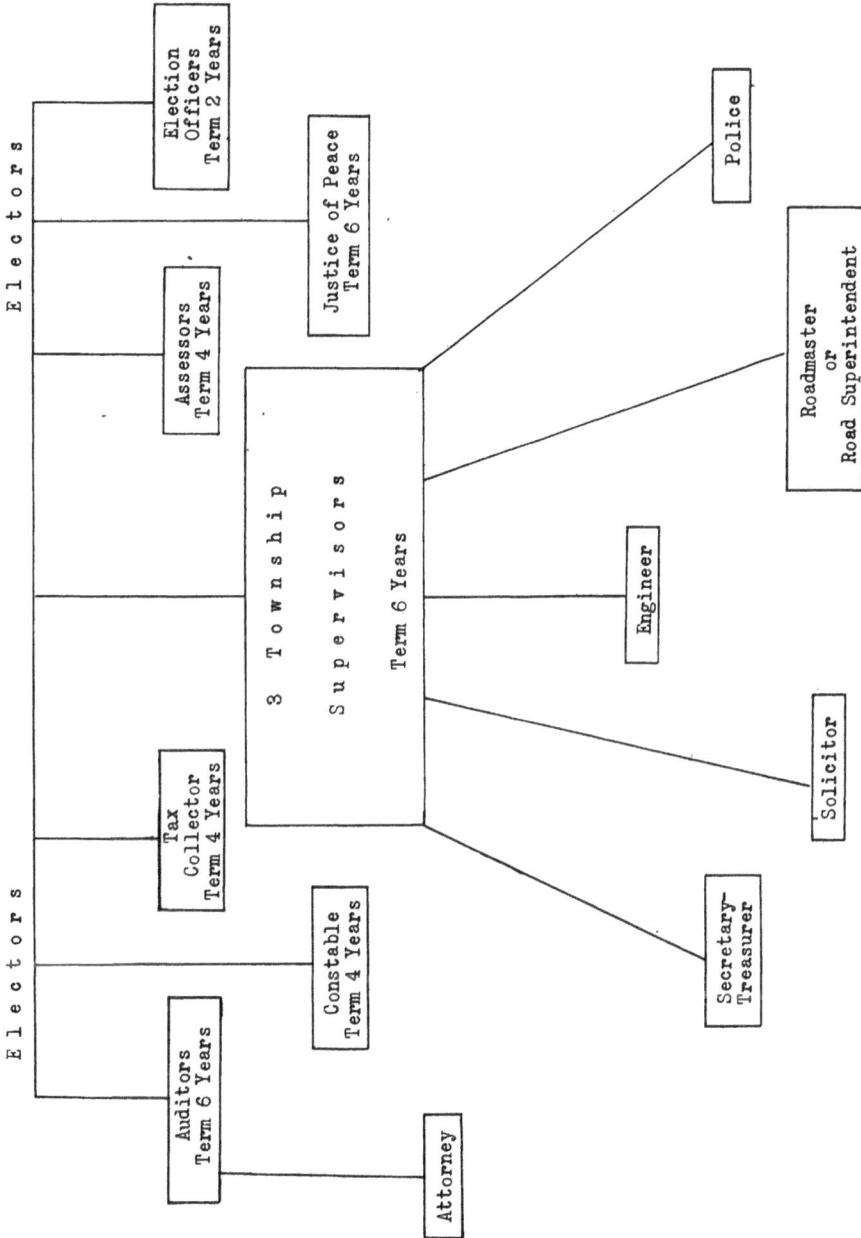

Electors

Election Officers Term 2 Years

Justice of Peace Term 6 Years

Assessors Term 4 Years

3 Township Supervisors Term 6 Years

Tax Collector Term 4 Years

Constable Term 4 Years

Auditors Term 6 Years

Electors

Police

Roadmaster or Road Superintendent

Engineer

Solicitor

Secretary-Treasurer

Attorney

Number of Township Officers - An estimate is here offered of the total number of persons engaged in public offices in second class townships. Conferences with township officers and other persons in various parts of the state were helpful in making the estimate. During the past few years a reduction has occurred in the number of roadmasters employed who were not township supervisors. This was confirmed during visits to individual townships. The estimated personnel may be summarized as follows:

Second Class Township Estimated Personnel

Office	1932	1934
Supervisors, 3	4,548	4,548
Roadmasters, (not Supervisors)	1,400	400
Treasurer-Secretary (a Supervisor)		
Tax Collectors, 1	1,491	1,491
Auditors, 3	4,548	4,548
Solicitors	700	300
Engineers	75	50
Police	100	75
Total	12,862	11,412

Compensation of Township Officers - Township officers for the most part receive inadequate compensation. Many of them are paid nothing whatever for their services. The following data on payments to officers were compiled largely from the 1932 Reports of Second Class Township Supervisors and furnished by the Pennsylvania Department of Highways:

Officer	Amount 1932(a)	Percent of Total Expenditures
Supervisors, Monthly Meetings	$127,814	1%
Roadmasters' Wages	873,079	7%
Secretary-Treasurer	199,176	2%
Tax Collectors	261,295	2%
Auditors	9,388	
Solicitor's Fees	17,129	
Engineers		
Police		
Total	$1,487,881	12%

(Data from Table 3 in Chapter 13)

The compensation of township officers appears to be based upon adequate remuneration for supervisors and tax collectors. Since four out of five supervisors are also roadmasters, the first three items on the list are for substantially the same persons. For attending monthly meetings the supervisors receive nearly $128,000. The supervisors acting largely as roadmasters receive $873,000. Supervisors acting as secretary-treasurer receive $199,176 in salaries. Together, these items comprised 10% of the expenditures of second class townships in 1932 (See Chapter 13, Table 3). It is characteristic of township expenditures that the amounts for particular officers of townships are ridiculously small, yet when extended over the unnecessarily large number of units (1,516 townships) the result is an expensive administrative system. This is the more striking in that a large number of supervisors and other officers are refusing to accept compensation other than for work performed as roadmaster. The average payment per officer listed below is based upon the estimated personnel and the amount expended in 1932, which data have been presented in the two preceding pages.

(a) For second class townships only. Road and Bridge labor, teamsters, etc. not included. Also not included were 3,260 local assessors who were paid $1,203,216 by the county commissioners in 1932. These were omitted because assessment is regarded as a county function.

Number of Officers	Officers Included	Purpose of Payment	Average Amount Per Officer 1932
5,948	Supervisors, roadmasters, secretary-treasurer	Monthly meetings, roadmaster's wages, secretary-treas.2%	$202.
4,548	Auditors	Annual audit	2.
1,491	Tax Collectors	Collection of Taxes	175.

Since the township tax is the smaller part of the tax collection (others being school, county and poor tax) the compensation per tax collector shown above is misleading. The tax collector a few years ago received an average of several times the amount shown per year. This has been reduced since 1932 due to reductions in township tax levies.

Compensation of Supervisors - Supervisors receive compensation for various purposes within limits fixed by the second class township code. These payments are as follows:

Purpose	Amount	Fixed
Attending Meetings	$2.50-$4.00 per diem	By township auditors
Inspection of roads & bridges	"Reasonable compensation"	By township auditors
Working on roads	$3.00-$6.00 per diem	By township auditors

Under the township code not less than 12 nor more than 16 meetings may be held yearly. The inspection of roads and bridges by supervisors is semi-annual. Unless the meeting is held at night, supervisors doing road work may not be paid for inspection, and for a meeting occurring on the same day (1933 P.L. 103, Sec. 515). In many townships supervisors accept no money for attending meetings or inspecting roads and bridges. The amounts for road work are small, less than $100 a year. In some cases a great deal of time is devoted to this duty without compensation. Often the cost in time and responsibility far exceeds the remuneration. Such work is usually undertaken from a sense of civic obligation.

The method of compensating supervisor-roadmasters varies greatly, some townships paying on a monthly basis, some by the week, some by the day, and others by the hour. These methods of payment can all be maintained as to amount within the legal minimum and maximum. In townships in which compensation was on an hourly basis, the common occurrence in reporting road work was for the roadmaster-supervisor to hand to the secretary-treasurer a list of days on which he worked. Some treasurers exhibited a penciled list of dates on the back of an old envelope. They are unable to tell from such a list on which day the work was done or for how many hours per day compensation was due. The list in some cases gave a total number of hours and in others it did not. This practice is not in conformity with the township code, which provides that "Records shall be kept, and reports made and filed, giving the names of persons employed, including supervisors, superintendents or roadmasters, dates on which work was done, with compensation paid to each person and the capacity in which he is employed" (1933 P.L. 103, Sec. 516 e). The conclusion was reached that a loose and unbusinesslike practice in such payments prevailed in many townships.

Duties of Township Supervisors - The more important duties of township supervisors as described by the second class township code (1933 P.L. 103) have been summarized below. The duties commence with the term of office on the first Monday of January following the election of the officer.

Organize as a "Board of Supervisors" on first Monday in January each year (511)
Hold monthly meetings (not to be paid for more than 16 meetings per year) to conduct township business, two of the three supervisors being a quorum (512)
Appoint township officers, including a secretary-treasurer (511), roadmasters, or a superintendent (514), township engineer, etc.
Keep minutes and records of proceedings. Such records must be open to inspection by taxpayers, by representatives of the Department of Highways, and by township auditors (513).
Fix rate of percentage, with approval of auditors, for compensation of township secretary-treasurer (531, 540). Also appoint (optional) and fix compensation

of solicitor (580), Engineer (585) etc.

Designate annually depositories for township funds (535).

May levy annual road taxes for the payment of debts incurred for road purposes (905).

May tax all property or all occupations or both (902).

Exercise corporate powers of second class townships (702).

Prepare a written budget or estimate on first Monday of January and file it with the treasurer (902).

Fix road tax (902) and other taxes.

Supervise, repair and improve highways and bridges in the township (510,516).

Divide the township into one or more road districts, (514) and into sections for road repairs and snow removal (516).

Fix wages and hours of roadmasters, superintendent, and laborers for work on roads and bridges (514).

Employ necessary persons, teams, and implements. Keep records of employees with dates on which work was done, compensation to each person, and capacity in which he was employed (516).

Inspect all highways and bridges annually in April and October. Construct and repair sluices, and culverts; remove all obstructions and snow. Remove loose stones from highways (516).

Attend road meetings and conventions individually (602) or at direction of board of supervisors (516).

Conform to rules and regulations of the Department of Highways (516).

Make contracts (optional) for the repair and improvement of roads of not more than ten miles of road or for over a period of four years (517).

Make an annual report to the State Department of Highways on or before February 1, which shall include the total taxes levied, amount expended on roads, etc. (518).

Make applications to county commissioners for county road aid when taxpayers petition for road improvement (519).

Not to violate, or refuse to carry out, any provision of township code ($50. fine); nor to be interested in any contract or purchase relating to roads and bridges (fine not in excess of $500.; imprisonment not in excess of six months, 520, 521).

The duties and powers of township supervisors in repairing roads are discretionary. They may decide what materials are to be used, where they are to be used, and at what time. They may divide the roads in sections of the townships among themselves and may apportion township funds to be expended in these districts.

Financial Authority of Township Supervisors - The Township Code entrusts broad financial powers to the supervisors. It enables them to levy taxes for various purposes including the following:

	Maximum Millage for Township Purposes
1. Annual road tax.	7
Indebtedness.	2
Upon appeal by the supervisors to the Court of Quarter Sessions and after a hearing the Court may order additional mills not exceeding a total of 16 mills. .	7
Other permissible taxes contingent upon approval by the voters, include:	
2. Construction and maintenance of a lockup...(not specified)	?
3. Lighting streets and highways.	5
4. Purchase of land and construction of townhouse, a tax not exceeding 50% of the road tax. The indebtedness for this purpose may not exceed one half of one per cent of the assessed value of township real estate . . .	8
5. Purchase and maintenance of fire apparatus and firehouse	2
Maximum	31

The above five taxes are specifically authorized by the Township Code (1933 P.L. 103, Sec. 905). The effect of the statutory restriction in mills may be observed as to 22 townships as follows:

1932 Tax Rate in Mills Chapter 13, Table 4.	Number of Townships
0	1
7	3
8	6
9	3
10	6
11	1
13	1
20	1

At least 10% of the annual road tax must be set aside for the payment of existing indebtedness. Taxes which are levied upon parts or districts of any township for particular purposes are not included in the restriction to types of taxes and limitations thereon mentioned above.

Auditors - The qualified voters of each township elect three auditors, one at each municipal election. The only legal qualification is that they may not hold any other township office (411).

The compensation of an auditor is fixed at five dollars per day for as many days as may be required to complete the audit (545). Although no limit is set on the number of days that may be spent in auditing the books in actual practice there is little abuse of this loosely framed provision. As has been pointed out, the average pay of an auditor was only $2.00 over the entire state in 1932. Many of them serve without compensation. The auditors meet each year to settle, audit and adjust the accounts of all township officers. They may issue subpoenas and compel attendance of township officers. They prepare an annual statement of township finances which is published. They may surcharge a township officer for deficiencies in his balance.

Tax Collector - Of the 2,774 tax collectors in Pennsylvania, 1,491 are collectors of second class township taxes. One is elected in each township for a term of four years (414). He acts as a collector of all township, school, county, poor and other taxes levied within the township by authorities empowered to levy taxes (570), except in a few counties in which special laws permit collection of county taxes in townships by the county treasurer. On the first day of each month, tax collectors pay to the township treasurer all moneys collected. Although the code requires the collector to forward a written statement to the secretary showing names, amounts, and total of taxes received each month (913), this is disregarded in many townships. The result in such cases is that the supervisors are deprived of knowledge of monthly collections and do not adjust their expenditures to monthly trends. This budget practice is unnecessarily loose and results in temporary borrowing for current expenditures. Where the collector fails to comply with these requirements or neglects to make complete settlement with the treasurer on July 1 each year he is guilty of a misdemeanor. The penalty is a fine of not less than $100 or imprisonment not exceeding one year or both (1933 P.L. 103, Sec. 911 and 913). The tax collectors' accounts are included in the annual audit conducted by the regularly elected auditors (1933 P.L. 103, Sec. 916).

Compensation of Tax Collector - The compensation of tax collectors is 2% for taxes collected during the period in which a discount is allowed, and 5% on all taxes collected thereafter (1933 P.L. 103, Sec. 912). The average compensation of tax collectors in 147 selected townships in 1932 was $306 per collector for township taxes only. In addition to a percentage on all collections, he is also paid printing and postage expenses incurred in the performance of his duties (1933 P.L. 103, Sec. 914). The cost of tax collection is treated further in the chapter dealing with fiscal problems.

Assessors - The qualified voters of every second class township elect one assessor for a term of four years (412) for the purpose of assessing real estate and other taxables. In view of the fact that the assessors, although responsible to the township electorate, perform duties under county supervision, the discussion of this subject will be treated in the section dealing with county government.

Secretary-Treasurer - The township supervisors at their organization meeting elect a secretary and treasurer, who should be the same person except where a banking institution acts as treasurer. The secretary-treasurer may or may not be a member of the board of supervisors (511). He acts as clerk of the board (540).

Compensation of Secretary-Treasurer - The township treasurer receives as compensation a percentage on money paid by him at a rate fixed by the supervisors with the approval of the township auditors. The combined amount paid to the secretary and treasurer may not exceed 2% of the money paid out by the treasurer, except when the amount would be less than $10.00. The calculation of this maximum compensation may not include a percentage upon money paid out by the treasurer for the repayment of loans, notes, certificates or indebtedness. The township secretary receives such compensation as may be fixed by the township auditors, within the 2% limitation stated above (1933 P.L. 103, Sec. 531). The practice is usually to allow the full 2% to the secretary-treasurer. The amount for both positions in 1932 averaged $132 per township. About 2% of all expenditures of 1,480 townships for road purposes were for compensation of Secretaries and Treasurers (Chapter Xlll, Table 3).

The Secretary-Treasurer of most townships is inadequately compensated under present legislation. The secretary in many instances was found to receive from $7 to $19 per year with a like amount for acting as treasurer. As clerk to the supervisors, he performs most of the detailed work, including preparation and recording of the annual report, warrant books, estimate forms of proposed expenses and cash book. Under the Parkinson Act, additional work was placed upon the secretary arising from correspondence over allocation of roads and agreements between the Highway Department and the supervisors.

In a number of cases the secretary-treasurer's wife was found to be much more familiar with the township finances than he was. She had filled in forms required by the Highway Department and handled township correspondence. Several secretary-treasurers had suffered loss of hearing, others were too advanced in years to leave their homes. Usually the supervisors continued them in office out of respect for their many years of service or because their wives had become proficient in township business.

Treasurer's Duties - Although the duties of the township treasurer and the township secretary are stated separately in the code, the offices are almost always conducted by the same person. The treasurer is required to furnish a bond, with two sufficient sureties or a surety company, subject to the approval of the auditors.

Among the duties of the treasurer are the receipt of moneys due the township, their prompt deposit in a bank and the keeping of distinct accounts of sums received from taxes and other sources. These accounts must at all times be open to the inspection of the supervisors. Orders for payment are on blanks furnished by the Department of Highways. At an annual accounting, vouchers are presented to the township auditors for settlement. Township money collected for a special purpose and paid to the treasurer may not be applied to a different purpose. Failure to observe this provision is a misdemeanor, and upon conviction the treasurer is punishable by a fine of not less than the amount misapplied or by imprisonment not less than three months and not more than one year. If the treasurer neglects or refuses to perform his duties he may be fined in a summary proceeding an amount not exceeding $100, and shall be disqualified from office.

The depository of township funds is selected by the supervisors on the first Monday of January, or as soon thereafter as possible, by a formal resolution. A treasurer and his surety are not liable for losses of township funds caused by the failure or negligence of depository, provided the funds were placed there in the township name at the request of the supervisors. The depository must be a banking institution located within the Commonwealth (1933 P.L. 103, Sec. 535).

Secretary's Duties - The secretary of the township is the clerk of the Board of Supervisors. He keeps a record of the proceedings of the township meetings in a minute book furnished by the State Department of Highways. The minute book and other records and documents of the township are open to inspection by any taxpayer of the township or his agent upon request (1933 P.L. 103, Sec. 540).

Roadmaster's and Superintendent's Duties - The duties of the roadmasters are to:-
 Report monthly to the supervisors on matters in the forms prescribed by the Highway Department (516).
 Inspect highways and bridges as directed by the supervisors, except in April and October (516).
 In the following instances, responsibilities imposed upon the supervisors may also be imposed upon roadmasters or superintendents. A bond may be taken to insure their faithful performance:
  Supervise, repair and improve highways and bridges in the township. Keep them free from obstruction and snow; give necessary directions for this purpose (516).
  Employ necessary persons, teams and implements for maintenance and repair of highways and bridges including snow removal. Provide for supervision of persons employed. Work on roads in person if directed by board of supervisors (516).
  Keep records of employees with dates on which work was done, and compensation to each person, and capacity in which he is employed (516).
  Construct and repair sluices and culverts and keep waterways, bridges and culverts open. Remove loose stones from the highways (516).
  Conform to the rules and regulations of the Department of Highways (516).

Road Districts and Roadmasters - Immediately after their organization it is the duty of the township supervisors to divide the township into one or more road districts, employing a road master for each district. They may appoint a superintendent for the entire township. A bond may be required for the faithful performance of the roadmaster's duties. Such officers are subject to removal by the supervisors, who fix their wages as well as the wages of all road laborers. Township supervisors usually employ themselves as roadmasters. The law specifically permits this and the extent to which it prevails is shown by counties in Chapter XlV, Table 2. No roadmaster may be the husband, father, brother, son, stepson, father-in-law or son-in-law of a township supervisor. Where desirable, two or more townships may join and appoint the same person as road superintendent (1933 P.L.103, Sec.514).

Township Supervisors as Roadmasters - The township supervisor acting as roadmaster is one of the outstanding points of controversy in township administration. In 1932, of 4,548 second class township supervisors, 3,148 were acting as roadmasters. At the organization meeting in January the three supervisors, having divided the township into road districts, elect themselves (in 70% of the cases) as roadmasters. This is the most remunerative township office, and it enables a politically minded officer directly to control patronage in the selection of labor and teams for road work. The principal criticism has been not so much from the abuse of patronage as from the desultory labor of some roadmasters, who have been known to take a scythe and cut back weeds from the road on days when they find it unprofitable or too inclement to work on their farm. This has not been well received by neighbors less happily situated, and it is the more irritating if the scythe is not diligently applied.

 Of the many supervisors who were interviewed, a number thought that the supervisor should not be permitted by law to act as roadmaster. Some agreed that in an enlarged township the roadmaster should be a man qualified by experience in road work. Although the time has not been reached when a qualified engineer can be required on township road work, the employment of only qualified men must come soon. Most of those interviewed believed the criticism unfounded, or at most confined to a few cases. Administratively, the best course is to separate the two offices, making the supervisors a policy-forming and ordinance-making body ineligible for road work of any kind.

Township Engineer - The township supervisors are authorized to appoint by majority vote a township engineer and fix his compensation. The appointee must be a registered civil engineer. He performs duties as designated by the supervisors relevant to the construction, maintenance and repair of roads, streets, sewers, bridges, culverts, etc. All complicated work of this type undertaken by the larger townships is based upon plans, specifications and estimates prepared for the supervisors by the engineer. The latter certifies the exact time of commencement and completion of improvements. The certificate is filed with the township secretary, who is required to enter it in a book kept for the purpose. This certificate is conclusive upon the parties in subsequent disputes (1933 P.L. 103, Sec. 585-87).

Township Solicitor - The township solicitor is elected by a majority of the supervisors on the first Monday of January, or as soon as possible thereafter. His compensation is also fixed by them. He is required to be a person learned in the law, and serves for a term of one year. No other counsel is provided by law for township purposes, except for the auditors. The solicitor prepares bonds, obligations, contracts, leases and conveyances, to which the township is a party as directed by resolution. He prosecutes and defends all actions for and against the township. It is also his duty to furnish the supervisors with his opinion in writing upon questions of law (1933 P.L. 103, Sec. 580-82).

Township Police - An elector, a resident of the county, may be selected as a policeman, upon a petition to the court of quarter sessions of 25 taxpayers of the township or of two or more adjacent townships. It must be represented to the court that the safety of citizens and security of property make the appointment necessary. If satisfied with the reasonableness of the application, the court directs the supervisors of the township to appoint the officer to serve at will. The number of policemen, their term of service and their compensation are fixed by the court. Where more than one township is concerned, the court fixes the contribution to be made by each.

Such officers possess the powers of policemen of the cities of the Commonwealth. When on duty, each is required to wear a badge with the words "township police" and the name of the township. Jailers are required to receive persons who may be arrested for any offense against the laws of the Commonwealth within the township. With the court's approval, the township supervisors may, out of the township funds, provide police with uniforms, equipment, and means of transportation. Such police may not accept any fees or compensation in addition to the salaries paid to them as police for services rendered pertaining to their offices. They may accept public rewards and legal mileage for travelling expenses. (1933 P.L. 103, Sec. 590-594.)

Relations Between Townships - The interrelations of townships with each other, while infrequent, usually evoke favorable comment from township officers and businessmen. Of the relationships observed, five were of considerable importance. These were joint retirement of indebtedness, care of roads and bridges on boundary lines, attendance of officers at conventions, rental of road machinery to other townships, and consolidated township school districts.

> The inter-relationships observed among townships were as follows:
> Appointment of township police by two adjacent townships.
> Joint contribution to indebtedness where a township has been divided.
> Care and maintenance of roads and bridges on boundary lines of
>   adjoining townships.
> Attendance of officers at county and state road conventions and
>   meetings.
> Connecting with water, sewer, street lighting, or fire protection
>   of adjoining township(occasionally with adjoining borough or
>   city).
> Relocation of roads and bridges by joint action.
> Rental of road machines, rollers and equipment from nearby town-
>   ships.
> Consolidated township school districts.

Only in connection with the controversies over highway maintenance on boundary lines was dissatisfaction expressed. Nearly every county appears to contain a township somewhere that is involved in a minor dispute concerning a short stretch of (boundary) earth road. In many cases state highway engineers appear to have exercised a good influence in the settlement of such disagreements. The determination of township road mileages was necessary to their work under the Parkinson Act (1933 P.L. 1520).

The lack of integrated structure on such common points is one of the principal causes of unnecessary waste in township administration. In recent years the rental of road machines, rollers and equipment from nearby townships has been of increasing importance.

This report recommends in a later chapter that the number of townships per county be greatly reduced in order to integrate further the existing functions which are now partly duplicated and partly co-operative. Relations between townships are treated under separate subjects in sections on Roads and Bridges, and Consolidations.

Township and County Relations - Township and county relations were more satisfactory before 1931 than thereafter because counties then had more funds for road and bridge improvements. A system of county-aid augmented the state-aid received by the township. Withdrawal of both county and state funds occurred about the same time. On the following points the relations of county and township seem closely identified:

> County road aid to townships.
> Poor relief in 51 counties on county-unit basis.
> Election officers in townships are paid by and
> responsible to the county.
> Assessors elected in townships but responsible to
> county commissioners.
> Filing and collection of delinquent township taxes
> by county treasurer.

The tax collection system, including filing of tax liens, is discussed elsewhere in this report. County aid is at present of relatively little importance. It is defined as a form of township revenue in chapter 13.

Modern Conditions Require Modern Administration - Township administrative methods are as ancient in origin as the township itself. In his opinion in Milford Townships Supervisors Removal (291 Pa. 46,) Mr. Justice Simpson indicated that the early township law was administrative law; that is, it dealt with township officers and their duties. (See Chapter 9)

During the period in which Pennsylvania's population has changed from a predominantly rural to a predominantly urban character, a remarkable transformation has occurred in the technique and methods of administration. Some of these new methods have already made their way into township management. Instances in which progress has occurred include the use of uniform reports furnished by the State Department of Highways, the partial supervision of bonded indebtedness by the Department of Internal Affairs, the requirement of minimum qualifications for public school employees, and the use of the secretary's minute book and the treasurer's voucher system for the recording of township business and township funds.

The township, despite the progress which has been noted, is still antiquated in methods, equipment, and treatment of personnel. As to methods, a good budget system, unit road cost records, and a competent auditing system have yet to be adopted. These must probably await the development of a larger township area if they are to be fully effective. The same may be said for equipment. The typewriter, adding machine, and in many instances the telephone, are seldom available. Townships are too small for the use of such equipment and in their present status it should not be expected. The objective should be to enlarge the township to a point where modern methods and their savings can be effected.

The township has made the least advance in the treatment of personnel. Too many officers (3 Supervisors, 3 Auditors, Tax Collector, Assessor, Election Officers, Constable, Justices of the Peace, School Board and, in some counties, the township Poor Board) are locally elected. Of the first three named only the supervisors need be elected. The audit should be made by a licensed municipal accountant. Whether the qualified accountant is selected for all the townships in a county by the Auditor General, the county commissioners, or by the township electors of the county seems of less importance. For the larger counties, 10 townships should be ample. Here again Mr. H. A. Thomson's recommendation of a total of not more than 500 townships in the state deserves endorsement. Although many township appointive officers are capable, no method exists for retaining them in office or for compensating them adequately for their services. In clerical offices, permanent employment, appointment based on qualifications, participation in municipal pension and retirement plans, and the wider use of county and state meetings by township employees other than supervisors, should be made applicable to the enlarged township. In federal, state and local government, including township government, the basic administrative problems are the same, They include leadership, co-operation, morale of employees, fiscal supervision, and the development of employee health, initiative and ability. Modern conditions have given rise to a tremendous desire for improvement and economy. This means that efficient local employees must be given a broader field of activity in an enlarged local community and a greater degree of security in their work.

<u>CONCLUSIONS</u>

ADMINISTRATION

<u>FINDING</u>: Corporate powers of township supervisors are broad enough to permit a high degree of urban and municipal development. Township administration in 1932 was composed of approximately 12,862 local officers. The cost of township administration in 1932 was 12% of the total expenditure for road purposes, as reported to the Pennsylvania Department of Highways.

<u>FINDING</u>: No qualifications are required for any office, and only electors of the township are eligible.
<u>RECOMMENDATION</u>: Certain offices, such as auditor and roadmaster, should be filled by applicants selected with qualifications appropriate to the work to be performed.

<u>FINDING</u>: Banking institutions acting as depositories of township funds are not specifically required to file a bond to secure payment of township deposits and interest.
<u>RECOMMENDATION</u>: A bond or collateral covering township deposits and interest should be required by law of banking institutions acting as depositories of funds of the township, except to the extent already covered by Federal Deposit Insurance.

<u>FINDING</u>: Too many township officers are elective, and township administration is thereby unnecessarily complicated.
<u>RECOMMENDATION</u>: Only township supervisors should be elected.

<u>FINDING</u>: Practically all of the township compensation for officers goes to supervisors and tax collectors.
<u>RECOMMENDATION</u>: Compensation of township officers should be readjusted so that the secretary-treasurer (to receive not less than $5 per month and not more than $50 per month) and the auditor (not to receive in excess of $10 per diem) are paid in a manner commensurate with services rendered.

<u>FINDING</u>: Roadmasters in a number of townships merely inform the secretary-treasurer how many days they worked each month without specifying the date or length of day, type of work, etc.
<u>RECOMMENDATION</u>: Roadmasters should specify the date, number of hours, and kind of work in their reports to the secretary-treasurer.

<u>FINDING</u>: The rapid advances in transportation require higher standards in township road work. Some formal training in engineering principles is essential for road work at least in the more densely populated areas. The time is rapidly passing when township supervisors may use road work as a fill-in to be done at times when weather and soil conditions are unsuitable for farming. Some roadmasters are now full time employees with many years of practical skill in road maintenance.
<u>RECOMMENDATION</u>: Township supervisors should not be permitted by law to act as roadmasters. All road work should be in charge of persons qualified by practical experience and, if possible, by engineering training. The experienced and trained roadmaster as a type must be increased in number, and made secure by more permanent tenure. In return, higher qualifications should be necessary to secure such positions.

<u>FINDING</u>: Relations between townships develop an integration of governmental services and administration in centers of growing population.

<u>FINDING</u>: Relations between townships and the county were more satisfactory before 1931 because of the county road-aid in greater use at that time. At the 1934 meeting of the State Association of County Commissioners, resolutions were adopted favoring the county unit for township road purposes. Since such proposals if adopted would abolish the second class township, the resolutions did not promote township-

county cooperation.

FINDING: Township administration based on locally elected officers has changed relatively little in Pennsylvania in the past 200 years. Under modern conditions new administrative devices are required which are too costly for the small township.

RECOMMENDATION: An enlarged township should be developed in which the saving due to unit road cost records, mechanical equipment and other modern devices together with a trained personnel can be economically utilized.

CHAPTER XII

FISCAL PROBLEMS

In the management of any business, private or public, certain operations are fundamental; among these are the accounting and recording of all transactions, the raising of funds, the development of budgetary control, the expenditure of money for authorized purposes and the independent audit of financial operations.

Accounting and Reporting. - In second class townships, books and records are kept by locally elected officers. In many cases this work is well done, but in a large number of townships the records are practically illegible. No qualifications are required by law for the office of township secretary-treasurer. In the great majority of cases he is a farmer but occasionally a miner, insurance agent, grocer, store clerk, school teacher, or undertaker is elected to the office. The lack of legal requirements for the office reflects itself in the township's financial records.

Annual Report To State Highway Department. - The township supervisors are required to make an annual report to the State Department of Highways on blanks furnished by the Department. The report must be forwarded (and usually is) before the first day of February, but a sufficient number are delayed until late in the year to prevent a complete compilation of township receipts and expenditures. The supervisor's report must be attested by a notary. It includes the amount of tax levied for road purposes, the amount of road taxes collected, and expenditures for maintenance, repairs, and new construction. Also included is the total number of miles of roads. The township code specifies that this be included, together with the names and addresses of the supervisors, the secretary-treasurer, and such other "matters and things as the State Department of Highways may require" (1933 P.L. 103, Sec. 518.).

The report should, but does not, contain separate mileages for road maintenance, repairs and permanent improvements. After specifying that the annual report form forwarded by the Highway Department shall include the amount expended for maintenance repairs, and new construction, the township code provides, "the number of miles of roads roads thus made" (Sec.518). The Highway Department has probably omitted this from the form because of the lack of generally accepted definitions of maintenance and new construction, and the manner in which these merge into repairs. Engineers do not have fixed, i. e., distinguishable standards for these types of work. Since township supervisors or secretaries who fill out the forms for the Highway Department are not engineers, further difficulty is encountered. It would seem that the mileage distinction should be attempted, however crudely. At the present time, maintenance costs per mile per township are figured by the Department and by township supervisors without reference to the number of miles maintained or newly constructed, although the latter greatly exceeds in cost. In some townships where only a third of the mileage was actually worked in 1933 the maintenance cost per mile was figured on the basis of the total number of miles in the township. In such townships the actual maintenance cost per mile would be three times the usual figures.

Minutes and Records. - The supervisors are required to keep minutes of their proceedings and such other books as they may find necessary. All books are open for the inspection of any taxpayer or any representative of the Department of Highways. They must be submitted to the township auditors at the close of the year. All books, papers and accounts must be delivered by the township supervisors to their successors (1933 P. L. 103, Sec. 513).

The Pennsylvania township officer is untrained but public spirited, and frequently serves without compensation or for a nominal sum. Invaluable assistance to the township officer has been furnished by the State Highway Department, which has done much to prevent errors in township records by supplying books and forms. Among the records used are:

Warrant Book   ---   This is a checkbook. Warrants are issued for township purchases, officers' compensation, and labor costs. Each warrant has a corresponding numbered numbered stub.

Annual Estimate form ---   This form budgets receipts and expenditures. Two copies are furnished, one being forwarded to the Highway Department, indicating estimated receipts and expenditures.

Monthly Payroll Report---   Shows type of road work, where performed, and names of persons paid, by months.

Annual Financial
 Report Blank   ---   A sworn statement of the itemized receipts and expenditures, as audited for the past fiscal year is shown on this report. It also shows the names and addresses of township officers, itemized list of township road equipment, township assessment and tax data. Miles of roads and number of roads and bridges, township assets and liabilities are also given.

Tax Collector's Warrant--   A grant of authority to the tax collector from the township supervisors advising him that the tax has been levied and authorizing him to proceed with the collection.

Treasurer's Bond Blank---   This is a blank form on which the township treasurer's bond is executed.

Treasurer's Book   ---   Record of cash receipts and expenditures, itemized by months.

Minute Book   ---   Contains minutes of the meetings of township supervisors. Kept by the township secretary.

The books and forms described above may be termed the records of the township. If the minutes of the meetings of the township supervisors are written in a manner to show the various motions and their sponsors, a record of the votes, a monthly statement of the receipts and expenses from the treasurer, and a statement as to the progress of tax collections from the collector, they are complete indeed. Exceptions sometimes occur, due to illegible handwriting of an otherwise competent secretary, and occasionally a person, unqualified by temperament for details and book work, is selected for the job.

Purchase orders for township supplies and materials are furnished by salesmen. Such purchase orders vary widely, as every firm has a different type of order blank.

The tax collector's monthly report blank could be furnished to the township supervisors in the envelope forwarded by the Highway Department in December containing the blanks and forms already mentioned.

Tax Collectors Monthly Report Blank --- Such a blank(suggested by several township supervisors) should be furnished by the Commonwealth, to enable the tax collector to forward to the township supervisors a formal statement of collections, in time for their monthly meeting.
From such a monthly statement, together with the treasurer's report, the supervisors could determine whether deposits were being regularly made, thus decreasing the possibility of negligence or default by the collector. The submission of this blank would assist the supervisors to adjust expenditures within probable revenues in accordance with changing conditions. In the township code (Sec. 913) such a written report is required monthly, but no special blank is specified. Some collectors omit the report entirely. Regularity of reporting is essential in the management of any business

in which the bill collection department is separated from the expending department.

Errors Made by Township Officers. - A number of errors were noted in the annual reports of township supervisors. Some of these appeared during tabulation by the Department of Highways. Others were ascertained from interviews.

Among those most frequently observed were:

1. Incorrect addition of receipts or expenditures on the Annual Financial Report to the state Highway Department. This was a natural error, and it must be remembered that township officers did not have adding machines and clerical assistance, such as a larger office might possess. Such errors total several hundred dollars each year in the 1516 townships.

2. Incorrect computation of 2% allowed by the township code to the secretary-treasurer of the township. In some cases, indebtedness is incorrectly included in the total on which the per cent is taken. Errors did not involve more than a few dollars in individual cases, but the total amounts to several hundred dollars each year for all townships 1933 P. L. 103, Sec. 531 and 540.

3. In complete write-up of secretary's minutes. In some townships only the briefest statement is made in the secretary's minute book, many of the essentials previously listed being omitted. The authors of the various motions and the vote thereon are not indicated. The treasurer's balances from the previous meeting, and copies of the resolutions fixing the tax rate, of the budget, and of the auditor's report are among the items more frequently excluded. In the case of taxpayers' appeals, auditor's surcharges or other litigation, the secretary's minutes become court records of great importance. In the same county there will be found townships in which the minutes are complete, concise, and well written, while in an adjoining township, there will be found inaccurate, illegible, and incomplete minutes.

4. Carrying forward temporary indebtedness by renewal each year. This is a violation of the purpose, if not the letter of the law which provides "payable on a certain date not exceeding one year from the date of issue" (1933 P. L. 103, Sec. 903). The construction placed upon this section of the code results in (1) borrowing in anticipation of taxes to be collected for the current fiscal year for highway purposes, (2) liquidation of the loan from collected taxes near the end of the fiscal year, (3) negotiation of a new loan based upon the uncollected taxes at the outset of the next succeeding fiscal year. When such a procedure is followed year after year, there is in effect a large outstanding floating indebtedness which is technically of a temporary nature, but is actually permanent. To this authority for temporary borrowing for operating expenses, there has been added a provision which further complicates the debt situation by permitting bonds to be issued for current operating expenditures. Bond issues are permitted during the next five years based upon uncollected taxes under the Mansfield Act (1933 P. L. 813).

5. Failure to use sinking funds or to keep them intact when established. Although many townships have no debts, either bonded or temporary, and many others have debts which they can discharge within the year, and require no sinking fund, there are some townships which have issued bonds but have failed to establish sinking funds for liquidation of the debt. This practice is contrary to the constitution of Pennsylvania which provides "Any. . .township. . .incurring any indebtedness shall, at or before the time of so doing, provide for the collection of an annual tax sufficient to pay the interest and also the principal thereof within thirty years" (Article 9, Section 10). This subject is discussed in Chapter 15. In several townships sinking fund moneys were utilized to supply the deficiency arising from uncollected taxes. In one instance $5,000 was withdrawn in 1933 from the township sinking fund for current expenses. On this subject, the township code provides "The supervisors may, by resolution, transfer moneys from one fund to another except from the fund allocated for the payment of debts" (1933 P. L. 103, Sec. 902).

Nearly all bonds issued are serial bonds which the township can retire out of current revenues, thus reducing both principal and interest at the same time. Of the larger townships that have financed macadam roads demanded by their more densely populated character, practically all have bonds outstanding. Some of the smaller townships could not set money aside out of revenues of the past few years regardless of what the statutory requirements might be. In such cases, as one township supervisor expressed it, "We just use common sense."

The Township Tax Procedure. - In Pennsylvania there are three steps in connection with local property taxes in second class townships. First the determination of the value of the property by the assessor; second the fixing of the tax rate or levy by the township supervisors or tax levying body; third the collection of the tax thus levied by locally elected tax collectors in each township.

Special Levies By Court - The court of quarter sessions is authorized by the township code to issue a writ of mandamus directing the supervisors to collect a special levy sufficient in amount to pay debts due. This may occur when it is shown to the court that the debts due exceed the amount which the supervisors may collect in any year by taxation (1933 P. L. 103, Sec. 907).

Budget Procedure. - The second class township operates on a fiscal year commencing the first Monday in January. All receipts, disbursements, contracts and purchases are chargeable to the fiscal year in which they are made. At the organization meeting held on the first Monday of January the supervisors make a written estimate of the money required for the ensuing year. The purpose of the estimate is to aid the board in determining how much road tax to levy. Such estimates are required to specify the amount of money necessary for:

1. Highway maintenance and improvements.
2. Repair and construction of bridges and culverts.
3. Purchases and repairs of machinery and tools.
4. Payment of debts or other miscellaneous purposes.

The township supervisors may, by resolution transfer money from one fund to another, but not from the fund allocated for the payment of debts (1933 P. L. 103, Sec. 902).

The estimate made up in January is a partial township budget for road purposes, and the items listed above may be characterized as the budgetary requirements. As we have seen, the township supervisors are authorized to levy separate taxes for roads and bridges, indebtedness, township building, lock-up, street lighting, and fire protection (1933 P. L. 103, Sec. 905). Why should the township code require a budgeting of the road tax and not include the other taxes listed? Although in most townships, the road tax is the only township tax, there is no reason why estimates and budgetary requirements should not apply to all types of taxes levied in townships having other taxes in addition to road taxes.

The township "written estimate", as required by the township code, is not a budget. A genuine budget would be prepared thirty days before the commencement of the fiscal year. The township estimate is adopted "before their organization meeting in January or as soon thereafter as practicable". A properly adopted budget should be posted in a public place prior to its adoption and a meeting of the tax levying body should be held, to which electors of the township should be invited by public notice. Changes in the budget should be permitted only during the last few months of the fiscal year. In emergencies, appropriations should be made for specifically authorized general purposes by a unanimous vote of the supervisors.

Unless the budget is properly coordinated with the accounting system and includes the use of control accounts, improvement in either the budget system or the accounting system cannot be given its full effect. It is the tying of the two systems together which affords assistance in keeping expenditures within revenues.

Budget Appeals by Taxpayers. - The clauses of the Township Code providing for appeals from the budget seem particularly onerous. The code authorizes taxpayers with property assessed as taxable amounting to 25% or more of the township valuation, to appeal to the court of common pleas. This appeal must be taken within fifteen days after the supervisors have completed the formation of the budget and fixed the tax levy. One of the petitioning taxpayers must file a bond of $500 with the Court for the payment of costs if the Court should assess them upon the petitioners. A copy of the petition requesting the Court to examine the necessity for and reasonableness of the various items in the township budget together with the specifications of objections must be served by the petitioning taxpayers upon the township supervisors (1933 P. L. 103, Sec. 908).

But if no budget has been adopted what has the taxpayer to appeal from? An illustration of this has occurred recently. When certain taxpayers of East Hanover Township, Dauphin County, attempted to appeal the tax levy of the township supervisors to the Court of Common Pleas of Dauphin County they found that the supervisors had not prepared a budget in 1934. The absence of the budget left the taxpayers with no legal basis for appeal. The taxpayers requested an injunction because they had received no notice of the 6 mill road tax which was levied. This the court denied on the ground that the taxpayers were given notice when the tax was levied. Assuming a budget had been adopted would the court have confined the appeal to the road tax? Or would it extend the appeal to taxes other than the road tax that might be involved and concerning which a budget does not seem to be required? The legal proceedings for taxpayers seem entirely too involved, costly and ineffectual to provide an adequate remedy.

Fiscal Year. - At present the fiscal year for townships, counties, cities and boroughs coincides with the calendar year. The fiscal year is the operating and accounting period of twelve months over which township receipts and expenditures are spread. To secure maximum benefits all local governments should operate on the same fiscal year, thus permitting all taxes to be mailed on the same bill so far as possible. The fiscal year for school districts is now from July 1st to June 30th. In eliminating the lack of uniformity among municipalities as to the fiscal year it would seem desirable to advance the beginning of the fiscal year of townships, counties, cities and boroughs from January 1st to July 1st. This could be done over a six year period, advancing the collection of taxes one month each year. To secure saving in mailing tax bills, either the date of collection of taxes or the date of assessment must be changed. If the tax collection date is advanced gradually the presend need for temporary borrowing in large amounts may be reduced. Such borrowing is now required to carry the township until taxes are collected six to nine months after the fiscal year which began on the preceeding January 1st. With a budget adopted January 1st and with the collection of taxes well under way prior to July 1st funds would be available for township purposes early in the proposed fiscal year, which would commence July 1st.

Mandatory Expenditures.-Although the maintenance, construction and improvement of roads and bridges in the township appears to be a mandatory duty imposed upon the township supervisors by the General Assembly, there is such a wide discretion in the powers exercised by the supervisors for such purposes that they cannot be deemed mandatory here. A mandatory expenditure, as used in this discussion of the subject, is an expenditure removed by statute or constitutional provision from the control of the township supervisors. Such expenditures in second class townships fall in general groups including salaries and commissions fixed by state statute, wages and salaries fixed by the township auditors, debt and sinking fund requirements as specified by the state constitution and statutes, and compensation insurance as required in the workmens' compensation insurance law.

Although the principal items of this type in second class townships are related to indebtedness, there are others of considerable importance. In 1932 the mandatory expenditures were as follows:

Purpose	Amount 1932	Per cent of Total
Sinking funds for bonds	$406,038	3%
Redemption of bonds	545,849	4
Interest on bonds	375,301	3
Notes Maturing	2,392,879	19

Purpose	Amount 1932	Per cent of Total
Interest on notes	$295,733	2
Supervisors monthly meetings	127,824	1
Compensation insurance	37,691	
Auditors fees	9,388	
Annual supervisors convention	26,411	
Tax Collectors' compensation	261,295	2
Secretaries' compensation	100,446	1
Treasurers' compensation	98,730	1
Roadmasters wages (largely supervisors)	873,079	7
TOTAL	$5,550,664	43%

As indicated above, 43% of the expenditures of Pennsylvania second class townships in 1932 were beyond the control of the township supervisors. These expenditures are shown by counties in Chapter XIII on Table 3.

Among the items which may be regarded as mandatory but which were not included in the above list were the following:

> Care of soldiers' memorials
> Burial plots, soldiers
> Maintenance of prisoners
> Advertising and printing
> Cost of views
> Police salaries
> State tax on bonds
> Tax Collector's printing and postage expenses

These items were not included either because the amount was less than 1% of the total expenditures for 1932 or because, like legal advertising and printing, they are included in "Miscellaneous".

In 1933 the State of New Jersey suspended so-called mandatory laws. In speaking of this, Governor (now U. S. Senator) Moore stated on March 27, 1934, that such expenditures fixed "a great many of the expenditures of local governments and dedicated a large part of incoming revenues to specific uses. By this action, the authority and responsibility for controlling local expenditures were restored to the local governing bodies, where it properly belongs."

It is important that the General Assembly of Pennsylvania restore to local governments a greater measure of control over their budgets. This should be done by a statute relaxing mandatory requirements as to the compensation, fees and percentages permitted various township officers. While debt requirements cannot and should not be relaxed, a measure of relief may be accorded on other points. In the 43% of total expenditures which were mandatory in 1932, debt items accounted for 31% and the remaining 12% were non-debt items.

Differences in Tax Rates - The diversity among townships in the tax-rate for the various purposes (township, schools, poor, etc.) is very great. While an extended analysis is given in the next Chapter at Table 4, a few typical illustrations may be noted here. They are expressed in mills.

		Tax Rate in Mills - 1932				
Township	County	Township	Schools	County	Poor	Total
East Goshen	Chester	8	9	4.	-	21.
Loyalsock	Lycoming	10	27	8.5	5	50.5
West Brunswick	Schuylkill	8	11	10.	-	29.
Osceola	Tioga	11	16	7.	6	40.
Hempfield	Westmoreland	8	28	8.5	-	44.5

Differences as to the tax rate in mills levied for township purposes only are equally striking. This millage is levied chiefly for the maintenance, construction and repair of roads and bridges, as follows:

Township	County	Tax Rate in Mills for Township Purposes 1932
Juniata	Blair	15
Cambria	Cambria	8
East Taylor	Cambria	10
East Caln	Chester	2.5
Middletown	Delaware	7
Eden	Lancaster	10
Manheim	Lancaster	6
Hunlock	Luzerne	8
Lake	Luzerne	10
Loyalsock	Lycoming	10
Pine	Lycoming	20
Brown	Lycoming	0
Upper Merion	Montgomery	6.25
Upper Salford	Montgomery	10
Monaghan	York	15

Among the many causes producing these differences are:

1. the difference in assessment practice; a high tax rate in mills may be offset by an assessment at a low percentage of the true property value;
2. the varying number of miles of road in different townships;
3. the difference in number of streams requiring more or fewer bridges.

Per Cent of True Value Assessed - Differences Among Townships: The percentage of true value assessed presents discrepancies in different townships which are to be expected in the present assessment system. Figures indicating this percentage are subject to greater error than others because they are secured from local tax collectors who may not in all cases be familiar with the data. In the following five townships the percentage varies not only between townships but between different types of taxes in the same township:

Township	County	Percent of True Value Assessed 1932			
		Township	School	County	Poor
East Goshen	Chester	75%	70%	50%	%
Loyalsock	Lycoming	33	20	33	33
West Brunswick	Schuylkill	90	90	40	-
Osceola	Tioga	50	100	50	66
Hempfield	Westmoreland	75	90	75	-

If we examine only the township tax the range in percent of true value assessed is equally great. The following were selected to indicate the range in the percent of assessment of township tax only:

Township	County	Percent of True Value Assessed for Township Purposes 1932.
Cambria	Cambria	100%
East Taylor	Cambria	50
Middletown	Delaware	35
Eden	Lancaster	100
Manheim	Lancaster	50
Hunlock	Luzerne	90
Lake	Luzerne	45
Loyalsock	Lycoming	33
Upper Merion	Montgomery	33
Upper Salford	Montgomery	90

When one considers that property is required by law to be assessed at sale value, the diversity in different sections of the state, within particular counties, and within the same township on the several taxes levied therein, seems striking.

Township Tax Delinquency By Counties 1932. The township tax delinquency varies greatly in different counties. While a township tax is technically delinquent if unpaid by October 1st (at which time a penalty of 5% is imposed) nevertheless, taxes may be regarded as delinquent for 1932 if they are unpaid and outstanding December 31, 1932. Among the counties in which the township tax showed a high and low delinquency were the following:

County	Percent of Unpaid 1932 Township Tax of Total Tax Levied	County	Percent of Unpaid 1932 Township Tax of Total Tax Levied
Beaver	35.7	Bradford	18.9
Blair	44.1	Elk	10.3
Center	37.1	Fayette	15.4
Clearfield	53.3	Greene	17.7
Cumberland	35.4	Lancaster	17.0
Jefferson	37.3	McKean	9.4
Lackawanna	38.3	Monroe	17.4
Mercer	35.8	Schuylkill	18.4
Mifflin	36.4	Pike	8.9
Snyder	39.1	Wyoming	3.9
Sullivan	38.8		
Tioga	38.1		

Township Tax Delinquency By Townships 1932. - The percent of unpaid 1932 tax (as of December 31, 1932) of the total tax levied was obtained for the township tax in each county. Figures include only the township tax levied, which is mainly devoted to road and bridge purposes. School, poor and county taxes are excluded. The townships on a selected list with the highest and the lowest percentage of unpaid 1932 taxes were the following:

Township	County	Percent of Unpaid 1932 Township Tax of Total Tax Levied	Township	County	Percent of Unpaid 1932 Township Tax of Total Tax Levied
Antis	Blair	51.6	Halifax	Dauphin	7.3
Freedom	Blair	50.8	Rush	"	6.6
Juniata	Blair	47.3	South Hanover	"	5.9
Logan	Blair	54.3	Lehigh	Lackawanna	4.5
Chest	Cambria	64.2	Newton	"	4.6
White	Cambria	49.6	East Earl	Lancaster	6.1
Gray	Greene	51.7	Plymouth	Luzerne	7.7
South Abington	Lackawanna	46.8	Loyalsock	Lycoming	5.7
			Lower Moreland	Montgomery	4.1
Sadsbury	Lancaster	46.5			
Pine	Lycoming	49.3	Hegins	Schuylkill	3.4
Beaver	Snyder	63.3	Porter	"	4.4
West Beaver	Snyder	47.6	Hillsgrove	Sullivan	2.9

The above statements of the percentage of unpaid 1932 tax of the total tax levied shows the stronger and the weaker groups of townships with respect to tax delinquency. Those familiar with the diversity of economic conditions among townships, will note that those listed have special significance. Economic conditions in particular townships as to local industries are of special importance in such a study. For example, the two townships in Schuylkill County which are on the list with the lowest unpaid tax of the total tax levy, are both fortunate in possessing an important mining undertaking which is in operation and paying its taxes. Most of the taxes in such townships come from a single corporation. Similarly, in townships in which the percentage of unpaid tax is high, the failure of local industries is equally important. With the collapse of one or several large operations, the real estate in the township suffers directly from the failure in taxes from those operations, and indirectly from the employees, storekeepers, merchants and others who were dependent upon it.

Weight of Property Tax Appears Greater in Rural Areas. - Studies of taxation in Pennsylvania lend credence to the view that real estate taxation is relatively heavier in rural farm, mining and forest areas. In 1926 an analysis of rural taxation in Pennsylvania† showed that agricultural communities over the entire state bear a proportionately larger share of the tax burden. The real property tax cannot be evaded, and in the case of unimproved lands, which cover a large part of the state, it cannot be easily shifted. The delinquency in second class townships is almost entirely a delinquency in real estate tax. It has been pointed out in an earlier chapter that the Pennsylvania townships' principal economic characteristics are farms, mines, and forests. In such areas the real estate tax appears to fall with greater weight upon the taxpayer in periods of declining revenue.

Other Tax Studies. Surveys were made in 1925 and 1927 in which it was concluded that this burden "ought not to be increased and should be decreased and equalized"‡. Nevertheless the tax burden continued to increase, and 1931 another inquiry was conducted on local taxes in Pennsylvania*. In this study, the sources of which were independent of previous inquiries, it was again shown that the rural farm sections of the state were weighted down by a heavy real estate tax which could be neither shifted nor evaded. In 1932 the collection of delinquent real estate taxes was thoroughly examined in a report§ in which the following conclusions were drawn: "in many counties, no real estate has been sold for delinquent taxes for a great many years. In some counties a sale is advertised and then adjourned, and sometimes readjourned in order to give the delinquent additional time in which to pay his taxes. In those counties where sales have been held regularly every other year, as provided by law, four things are noticeable:

(1) the number of properties sold is increasing with every sale held;
(2) the amount of taxes due on each property is increasing every year;
(3) the percentage of property bought by the county commissioners, because there is no private bidder, is increasing with every year;
(4) the number, or percentage of redemptions, on the part of the owners is decreasing every year."

Attention to the growing difficulty of real estate tax collection has not been confined to Pennsylvania. In 1934 a survey of taxation in New York state★ indicated that, as in Pennsylvania, there has been a steady decline in the percentage of collectible real estate taxes. The decline began in 1927, and had continued with increasing rapidity up to 1934.

Causes of Tax Delinquency - It is not intended to make an extended analysis of the causes of tax delinquency. This subject was dealt with in detail at the 46th Annual Meeting of the American Economic Association at Philadelphia in December 1933, and also at recent meetings of the National Tax Association. The following causes were advanced at these meetings:

1. Expenditures beyond the means of the community. In some communities expenditures have been wasteful, extravagant, or due to unwise assumption of public services, such as construction of roads, sewers, public buildings, etc.

2. Incorrect assessment procedure. This is also regarded as a fundamental cause. Where property is assessed above its true value (as is generally the case at the present time) the resulting levy is more onerous, and often leads to foreclosure.

3. Tax collection procedure. This is the chief cause of tax delinquency. Inefficient collection encourages postponement of payment on the part of the taxpayer.

---

† F. P. Weaver and Clyde L. King "Some Phases of Taxation in Pennsylvania" Bulletin of the Pennsylvania Department of Agriculture, December 1926, page 4 (out of print).
‡ Pennsylvania Tax Commission reports to the General Assembly, 1925 report, page 28.
* F. P. Weaver "The Rural Tax Problem in Pennsylvania", Bulletin 263, Pennsylvania State College, Agricultural Experiment Station, March 1931, pages 6 and 8.
§ Blake E. Nicholson "Collection of Local Taxes in Pennsylvania", Bulletin of the Pennsylvania Department of Revenue 1932, page 214.
★ New York State Commission for the Revision of The Tax Laws, Fourth Report February 15, 1934, page 51.

In some counties, prior to the depression, sales had not been held for several years, as Dr. Nicholson has pointed out.

4. Popular election of tax collectors. Although one of the best paying local public offices, the duties involved are decidedly unpopular. The collector is beset by requests for time extensions, remission of penalties, and withholding of names from advertised lists, especially before election. Such requests are often difficult to refuse.

5. Ineffective lien system. Where the difficulties involved in final settlement of tax liens are great, the procedure expensive and time-consuming, and the titles secured at foreclosure sales questionable, the system requires fundamental change.

6. Delinquencies cause further delinquencies. When a larger percentage of taxes becomes delinquent, a heavier burden is imposed upon those who pay promptly. A sufficient tax must be levied upon those able to pay to meet governmental expenditures despite the shortage caused by the falling off in tax collections. In this manner, as the burden becomes heavier, additional delinquencies result.

7. Relation to income value of the property. Although property taxes are assessed by law on a (fictitious) sale value, they were designed to be paid from the rents or income value of the property. A readjustment of assessments nearer to an income basis would revitalize property values. The lack of relation of present assessments to the income derived from the property makes the tax burden unduly heavy for the taxpayer.

Should Assessments Be Related To Income Value? - Property taxes were not intended to be paid as is so frequently the case today out of capital savings. They are a direct tax and are levied upon a fictitious sale or capital value, but intended to be paid from rentals. Probably there is no tax in which the relation to income is of so little significance as in the present local tax on real estate. If the tax were adjusted to the income value of the property, the problem of tax delinquency would assume a more reasonable ratio, properly 5% or 10% of the tax levy. Property taxes in England were placed on an income value basis in 1601. The "Statute of Elizabeth", otherwise known as the Poor Relief Act of 1601, placed upon every real property owner a tax which today is the foundation for an Englishman's entire contribution to public local expenditure. The purpose of the English assessment is to find the annual value of the property; not what it will sell for, but what it will produce in annual rentals, averaging the years and deducting repairs and insurance necessary to keep the property in rental condition. This annual value is an average, and represents a theoretical net income. Under English acts of recent date, industrial and freight transportation properties are given a rebate of 75% of the tax. Agricultural real estate other than farm houses is tax exempt. Railway properties privately owned are assessed by a specially constituted Railway Assessment Authority.

This brief note of the English assessment system⊙ is included to indicate a different approach to the delinquency problem which has become acute in recent years. It opens an interesting avenue of thought concerning the proper relationship of property values and tax levies.

Should There Be a Real Property Tax Limit? - Studies of the tax limit on real estate are in substantial agreement as to its harmful effects. The limit usually adopted is a flat maximum millage for each type of municipality. The average millage for 1,481 second class townships in Pennsylvania was 9.88 in 1923. Nine years later (1932) the average millage for 1,435 townships was 9.33. Between 1927-1930, figures compiled by the Pennsylvania Department of Highways indicate that millage averaged between 11 and 11.4. The effect of a flat limitation can be observed in Chapter 13, Table IV. A Lycoming County township in 1932 has levied a township tax of 20 mills, its assessment being at 33% of true value. Many other townships shown on this Table have levied township taxes of 10 mills. A Snyder County township levied a 4 mill tax in the same year,

---

⊙ See John J. Clarke, "The Local Government of the United Kingdom", new edition, 1934.

but assessed at 100% true value. The average rate for the townships visited was 8 4/10 mills, and the average valuation was 64% of true value. If we assume the enactment of a flat rate of 8 4/10 mills by the Legislature as a maximum township tax, many townships with complicated budgetary problems or a large bonded debt would face such immediate and drastic curtailment of services that actual har : might result. In some cases excessive temporary borrowing has been resorted to by municipalities in states where the tax limit has been adopted. A limitation on the tax rate can be partially evaded by raising assessments. This enlarges borrowing capacity and permits bonds to be sold to pay for the temporary borrowing. The sale of bonds would be permissible in Pennsylvania under the terms of the Mansfield Act already referred to. The tax limit makes it almost impossible for a thrifty municipality to include a capital improvement in current operating expenses. It runs counter to the usually accepted maximum that in a state as diversified as Pennsylvania no general rule can be made applicable to all townships. The possibility of adjusting such a tax limit to various types or population groups of second class townships is a very complicated matter. The enactment of a limitation would probably result in an increase of interest rates for municipal borrowing, since it imposes a hard and fast restriction where formerly there had been support for municipal credit. Where the limit results in drastic curtailments of service, further depreciation of property may result, due to the lack of adequate police and fire protection. Despite these objections, the tax limitation has strong support because it is easy to understand. Its operation is regarded as simple and its scope is so broad that it is generally accepted as a measure conducive to sound economy on a state wide scale.

Procedure In Collecting Tax Levy. - Although the initial steps for the collection of the tax levy are relatively simple, the later stages are involved in a maze of complications. The township supervisors make a duplicate or list indicating the amount of road and other taxes levied against each taxpayer of the township. This duplicate is delivered to the tax collector on or before May 1; together with a warrant authorizing him to collect the same. Taxpayers who pay their taxes before June 1 receive an abatement of 5% of the amount of tax.

For taxes paid to the collector between June 1 and October 1 full payment must be made. All taxes unpaid on October 1 are regarded as delinquent and a 5% penalty is added to the amount of tax. The collectors' first step is to give public notice as soon as possible, posting it in ten different places. The notice states that the duplicate has been delivered to him for collection. Within 30 days after receiving the duplicate, each tax collector must notify persons on his list of the rate of taxation, property valuation, occupation valuation, and full amount of tax. The notice must include a statement of the time and of payments; also the place, periods for abatement and penalty (1933 P. L. 103, Sec. 911 & 913).

What Constitutes Good Tax Collection Procedure? - The importance of an efficient collection program can not be emphasized too strongly. Each step should be clear and well defined for the taxpayer's convenience. The following brief resume of a sound collection procedure is suggested: it is in general harmony with views expressed in recent examinations of the subject. ①

1. The budget should be adopted and the tax rate determined within one month of the time of the completion of the assessment.
2. Bills should be sent to the taxpayers as soon as they can be prepared. One month should be adequate in most cases.
3. The tax or its first installment should be due and payable within one month of the receipt of the tax bill. Taxes not paid at this time should be deemed delinquent. In ordinary times a penalty of 10% and interest at 7% should be added to the bill.
4. Some months should be permitted to elapse to enable the delinquent taxpayer to make payment. Before twelve months have elapsed (nine months from the date taxes are due) steps should be taken for foreclosure.
5. Where the number of foreclosures mounts in times of emergency the court of common pleas should be authorized to stay the sale for limited periods. (1933 P. L. 826) This Act should be extended, since it expires March 31, 1935.

---

① This follows in part the procedure outlined by Prof. F. R. Fairchild in "The Problem of Tax Delinquency", American Economic Review Supplement, March 1934. Also see Prof. H. D. Simpson "Tax Delinquency, Economic Aspects", Illinois Law Review, June, 1933; M. H. Hunter, "Legal Provisions Affecting Real Estate Tax Delinquency, Tax Sales, and Redemption" in Bulletin of the Bureau of Business Research, University of Illinois, June, 1933.

6. Foreclosure procedure should be the same as for an ordinary mortgage foreclosure and the property should be sold. This last is the present practice in Pennsylvania. In some states a tax certificate or tax lien of doubtful value is issued.

7. A period of one year during which a property might be redeemed by the owner upon payment of sale costs interest, penalties, taxes, should be permitted. Where a taxpayer is solvent, but finds difficulty in paying his taxes for lack of cash on hand, the court should be authorized to extend tax payments over a five-year period. Payments should be made with interest at regular intervals where such an extension is granted.

8. If no purchaser bids an amount equal to taxes, penalties, interest, and other charges at the foreclosure sale the local unit should purchase the property in the same manner as a purchase by a private person.

Annual Settlement By The Collector - The tax collector keeps an account of all money collected, marking "paid" on his duplicates at the name of each taxable, together with the amount of tax and date on which paid. A complete settlement of all taxes is made by the collector with the secretary-treasurer not later than July 1st of the year following the year of the levy. The collector is credited in this settlement for unpaid taxes which have been returned or for which liens have been filed. Until this settlement is made the collector is not entitled to the duplicate of any succeeding year, and the supervisors must appoint a different collector where settlement has not been secured. Professor Nicholson states that in 1930 he found "one instance where the (school) collector admitted that no duplicate of his had been finally settled for seven years.....and townships.....have had similar experiences."♦ A fine of not less than $100, and imprisonment not exceeding one year, or both, are the penalties provided for a collector who neglects or refuses to meet requirements (1933 P. L. 103, Sec. 913).

Tax Collector's Commission. - There were 2,774 tax collectors in Pennsylvania in 1932. Of this number 1,491 were in second class townships. The commissions received by all collectors in the state in 1932 amounted to $4,500,000. This includes commissions on both the current and delinquent taxes. It also includes salaries paid tax collectors who had been put upon a salary basis. Of this total $268,000 came from second class township taxes. In 1931-1932, tax collectors' commissions declined 10 and 28 per cent respectively. The commissions in selected townships for township taxes only follows:

Township Tax Collectors' Commissions 1932

Township	County	Total
Tredyffrin	Chester	$1,694
Hempfield	Westmoreland	1,332
Whitemarsh	Montgomery	1,298
Manheim	Lancaster	1,216
Upper Dublin	Montgomery	1,069
Logan	Blair	1,010

The list of second class townships in which collectors receive the smallest remuneration of any on the list of townships visited is as follows:

Township Tax Collectors' Commissions 1932

Township	County	Total
Beaver	Snyder	$ 7
Portage	Potter	9
Hillsgrove	Sullivan	11
McHenry	Lycoming	16
Putnam	Tioga	16
Rush	Dauphin	18
Brown	Lycoming	20
Pine	Lycoming	20
East Caln	Chester	23

♦ Blake E. Nicholson "Collection of Local Taxes in Pennsylvania", Pennsylvania Department of Revenue, Bulletin No. 1, 1932, page 43.

As has been previously pointed out, the commissions discussed above arose only from collection of township tax and this is a smaller part of the commission. The following computation shows the commissions received by township collectors on various taxes:

Tax Collectors' Commission 1 9 3 2

Tax	Tredyffrin Township Chester County	Whitemarsh Township Montgomery County	Hempfield Township Westmoreland Co.
Township	$1,694	$1,298	$1,332
School	3,811	3,046	3,684
County	452	804	1,249
Poor	---	---	---
Total	$5,957	$5,148	$6,265

The above commissions are shown for all types of taxes levied in the second class township, including township, school, county, and poor taxes in Chapter 13, Table IV, Column 17. For purposes of contrast, the collectors' commissions are shown for the same types of tax in three townships in which it was very low:

Tax	Commission 1 9 3 2		
	Brown Township Lycoming County	Portage Township Potter County	Putnam Township Tioga County
Township	$20	$ 9	$16
School	23	25	81
County	31	10	41
Poor	19	9	16
Total	93	53	154

The widespread dissatisfaction which exists with regard to the collection system in Pennsylvania has been a subject of comment in every tax inquiry. In expressing this view in connection with townships, the Township Law Revision Commission stated ▲.

"The Commission heartily favors the modernizing of the local tax laws and the adoption of sound business and administrative methods. The commission calls attention to the fact that more than $5,000,000 is expended annually for the collection of local taxes."

In Cameron County in 1932 the average income of the five collectors for collecting township tax was about $28.40 per year, while in Westmoreland County the average commission received by the 23 collectors from township tax was about $476. The average commission of the collectors of the entire state for collecting township tax was about $180 per collector. If the assumption be correct that commissions earned by the collector for collecting township tax constitute 29% of his earnings from all types of taxes, then the collectors in Cameron County would have a total income of approximately $98 each; those in Westmoreland County, $1,650 each. The average for collecting all second class township taxes (county, school, township, poor) for the entire state would be $620. This amount is not unduly high, it is true, but the real question is as to the need for the large number of collectors now employed under an indefensible system. The reasons for a centralized plan to be operated by the county treasurer have already been discussed in the chapters on the County.

Is A Qualified Auditor Required? - The elected auditor is often not fitted by experience for a careful checking of local receipts, expenditures, balances, etc. It is believed to be to the best interest of the township to have an independent audit by a qualified accountant. The objection to such a proposal is mainly one of cost. A reduc-

▲ Township Law Revision Commission, Final Report to General Assembly, page 10

tion in the number of townships per county would require fewer audits. Auditors could be selected from lists of licensed municipal accountants (licenses are now required of municipal auditors in New Jersey and other states) in the same or nearby counties, and their audit should include all townships in a particular county. This would require the appointment or selection of the auditor by an outside authority such as the State Auditor General, the county commissioners, or by popular vote of the electors of the townships affected.

Audit Procedure - There has been considerable discussion as to what kind of an audit is necessary, and how extensive an audit may be expected from local officers without training engaged at a salary of $5 per diem. Although some of the steps indicated below as part of audit procedure are inapplicable to local units as small as many Pennsylvania second class townships, the same principle has been used here that was employed so well by the Township Law Revision Commission in drafting the Code for second class townships. The Commission drafted a code adequate for all townships including the most advanced in municipal functions. In the present discussion of accounting, bookkeeping, budget, and audit procedure, the same plan has been used. The plan is to suggest fundamental requirements which might be applicable to all--even the most advanced.

   Audit procedure is divided into two parts (1) the audit of receipts, (2) the audit of disbursements, as follows:♦

Audit of Receipts:

   Compute assessed valuation.
   Determine if current tax list is closed before end of fiscal year, eliminating errors duplicated entries, and tax sales.
   Note if accounts receivable were actually received.
   Examine accounts of any improvements made, to see if collections were secured wherever possible.
   Read minutes of proceedings of township supervisors, to see if the budget was adopted and taxes levied as required by law.
   Examine records of county treasurer for collections of any money, fees, or taxes due during fiscal year.
   Examine dockets of Justice of Peace, to see if fines payable to the township were fully paid to township treasurer.
   Check any fees paid to county or township officers for township purposes.
   Check interest on bank deposits (2%) with bank statements and with passbook entries.
   Check receipts from temporary borrowing and from bond issues with the minutes of the proceedings of township supervisors and with bank deposits, passbook and cash book entries.

Audit of Disbursements:

   Check all expenditures as entries in the Township Disbursements Book.
   Examine vouchers indicating expenditures by township officers, noting whether items indicated are for township purposes.
   Total bills on file, to see if charges to township are properly computed.
   Examine warrants drawn, to determine if they check with bills rendered.
   Check tax collectors' commissions and penalties by actual computations.
   Examine police records for any expenditures.
   Bills and Claims:
     Verify amount, purpose, and claimant.
     Is signature of claimant on bills?
     Is signature of township officer acknowledging the claim included?
     Where bills are rendered in corporate or firm name, does the individual name also appear?
     Are claims itemized as to quantity, price, grade, discounts, etc.?

─────────────
♦ The audit of receipts and disbursements here proposed is based in part upon the favorable results secured from the instructions issued for the local government audits in New York State as performed by the State Comptroller's office.

Traveling Expenses:
Does bill rendered show places visited, purpose of trip, and number of miles each way?
Are hotel bills accompanied by a receipt made out by the hotels?
Is the amount due each person made out on a separate claim?
Are various kinds of expense including hotel, mileage, and meals distinguished and the amount for each itemized?
Bills submitted in lump form should be rejected by the auditor, who has power to surcharge officials permitting payment of improper bills.
Prepare a list of outstanding warrants.
Examine principal and interest payments on both funded debt and temporary loans to to see if they are paid in accordance with the record of maturity indicated and in the proper amount due.
Note if bonds paid during the fiscal year are cancelled and filed.

Audit of Contracts and Purchases. - Before concluding the audit, township contracts should be examined with reference to provisions of the Township Code prohibiting township officers from being directly or indirectly interested. Materials and supplies are occasionally ordered from firms or stores of which officials of the township are owners or interested partners. These transactions are often indirect, as with a firm in which the official is a silent partner or is owner of the property. Such practices are prohibited by law, and should be discontinued.

Audit of Surety Bonds - Some bonding companies prefer to insure only for the highest amount held by the collector or treasurer at any one time. This has been carried so far as to bond tax collectors for the total tax delinquencies over a period of years, with consequent increase in bond premium. A number of townships reduced the cost of bond premiums by deputizing several fiscal agents, thus dividing the amount of the bond. Surety bonds covering the estimated amount of funds on hand are required of both the township treasurer and tax collector.

Bonds of tax collectors and township treasurers are subject to numerous errors, of which the following may be said to be most frequent:

1. Failure to renew the bond upon re-election, by a continuation certificate indicating new period of liability. The township should require a new bond covering a re-elected officer; in case the amount of liability may have been reduced, the charge will be correspondingly lowered. If the liability of officer has increased, the township will be protected by the new amount instead of to the extent of the old figure in the original bond.
2. Omission of proper title or office of person bonded.
3. Failure to have signature of the principal affixed to the bond.
4. Misspelling of names and initials of the bonded employee.
5. Misdating the period of liability which should coincide with the term of office.

A few townships have made a local bank the fiscal agent. Such institutions should be required to give bond, and a complete audit would require the examination of the bonds and of the securities. The bonds should have affixed the seal of the institution, and should state the securities offered. These should be listed and described by the auditors. They should be renewed each time the appointment as fiscal agent is renewed.

CONCLUSIONS

ACCOUNTING SYSTEM, ANNUAL REPORT, RECORDS

FINDING: Accounting is not performed in a manner calculated to yield the best results. Although township accounts are usually classified by separate funds, there is no use of such simple devices as control accounts, accrual basis, double entry bookkeeping, etc.

RECOMMENDATION: That a simple but uniform system of municipal accounts, similar to those now employed for rural municipalities in New York, be installed under the guidance of a state agency.

RECOMMENDATION: That a state agency be required to publish an annual report on the finances of local government, including townships, based upon uniform accounting and adjusted to types of municipalities.

FINDING: The Annual Report form of the State Department of Highways does not contain separate mileages for road maintenance, repairs, and permanent improvements performed in the township.
RECOMMENDATION: It is believed that more accurate cost per mile figures would be obtained if based upon the actual number of miles. In some instances the cost per mile would greatly exceed costs as now computed.

FINDING: The Township Code requires a monthly report from the tax collector.
RECOMMENDATION: That a form be provided by a state agency on which this report should be made.

FINDING: The minutes of township supervisors' meetings show marked defects in contents, quality of reporting, and legibility.
RECOMMENDATION: Minutes may become court records and therefore should show the various motions, their sponsors, a record of the vote, a monthly statement from the treasurer showing receipts and expenditures, and a statement from the tax collector showing the currently collected taxes.

BUDGET

FINDING: A "written estimate" is required for township road purposes.
RECOMMENDATION: That a genuine budget including all township taxes and expenditures be required, and that this budget be coordinated with a simple accounting system on an income and expenditure basis for purposes of budgetary control. An amendment to the Township Code would be required to extend the budget to other township taxes than the road tax (1933 P.L. 103, Sec. 905).

FINDING: Appeals from township budgets are difficult, delayed and expensive. They can only be had in those townships in which a budget is adopted.
RECOMMENDATION: That a simple and inexpensive procedure be devised for taxpayers' appeals. Such appeals should be originally taken to a state administrative agency instead of being involved in the formality and delay of judicial proceedings.

FISCAL YEAR

FINDING: The present fiscal year as of January 1st necessitates temporary borrowing in many townships until sufficient taxes are collected for operating purposes. The lack of uniformity in the local government fiscal year prevents consolidation of tax bills for purposes of economy.
RECOMMENDATION: That the fiscal year be advanced six months, one month per year for six years. The fiscal year should commence on July 1st when sufficient funds from tax collections may be substituted for loans.

MANDATORY EXPENDITURES

FINDING: 43% of the expenditures (non-debt items 12%, debt items 31%) in 1932 were beyond the control of township supervisors.
RECOMMENDATION: That state legislation requiring specific use of township funds be suspended, permitting local control over purposes of expenditure other than indebtedness

TAX DELINQUENCY

FINDING: The rapid increase in delinquency (almost doubled between 1926 and 1932) discloses fundamental weaknesses requiring immediate correction.
RECOMMENDATION: Steps should be taken to adjust assessments nearer to the income value of the property. More efficient tax collection and more effective execution of lien procedure are urgently required. Adequate relief from this distressing situation can be secured only by abandonment of the present extremely decentralized tax collection and assessment systems.

REAL PROPERTY TAX LIMIT

FINDING: The adoption of a fixed number of mills as a flat limitation on the amount of real property tax would probably apply equally to townships accustomed to a tax of 15 mills and to those accustomed to a tax of 4 mills. The average millage for all second class townships in 1932 was 9.33 mills.
RECOMMENDATION: A flat limitation on real property taxes should not be adopted. It is entirely too simple and arbitrary a solution for this complicated problem. It lacks flexibility and adjustment to many diversified conditions.

TAX COLLECTION

FINDING: The present method of collecting property taxes by 2,774 independent collectors results in extreme diversities in methods, efficiency of operation, and costs.
RECOMMENDATION: That tax collections as recommended in the County section of the Report be cleared through a centrally located county office. This plan is followed in Ohio and other states. This should result in a substantial saving in the $4,500,000 spent yearly for commissions. Payment of a county tax collector should be put upon a salary basis. A greater saving would be effected if the change in fiscal year were made as recommended, thus permitting the mailing of as many taxes (county, school, poor, township) as possible in one envelope.

ANNUAL AUDIT

FINDING: The annual audit is performed in a careful and exacting manner in many townships. In others, it lacks the essentials of an audit. Technical training is needed for the proper performance of this function.
RECOMMENDATION: That qualified accountant be selected by a state agency to act as auditor for the entire county, and in some areas for several counties. This recommendation should, for reasons of cost, be contingent upon township reorganization into larger areas.

CHAPTER XIII

FISCAL PROBLEMS CONTINUED

I. Tax Levies

The rapid increase in expenditure of local governments since the War has placed a heavy burden upon their revenue systems. The townships which, like the federal and state governments, attempted to develop new sources of revenue, have had only scant success. In general their avenues of income have changed but little. The main source of local funds for the past 200 years has been the general property tax.

So true is this that property tax levies are regarded in Pennsylvania as the best indication of the cost of government. The tax levies are here presented in data which are comparable with those for: bonded debt, tax delinquencies, taxable valuation, tax collectors' commissions, etc. These data for the state as a whole were made available through the courtesy of the Department of Internal Affairs, Bureau of Statistics.

Township Tax Levies Relative To Other Local Governments, 1932. Since in Pennsylvania nearly $363,000,000 in taxes are levied for local government purposes each year, the place of the township is a relatively inconspicuous one. Second class township tax levies were $11,570,631 or only 3.2% of the total taxes levied for local government purposes. This is an average of $7,600 for each of the 1,516 townships. Taxes levied by local governments in 1932 were as follows:

TOTAL TAXES LEVIED BY LOCAL GOVERNMENTS-1932

Data from Bureau of Statistics, Dept. of Internal Affairs

Local Government	Number of Local Governments	Population 1930	Total Taxes Levied Amount	Per Capita	Percent of Total
Cities					
First Class	Philadelphia	1,950,961	65,946,879	33.8	18.2
Second Class	Pittsburgh	669,817	20,651,587	30.8	5.7
Second Class A	Scranton	143,433	2,241,207	15.6	.6
Third Class	44	1,543,602	21,169,951	13.7	5.8
Boroughs	936	2,573,820	22,750,387	8.8	6.3
First Class Townships	61	472,120	5,569,621	11.8	1.5
SECOND CLASS TOWNSHIPS	1516	2,277,597	11,570,631	5.1	3.2
School Districts	2584		151,121,340	15.7	41.6
Counties	67		52,488,225	5.4	14.5
Poor Districts	424		9,461,096	1.0	2.6
Total	5635	9,631,350	362,970,924	37.69	100.0

The 1,516 second class townships with a population of 2,277,597 levied taxes amounting to $5.10 per capita. The tax levy for all local government was $37.69 per capita in 1932. The tax for local government purposes as we have seen is a general property tax levied upon both real estate and personal property. As in other states, the personalty almost entirely escapes the levy. Personal property is levied upon for county purposes at the rate of four mills. The amount of the personal tax levied in 1932 was $6,794,563 outside of Philadelphia and $4,205,151 in Philadelphia.

Decrease 1926-1932 in Township Tax Levies Based on 1930 The 1,516 second class townships levied local taxes amounting to $13,846,519 in 1926. Since that time the trend of township taxes has been, first to increase to a peak of $15,056,277 in 1929 and second to decline substantially throughout 1931-1934. In 1932 the figure dropped to $11,750,631 and in later years may be expected to go much lower. The average tax levied per township in 1932 was $7,632 for 1,516 townships. The amount of tax levies with the percent of the increase or decrease based on 1930 was as follows:

Second Class Township Tax Levies Compared With All Local Governments 1932.

Data from Bureau of Statistics, Dept. of Internal Affairs

	Second Class Township		All Local Governments	
Year	Amount	Percent Increase or Decrease Based on 1930	Amount	Percent Decrease Based on 1930
1932	$11,570,631	-22.7	$362,970,924	- 4.0
1931	13,836,395	- 7.6	372,984,774	- 1.3
1930	14,967,007		377,963,391	
1929	15,056,277	.6	368,719,293	- 2.5
1928	14,790,741	- 1.2	358,674,018	- 5.1
1927	14,177,185	- 5.3	330,468,399	-12.6
1926	13,846,519	- 7.5	321,937,872	-14.8
1925			300,316,507	-20.5
1924			272,259,893	-28.0
1923			250,151,933	-33.8

A decline in tax levies of nearly 23% for townships over the entire state was indicated for 1932. A more recent study, also based on 1930, showed a decline of 42% in actual expenditures for the 52 townships in Chester County in 1933. Although comparative tax figures were not available for 1933 and 1934, interviews with township supervisors in many sections of Pennsylvania confirm the opinion that an even greater reduction over 1930 will appear when the figures for those years have been compiled. It should be noted that the 1934 reduction in taxes levied for road purposes is attributable to a different cause than that of previous years. During 1931, 1932 and 1933 the reductions were a result of economies in township administration. In 1934, however, the reduction will be attributed to an allocation by the state of $11,352,000 on a mileage basis to be spent by the state for road maintenance in second class townships. This allocation substantially reduces the outlay of the township although it does not entirely replace it because the Parkinson Act (1933 P.L. 1520) has been interpreted to place upon the township the care of bridges, construction of new roads and other road expenses which are not strictly "maintenance".

Per Capita Tax Levy by Townships, 1932. - In an analysis of governmental costs, the per capita tax levied is the most significant index. This figure was secured for 147 townships selected from all sections of the state; some of the highest from this list are presented below:

Township	County	Per Capita Township Tax Levied 1932		Economic Characteristic
Newton	Bucks	$10.94		Farming
Solebury	Bucks	11.00		"
Upper Makefield	Bucks	11.28		"
Middle Taylor	Cambria	11.99	1/2	" & Bituminous
Charlestown	Chester	10.86		"
East Bradford	"	11.05		"
East Goshen	"	10.35		"
Easttown	"	15.03	1/5	"
Kennett	"	13.21		"
Tredyffrin	"	10.13	1/3	"
West Goshen	"	10.02		"
Willistown	"	12.64		"
Marple	Delaware	10.19	2/7	"
Newtown	"	10.68		"
Upper Providence	"	11.39	1/4	"
Lehigh	Lackawanna	11.10	1/4	" thickly wooded

Township	County	Per Capita Township Tax Levied 1932	Economic Characteristic
Lower Moreland	Montgomery	$10.61	1/5 Farming
Dingman	Pike	10.29	1/4 " ; heavily wooded
Upper Burrell	Westmoreland	10.34	Farming, Gas, Bituminous, Wooded

The principal economic characteristic of the township plays an important part in its cost of government. Where farming predominates, better roads are required to take produce to markets. In wooded sections, there is less travel, especially during the winter. Earth roads in such sections do not receive continuous maintenance. Comparisons may be made as to whether roads are well-maintained, whether bituminous macadam surfaces have been used, and the number of miles of road in the particular township. All these factors affect the per capita township tax levy to a considerable degree. For example, Easttown Township, Chester County (population 2,570) had a per capita tax levy of $15.03. This was the highest per capita township tax levied in 1932 in a list of 147 townships selected for special study. This tax was levied almost exclusively for road purposes. Easttown Township has more than 14 miles of bituminous surface macadam road. This is a high cost surface, and nearly all of it is in good condition. In contrast to this, Hillsgrove Township, Sullivan County (population 216) levied a per capita township tax of $1.75. Hillsgrove has not quite four miles of road, all of which is unimproved dirt surface, and most of which is in poor condition. It would seem, therefore, that the difference in per capita costs is well accounted for by the needs of the two communities mentioned, by the population served, and by the kind of service rendered.

In contrast with the previous list, there is shown below a group of townships having a low per capita township tax:

Township	County	Per Capita Township Tax Levied 1932	Economic Characteristic
Catherine	Blair	$2.11	Farming, wooded area
Center	Butler	2.57	" , oil, bituminous
Derry	Dauphin	2.75	" , wooded
Newton	Lackawanna	2.23	" "
Hunlock	Luzerne	2.55	1/3 " , partly wooded
Hepburn	Lycoming	2.73	" " "
McHenry	"	1.90	All wooded, bituminous -
Hillsgrove	Sullivan	1.75	1/4 Farming, wooded, anthracite

The above tabulation includes the township tax levy only, and is taken from tables which were prepared for but not printed with this report. Again it may be observed that the list consisting of townships with a high per capita tax levy is one on which farming characteristics predominate. In the list with a low per capita tax, the wooded and mineral sections are represented to a greater extent.

II. Expenditures

Township Expenditures by Townships, 1932 - For 161 selected second-class townships, the expenditures were secured from the Pennsylvania Department of Highways. They are based upon the Annual Reports of Township Supervisors. The total expenditures in 1932 in the 161 townships, were $2,257,998. Since substantially the same townships were studied as to costs and expenditures in the Highway Department, as to tax and assessment statistics, in the Department of Internal Affairs, and by personal observation and travel, it was possible to compare the various kinds of information. The townships with the highest and lowest expenditures in 1932 were as follows:

Township	County	High Total Expenditures 1932	Township	County	Low Total Expenditures 1932
Hempfield	Westmoreland	$ 68,670	Portage	Potter	$ 370
Unity	Westmoreland	63,611	East Caln	Chester	677
Mt. Pleasant	Westmoreland	61,651	Beaver	Snyder	707
Whitemarsh	Montgomery	55,064	Hillsgrove	Sullivan	958
West Goshen	Chester	54,149	Aleppo	Allegheny	973
Manor	Lancaster	53,375	Rush	Dauphin	1,071
Tredyffrin	Chester	52,197	Jefferson	Dauphin	1,083
Manheim	Lancaster	51,738	Putnam	Tioga	1,171
Upper Merion	Montgomery	38,173	Bridgeton	Bucks	1,307
Upper Dublin	Montgomery	36,981	Lehigh	Lackawanna	1,317

Expenditures for 147 individual townships were analyzed on a per capita basis, using the population figures on Table I at end of this Chapter. Per capita figures are the more reliable but do not take into consideration types of service rendered, geographic conditions including number of streams and bridges, and condition of the soil as it affects cost of road maintenance. The same objections apply to figures on a per mile basis with the additional difficulty that only a proportion of the roads in a township may be worked in one year. This is particularly true in those few townships where there is considerable mileage of high cost (bituminous macadam) surface. Patching of this surface is not done every year, and if the cost per mile were based only on the mileage worked, instead of on the total number of miles in the township, an entirely different result is obtained.

In pointing out that individual township expenditures may be analyzed on a per capita basis the unreliable nature of conclusions which may be drawn should be indicated. The absence of any continuous state agency for recording township statistics and of any uniform accounting procedure among townships prevents comparisons of the type so frequently made in New York and other states where these conditions have to some extent been dealt with.

Classification of Road Expenditures. Expenditures of second class townships for 1932 may be classified into three groups for further analysis. These groups include costs directly connected with the construction and maintenance of roads, cost of debt service, and overhead expenditures. The first group contains 57% of the total expenditures. Among the items included were:

Maintenance of earth roads	22%
Maintenance of improved roads	4%
Repairs state reward roads	1%
Opening and building new roads	1%
Permanent improvements of roads	14%
Permanent improvements of bridges and culverts	3%
New tools and machinery	2%
Repairs to tools and machinery	2%
Snow removal	1%
Roadmaster's wages	7%
Total	57%

The second group consisting of debt service costs, consumed 31% of the total township expenditure. It included:

Interest on notes	2%
Notes maturing	19%
Interest on bonds	3%
Sinking fund	3%
Redemption of bonds	4%
Total	31%

The last group absorbed the remaining 12% of second class township road expenditures. Items included as overhead were:

Treasurers' Compensation	1%
Secretaries' Compensation	1%
Tax Collectors' Compensation	2%
Supervisors' monthly meetings	1%
Miscellaneous	5%
Premiums Treasurers' & Tax Collectors' Bonds	
Annual Supervisors' Convention	
Rent, heat and light for meetings	
Stationery	2%
Signs and index boards	
Compensation insurance	
Attorneys' & Auditors' Fees	
Gas and Oil	
Total	12%

While the miscellaneous expenditures are here included among the overhead, it often happens that costs of special municipal services are included under these expenditures. The Township Code definitely provides that a special tax shall be levied for municipal services such as street lighting, and fire protection.

In some cases the amount spent for such services is so small that it does not warrant a special tax levy and the township auditors therefore include them under miscellaneous.

Maintenance of Earth Roads - We have seen that the expenditures directly connected with road construction and maintenance represent by far the largest portion of township outlay. The leading item for road purposes in second class townships was "Maintenance of Earth Roads" (Table III, Column 1 at end of this Chapter). This item required $2,862,000 in 1932 and comprised 22% of all expenditures of second class townships. Since 43,000 of the 48,479 miles of second class township roads are unimproved earth, it can be readily understood that this amount should play a large part in township road expenditure. It provides for the care and upkeep of nearly 90% of the township roads.

Among 161 selected townships studied for the purpose the amount expended on earth roads varied from $53.00 in Lower Moreland Township, Montgomery County, to $23,484 in Mount Pleasant Township, Westmoreland County. Townships with the highest and lowest expenditures for earth roads maintenance were as follows:

Township	County	- High - Maintenance Earth Roads 1932	Township	County	- Low - Maintenance Earth Roads 1932
Mount Pleasant	Westmoreland	$23,484	Lower Moreland	Montgomery	$ 53
Hempfield	Westmoreland	23,067	Montgomery	Montgomery	74
Hegins	Schuylkill	14,952	Summit	Butler	77
Unity	Westmoreland	12,580	Derry	Dauphin	119
Manor	Lancaster	11,001	Bridgeton	Bucks	126

If the township expenditures for earth roads in 1932 are totaled by counties, the amount varies from $7,145 for Cameron County to $156,729 in Westmoreland County. The five counties in which townships had the highest and lowest expenditure in 1932 for maintenance of earth roads were as follows:

County	High Maintenance Earth Roads, 1932	County	Low Maintenance Earth Roads, 1932
Westmoreland	$156,729	Cameron	$7,145
Fayette	125,868	Fulton	8,261
Allegheny	99,145	Forest	8,735
Chester	91,527	Union	8,931
Crawford	86,426	Montour	9,203

It should be noted that the items presented above, and similar enumerations in other sections of this study, are not so much comparisons of actual cost per unit of work done as they are lists of amounts expended on the basis of the best figures obtainable. It is doubtful if comparable cost figures can be presented until there is either (1) uniform local accounting, or (2) a special reclassification of all items in each municipality on a uniform basis.

Permanent Improvement of Roads - Expenditures for the construction of new roads were exceeded in amount by only two items, the cost of maintenance of earth roads and the cost of indebtedness. The total expenditures for new construction (i.e. permanent improvements) in 1932 were $1,763,133 (Table III Column 6). This was 14% of the total expenditures of second class townships in that year. The number of miles of new road constructed in a particular township varies greatly from year to year. Numerous townships spend nothing for this purpose. Expenditures of townships which have undertaken permanent improvements on their road systems range in amount from $38 in Bridgeton Township, Bucks County, to $42,846 in Manheim Township, Lancaster County. The townships which were lowest and highest in expenditures for permanent improvements for 1932 were as follows:

Township	County	- High - Permanent Improvement of Roads 1932	Townships	County	- Low - Permanent Improvement of Roads 1932
Manheim	Lancaster	$ 42,846	Bridgeton	Bucks	$ 38
Tredyffrin	Chester	40,357	North Londonderry	Lebanon	40
Whitemarsh	Montgomery	27,228	East Caln	Chester	85
Easttown	Chester	22,710	Lower Makefield	Bucks	106
Derry	Dauphin	19,313	Horsham	Montgomery	131

Expenditures for permanent improvements undertaken on township roads were totaled for 1932 by counties. These expenditures are shown in Chapter XIV Table 7, column 6. The counties in which township expenditures were highest and lowest for this purpose were as follows:

County	- High - Permanent Improvement of Roads 1932	County	- Low - Permanent Improvement of Roads 1932
Lancaster	$ 175,279	Cameron	$ 49
Chester	169,064	Fulton	103
McKean	161,451	Mifflin	150
Washington	118,048	Wyoming	648
Bucks	99,702	Bradford	727

A permanent road improvement cannot at present be deemed significant other than as an indication of the range of expenditure which prevails on this subject. It is evident that no large permanent improvement could be accomplished for $38. An earth road may be permanently improved by laying a drainage pipe beneath the road bed. Due to the diversity of practice among township officials as to the classification of road costs, the analysis of particular townships on a per capita or per mile basis leads to difficulty which can best be avoided and which probably should not be undertaken with present figures for comparative purposes.

Roadmasters' Wages - The wages of roadmasters occupy an important place in township road expenditures. Of the total expenditure of second class townships in 1932, 7%, or $873,079 was spent for this purpose. Other aspects of this subject appear in the chapters on "Administration" and "Roads and Bridges". It should be pointed out, however, that the wages vary greatly in different townships. Some townships record no expenditure. In others, the amount ranges from $20.00 in Portage Township, Potter County, to $4,615.00 in Plymouth Township, Luzerne County. Of 161 townships studied as samples the five highest and five lowest in amount of wages paid roadmasters were as follows:

Township	- High - County	Roadmasters' Wages 1932	Township	- Low - County	Roadmasters' Wages 1932
Plymouth	Luzerne	$ 4,615	Portage	Potter	$ 20
Mount Pleasant	Westmoreland	4,132	Bell	Westmoreland	26
Unity	Westmoreland	3,838	Aleppo	Allegheny	30
Hempfield	Westmoreland	3,584	Beaver	Snyder	70
Adams	Cambria	3,389	Middle Taylor	Cambria	75

The expenditures of second class townships for roadmasters' wages are indicated by counties on Table III, Column 10. These figures were secured from the Annual Financial Reports of Township Supervisors for the year 1932.

Miscellaneous Road Expenditures. The item "miscellaneous" originates in the list of expenditures on the Annual Financial Report Form forwarded by the Highway Department to the township supervisors. Since this item contained nearly 5% of the township road expenditures in 1932, an effort was made to secure information as to the kind of expense listed under this item. The results of this inquiry show that the following kinds of expense are frequently included in the item "miscellaneous":

> Gasoline and oil
> Fire hydrants and fire companies
> Printing and advertising
> Postage and affidavits
> Workmen's compensation insurance
> Telephone and stationery
> Expenses, county road supervisors association
> Subscriptions to "Road Builder" and other road magazines
> Insurance on township automobile
> Janitor service
> Refunds of tax collected by mistake
> Coal and wood

The township in which the largest amount was listed in the auditor's report as "miscellaneous" was Willistown Township, Chester County. In 1933 the total for this item was $1,148.66. Auditors lump small items together in this manner for reasons of convenience, despite the fact that such a practice defeats the purpose of requiring annual publication of the report. There was no hesitation in responding to requests for itemization of "miscellaneous" in the various townships. The impression was gathered that township officers welcomed inquiries concerning township business and were almost without exception willing to explain and discuss details of township practice.

III. Receipts

For particular items of receipts and expenditures, the cost of government is best analyzed in figures compiled by the office of the township engineer, Department of Highways. They are given in Tables II and III at the close of this chapter, for a total of 1480 of the 1488 second class townships. The itemized receipts for 1932 by counties are given in Table II columns one and six, the itemized expenditures in Table III at columns one and thirty-two.

Computations of receipts and expenditures originate in the annual financial reports submitted by Township Supervisors to the Department of Highways. In 1932 the receipts of second class townships totaled $14,202,915.

	Amount 1932	Percent
Cash Road Tax	$9,614,650	68
State Reward	283,664	2
Forest Road Aid	55,111	1
County Aid	344,657	2
Other Sources	3,904,833	27
Total	$14,202,915	100

Cash Road Tax. The cash road tax in second class townships is obtained by the tax collector in individual payments and forwarded monthly to the township treasurer.

State Reward - In 1919 the General Assembly passed the State Reward Act which was designed to form a new basis for state assistance to second class townships and to induce them to do better road construction. Under the State Reward Plan the Highway Department apportioned the funds appropriated by the legislature on the basis of the township mileage in each county. The allotment was divided according to the applications filed by townships in order of filing. No township was eligible to receive more than 15% of the total reward allotted to its county. Where the townships in a county did not apply for funds the surplus was divided among the counties having applications in excess of the original 15% limit.

Applications for the state reward were made by township supervisors on forms supplied by the State Highway Department. Such applications numbered 2,537 in 1920 and were filed at a fairly steady rate, terminating with 3,107 in 1928. The applications on file by years during the period were as follows:

STATE REWARD APPLICATIONS ON FILE

1928	3,107
1927	3,072
1926	2,846
1925	2,487
1924	2,114
1923	1,199
1922	2,521
1921	2,263
1920	2,537

In filing applications for state reward, township supervisors could choose from eight different types of road construction. Each type of road had a maximum state contribution. Payments were based an improved surface road having a width of 16 feet. Townships with an assessed value per mile in excess of $20,000 were eligible to receive 50% of the cost from the state. Townships in which the assessed value per mile was less than $20,000 were entitled to 75% of the cost. During the period 1919-1929 $16,000,000 was appropriated for State Reward purposes. The appropriations were as follows:

STATE REWARD APPROPRIATIONS AND PAYMENTS 1919-1929

	State Appropriation	Reward Payments	Total Cost of Work
1919	$1,000,000	$ 971,466	5,130,126
1921	1,000,000	1,000,000	4,103,985
1923	2,000,000	2,000,000	5,647,439
1925	3,000,000	3,000,000	7,751,355
1927	3,000,000	2,846,600	7,671,154
1929	6,000,000	-	-
Total	$16,000,000		

In 1929 the status of State Reward was as follows:

STATE REWARD STATEMENT

1929 Appropriation	$6,000,000
Balance from 1927 Appropriation	153,399
Total	6,153,399

Unobligated Balance, Apr. 7, 1933 $ 116,009
Apportionment   6,153,399
Amount Taken Up   6,036,968
Agreements in Force   29,165
Balance   116,431

In 1931 the State Reward System became inoperative, due to lack of funds. The General Assembly in the 1933 session passed a measure providing that all appropriations for township reward which had been made from the Motor License Fund should revert to and become part of that Fund. By this enactment the Reward System was left without funds although the law itself remained in force (1933 P.L. 28). The System was a means of educating township officials in proper road maintenance methods. In 1930 more than 1,600 miles of roads were repaired in townships under this plan. In 1931, 678 miles were constructed through state reward. During 1930 and 1931 the maintenance costs on such projects were as follows:

MAINTENANCE COSTS -- STATE REWARD PROJECTS

	1930	1931
Stone Roads		
Length, miles	967	447
Cost	$250,490	$81,640
Cost per mile	$259	$183
Bituminous Macadam		
Length, miles	627	231
Cost	$317,991	$83,438
Cost per mile	$507	$362
Concrete, Brick, etc.		
Length, miles	13	--
Cost	$2,614	--
Cost per mile	$194	
Total		
Length, miles	1,607	678
Cost	$571,095	$165,078
Cost per mile	$355	$244

Under the Reward plan applications were filed without obligation on the part of the township. If, at the time the application was considered, the township was unable to proceed with the work, the application could be cancelled. Bridges were frequently built under State Reward, and most of the concrete structures now observed on township roads were constructed on this basis.

The State Reward System was at first inserted as an amendment to the Township Code of 1917. Upon the revision of the laws affecting second class townships by the Township Law Revision Commission in 1931, a separate bill was drafted in order to separate the proposed township code from controversy over State Reward. A revised State Reward Bill was proposed by the commission at the same time.

In 1932 nearly $30,000 was received from this source by 12 of the 161 townships studied as a representative list. The amount received by all second class townships in that year totaled $283,664, or 2% of the total receipts of second class townships (Table II, Chapter XIII). Incompleted contracts accounted for income from State Reward in 1932. As has been previously noted, such reward terminated from lack of funds in 1931 and any remaining balances were lapsed by the Act of 1933. The State Reward policy from 1919-1929 has been presented in some detail because many township officers look to a resumption of this policy.

Forest Road Aid. In 1932 the second class townships received $55,111 in Forest Road Aid from the state. This was approximately 1% of all expenditures. Among the 161 selected townships studied, 16 had receipts from this source. The largest income of any of the selected townships was $477, received by Brown Township, Lycoming County which is about 80% state forest. It has approximately 6½ miles of road and no indebtedness of any kind. Roads in the state forest run through the township. For the maintenance of these roads the state pays 2¢ an acre. The state has taken so much land in the township for forest purposes that tax receipts, which some years ago amounted to $1,500, would now amount to $300 per year at the same rate of levy.

One of the most interesting roads in the entire state runs through Brown township. It is located on a high shelf and is so narrow that vehicles may pass only at infrequent intervals. This is one of the few remaining earth surfaced roads on the old State Highway System of which it has been a part for more than thirty years. The township tax levy was 20 mills in 1932. In 1934 no tax was levied. Taxpayers in the township formerly engaged in lumbering, farming, railroad work, wood mills, and summer camps. By 1934 only a few farms and summer camps remained as taxpayers and places of employment.

In such districts, a State Administrative area would be most applicable. Peculiarly enough, it is in the most rural and wooded areas that local sentiment is strongest in favor of local control. The sentiment for centralization of local functions develops in more densely populated, almost suburban-rural regions.

County Aid. - County aid like state reward is a form of subsidy to townships. The county commissioners are authorized by the county code to make contracts with the township supervisors for the construction, maintenance and repair of roads in which they are jointly interested. New construction is usually undertaken by the county. Maintenance is "under such terms and conditions as may be mutually satisfactory" (1929 P.L. 1278 Sec. 957). Contracts are made whereby the county and township agree to contribute jointly on a fifty-fifty basis.

When a township desires to improve a stretch of road connecting county roads or located as a feeder road to a main county road, it makes application to the county commissioner for the improvement at joint expense. Many miles of township road have been built in which State Reward paid for half of the construction costs and county aid for 25% with the township paying the 25% remaining. It was to the interest of both the state and the county to have such roads built. In 1932 county aid amounted to $344,657, approximately 2% of the receipts of all second class townships. Among the 161 townships listed 33 received $53,010 from the county for road purposes.

Applications for County Aid. Applications in the form of a petition may be made by the owners of the majority of assessed value of real property in the township for the improvement and maintenance of any principal road at the joint expense of the county and township. This petition is presented to the township supervisors requiring them to make application to the county commissioners for county aid. If the township supervisors refuse or delay action, any citizen taxpayer of the township or county may, by petition, present the facts to the court of quarter sessions. If the court finds the allegations in the petition are sustained, it may order the supervisors to act upon the application and require it to be forwarded to the county commissioners (1933 P.L. 103-Sec. 519).

Other Sources. - Since the item "other sources" constituted 27% of the receipts of second class townships in 1932, an effort was made to determine what types of income were included. Nearly $4,000,000 of the $14,200,000 available for road purposes was included in this item. Similarly the same items accounted for 27% of the receipts of the 161 townships studied. Income from the following items is included among "other sources."

Balance from previous year
Sale of bonds
Temporary loans
Rental of road machines
Fines by justice of peace
Sale of road material
Donations
Rebate on compensation insurance.

The first three items listed above are by far the most important. Many townships carry heavy balances from year to year. In Easttown Township, Chester County, one of the larger townships in the state, balances in the treasury at the end of the year were as follows:

1931	$4,753
1932	8,610
1933	7,275

Since the expenditures of second class townships for road purposes in 1932 were $12,700,000 and the receipts were $14,200,000 it would appear that a cash balance of $1,445,000 was carried forward as a balance into the next year. This is an average balance of $980 per township. In numerous townships several warrants are left outstanding and unpaid at the end of the fiscal year. It is probable that a considerable amount remains to be deducted from the cash balance of $1,445,000 shown above for the payment of these warrants. Amounts received by second class townships from temporary loans and the sale of bonds appear to have been merged with the other items listed above in Highway Department computations. Expenditures in the form of interest costs, balance to sinking fund, balance for the retirement of bonds, and for temporary loans are discussed at length later in this chapter under the heading of indebtedness.

Proposed State Sources of Revenue for Township Road Purposes. - The necessity of lessening the burden of the township road tax on rural real estate is generally admitted. The methods by which this should be done, if it can be done at all, are most debatable. The state has been returning large sums every year to localities and the constant expansion of state assistance has in some cases tended to expose real estate to additional burdens rather than to relieve it. Among the recent proposals of state sources of revenue for townships are the following:

1. Amendment of the Parkinson Act to return 1935 allocation to township treasuries. From the standpoint of a large number of township officials this is the most advantageous proposal. It aims by a simple amendment to the Parkinson Act to transfer the appropriation from the maintenance forces of the State Highway Department to the road supervisors of the various townships. The data collected in this Survey do not warrant a judgment on the engineering aspects of this problem.

2. Allocation of ½¢ of the gasoline tax to townships for road purposes. This proposal has more merit than any of the other suggestions advanced. It should be made a part of a planned allocation distributed among all types of municipalities and devoted by them exclusively to road and street expenditures. Such an allocation has been described in highway studies as a "Public Highway Program". (See Chapter XIV, "Roads and Bridges")

3. Continuance of beer license funds to township road districts. Approximately a quarter of a million dollars annually is expected to be divided among townships of the second class for road purposes. These funds were allocated to municipalities in which the licensed places are situated at the special session of the General Assembly in November 1933 (1933 Sp.S. 15, Sec. 701). Since the amount is not quite 2% of the total expenditures of second class townships in 1932 it can hardly be compared as a property tax relief measure on a broad scale with the Parkinson Act. The expenditure under that act, ($5,676,000) in 1934, amounted to 42% of the township expenditures in 1932.

4. Resumption of state reward for township roads. The statement concerning the state reward policy which appeared earlier in this chapter showed that $16,000,000 had been paid to townships during the period 1919-1929. The good feature of state reward was that it enabled the state to impose standards for construction and materials as conditions upon which the money was accepted and spent by townships. Many township road masters became better acquainted with improved methods of road construction through this policy. The disadvantage of state reward was that in many townships taxes were raised and debts were incurred to secure funds to match the state reward allocation. Numerous

townships at the present time labor under indebtedness incurred for such roads which became part of the state highway system in 1931. The resumption of state reward at the present time would be unwise and it does not appear to be actively proposed by any large group. Construction, under present conditions, is at a minimum and townships would be unable to raise their share of a state reward contribution. Many of them could not do so even in prosperous times.

Alternatives to Increasing Township Road Revenue. - A number of alternatives may be considered which do not directly increase the township road funds. These are briefly summarized below in order to present a concise view of what should properly be made the subject of an extensive study. Among the principal alternatives are the following:

1. Continuation of the Parkinson Act. Although the Parkinson Act appears to have brought financial relief to many townships it is almost everywhere severely criticized. Responsibility for township roads is unhappily undivided. The township incurs the legal responsibility, but the state does the road maintenance work. The township maintains the bridges and its road forces perform all new construction on both roads and bridges. The dissatisfaction with this compromise has been so widespread that a continuance of the Parkinson Act in its present form hardly seems feasible.

2. Place responsibility for the 48,461 mile township road system on the county. The counties apparently desire to make this change. At the state convention of county commissioners at Pottsville, September 1934, a resolution was adopted advocating the county unit for road purposes. It was pointed out in Chapter 14 "Roads and Bridges" that counties now maintain only 2% of the road mileage of the state. The total county road mileage is less than 1,800 miles. At the present time counties are not prepared to perform this work and a transfer of the township system would require the acquisition of additional road forces and equipment. The counties do not propose to adopt township bridges and there thus looms the possibility of three duplicating forces: the state maintenance unit, the county maintenance unit, and the township bridge maintenance unit. It should be noted that no tax relief for real estate in townships would be secured by this plan because the tax formerly levied for township road purposes would be replaced on the same tax bill by an increased tax for county road purposes.

3. State control of the township road system. Ultimately all roads in the Commonwealth excepting those in cities should be expected to come under state control. In other states township and county road systems have been placed on the state highway system. It is believed that the savings in administrative costs under state control are greatly over-estimated. But that there are very great savings through central purchasing which such a large system permits, is confirmed by the experience of Virginia and North Carolina. Although it is frequently objected that the state would be unable to finance such an enormous addition, it should be noted that in both Virginia and North Carolina, rural road systems were assumed by the state in a period of financial stringency.

In regard to highway centralization Pennsylvania should work out a solution which fits her peculiar needs. Certainly the state should not attempt to assume the burdens of public school support and of the rural road system at the same time. Since the township school tax and the township road tax are the two most important taxes on rural real estate, the question of property tax relief is a major consideration.

4. Return township road maintenance to the townships. The township 48,461 mile road system will automatically return under the supervision of township road maintenance forces in 1936, unless new legislation is enacted. At that time the Parkinson Act which provides for state maintenance of the township system will cease to operate and the funds appropriated for the purpose will be exhausted. The townships would be expected to increase the township road tax by amounts

sufficient to enable them to resume work. It would appear that whether this should occur ought to depend on whether it is desired to continue local control of the largest highway system in the state and on whether real estate delinquencies have subsided sufficiently to permit the necessary increase in township tax. In the latter case township participation in any proposed "Public Highway Program" would be a township tax reduction factor.

IV. Conclusions

Tax Levies

FINDING: Township tax levies were a relatively small proportion, only 3.2%, of the total taxes levied for local government purposes in 1932. The township tax in that year amounted to $11,570,631. This is an average levy of $7,600 for each of 1,516 townships.

FINDING: The per capita township tax levy was $5.10 per person.

FINDING: During the nine-year period 1923-1932, the township tax levy was reduced from 4.3% of the taxes levied by all local governments in Pennsylvania in 1926, to 3.2% in 1932. Still more striking is the reduction effected in the township tax levy in 1930 to 1932. During these three years a reduction of nearly 23% was effected while the reductions accomplished by all local governments amounted to only 4%.

Expenditures For Road Purposes

FINDING: The trend in township expenditures during the past 12 years confirms the substantial reduction already indicated by the study of the township tax levy. Since a longer period is included in the study of road expenditures a greater range is apparent. The figures are further influenced by large state reward and county aid appropriations which as they appeared and disappeared tended to increase the fluctuation in expenditures. These did not influence greatly the tax levies which were relatively stable.

FINDING: In 1920 township expenditures for road purposes were $14,800,000. In 1930 expenditures had increased to $23,700,000. By 1932 expenditures for road purposes declined 46% from those of 1930, reaching a total of only $12,700.000.

FINDING: In 1932 the three largest items of expenditure for township road purposes were maintenance of earth roads 22%, notes maturing 19%, and permanent improvement of roads 14%. These three items comprise 55% of township expenditures.

FINDING: A classification of 1932 road expenditures indicated that costs directly connected with the construction and maintenance of roads required 57% of township expenditures; debt service costs took 31% and township overhead containing the remaining items required 12%.

Receipts

FINDING: Receipts of second class townships from all sources in 1932 amounted to $14,202,915. Of this amount 68% was received from the Cash Road Tax, 2% from State Reward, 1% from Forest Road Aid, 2% from County Aid, and 27% from other sources.

Proposed State Sources of Revenue for Township Road Purposes

FINDING: The amendment of the Parkinson Act to return its allocation for 1935 to township treasuries is the most desirable proposal for additional township revenue as viewed by township officials. As previously mentioned the Parkinson Act was a compromise and it is still an experiment. State Highway Department

maintenance although encountered in many sections of the state was too broad in
scope and too brief in duration to enable a conclusion to be reached upon this
subject.

FINDING: Although some township officers desire ½¢ of the gasoline tax allocated directly to township road purposes to be expended under township supervision
it is
RECOMMENDED: That the distribution of any such funds be made a part of a general program in which all municipalities participate to the extent that funds may
be available for such purposes.

FINDING: If the distribution of beer license and liquor license funds is continued to the localities in which the licenses are purchased it is
RECOMMENDED: That the township road district should continue to share in this
distribution.

FINDING: State reward as a source of revenue for township road purposes is not
feasible at the present time, and it is
NOT RECOMMENDED that any such policy be resumed.

Alternatives to Increasing Township Road Revenue

FINDING: The Parkinson Act appears to have brought about substantial tax reduction despite the dissatisfaction arising from it on all sides.
RECOMMENDATION: That the continuance of the Parkinson Act for another two years
should not be undertaken. The Act was a compromise in its inception and its
continuance merely postpones the adoption of a settled policy.

FINDING: One alternative to increasing township road revenue would be to place
the responsibility for the township road system on the county. This is
NOT RECOMMENDED since counties at the present time are equipped to maintain only
2% of the road mileage in the state. The township system constitutes 49% of the
road mileage. The proposed county-road unit merely transfers the township road
tax to the county-tax on the same property. It offers little if any tax relief.

FINDING: Adoption by the state of the township road system would operate as a
substantial local property-tax relief. Maintenance of township bridges by township officers would require the continuance of the township road tax. Ultimately
all roads and bridges except those in cities, may be expected to become part of
the State Highway System. The placing of this additional burden directly upon
the state is
NOT RECOMMENDED at this time.

FINDING: The lack of taxable resources in the local units, their small area,
inefficiency and long continued tax delinquency have led Virginia and North Carolina to adopt a highly centralized system of state road control. In Pennsylvania, however, centralized state road control has been shown to possess grave
political dangers. The fiscal condition of the state also makes it more than
doubtful if the burden of such control should be continued.
RECOMMENDATION: That the enlarged and strengthened township organization and the
supervision of township road maintenance by qualified roadmasters and road
supervisors, as recommended elsewhere in this report, should first be given a
fair trial before a centralized state road system is attempted.

TABLE I. SUMMARY OF ASSESSMENTS AND TAXES FOR TOWNSHIP PURPOSES BY COUNTIES 1932

County	Number of Townships	Township Square Miles Per County	Township Population		Taxable Valuation			Total Taxes Levied		
			Number	Per Square Mile	Amount	Per Capita	Per Cent of Total	Amount	Per Capita	Per Cent of Total
	1	2	3	4	5	6	7	8	9	10
Adams	21	521	23,022	44	$11,659,110	506	.8	$106,602	$4.63	.9
Allegheny	29	419	75,312	180	80,856,465	1,074	5.8	423,505	5.62	3.7
Armstrong	28	574	46,225	81	30,485,710	660	2.2	182,346	3.94	1.6
Beaver	22	400	26,569	66	16,943,253	638	1.2	164,742	6.20	1.4
Bedford	25	1,024	28,006	27	12,009,979	429	.9	88,384	3.16	.8
Berks	44	841	71,040	84	46,013,524	648	3.3	371,495	5.23	3.2
Blair	15	526	32,525	62	12,340,228	379	.9	109,908	3.38	.9
Bradford	38	1,132	25,854	23	11,945,316	462	.9	131,653	5.09	1.1
Bucks	31	594	52,432	88	46,058,927	878	3.3	393,378	7.50	3.3
Butler	32	756	34,675	46	22,183,142	640	1.6	217,769	6.28	1.9
Cambria	29	691	67,015	97	41,470,976	619	3.0	438,182	6.54	3.8
Cameron	5	391	2,096	5	787,285	376	.1	5,752	2.74	.1
Carbon	12	385	20,393	27	9,088,644	446	.7	49,875	2.45	.4
Center	25	1,144	29,003	25	8,318,453	287	.6	52,621	1.81	.5
Chester	57	757	64,520	85	61,340,835	951	4.4	523,949	8.12	4.5
Clarion	22	593	22,236	37	6,593,162	297	.5	102,231	4.60	.9
Clearfield	30	1,117	51,575	46	10,255,193	199	.7	138,066	2.68	1.2
Clinton	21	874	12,916	15	4,553,005	353	.3	36,960	2.86	.3
Columbia	24	467	20,347	44	12,295,461	604	.9	68,427	3.36	.6
Crawford	35	1,016	30,245	30	17,610,665	582	1.3	191,311	6.33	1.7
Cumberland	22	517	27,741	54	17,046,485	614	1.2	121,516	4.38	1.1
Dauphin	22	489	34,595	71	20,432,079	591	1.5	156,672	4.53	1.4
Delaware	11	121	21,174	175	17,870,689	844	1.3	124,248	5.87	1.1
Elk	10	800	14,948	19	3,399,435	227	.4	50,514	3.38	.4
Erie	21	738	30,130	27	24,444,424	811	1.8	252,532	8.38	2.2
Fayette	24	790	136,582	173	61,664,553	451	4.5	591,410	4.33	5.1
Forest	8	422	4,516	11	1,658,397	368	.1	16,152	3.58	1.1
Franklin	15	744	35,616	48	19,728,861	554	1.4	130,479	3.66	1.1
Fulton	11	402	8,463	21	2,120,398	251	.2	18,274	2.16	.2
Greene	20	573	32,618	57	46,011,812	1,412	3.3	356,613	10.93	3.1

	1	2	3	4	5	6	7	8	9	10
Huntingdon	30	910	20,297	22	9,777,960	482	.7	61,809	3.05	.5
Indiana	24	823	51,589	63	33,528,963	650	2.4	305,074	5.91	2.6
Jefferson	23	653	26,103	40	10,771,261	413	.8	98,691	3.78	.9
Juniata	13	391	11,481	29	3,787,356	330	.3	23,848	2.08	.2
Lackawanna	20	336	20,540	61	10,858,609	529	.8	109,257	5.32	.9
Lancaster	41	927	94,345	102	74,027,087	785	5.3	569,784	6.04	4.9
Lawrence	17	347	31,232	90	17,024,048	545	1.2	113,530	4.60	1.2
Lebanon	16	346	24,004	69	17,537,616	731	1.3	129,602	5.40	1.1
Lehigh	14	326	32,521	100	20,132,537	619	1.5	171,449	5.27	1.5
Luzerne	32	786	64,412	82	25,025,487	389	1.8	207,529	3.22	1.8
Lycoming	42	1,202	25,135	21	7,546,219	300	.5	82,500	3.28	.7
McKean	15	974	22,369	23	14,813,402	752	1.2	250,431	4.20	2.2
Mercer	32	682	30,368	45	14,672,926	483	1.1	151,019	4.97	1.3
Mifflin	10	396	22,457	57	8,052,127	359	1.6	57,170	2.55	.5
Monroe	16	617	15,317	25	9,311,673	608	.7	105,863	6.78	.9
Montgomery	30	384	57,429	150	48,234,898	840	3.5	411,062	7.21	3.6
Montour	9	129	7,150	55	2,611,320	365	.2	21,439	3.00	.2
Northampton	17	337	34,945	104	21,749,881	622	1.6	220,246	6.30	1.9
Northumberland	23	414	29,219	71	26,125,031	894	1.9	89,206	3.05	.8
Perry	21	561	13,570	24	6,701,479	494	.5	44,763	3.30	.4
Philadelphia	--	--	--	--	--	1,258	--	--	--	--
Pike	11	543	4,813	9	6,055,856	302	.4	53,569	11.13	.5
Potter	25	998	9,406	9	2,836,017	589	.2	43,667	4.64	.4
Schuylkill	37	748	77,011	103	45,397,069	520	3.3	222,775	2.89	1.9
Snyder	15	309	13,282	43	6,902,669	520	.5	35,922	2.70	.3
Somerset	25	1,021	49,562	49	23,131,984	467	1.7	216,230	4.36	1.9
Sullivan	9	452	6,304	11	2,028,385	322	.1	20,082	3.19	.2
Susquehanna	27	808	17,880	22	11,708,718	655	.8	112,642	6.30	1.0
Tioga	30	1,126	19,790	18	9,264,944	468	.7	118,421	5.98	1.0
Union	10	302	11,573	38	5,792,163	500	.4	40,826	3.53	.4
Venango	21	647	23,975	37	12,621,361	526	.9	168,527	7.03	1.5
Warren	23	890	21,897	25	8,765,246	400	.6	97,948	4.47	.8
Washington	32	813	85,640	105	81,646,717	953	5.9	578,355	6.75	5.0
Wayne	22	723	19,534	27	9,626,574	493	.7	90,255	4.65	.8
Westmoreland	23	1,029	129,996	126	66,430,320	513	4.8	604,540	4.67	5.2
Wyoming	18	394	10,737	27	3,490,752	325	.5	54,125	3.03	.5
York	34	848	64,586	76	27,898,892	432	2.0	241,809	3.74	2.1
Total	1516	43,010	2,277,597		1,385,117,013		100.1	11,570,631		100.3
Average Per Township	23	642	33,994	53	$20,673,388 $ 913,665	608	2.0	$172,696	$5.08	

TABLE I. SUMMARY OF ASSESSMENTS AND TAXES FOR TOWNSHIP PURPOSES BY COUNTIES 1932

County	Discounts and Rebates Allowed	Unpaid 1932 Taxes outstanding Dec. 31 1932	Percent Unpaid 1932 (Delinquent) Taxes of Total Tax Levy	Current Delinquencies Exceed Current Collections of Delinquent Taxes For Prior Years		Exonerations and refunded taxes on 1932 duplicate	Penalties on 1932 duplicate	Number of Tax Collectors	Tax Collectors' Commissions
				Amount	Per Cent				
	11	12	13	14	15	16	17	18	19
Adams	$2,653	$31,824	29.9	$9,710	43.9	$684	$ 237	21	$2,509
Allegheny	12,926	115,993	27.4	77,620	202.3	5,479	198	29	8,299
Armstrong	4,889	46,011	25.3	32,350	236.3	10,490	638	28	3,820
Beaver	4,390	58,887	35.7	43,494	282.6	1,474	713	21	3,194
Bedford	2,313	20,076	22.7	6,997	53.5	616	393	25	2,185
Berks	10,494	107,629	29.0	38,542	55.8	3,243	1,210	44	8,353
Blair	2,022	48,512	44.1	25,190	108.0	1,283	53	15	3,303
Bradford	2,916	24,938	18.9	12,539	101.1	8,474	749	38	3,062
Bucks	10,481	80,246	20.4	33,130	70.3	3,398	2,267	31	10,227
Butler	5,247	63,409	29.1	31,074	96.1	11,762	3,327	32	4,789
Cambria	13,250	111,821	25.5	69,611	164.9	10,386	2,632	29	9,434
Cameron	----	1,269	22.1	1,205	1882.8	18	137	5	142
Carbon	1,315	11,883	23.8	5,281	80.0	1,167	233	12	1,167
Center	1,371	19,527	37.1	10,006	105.2	452	820	24	1,439
Chester	14,011	126,352	24.1	55,819	79.1	3,265	1,189	57	12,991
Clarion	2,278	30,301	29.6	15,834	109.4	2,508	672	22	2,410
Clearfield	2,241	73,658	53.3	53,001	256.6	1,612	2,121	29	1,659
Clinton	981	11,434	30.2	791	7.4	734	646	21	1,564
Columbia	1,591	15,306	22.4	10,550	221.8	3,250	1,192	24	1,782
Crawford	3,938	51,496	26.9	13,418	35.2	10,521	2,532	35	5,355
Cumberland	3,494	43,071	35.4	9,047	26.6	1,195	1,346	22	3,446
Dauphin	4,714	36,001	23.0	20,579	133.4	1,417	1,167	22	3,821
Delaware	3,424	32,328	26.0	15,862	96.3	5,347	546	11	2,956
Elk	1,261	5,194	10.3	1,272	32.4	3,441	166	10	1,105
Erie	4,907	66,553	26.4	44,612	203.3	9,906	2,615	21	5,400
Fayette	20,798	91,201	15.4	77,120	547.7	37,537	2,773	24	10,354
Forest	436	4,503	27.9	3,191	243.2	638	193	8	343
Franklin	3,649	39,547	30.3	2,698	7.3	807	1,125	15	3,692
Fulton	506	4,362	23.9	34	7.8	747	237	11	507
Greene	13,227	63,034	17.7	52,562	501.9	10,380	187	20	6,517

	11	12	13	14	15	16	17	18	19
Huntingdon	1,694	18,955	30.7	9,454	99.5	620	527	26	1,800
Indiana	9,201	90,659	29.7	44,893	98.1	7,548	1,363	24	6,386
Jefferson	2,278	36,810	37.2	15,588	73.5	1,864	890	23	2,869
Juniata	626	5,064	21.2	2,677	112.1	666	268	13	654
Lackawanna	2,319	41,804	38.3	25,998	164.5	1,512	37	18	2,494
Lancaster	20,870	96,662	17.0	25,457	35.8	1,970	1,899	41	13,237
Lawrence	3,480	48,642	33.9	32,765	200.1	1,677	1,041	17	3,252
Lebanon	3,939	31,624	24.4	11,968	60.9	434	457	15	3,222
Lehigh	5,220	38,543	22.5	6,567	20.5	998	484	14	4,287
Luzerne	4,399	56,719	27.3	44,482	363.9	5,633	1,219	29	5,343
Lycoming	2,033	21,282	25.8	15,006	239.1	1,184	632	42	2,379
McKean	8,216	23,437	9.4	34,476	463.9	6,551	1,997	15	6,078
Mercer	3,359	54,063	35.8	8,270	662.0	3,002	591	32	3,751
Mifflin	1,250	20,796	36.4	5,502	43.9	175	164	10	1,597
Monroe	3,118	18,027	17.4			1,200	945	16	2,446
Montgomery	10,914	85,030	20.5	36,758	76.1	6,891	6,183	30	10,659
Montour	572	6,750	31.5	4,158	160.4	224	107	9	579
Northampton	5,712	69,798	31.7	28,667	69.7	3,974	1,402	17	5,198
Northumberland	2,724	24,119	27.0	10,922	82.8	246	1,053	23	5,055
Perry	1,276	13,101	29.3	2,356	21.9	170	452	21	1,307
Philadelphia	---	---	---	---	---	---	---	---	---
Pike	1,383	4,790	8.9	4,790	20.0	647	222	11	1,617
Potter	1,125	13,512	30.9	2,250	85.4	744	508	24	1,370
Schuylkill	7,256	41,031	18.4	18,903	141.4	5,085	509	27	6,068
Snyder	841	14,047	39.1	8,227		29	562	15	734
Somerset	5,835	46,241	21.4	26,131	129.9	14,405	2,409	25	4,910
Sullivan	410	7,790	38.8	4,815	161.8	136	186	9	456
Susquehanna	2,508	31,662	28.1	19,874	168.6	1,447	1,238	27	3,048
Tioga	3,245	45,162	38.1	17,458	63.0	2,369	2,114	30	2,863
Union	1,131	13,537	33.2	7,623	128.9	201	53	10	885
Venango	4,222	44,775	26.6	24,482	120.6	2,886	1,290	21	4,064
Warren	2,818	26,234	26.8	7,375	129.1	4,037	863	22	2,601
Washington	20,601	128,125	22.2	99,290	344.3	7,098	325	32	10,256
Wayne	2,286	18,957	21.0	5,336	39.1	2,021	751	22	2,364
Westmoreland	23,920	122,948	20.3	95,165	342.5	15,035	3,520	23	10,921
Wyoming	1,351	2,097	3.9	2,097		949	79	18	2,136
York	7,789	53,793	22.2	22,225	70.4	2,026	2,171	34	5,657
Total	$335,644	$2,852,962		$1,505,144		$257,881	$70,825	1,491	$268,380
Average	$5,010	$42,582	24.7	$22,465		$3,849	$1,057		$4,006 per county collector
									$ 180 per collector

TABLE II SECOND CLASS TOWNSHIP ROAD RECEIPTS 1932

County	Cash Road Tax	State Reward	Forest Road Aid	County Aid	Other Sources	Total Available for Road Purposes
	1	2	3	4	5	6
Adams	$ 91,902	$ 2,175	$ 515	$ 3,359	$ 22,500	$ 120,451
Allegheny	319,982	6,370		12,000	109,549	447,901
Armstrong	145,766	11,827		2,939	101,723	262,255
Beaver	125,646	5,274	6,866	5,998	76,907	220,691
Bedford	80,041	4,353	464	855	21,404	107,117
Berks	321,040	5,690	4	8,657	105,718	441,109
Blair	74,375	1,784	6	680	34,189	111,034
Bradford	111,860	3,185	487	596	55,224	174,352
Bucks	348,844		24	26,714	131,241	506,823
Butler	173,891	5,407	13	8,111	115,591	303,013
Cambria	334,990	8,884	1,634	438	192,408	538,354
Cameron	6,089		3,093	3,974	6,668	19,824
Carbon	37,929	3,138	12		15,311	56,390
Center	50,604	7	2,712		32,223	85,546
Chester	423,778	6,505	1,502	37,675	144,200	613,660
Clarion	80,861	2,042	208	2,500	26,433	112,044
Clearfield	87,717		1,997	1,667	44,558	135,939
Clinton	38,556		4,525	2,873	17,229	63,183
Columbia	64,412	1,706	859	5,217	25,838	98,032
Crawford	155,224	2,921	1,936	533	80,416	241,030
Cumberland	110,239		568	740	60,361	171,908
Dauphin	131,146	2,899	78	4,137	42,660	180,920
Delaware	104,960	1,719			16,958	123,637
Elk	48,053	8,129	3,689	150	32,813	92,834
Erie	200,694	7,621	6		60,954	269,275
Fayette	463,035		144		82,880	546,059
Forest	16,253	1,969	2,456		7,002	27,680
Franklin	123,505		782	16,120	48,031	188,438
Fulton	16,970	877	506		10,022	28,375
Greene	274,736	16,481		22,095	178,550	491,862
Huntingdon	52,266	6,466	778	3,700	30,328	93,538
Indiana	244,390	10,374	13	1,947	143,971	400,695
Jefferson	83,185	1,155	518	2,416	39,969	127,243
Juniata	23,503		177		7,478	31,158
Lackawanna	70,510	3,135	106	500	49,851	124,102
Lancaster	535,700	3,119	54	2,535	228,313	769,721
Lawrence	111,200	13,381	7	250	52,504	177,342
Lebanon	114,188	2,584		11,029	70,604	198,405
Lehigh	158,117	2,625		1,313	63,783	225,838
Luzerne	153,803	5,091			44,558	203,452
Lycoming	78,022	5,941	2,466	2,990	41,341	130,760
McKean	238,557	3,846	3,215	34,177	182,413	462,208
Mercer	123,686	2,084	168	14,360	74,291	214,589
Mifflin	47,155		1,024		29,214	77,393
Monroe	94,688	2,321	198	5,785	37,651	140,643

TABLE II (CONTINUED) SECOND CLASS TOWNSHIP RECEIPTS 1932

County	Cash Road Tax	State Reward	Forest Road Aid	County Aid	Other Sources	Total Available for Road Purposes
	1	2	3	4	5	6
Montgomery	$358,872	$6,963	$ 500	$42,055	$116,693	$525,083
Montour	17,821	3,177		893	6,732	28,623
Northampton	182,896	1,128	19		85,680	269,723
Northumberland	66,051	4,876	40	10,645	29,295	110,907
Perry	40,732	2,686	646	1,485	15,287	60,836
Philadelphia						
Pike	53,723	1,554	901	650	19,770	76,598
Potter	53,279	10,506	490	333	8,091	72,699
Schuylkill	156,048	69		2,265	30,164	188,546
Snyder	32,351	2,317		308	4,159	39,135
Somerset	186,219	12,822	261	60	56,113	255,475
Sullivan	21,373	1,211	1,383	125	4,593	28,685
Susquehanna	93,321	2,364	135	74	46,159	142,053
Tioga	98,363	859	1,774		48,083	149,079
Union	36,712	1,733			2,485	40,930
Venango	154,521	189			26,749	181,459
Warren	85,236	3,970	1,080	2,651	20,769	113,706
Washington	452,564	23,733		12,472	189,820	678,589
Wayne	76,255	8,070	4,072		18,942	107,339
Westmoreland	482,576	19,645		5,403	117,271	624,895
Wyoming	47,981	994			10,081	59,056
York	222,688	1,713		16,208	52,067	292,676
TOTAL	$9,614,650	$283,664	$55,111	$344,657	$3,904,833	$14,202,915
Per Cent of Total	68%	2%	1%	2%	27%	100%

TABLE III SECOND CLASS TOWNSHIP EXPENDITURES BY COUNTIES 1932

County	Maintenance Earth Roads	Maintenance Improved Roads	Repairs State Reward Roads	Repair State Aid Roads	Opening and Building New Roads	Permanent Improvements of Roads	Permanent Bridges and Culverts	New Tools and Machinery	Repairs Tools and Machinery	Roadmasters' Wages	Treasurers' Compensation
	1	2	3	4	5	6	7	8	9	10	11
Adams	$28,620	$2,405	733	$2,706	$63	$9,338	$2,661	$6,733	$2,592	$11,261	$1,189
Allegheny	99,145	5,228	9		7,067	56,245	10,989	11,487	10,637	70,530	2,968
Armstrong	76,877		4,818		5,413	7,211	7,912	1,882	2,532	16,503	1,531
Beaver	33,447	3,983	107	440	5,660	23,827	4,152	2,871	1,736	9,243	1,361
Bedford	36,420				1,423	5,435	3,206	1,403	1,795	10,684	1,216
Berks	69,811	24,301		799	2,508	76,203	13,258	9,374	8,637	34,157	3,356
Blair	19,365	1,219			525	15,272	7,464	1,432	4,133	3,827	942
Bradford	56,326	1,499	522	2,500	88	727	10,350	2,188	2,441	3,880	1,448
Bucks	80,957	37,270		1,222	8,569	99,702	13,159	9,323	9,843	32,306	3,757
Butler	44,800	1,288		581	5,512	42,047	5,072	4,569	3,223	114,783	1,349
Cambria	83,713	1,889	250		22,299	21,493	2,126	11,468	4,991	25,856	1,945
Cameron	7,145	441			455	49	2,024	381	335	2,648	151
Carbon	15,031	1,636	200			2,314	1,952	107	614	3,170	578
Center	22,945	4,375	955	16		1,382	2,671	2,560	2,324	8,040	837
Chester	91,527	18,914		1,196	246	169,064	12,834	12,831	5,071	27,910	5,213
Clarion	25,213	1,070	16	1,069	251	7,137	2,608	6,563	1,133	9,361	761
Clearfield	41,963				113	1,256	1,602	2,028	1,750	17,547	882
Clinton	13,533	5,116	126	151	3,310	10,377	1,335	847	427	5,192	658
Columbia	32,351	21	73	1,011	530	1,007	10,162	654	727	10,104	1,040
Crawford	86,426				5,394	21,348	12,942	2,672	1,631	6,524	1,813
Cumberland	25,209	11,114	5	1,218	1,180	15,347	2,355	685	3,159	10,861	1,175
Dauphin	31,011	12,792			2,842	48,323	7,333	1,580	2,341	10,165	1,449
Delaware	15,404	5,671	5,780	65	9,958	14,345	1,203	172	872	8,523	1,039
Elk	25,927				171	1,185	1,134	6,017	1,656	7,771	843
Erie	48,051	1,166				16,442	6,552	4,777	3,675	5,648	1,875
Fayette	125,868	29,758			5,803	45,297	5,285	17,939	10,287	39,447	2,704
Forest	8,735					6,100	925	84	311	1,235	254
Franklin	48,399	12,145		220	297	21,422	4,176	532	3,207	10,749	1,236
Fulton	7,261		165	219		103	2,674	200	64	2,561	282
Greene	71,652	3,176			159	82,245	13,090	18,499	15,115	21,856	2,818

	1	2	3	4	5	6	7	8	9	10	11
Huntingdon	22,585	240	235	126	18	12,416	2,049	801	448	5,694	217
Indiana	55,233	1,387		1,000	4,551	11,250	4,554	3,873	3,206	16,153	2,343
Jefferson	25,099			273	5,684	6,352	2,458	809	916	10,205	520
Juniata	15,566						548	56	62	3,007	155
Lackawanna	31,931	890			5,005	5,575	2,510	1,193	916	7,511	1,078
Lancaster	65,266	64,208	4,543	4,950	9,306	175,279	21,043	17,779	11,057	46,314	6,524
Lawrence	43,330	117	207	274	150	30,838	8,658	4,381	2,338	12,281	1,409
Lebanon	17,951	18,934			1,681	29,040	3,691	3,531	2,709	14,384	1,194
Lehigh	55,102	4,928			10,447	21,141	3,028	3,812	2,231	17,267	1,927
Luzerne	70,263	2,521			641	10,833	3,537	2,728	1,422	22,194	1,894
Lycoming	33,399	37	25	7	460	17,078	5,237	3,633	2,208	12,522	1,203
McKean	26,819	4,949	1,767	47	11,432	161,451	11,900	12,430	5,575	15,039	3,547
Mercer	36,409	3,333	109	7,139	1,673	46,423	1,680	5,965	1,158	6,794	897
Mifflin	14,932	5,969	839	100	266	150	2,893	159	612	4,217	259
Monroe	46,036				1,367	9,821	7,559	4,854	2,562	14,052	1,422
Montgomery	51,608	86,149	2,046	250	6,851	93,557	13,311	3,265	6,872	23,920	2,975
Montour	9,203	1,138			664	1,812	1,560	59	90	3,593	214
Northampton	43,498	12,377	6,891	3,150	671	39,217	3,559	2,506	2,073	17,006	1,943
Northumberland	30,631				1,228	2,859	12,361	1,045	843	9,260	1,132
Perry	19,854		88		6	4,661	2,106	156	356	5,259	344
Philadelphia	28,743	846	431	556		2,886	2,500	809	275	3,325	611
Pike	23,217						2,548	608	488	2,878	836
Potter	53,473	3,513	395	197			2,548	608	488	2,878	836
Schuylkill	16,556	49	191			23,295	10,696	10,572	4,810	17,581	2,386
Snyder						1,205	4,609	1,192	292	4,422	609
Somerset	61,474	1,339	20,973	1,066	6,264	13,428	6,689	5,116	5,792	16,611	1,600
Sullivan	10,669	25			440	5,789	2,377	78	74	1,618	225
Susquehanna	46,316		217		215	2,176	7,931	1,761	959	9,793	1,217
Tioga	44,949	877	5,154		268	2,984	4,617	2,039	2,722	12,062	1,448
Union	8,931		2,462				2,545	142	270	4,219	323
Venango	61,091	1,770	320	375	158	14,794	14,316	4,318	6,642	8,914	711
Warren	34,551	5	684			13,355	3,546	2,150	2,038	5,980	650
Washington	76,497	26,479			9,662	118,048	15,264	25,192	17,132	40,298	2,482
Wayne	41,202			2,098	1,000	8,868	2,963	4,236	515	6,243	1,029
Westmoreland	156,729	14,861	98	795	531	16,278	28,340	29,992	19,294	40,779	3,526
Wyoming	25,689	7,821	1,140	4	233	648	2,503	969	477	6,650	658
York	81,110				10,438	37,383	11,590	8,815	3,148	23,392	2,542
Total	$2,862,044	$451,239	$62,573	$35,820	$181,146	$1,763,133	$413,685	$310,882	$219,761	$873,079	$98,730
	22%	4%	1%		1%	14%	3%	2%	2%	7%	1%

TABLE III (CONTINUED) SECOND CLASS TOWNSHIP EXPENDITURES BY COUNTIES 1932

County	Secretaries' Compensation	Tax Collectors' Compensation	Premium Treasurers' Bond	Premium Tax Collectors' Bond	Snow Removal	Annual Supervisors' Convention	Interest on Notes	Notes Maturing	Interest on Bonds	Sinking Fund for Bonds	Redemption of Bonds
	12	13	14	15	16	17	18	19	20	21	22
Adams	$960	$2,675	$15		$465	$442	$5,308	$26,066	35		9,000
Allegheny	3,210	7,154	680	920	644	314	3,963	60,572	5,460	5,772	31,285
Armstrong	1,635	4,231	235	133	1,213	783	4,787	30,735	5,852		13,000
Beaver	1,638	3,076	115	54	484	353	2,531	31,284	21,374	11,089	
Bedford	949	2,068	72		1,197	19	4,368	13,331	496		500
Berks	3,339	9,255	215	175	2,217	851	9,180	80,857	712		
Blair	920	2,636	70	157	1,594	301	2,732	12,407	816		1,300
Bradford	1,699	3,839		130	7,860	831	6,950	38,450		6,397	652
Bucks	3,407	9,357	207	72	988	584	8,208	80,018	4,326		4,500
Butler	1,679	5,014			1,478	530	4,730	66,160	14,508	19,678	17,644
Cambria	2,830	9,753	105	365	9,205	524	9,146	24,496	49,064	45,525	84,000
Cameron	267	10			20	148	28	1,500			
Carbon	341	1,228		2	305	122	1,401	14,700	27	1,285	
Center	615	1,561	65	176	2,694	833	1,517	4,763	54		105
Chester	3,569	13,502	468	371	383		13,317	131,783	9,604	9,538	26,000
Clarion	1,167	2,807	13	35	1,689	301	2,314	14,940	4,574	1,644	
Clearfield	1,265	2,377	25		1,956	551	2,723	6,835	8,026	2,735	7,500
Clinton	481	1,271	35	36	1,169	161	1,705	5,234	3		14,100
Columbia	637	1,860	27		2,755	430	2,078	12,777			
Crawford	1,961	5,049	120	50	2,364	712	10,043	43,759	3,049	1,143	2,036
Cumberland	1,618	4,450	71	80	803	429	4,154	59,854	1,196	500	
Dauphin	1,553	3,733	120	81	635	498	2,943	19,136	643	1,000	2,500
Delaware	908	2,779	110	206	123	153	2,126	24,205	3,948	3,994	1,500
Elk	691	1,168		87	1,334	151	597	11,180	1,463		800
Erie	2,200	5,617	65		4,116	536	9,874	45,390	22,434	17,903	33,400
Fayette	4,751	11,246	996	143	2,331	385	17,785	92,038	12,795	31,298	29,458
Forest	277	400	25		18	65	238	1,221	165		1,531
Franklin	1,163	3,441	130	20	2,992	310	6,849	47,114	951		5,065
Fulton	274	507			355	233	444	4,594	90	2,000	
Greene	2,896	6,381	1,412	157	1,184	411	4,958	74,288	19,221	37,800	7,000

	12	13	14	15	16	17	18	19	20	21	22
Huntingdon	1,010	1,565	15		1,050	641	2,776	13,814	3,080		4,500
Indiana	2,345	6,966	188	127	2,764	370	7,652	46,918	35,047	40,987	31,212
Jefferson	777	2,536	35	156	799	652	8,238	19,005	7,686	2,498	11,500
Juniata	221	828			323	262	862	6,500			
Lackawanna	805	1,850	140	409	4,858	315	3,660	33,492	275	1,066	1,000
Lancaster	3,621	11,358	525	41	460	730	8,965	40,538	18,325	9,351	5,235
Lawrence	1,145	2,942	110	29	340	267	2,594	18,997	2,198	6,445	10,522
Lebanon	1,194	3,172	55		328	266	2,382	56,209	1,068	1,600	
Lehigh	1,009	4,286	95		178	284	2,884	41,980	150	2,223	
Luzerne	1,369	5,038	200	83	4,435	369	3,532	29,902	957	1,000	
Lycoming	791	2,317	80	25	3,236	818	2,136	13,292	1,042	7,632	2,039
McKean	5,610	6,491	50	20	619	268	4,855	68,291	15,747	2,841	29,500
Mercer	1,640	3,584	59	82	1,498	841	7,586	57,627	300	2,804	
Mifflin	1,013	1,660	35		724	169	2,525	18,887	1,078	800	5,000
Monroe	886	2,596			1,287	276	1,217	28,001	761		3,000
Montgomery	2,985	9,479	494	216	351	604	11,005	114,499	1,807	6,209	8,108
Montour	208	497			1,828	155	627	2,150			
Northampton	1,848	4,883	40		1,262	284	5,163	44,721	12,521	5,882	5,000
Northumberland	873	1,813			2,690	288	1,870	12,886	445		500
Perry	529	1,156	15	1	668	338	988	13,964			
Philadelphia											
Pike	524	1,417	25		338	181	1,101	20,940	235		500
Potter	640	1,503	26	80	865	23	2,700	22,361	894		
Schuylkill	1,386	4,552	2		3,292	289	8,704	34,546	100		1,500
Snyder	328	811			758	281	768	5,080			
Somerset	1,662	4,872			4,774	795	7,218	31,671	6,441	9,462	6,992
Sullivan	318	277			2,221	177	1,624	4,964		28	
Susquehanna	1,132	2,633	20	1	1,694	570	4,842	28,418	548		3,800
Tioga	994	3,111	49		4,189	525	5,282	35,913	3,606	1,609	3,000
Union	334	811		22	438	183	644	7,881			
Venango	1,511	4,677	213	87	1,788	439	1,874	8,787	9,549	2,302	24,000
Warren	1,264	2,708			740	446	3,859	16,549	2,225	5,049	2,500
Washington	3,846	10,386	700	612	4,025	553	11,122	111,325	21,803	11,000	38,000
Wayne	836	2,214	23		288	211	5,057	30,833	78		
Westmoreland	5,776	10,212	115		3,820	541	6,849	92,294	35,197	85,946	51,565
Wyoming	2,744	1,574	60	138	3,848	330	2,114	11,416	1,252		4,000
York	2,342	6,175			3,773	699	3,908	67,541			
Total	$100,446	$261,295	$8,765	$5,579	$122,150	$26,411	$295,733	$2,392,879	$375,301	$106,038	$545,849
	1%	2%			1%		2%	19%	3%	3%	4%

TABLE III (CONTINUED) SECOND CLASS TOWNSHIP EXPENDITURES BY COUNTIES 1932

County	Supervisors' Monthly Meetings 23	Rent, Heat, Light for Meetings 24	Stationery 25	Signs and Index Boards 26	Miscellaneous 27	Compensation Insurance 28	Attorneys' Fees 29	Auditors' Fees 30	Gas and Oil 31	Total Expenditures 32
Adams	$ 1,435	$ 64	$ 112	8	$ 3,088	$ 126	$ 265	$ 111		$ 109,476
Allegheny	3,540	438	1,665	19	25,944	3,921	734	498	3,070	366,051
Armstrong	2,155	124	245		10,790	588	204	263		225,709
Beaver	1,719	83	376	68	19,683	393	175	183	1,304	195,369
Bedford	1,592	102	234		11,771	539	177	193		99,630
Berks	3,591	496	1,487	302	30,707	1,289	976	432	1,305	389,790
Blair	2,169	313	148	86	9,362	487	111	173	664	92,605
Bradford	4,175	197	283	2	8,562	178	80	267	114	155,791
Bucks	2,771	593	1,478	56	17,663	1,611	109	120	2,126	440,699
Butler	2,613	136	243	266	14,269	196	230	149	158	272,905
Cambria	2,757	385	1,376	212	16,150	926	965	113	920	435,147
Cameron	248	30	32		418	913		10	28	17,173
Carbon	1,293	94	160		2,194	88		50	24	48,642
Centre	1,099	89	127	10	7,679	179	30	86	259	66,583
Chester	4,842	589	1,356	250	23,957	1,575	2,044	460	822	591,024
Clarion	1,103	173	212	56	4,595	31	63	35		98,371
Clearfield	1,986	131	171	112	7,430	623	49	239	1,264	119,713
Clinton	999	48	103	24	2,793	108	338	65		57,230
Columbia	1,229	311	76	46	2,488	460	236	105	516	86,349
Crawford	3,753	196	71	101	10,504	600	236	243		225,329
Cumberland	1,407	257	270	7	5,413	532	410	195	454	154,408
Dauphin	1,922	178	437	43	3,588	199	56	70	1,141	158,312
Delaware	1,176	146	716	3	11,471			15		110,791
Elk	540	30	92	2	2,398	153	61	42	752	71,965
Erie	3,227	162	196	82	10,207	291	227	65		244,208
Fayette	2,236	974	734	27	21,655	779	787	170	2,466	515,442
Forest	694	15	67	286	1,156	145	10	40		23,997
Franklin	1,196	343	208	5	3,600	201	100	89	1,364	177,524
Fulton	736	26	90	5	1,509	98	20	55		24,181
Greene	1,396	255	484	1	39,832	1,910	91	245	2,811	431,727

	23	24	25	26	27	28	29	30	31	32
Huntingdon	1,333	171	313	31	5,583	526	100	165	46	81,548
Indiana	2,163	132	296	13	7,393	345	250	156	2,543	291,417
Jefferson	1,015	119	184	3	5,362	158	155	123	16	113,343
Juniata	688	13	89		1,774	113	15	38		31,137
Lackawanna	6,359	453	157	17	4,579	143	28	59	425	116,682
Lancaster	3,478	839	1,036	236	18,860	1,937	266	165	1,755	652,990
Lawrence	1,321	151	352	28	4,767	335	109	115	440	156,983
Lebanon	1,106	213	355	16	9,402	232	70	99	1,917	173,006
Lehigh	1,541	451	686	66	17,180	312	78	45	351	190,982
Luzerne	1,799	335	375	4	17,569	522	187	85	167	185,959
Lycoming	1,909	171	205	53	4,928	467	455	309	222	110,282
McKean	1,736	211	263	114	13,374	1,219	25	110	46	408,355
Mercer	3,582	55	188	2	5,268	638	400	326	308	203,063
Mifflin	1,299	63	317	26	3,741	219	45	55		69,326
Monroe	1,339	145	293	23	4,851	441	160	63	530	135,181
Montgomery	2,740	839	780	385	40,671	1,326	308	120	951	494,681
Montour	295	38	14	15	894	222	50	53		28,529
Northampton	2,089	384	317	48	7,047	260	791	66	150	222,497
Northumberland	1,839	134	283	19	9,321	214	210	137	82	93,963
Perry	1,392	57	102	.	1,567	243	55	89		53,994
Philadelphia	942				4,358	77	50	33	94	72,041
Pike	242	56	71	37	3,120	102	114	59	5,014	73,381
Potter	242	206	140	100	3,143	1,257	647	162	2,048	191,766
Schuylkill	1,910	259	554	194		205	100	73	24	39,992
Snyder	743	34	144	24	1,494					
Somerset	1,801	92	332	6	7,703	705	226	266	1,127	226,498
Sullivan	560	80	71		963	165	50	74		26,613
Susquehanna	2,026	282	94	3	5,743	85	10	74		126,540
Tioga		97	251	87	8,313	258	66	115	12	144,917
Union	860	126	128	4	3,020	162	280	45		37,937
Venango	2,661	209	300	4	13,212	309	132	80	121	184,999
Warren	2,329	273	240	15	9,276	426	2	115	1,000	111,611
Washington	3,772	368	877	79	28,664	2,548	745	181	6,185	588,904
Wayne	1,413	152	169	78	4,821	168	253	60	112	114,858
Westmoreland	2,137	664	799	195	32,868	1,652	1,805	623	3,738	648,019
Wyoming	1,284	87	74	7	1,954	25		45		61,333
York	2,368	414	857	149	8,193	536	344	246		290,374
Total	$127,814	$15,351	$24,955	$4,098	$657,852	$37,691	$17,129	$9,388	$50,986	$12,757,802
	1%				5%					100%

CHAPTER XIV

ROADS AND BRIDGES

The principal activities of the second class township in Pennsylvania are the maintenance and construction of township roads and bridges. The township as a district also has other functions such as the support of public schools, entrusted to the school directors, and in 16 counties the township provides for the poor through poor directors. In a few townships the supervisors levy special taxes for street lighting and police and fire protection. The most important duty of the township supervisors throughout the state is in connection with their highway services.

Mileage of Public Roads in Pennsylvania - The place occupied by the township in the public road system of the Commonwealth exceeded that of any other governmental division in 1933 with regard to the number of miles of highways. In second class townships 49% of the public road mileage or 48,462 of a state total of 99,175 miles is maintained, constructed and repaired. The miles of highways are presented by governmental divisions on Table 1 and may be summarized as follows:

MILEAGE OF PUBLIC ROADS IN PENNSYLVANIA [†]

	Miles of Road	Percent of Total
State Highways*	35,516	36%
SECOND CLASS TOWNSHIP	48,461	49
County	1,798	2
City	6,076	6
Borough	6,938	7
First Class Township	376	-
Total	99,165	100%

The second class township mileage varies greatly from county to county. In 16 counties the township mileage exceeds 1,000 miles. In 16 other counties the township mileage is less than 400 miles per county. Cameron County has the smallest number of miles (70) of township roads.

[†] Figures are approximate. New figures adjusted to 1933 session changes have since been compiled by the Pennsylvania Highway Department. Former compilations of public road mileage show 103,590 miles in Pennsylvania. The new total is 4,425 miles less, probably due to a new figure of 48,461 miles of second class township roads. The new figure is based on mileage certified by township supervisors in 1933 instead of a former figure of 53,874 miles compiled from the annual reports from township supervisors.

* State highway mileage may be allocated as follows:

Old state highway system	13,451
1931 rural state highway system	20,172
Other	
City streets connecting state highways	302
State aid in boroughs	65
1933 legislation, State, city and borough	831
Abandoned turnpikes	695
Total	35,516

PENNSYLVANIA LOCAL GOVERNMENT SURVEY

TABLE I

HIGHWAY MILEAGE BY GOVERNMENTAL DIVISIONS①

County	State ②	Second Class Townships ③	County ④	City ⑤	Borough⑥	First Class Townships	Total
	1	2	3	4	5	6	7
Adams	481	670	3		32		
Allegheny	788	451	547	1450	875		
Armstrong	608	1,110			63		
Beaver	536	572	19		244		
Bedford	718	914	16		33		
Berks	880	1,458		55	143		
Blair	348	310	73	285	54		
Bradford	880	1,533			88		
Bucks	817	1,034	26		140		
Butler	678	1,094	5	60	81		
Cambria	540	590	15	82	152		
Cameron	100	70			8		
Carbon	228	305			97		
Center	538	499	57		39		
Chester	990	1,347		25	153		
Clarion	472	889			42		
Clearfield	726	960	68	25	109		
Clinton	266	300		19	27		
Columbia	491	731	2		91		
Crawford	821	1,383		67	58		
Cumberland	496	640	31		119		
Dauphin	423	595	4	133	97		
Delaware	374	110		62	234		
Elk	240	314	2		44		
Erie	693	969		246	82		
Fayette	612	1,085	61	89	143		
Forest	173	193			3		
Franklin	519	739	6		58		
Fulton	302	315			2		
Greene	528	884	11		48		
Huntingdon	614	611	10		45		
Indiana	806	1,141			94		
Jefferson	496	782	6		76		
Juniata	295	405			6		
Lackawanna	373	361	45	183	373		
Lancaster	1,046	1,854		79	104		
Lawrence	351	516		87	64		
Lebanon	345	475		70	58		
Lehigh	430	715	5	84	66		
Luzerne	672	680	173	257	268		
Lycoming	750	900	6	135	73		
McKean	389	347		22	45		
Mercer	624	1,056		61	117		
Mifflin	228	282			25		

TABLE I

HIGHWAY MILEAGE BY GOVERNMENTAL DIVISIONS ①

County	State ②	Second Class Townships ③	County ④	City ⑤	Borough ⑥	First Class Townships	Total
	1	2	3	4	5	6	7
Monroe	423	525	37		50		
Montgomery	745	785	62		291		
Montour	153	197			15		
Northampton	417	657		159	150		
Northumberland	481	676		31	99		
Perry	389	537			23		
Philadelphia		-		1995			
Pike	283	313	21		24		
Potter	452	595			40		
Schuylkill	557	925		36	186		
Snyder	307	444			18		
Somerset	816	1,252	33		113		
Sullivan	262	268			10		
Susquehanna	815	980	81		84		
Tioga	682	1,012	2		62		
Union	238	270	9		23		
Venango	436	741		70	29		
Warren	443	665			63		
Washington	858	1,221	142	71	221		
Wayne	618	719	125		57		
Westmoreland	1,115	1,389	90	75	393		
Wyoming	328	383	5		24		
York	1,013	1,718		63	190		
Total	35,516	48,461	1,798	6,076	6,938	376	99,165
Percent	36%	49%	2%	6%	7%		100%

① Figures obtained through the courtesy of Honorable Samuel S. Lewis, Secretary, and Harold G. Van Riper, Township Engineer, Pennsylvania Department of Highways.
② As of November 24, 1933.
③ This is termed "Certification Mileage" by the Highway Department. It is the mileage certified by the township Supervisors to the Department in the Fall of 1933 for Parkinson Act purposes.
④⑤⑥ As of July 1, 1929.

Differences in Total Mileage of Township Roads - Since there are different figures for the total mileage in the Township Highway System an explanation indicating the manner of their origin is appropriate. The various mileage figures may be summarized as follows:

Type of Mileage	Number of Miles
Certification Mileage (1933)	48,461
Speedometer Survey Mileage (911 Mileage) (1933)	48,479
1932 Mileage (Certification Mileage adjusted to 1932)	
Township Highway System Mileage (approximate) (Before 1933)	53,874

The "Certification Mileage" is the mileage determined by the township supervisors and forwarded by them to the State Department of Highways for Parkinson Act purposes.

This Act was adopted in 1933 and it provided for the payment of State money to the Highway Department for maintenance work in second class townships at a fixed rate per mile of road. Provisions of the Act required that the mileage be attested and sworn to by municipal officials and that the Department be governed by the certification (1933 P.L. 1520 Sec. 6).

When the allocation for various townships had been made in the fall of 1933 controversies arose as to the correctness of certification mileage. To dispel these doubts the Secretary of Highways ordered a recheck by State Highway engineers. The result was the second type of mileage indicated above. The engineers traversed every mile of township road making ratings as to the condition of the road as well as to its length. The ratings were entered upon a mimeograph form and were termed "911 Mileage". This mileage is also termed "Speedometer Survey Mileage" because distances were measured upon the speedometers of State cars used by the engineers.

Mileage described as "1932 Mileage" employed the Certification Mileage as a basis but various changes especially those made by legislative enactment in 1933 were eliminated from its computation in order to get the actual mileage obtaining in 1932. This mileage is used to determine maintenance costs per mile in individual townships, and the 1932 Annual Reports of Township Supervisors supply the figures upon which these costs are based. In this manner the actual mileage prevailing in 1932 is used with the expenditures for the same year.

Township highway system mileage was formerly estimated by the Highway Department at approximately 53,000 miles. These estimates were based upon figures entered by supervisors of each township on their annual reports. Township roads were more frequently termed "The 53,000 Mile System". The township roads are still so described by many observers and this was the only figure for the system until the two computations in the fall of 1933 arising from the Parkinson Act.

Township Roads Mainly Unimproved - The road mileage in second class townships has been classified by the Department of Highways. This classification was secured as to type of road for all roads in the township system. Ratings of the surface type distinguished between 4 groups as follows:

Type of Road 1933	Miles
Unimproved Earth	42,989
Stone	4,040
Bituminous macadam	1,148
Concrete, brick, etc.	98
Total	48,275 ‡

Not only do the townships have nearly half of the road mileage of the State but also the mileage which they have is almost entirely dirt road.

---

‡ These are preliminary figures. Later corrections increased the total mileage to 48,479.

Township Roads in "Fair" Condition - During the speedometer survey made by the State Highway Department in 1933, each township road was checked on a field sheet as to its condition. While individual errors will occur in such a computation, the ratings in general appear to have been accurately listed.

Condition of Road 1933	Miles
Good	4,154
Fair	21,707
Poor	20,757
Impassable	1,924
Unclassified	3
Total	48,545 §

Classification by type of road by counties is based on road computations in individual townships. There are 161 townships scattered throughout the State of which special study was made including road classification as unimproved, stone, bituminous macadam, concrete, and brick. They have also been classified as good, fair, poor, and impassable. As was indicated in an earlier chapter, a proportionately large mileage of bituminous macadam road may usually be taken as an indication of suburban development.

Readjustment of Mileage Under State Control of Township Roads - If the Commonwealth continues to perform maintenance on second class township roads, a considerable readjustment of road mileage is involved. Under state control maintenance mileage would be altered as follows:

	Miles of Road
State	83,977
County	1,798
City	6,076
Borough	6,938
First Class Towship	376
Total	99,165

Since the State Department of Highways performed road maintenance work in second class townships during 1934 under the provisions of the Parkinson Act (1933 P.L. 1520) the manner in which the township road system was divided for the purpose is significant. The Department grouped the township system into 6 main divisions as follows:

Highway Department Divisions for the Township Road System

Highway Department Division	Number of Townships	Miles of Townships Roads ★
1	271	8,093
2	329	9,468
3	269	7,861
4	215	7,913
5	239	7,757
6	192	7,371
	1,515	48,462

Range of Township Road Mileage - The range of road mileage in second class townships is as diverse as the range of population noted in Chapter 10. The Department of Highways has grouped townships according to number of miles of roads. This grouping is summarized below:

Miles	Number of Townships
Less than 1 mile	3
1-10	162
10-35	781
35-100	557
100-200	11
Total	1,514

§ These are preliminary figures. Later corrections reduced the total mileage to 48,479.
★ This is certification mileage adjusted to nearest even mile.

The three townships having less than one mile of road in 1933 were Aleppo Township, Allegheny County, .7 miles; Buck Township, Luzerne County, and Portage Township, Potter County, no miles of roads.

Miles of Township Road in Counties - Average Per Township - The average number of miles of road for each second class township is 32 miles. This is based upon a total road mileage of 48,461. The counties in which the average miles per township were highest and lowest in 1933 were as follows:

County	Highest Average Miles per Township	County	Lowest Average Miles per Township
Westmoreland	60	Delaware	10
York	51	Clinton	14
Lehigh	51	Cameron	14
Somerset	50	Allegheny	16
Franklin	49	Center	20
Indiana	48	Lackawanna	20
Erie	46	Cambria	20
Fayette	45	Huntingdon	20
Lancaster	45	Blair	21
Greene	44	Luzerne	21
		Lycoming	21
		Wyoming	21

The low mileage in Allegheny and Delaware counties may be attributed to the large number of first class townships in those counties. The average mileage per township would have to be doubled or even trebled to secure an efficient operating unit, appropriate for trained supervision such as that of a qualified road man. At the present time road supervisors are paid by the day and possess no special training.

Classification of Township Roads - The roads in a township should be classified by the supervisors to designate as primary the roads that are most important. These roads connect schools and populated centers and carry most of the township traffic. A less traveled group of roads may be designated as secondary highways. These are necessary but not subject to frequent use. They are needed by the postman, doctor, fire company, school bus and occasional visitor. Such classification enables the supervisors to put their major efforts on the primary system without neglecting the secondary roads.

Vacation of Roads - Some roads are travelled so infrequently that they cannot be placed in either of the above classes. These roads are unimportant and should be vacated. There are many miles of useless roads in Pennsylvania and each county contains its share. The vacation of unused roads not only relieves the township of maintenance expense but it also reduces the township liability for accidents which may be caused if someone wanders from the usually travelled route. The greatest possibility for economy in township road cost (other than reduction of indebtedness) lies in diminishing the number of miles of road on which expenditures may be made. Progress might be made in this direction through a study by the township supervisors or by a committee upon which they were well represented. Inquiries in particular townships in mountainous districts where hunting camps and mountain wood lots provide almost the sole highway users indicated the very great difficulty attendant upon efforts to vacate roads. The owner of a hunting camp, for example, is a taxpayer who wants roads maintained in return for his taxes. Many such roads should be private roads and maintained at the expense of the property owner rather than the public.

Township Road System Divided Into Road Districts; Roadmasters' Wages - The township road system consists of 1,516 townships which are in turn subdivided, each independent of the other, into 3,633 road districts. The maintenance, repair and construction of the township highway system in Pennsylvania is conducted in these districts. As was pointed out in the chapter on "Administration" (Chapter 11) the supervisors in a great majority of cases select themselves as roadmasters. In 1932, 3,148 of the 4,548 township supervisors had done so and the 3,633 road districts constituted the administrative areas for maintenance purposes. The average wages of township roadmasters per capita of population ranged in 1932 from $.15 in Bradford County to $1.26 in Cameron County. There was an average of 3 road districts per township within the state, and an average of 2.56 roadmasters per township.

It is evident that too many townships have been established and that the present units are too small in area, population and taxable resources to function effectively. A subdivision of the small township unit into independent road districts, (usually a district for each supervisor) has been effected throughout the state. Such a system is wasteful and ineffective from an administrative viewpoint. In order to pass judgment on township road maintenance and construction a jury of engineers would be required and their verdict in any event would be warmly disputed. But a jury of highway experts is not required to find that the supervision and management of township roads is divided, disorganized and uncoordinated by numerous independent managers and management areas. The supervisor-roadmaster is a per diem worker, and he frequently employs several subordinates who are also paid on a per diem basis. Such supervisors naturally receive a higher rate of compensation than their subordinates because of the greater responsibility which is theirs. Since the number of road districts and supervisors is at least three times in excess of present needs the excess cost involved must reach a substantial sum each year.

The compensation of the various township officials was presented in earlier chapters, showing that 7% out of the 12% of township expenditures allocated to township officials in the form of salaries, wages and fees went to the roadmaster. In view of the discussion of compensation in Chapter 11 it need not be further dealt with here.

Roads in Losch Act Townships - There are 26 so-called "Losch Act" second class townships in Pennsylvania. The legal basis under which such townships operate was described in Chapter 11. In these townships contracts are made between the township supervisors and the private corporations for the maintenance and construction of roads and bridges.

The contract obligates the corporation to maintain, construct and repair all township highways and bridges at its own expense. The township is protected from claims for damages arising from such work. Salaries of township officers are paid by the corporation as follows:

3 Supervisors	$250. each
3 Auditors	25. each
Township Secretary	50.
Attorney	50.

The township supervisors agree on behalf of the township not to levy or to collect any tax for road purposes. They are required to inspect the roads once a month and to notify the corporation of any portions in need of repairs. Where the corporation fails to perform the necessary improvement within five days the supervisors are authorized to do the work at the expense of the corporation (1933 P.L. 103 Sec.1194-1196). The following table indicates the type of road and its condition in the 26 Losch Act townships in 1933.

The condition of roads in Losch Act townships was reported by engineers of the Department of Highways during the fall of 1933 when speedometer survey was made of the township system over the entire state. The purpose of this survey was to serve as a recheck on the number of miles in Losch Act and other townships in order to facilitate allocations and state maintenance on the township road system under the Parkinson Act. Each road was analyzed individually in the township. Reports were made as to the condition of the road, the type of surface and the mileage. The percentage of improved road in Losch Act townships compares favorably on a percentage basis with the number of miles of road improved in second class townships throughout the state.

Township Bridges 1932 - Probably too much stress is placed upon the road functions of second class townships in Pennsylvania in relation to the township highway system. An important part of this system that is frequently overlooked, is to be found in the 17,922 bridges which it contains.

This is an average of more than 13 bridges per township for the 1,241 townships which reported on this subject in 1932. Only bridges with a span in excess of 10 feet are included in these figures. Since there were 1,515 townships, according to the Highway Department in 1932, there were 274 townships which did not report bridges in the space provided on the annual report of the township supervisors to the Department in that year. The majority of township bridges are constructed of wood. An analysis

TABLE II

LOSCH ACT TOWNSHIPS - ANALYSIS BY TYPE OF ROAD - 1933

County	Township	Unimproved Roads	Stone Roads	Bituminous Macadam	Concrete & Brick	Total Mileage	Good	Fair	Poor	Impassable
Carbon	Banks	1.5		1.2		2.7	1.	1.6	.1	
	Mauch Chunk	9.31	.53	5.02	5.7	20.56	10.8	5.73	4.03	
Columbia	Conyngham	5.3		.3		5.60		3.7	1.9	
Lebanon	Cold Spring	3.6				3.60			1.4	2.2
Luzerne	Foster	21.7	.4	1.9		24.00	3.3	18.9	1.8	
	Hazle	12.1		9.3		21.40	3.8	14.4	3.2	
	Jenkins	2.95		1.66		4.61		.96	3.65	
	Pittston	9.43	.73	3.08		13.24		3.04	10.2	
Northumberland	Mt. Carmel	4.		2.25		6.25	.9	3.05	1.1	1.2
	Zerbe	6.57	.23	1.98		8.78	3.8	3.88	.9	.2
Schuylkill	Blythe	16.1		4.4		20.50		10.3	6.	4.2
	Branch	12.5				12.50		9.5	1.2	1.8
	Butler	18.4		8.1		26.50		9.4	17.1	
	Cass	7.7		7.1		14.80		9.9	1.6	3.3
	E.Norwegian	2.6		1.6		4.20		1.8	2.4	
	Foster	4.		1.5		5.50		5.5		
	Fraley	7.53				7.53		5.2	1.63	.7
	Kline	5.2	1.9			7.10	1.9	4.4	.8	
	Mahanoy	6.1		7.6		13.70		8.9	4.8	
	New Castle	8.		.8		8.80		.6	2.	6.2
	Norwegian	11.		2.1		13.10	.5	4.5	7.9	.2
	Rahn	5.				5.00		5.		
	Reilly	9.				9.00		9.		
	Schuylkill	1.2		2.6		3.80		3.8		
	Tremont	10.7				10.70		6.2	2.8	1.7
	West Mahanoy	7.4		2.2		9.60		4.4	5.2	
Total		208.89	3.79	64.69	5.7	283.07	26.00	153.66	81.71	21.7

of bridges maintained by townships with a span in excess of 10 feet follows:

Wood	10,154½
Concrete	5,899½
Steel	1,868
Total	17,922

Township bridges in many sections of Pennsylvania were built with the assistance of state and county funds under what is described in an earlier Chapter as the Township Reward System. In 1932, 33 bridges (with a span exceeding 10 feet) per township were reported in York County. This was the largest number of township bridges in the state. In the same year only three bridges per township were reported in Cameron and Luzerne Counties. This was the smallest number reported. An analysis of the bridges per township, according to counties is given below in Table II. The number of bridges per township has an important affect on the amount of tax which must be levied for road purposes. Throughout the Township Code the tax for road purposes is interpreted to include the maintenance and construction of both roads and bridges. Geographic conditions, including nearness to streams and rivers, and elevation above sea level, are important factors in determining the extent to which township funds must be expended. In the 1933 Session of the General Assembly a bill was introduced to amend the 20,000 mile state rural road system (1931 P.I. 594) so as to place upon the state the obligation to construct and maintain all township bridges on that system. The bill appears to have been passed unanimously by the General Assembly. It would have relieved second class townships of approximately 8,000 bridges on roads which have been part of the State Highway System since 1931, as the Baker (Union County) bill included bridge structures of both counties and townships on the 20,000 mile system. Although it was unanimously passed in this form, it was reconsidered at the Governor's request and an amendment was adopted eliminating county bridges from the bill. The bill passed both Houses but was vetoed by the Governor.

Purchases of Material and Equipment Approved by the Highway Department - Purchases of materials and equipment by second class townships are required by the township code to be approved by the Pennsylvania Highway Department where the amount of the purchase is in excess of $200 (1933 P.L. 103, Sec. 802). The amount of such purchases has steadily declined over the years 1930-1933.

Year	Amount of Purchases	Percent of Decrease Based on 1930
1930	$2,344,684	
1931	1,583,745	32%
1932	1,277,514	46
1933	747,669	68

In 1931, purchases were made by townships in every county, the lowest being $448, approved for townships in Fulton County. In 1932, every county was again represented as making purchases, the lowest being $250, made by townships in Juniata County. In 1933, a somewhat different situation was presented. In that year, no purchases in excess of $200 were made by townships in 7 counties: Cameron, Center, Clearfield, Juniata, Potter, Tioga and Union.

The trend of township expenditures during the four-year period 1930-33, may be further illustrated by an analysis of the items purchased with the approval of the Department of Highways. As the expenditures declined approximately 68% during the period, a corresponding decline occurred in the various businesses which would ordinarily supply the materials indicated.

Iron-pipe, road machinery, asphalt, oil, lumber and other industries lost several million dollars in sales of these materials as the depression progressed. Since it became increasingly difficult for industries to find purchasers for their products, the corresponding increase in tax delinquency noted during the same period in a previous chapter, can readily be understood.

TABLE III

SECOND CLASS TOWNSHIP BRIDGES 1932

County	Number of Townships	Number of Townships Reporting Bridges 1932	Bridges Per Town-ship	Bridges Maintained by Townships Over 10' Span 1932			
				Wood	Concrete	Steel	Total
Adams	21	16	16	181	62	6	249
Allegheny	29	23	7	79	84	7	170
Armstrong	28	24	23	332½	193	24	549½
Beaver	22	16	9	101	39	11	151
Bedford	25	23	19	387	39	10	436
Berks	44	35	14	266	180	47	493
Blair	15	13	9	53	52	6	111
Bradford	38	32	23	575	70	97½	742½
Bucks	31	25	23	228	277	81	586
Butler	32	27	8	118	54	19	191
Cambria	29	20	6	32	83	5	120
Cameron	5	4	3	8	3	1	12
Carbon	12	10	7	35	30	4	69
Center	25	21	13	167	68	37	272
Chester	57	49	12	339	180	80	599
Clarion	22	19	20	211	100	62	373
Clearfield	30	27	9	122	101	29	252
Clinton	21	17	6	51	39	17	107
Columbia	24	20	9	101	71	17	189
Crawford	35	34	25	344	335	167	846
Cumberland	22	22	9	125	67	9	201
Dauphin	22	17	11	95	96	3	194
Delaware	11	4	11	23	12	7	42
Elk	10	9	9	11	32	42	85
Erie	21	19	31	106	313	171	590
Fayette	24	22	21	294	162	11	467
Forest	8	6	9	19	3	10	32
Franklin	15	13	21	140	120½	6	266½
Fulton	11	11	10	92	11	3	106
Greene	20	18	18	227	83	15	325
Huntingdon	30	28	9	205	26	7	238
Indiana	24	18	14	142	90	24	256
Jefferson	23	18	8	90	47	9	146
Juniata	13	12	18	163	47	-	210
Lackawanna	19	11	5	27	20	6	53
Lancaster	41	38	19	278	382	54½	714½
Lawrence	17	14	11	55	63	37	155
Lebanon	16	15	12	121	61	-	182
Lehigh	14	13	20	95	140	20	255
Luzerne	32	3	3	3	4	3	10
Lycoming	42	34	10	208	88	58	354
McKean	15	13	10	16	44	70	130
Mercer	32	25	9	134	93	5	232
Mifflin	10	9	15	58	61	14	133
Monroe	16	13	11	70	64	3	137

TABLE III

SECOND CLASS TOWNSHIP BRIDGES 1932

County	Number of Townships	Number of Townships Reporting Bridges 1932	Bridges Per Township	Bridges Maintained by Townships Over 10' Span 1932			
				Wood	Concrete	Steel	Total
Montgomery	30	27	16	184	248	7	439
Montour	9	4	7	15	9	3	27
Northampton	17	11	14	107	43	-	150
Northumberland	23	18	9	128	37	-	165
Perry	21	15	10	118	20	5	143
Philadelphia	-	-	-	-	-	-	-
Pike	11	9	9	68	8	1	77
Potter	25	21	13	93	63	107	263
Schuylkill	37	24	17	317	70	26	413
Snyder	15	15	10	128	22	1	151
Somerset	25	21	19	338	56	14	408
Sullivan	9	9	10	37	14	35	86
Susquehanna	27	24	10	198	22	12	232
Tioga	30	24	18	190	135	113	438
Union	10	10	12	73	43	-	116
Venango	21	20	15	181	87	25	293
Warren	23	20	13	102	58	97	257
Washington	32	28	12	151	184	8	343
Wayne	22	17	13	136	71	15	222
Westmoreland	23	20	30	278	241	75	594
Wyoming	18	14	7	82	15	1	98
York	34	30	33	703	264	8	975
Total	1,515	1,241	13 1/3	10,154½	5,899½	1,868	17,922

The number of bridges per township has an important affect on the amount of tax which must be levied for road purposes. Throughout the Township Code the tax for road purposes is interpreted to include the maintenance and construction of both roads and bridges.

Geographic conditions, including nearness to streams and rivers, and elevation above sea level, are important factors in determining the extent to which township funds must be expended.

TABLE IV

SECOND CLASS TOWNSHIP PURCHASES OF MATERIALS AND EQUIPMENT BY ITEMS

1930-1933

	1933 ♦	1932 ▲	1931 ♦	1930 ☉
Pipe				
Feet	44,297	130,564	132,906	268,681
Cost	$53,725	$170,050	$186,677	$414,387
Average	$ 1.2128	$ 1.30	$ 1.40	$ 1.54
Road Machinery				
Pieces	35	180	281	398
Cost	$38,411	$249,138	$456,405	$619,006
Road Material				
Tons	350,394	421,997	428,696	519,371
Cost	$410,463	$534,027	$582,562	$742,477
Average	$ 1.17	$ 1.27	$ 1.36	$ 1.43
Road Oil				
Gallons	1,594,479	2,462,505	2,396,669	3,399,979
Cost	$202,908	$285,715	$306,556	$493,817
Average	.1272	$ .12	$ .13	$ .15
Snow Fence & Posts				
Feet	31,500	16,600	87,800	
Cost	$2,719	$2,168	$10,518	
Average	$ .086	$ .13	$ .12	
Surfacing				
Yards	11,740	5,200	32,889	
Cost	$ 5,064	$11,612	$11,320	
Average	$ .431	$ 2.23	$ .34	
Bridge Material				
Number	20	9		
Cost	$22,803	$ 5,367		
Miscellaneous				
No. Orders	27	68	87	
Cost	$11,574	$19,436	$29,707	$74,997 ⊕
Total Cost	$747,667	$1,277,513	$1,583,745	$2,344,684
Percent of Decrease Based on 1930	68%	46%	32%	

♦. From Dec. 1, 1932, to Dec. 31, 1933 or 13 months, due to change in fiscal year for Townships. Total cost itemized by counties, see Table 5. This was a period of relatively few purchases due to (1) depression, (2) Parkinson Act, then pending. The act created uncertainty, some township officers thinking the State might bear such expenses.
▲ From Dec. 1, 1931 to Dec. 1, 1932
♦ From Dec. 1, 1930 to Dec. 1, 1931
☉ From Dec. 1, 1929 to Nov. 30, 1930
⊕ Includes bridge materials, snow fence, small tools and repair parts, wheels; tires, tool and equipment sheds, etc.

Township Road Machinery - A large amount of expensive equipment is necessary in order that a road system may be maintained. Some indication of the scope in terms of road machinery which a rural road system requires may be secured from the following tabulation. The information listed was taken from annual reports of second class township supervisors of 1932 and compiled by the Pennsylvania Department of Highways.

Steam or Gas Rollers	318
Traction Engines and Tractors	593
Road Machines	2,343
Motor Trucks	549
Crushers	373
Concrete Mixers	122
Wheel Scrapers	906
Pumps	87
Dump Wagons	267
Road Drags	7,200
Horse Rollers	45
Sprinklers	74
Scarifiers	163
Plows	542
Scoops	1,066
Storage Houses	527
Air Compressors	16
Steam or Gas Shovels	13
Power Graders	113
Snow plows	68
Conveyors	8
Steam or Air Drills	11
Jack Hammers	1
Road Maintainers	8
Stone Spreaders	8
Mowing Machines	1
Power Scrapers	4
Water or Tank Wagons	3
Gas Engines	2
Chain Hoist	1
Unloaders	4
Gravel Plant	1

Purchases in Excess of $200 Must be in Writing and Approved by Highway Department - All purchases of materials, equipment, and machinery for the construction and improvement of roads and bridges must be by contract in writing if they involve an expenditure in excess of $200. Purchases of less than $200 must be recorded for the Department of Highways in the supervisor's annual report. It is specifically forbidden to evade this provision by making two or more contracts for small amounts and the auditors are directed to disapprove such contracts. Contracts for the construction, reconstruction, and improvement of roads and bridges involving an expenditure in excess of $200 are valid only when approved by the Department of Highways.

Contracts for Public Work in Excess of $500 - Where contracts are for a public work in excess of $500, the township must require the execution of a bond not less than 50% nor more than 100% of the contract price for the prompt payment of labor and material. The bond must have as surety one or more surety companies authorized to do business in Pennsylvania. It must be deposited with the township for the use of any interested parties. A provision of the bond permits action in assumpsit in the name of the township by labor or material men.

Competitive Bids Not Required - In the erection of public buildings, costing in excess of $1,000, separate specifications must be prepared for plumbing, heating, ventilating and electrical work. Separate bids are required on each of these branches of the work and the contract must be awarded to the lowest responsible bidder. There seems to be no good reason why the township code should not require public buildings costing between $200 and $1,000 to be let to the lowest responsible bidder. In addition, road and bridge contracts for more than $200 and now subject to approval by the Department of Highways, should be let to the lowest responsible bidder, instead of being left to the discretion of the township supervisors as is the present practice. As observed in an earlier chapter, public bidding, if made a legal requirement, need not become a substitute for contract approval by the Highway Department. Both conditions could properly be imposed. The practice of requiring public bids is followed in many townships despite the failure of the township code to specify its use other than in the limited sense already indicated. Desire for economical administrative methods and good management compel a resort to competitive bidding where these principles are given primary consideration. There seems to be no sound reason for not making the practice obligatory by statute on a much larger scale than at present.

Workmen's Compensation - All contracts executed by the township involving the employment of labor must contain provisions that the contractor accepts in such work the provisions of the Workmen's Compensation Act of 1915, as amended, and that he will insure his liability thereunder. Township officers signing such contracts must require proof from the contractor that this liability has been insured. Where a contract is executed in violation of these requirements, it is null and void.

Interest in Contracts - Supervisors and Roadmasters - Township supervisors and roadmasters are prohibited from being interested directly or indirectly in any purchase or contract relating to roads and bridges, or to the furnishing of materials for such purposes. Those violating this provision knowingly, are guilty of a misdemeanor. Upon conviction, the fine shall not exceed $500, the imprisonment shall not exceed six months or both. The office held by any person violating this provision is forfeited (1933 P.L. 103 Sec. 520).

Interest in Contracts Prohibited by Engineers and Architects - Should Include All Officers - Engineers and architects in the employ of the township and engaged in preparing plans, specifications, or estimates are not permitted to bid on any public work for the township. The township code makes it illegal for township officers in charge of letting public work to award such a contract. It is not lawful for any architect or engineer in the employ of the township to be "in any wise interested in any contract for public work in such township, nor receive any remuneration or gratuity from any person interested in such contract."

This above provision should be extended to apply to all purchases of materials and supplies. It should not be restricted to contracts for public work. It should of course apply to all township officers elective and appointive, not merely to architects and engineers. The penalty for persons violating provisions of the township code prohibiting engineers and architects from being interested in contracts is a fine not exceeding $500, imprisonment not less than 6 months, or both, and forfeiture of office (1933 P.L. 103 Sec. 806). Any extension of the prohibition indicated above should be accompanied by an appropriate extension of the penalty.

Record of Public Roads Kept by Court of Quarter Sessions - Township roads are laid out, widened, changed and vacated by the court of quarter sessions, as provided by the general road law and its amendments. It is the purpose of this procedure to provide a permanent record of public roads in the township. Roads not of record used for public travel, and maintained and kept in repair through township funds for at least 21 years are deemed to be public roads for a width of 33 feet.

Roads May Be Temporarily Closed by Supervisors - It is the duty of the township supervisors to see to it that all public roads in the township are "effectually opened and constantly kept in repair." A road may be temporarily closed when unfit or unsafe for travel and a proper marking may be placed at its extremities. Anyone using a road designated as closed, without a permit from the supervisors, is subject to a penalty of not more than $100.

Road Contracts with Private Corporations - The township code permits township supervisors to contract for road improvement and repair with private persons or corporations. Such contracts are limited to four years and must be signed by at least two township supervisors. Not more than ten miles of road may be included in a contract, but no restriction appears on the number of contracts. In such cases contractors are required to give bond for the amount of the contract and to sign specifications furnished by the supervisors (1933 P.L. 103 Sec. 517).

Township Highway System Compared 1923 and 1932 - What are some of the outstanding points in the development of the Township Highway System during the past nine years? If it were possible to have access to a central agency which recorded important developments during this period the answer to this question would be relatively simple. The township road system is based upon thousands of independent districts and the principal source from which information can be obtained comes from these districts in the form of a brief annual report to the Highway Department. On a few points certain comparisons between the years 1923 and 1932 are possible, and official figures for this purpose were made available through the courtesy of the Highway Department. A significant trend may be noted from the following figures given by this Department:

	Second Class Township Assessed Valuation	Tax Collected	Mileage of Township Roads	Tax Collected Per Mile	Assessed Valuation Per Mile
1923	$1,387,488,348	$11,235,041	74,875	$150	$18,531
1932	$1,278,409,095	9,133,877	48,461	188	26,380

The principal changes may be summarized briefly in the order in which they were presented.

    1. Decline of nearly 8% in township assessed valuation due to (a) depreciation of farm values, (b) exhaustion of timber and natural resources, and (c) annexations of territory by boroughs and cities.

    2. Decrease of 18% in total tax collected due to (a) tax delinquencies and (b) the depression.

    3. Reduction in mileage of 35% because of assumption of road mileage by the state (20,000 miles in 1931) and remeasurement of township systems.

    4. Increase of 20% in tax collected per mile attributed to (a) better earth road construction and maintenance than in 1923, (b) construction of brick, concrete and macadam surface and (c) increase in debt requirements.

    5. Increase of 30% in assessed value per mile of road due to 35% reduction in mileage already noted.

The decrease in general assessed valuation in second class townships was discussed in some detail in previous chapters. The extent to which this decrease may be attributed to the extraction of lumber, minerals, oil, and other natural resources cannot be determined on facts now available. There is no doubt, however, that a substantial part of the decline should be charged to exploitation of forest and mineral areas.

The township assessed valuation is shown by counties on Table 5. In addition, the assessed value per mile of township road, indicating the wide diversity which prevails in Pennsylvania in taxable resources, is also presented.

Effect of Parkinson Act on Township Tax Levies - Many townships have made substantial reduction in their tax levy. The following illustrations indicate the trend of tax levies in four townships:

		Tax Rate In Mills		
Township	County	1932	1933	1934
Pine	Lycoming	20	18	0
East Bradford	Chester	8	5	3
Easttown	Chester	8	4	4
Willistown	Chester	-	8	4

This is due in part to economies affected by the township supervisors, but the principal influence has been the Parkinson Act. The passage of the Act permitted most of the townships to devote their reduced tax levies mainly to debt retirement and the care of bridges. Road work was performed at state expense by state employees.

How was the Parkinson Act Allocation Estimated? - In 1933 estimates were requested from

TABLE V. ASSESSED VALUE PER MILE AND EXPENDITURES PER CAPITA--1932

DATA FROM PENNSYLVANIA DEPARTMENT OF HIGHWAYS

County	Number of 1932 Townships	Assessed Value	Total Tax Collected	Miles of Township Road	Tax Collected Per Mile	Assessed Value Per Mile	Expenditures Per Capita	Expenditures Per Mile Of Road	Townships Not Reporting
	1	2	3	4	5	6	7	8	9
Adams	21	$11,618,988	$ 92,124	670	$137.50	$17,342	4.76	$163.40	
Allegheny	29	79,903,950	316,560	451	701.91	177,171	4.86	811.64	2
Armstrong	28	26,530,698	127,194	1,110	114.59	23,902	4.88	203.34	1
Beaver	21	14,243,299	102,656	572	179.47	24,901	7.35	341.55	
Bedford	25	12,201,547	79,315	914	86.78	13,350	3.56	109.00	
Berks	44	45,997,071	321,972	1,458	220.83	31,548	5.49	267.35	
Blair	15	12,340,228	79,612	310	156.81	39,807	2.88	301.95	2
Bradford	38	11,712,957	111,673	1,533	72.85	7,641	6.03	101.62	
Bucks	31	45,802,246	345,156	1,034	333.81	44,296	8.41	426.21	
Butler	32	21,114,138	161,614	1,094	147.73	19,300	7.87	249.46	1
Cambria	29	41,256,535	336,588	590	570.49	69,926	6.49	737.54	2
Cameron	5	901,330	4,542	70	64.89	12,876	8.17	244.76	1
Carbon	10	3,495,860	28,876	305	94.68	11,462	2.39	159.48	2
Center	25	8,439,539	43,538	499	87.25	16,913	2.30	133.43	
Chester	57	52,836,509	399,085	1,347	296.28	39,224	9.16	438.77	
Clarion	22	6,351,674	77,338	889	86.99	7,145	4.42	110.65	1
Clearfield	29	9,210,828	74,228	960	77.32	9,595	2.32	124.70	2
Clinton	21	4,252,864	30,500	300	101.67	14,176	4.28	184.10	1
Columbia	23	8,388,142	59,411	731	81.27	11,475	4.24	118.12	2
Crawford	35	16,184,792	146,095	1,383	105.64	11,703	7.45	162.93	
Cumberland	22	17,407,196	108,109	640	168.92	27,199	5.57	241.26	1
Dauphin	22	18,827,332	120,708	595	202.87	31,643	4.58	266.07	2
Delaware	11	12,356,981	89,012	110	809.20	112,336	5.23	1,007.19	1
Elk	10	2,839,546	34,272	314	109.15	9,043	4.81	229.19	
Erie	21	24,689,381	191,751	969	197.89	25,479	8.11	252.02	
Fayette	24	58,276,939	432,520	1,085	398.64	53,711	3.77	475.06	1
Forest	8	1,370,837	11,486	193	59.51	7,103	5.32	124.34	
Franklin	15	19,754,267	125,543	739	169.88	26,731	4.98	240.22	1
Falton	11	2,126,014	17,170	315	54.51	6,749	2.86	76.77	
Greene	19	45,762,655	275,129	884	311.23	51,768	13.24	488.38	

	1	2	3	4	5	6	7	8	9
Huntingdon	30	$10,424,295	$ 51,683	611	$ 84.59	$17,061	4.02	$133.47	1
Indiana	24	32,861,871	250,043	1,141	219.14	28,801	5.65	255.40	
Jefferson	23	10,772,927	75,601	782	96.68	13,776	4.34	144.94	
Juniata	13	3,826,259	23,439	405	57.87	9,448	2.71	76.88	1
Lackawanna	18	10,026,206	62,544	361	173.25	27,773	5.68	323.22	
Lancaster	41	73,656,134	530,476	1,854	286.13	39,728	6.93	352.75	1
Lawrence	17	16,907,659	110,215	516	213.59	32,767	5.03	304.23	
Lebanon	15	17,589,286	113,162	475	238.24	37,030	7.21	364.22	
Lehigh	14	20,022,637	159,205	715	222.66	28,004	5.87	267.11	
Luzerne	28	16,887,585	138,214	680	203.26	24,834	2.86	270.53	5
Lycoming	42	7,215,651	67,119	900	74.58	8,017	4.39	122.54	2
McKean	15	15,817,193	212,000	347	610.95	4,558	18.26	1,176.82	2
Mercer	32	14,820,804	120,774	1,056	114.37	14,035	6.69	192.29	
Mifflin	10	8,048,910	46,298	282	164.18	28,542	3.09	245.84	
Monroe	16	8,646,845	84,818	525	161.56	16,474	8.83	259.39	1
Montgomery	30	48,202,085	360,262	785	458.93	61,404	8.61	630.17	
Montour	9	2,603,099	17,046	197	86.53	13,214	3.99	144.82	
Northampton	17	21,826,760	175,307	657	266.83	33,222	6.37	338.66	
Northumberland	21	8,056,143	62,206	676	92.02	11,917	3.22	139.00	
Perry	21	6,696,745	42,013	537	78.24	12,471	3.98	81.93	
Philadelphia	11	6,069,143	48,301	313	154.32	19,390	14.97	230.16	1
Pike	24	2,766,639	35,478	595	59.63	4,650	7.80	123.33	1
Potter	21	14,317,714	127,383	925	137.71	15,479	2.49	207.31	
Schuylkill	15	6,919,357	31,217	444	70.31	15,584	3.01	90.07	
Snyder	15								
Somerset	25	23,104,107	168,779	1,252	134.81	18,454	4.57	180.91	1
Sullivan	9	1,371,860	11,013	268	41.09	5,119	4.22	99.30	1
Susquehanna	27	11,421,977	82,010	980	83.68	11,655	7.08	129.12	2
Tioga	30	8,593,849	87,901	1,012	86.86	8,492	7.32	143.20	
Union	10	5,753,734	36,530	270	135.30	21,310	3.28	140.51	
Venango	21	10,316,955	112,744	741	152.15	13,923	7.72	249.66	1
Warren	23	8,694,116	75,904	665	114.14	13,074	5.10	167.84	2
Washington	32	81,916,387	460,769	1,221	377.37	67,090	6.88	482.31	
Wayne	22	8,050,698	67,823	719	94.33	11,197	.59	159.75	3
Westmoreland	28	66,909,461	476,277	1,389	342.89	48,171	5.01	466.54	
Wyoming	18	3,541,155	45,741	383	119.43	9,246	5.70	160.14	1
York	34	27,556,506	219,805	1,718	127.94	16,040	4.50	169.02	
TOTAL	1,484	$1,278,409,095	$9,133,877	48,461	$188.48	$26,380		$263.26	49

the Highway Department as to how much money would be required for the maintenance of
the 53,000 mile Township System during 1934 and 1935. The Annual Reports of township
supervisors furnish a starting point for the estimate. These reports are forwarded to
the Highway Department each year. From them the expenditures in each township on matters of maintenance can be totaled for the entire state. Selecting items dealing with
roads and omitting those costs attributable to bridges also omitting all costs arising
from new surfacing, reconstruction, and permanent improvement, there would be left only
the cost of dragging and scraping the roads, filling holes and ruts, providing ditches
and drains, and cutting back weeds and brush. This constitutes maintenance. It is the
work performed by the State Highway Department under the terms of the Act. The totals
on these points from the Annual Reports in 1932 were as follows:

	Chapter 13, Table 3, Column No.	Second Class Township 1932 Expenditures for Maintenance
Maintenance Earth Roads	1	$2,862,044
Maintenance Improved Roads	2	451,239
Repairs, State Reward Roads	3	62,573
Repairs, State-Aid Roads	4	35,820
New Tools and Machinery (20%)	8	62,176
Repairs to Tools and Machinery	9	219,761
Roadmasters' Wages	10	873,079
Snow Removal	16	122,150
Signs and Index Boards	26	4,098
Miscellaneous (50%)	27	328,926
Gas and Oil (50%)	31	25,493
Total		$5,047,359

The total of the above items equals $5,047,359 which would be required to maintain
the township roads (but not the bridges) for one year. If this amount is doubled the
approximate appropriation for a biennieum or $10,094,718 would be the result.

The allocation of this amount to particular townships followed the procedure designated by the statute. The Parkinson Act provided "there shall be expended by the
Department of Highways.....such proportion of the total amount herein before provided..
....as the miles of streets or roads.....within its municipal limits......bears to the
total number of miles of such streets or roads within the municipal limits of all townships of the second class" (1933 P L. 1520 Sec. 3). What was needed, therefore, was
the township road mileage for each township and the total mileage for the Township
Road System. These were supplied by "Certification Mileage" (1933) and "Speedometer
Survey Mileage" (1933), and were described earlier in this chapter. The former served as the basis of allocation while the latter served as a check to assure accuracy.

Mileage Basis for Parkinson Act Maintenance Allocation - One of the principal points
upon which the Parkinson Act has proved unsatisfactory involves the relation between
the amount of maintenance required by the roads of the particular township and the
allocation for that maintenance on the basis of mileage. In brief, the mileage may or
may not have a reasonable relation to the maintenance cost. We have seen how the allocation was based upon a flat mileage figure.

Where Maintenance Costs Exceeded 1934 Parkinson Act Allocation - In 466 townships the
cost of maintenance per mile in 1932 averaged $238 for a total of 12,077 miles of roads.
This mileage received a 1934 Parkinson Act allocation of $1,400,000 to replace a town-
ship maintenance in 1932 in excess of $2,800,000. The counties containing more than
19 townships whose 1932 maintenance cost exceeded their allocation under the Parkinson
Act were as follows:

County	Number of Townships in Which 1932 Maintenance Exceeded Parkinson Act Allocation
Allegheny	26
Chester	22
Cambria	22
Bucks	20
Luzerne	20

In the above counties a substantial 1934 road tax would be required to maintain the services, because the allocation from the Parkinson Act did not equal what was spent in 1932 per mile for maintenance purposes. Some tax for 1934 township road purposes would be required in nearly all townships to maintain bridges, retire indebtedness, and for other road purposes not included in "maintenance".

Where 1934 Parkinson Act Allocation Exceeded 1932 Maintenance Cost - In 1,049 townships the 1934 Parkinson Act allocation, which was computed as a payment of $117 per mile, exceeded what was spent for maintenance purposes in 1932. It would be expected that state maintenance in such townships would be superior to the former township maintenance. In 1932 these townships, about two-thirds of the entire number, had an average expenditure of $61 per mile for maintenance. The expenditures in the same townships during the two-year period, 1934-1935 will be at the rate of $117 per mile. The counties having the most townships in this group are as follows:

County	Number of Townships in Which 1934 Parkinson Act Allocation Exceeded 1932 Maintenance Cost
Lycoming	38
Chester	35
Berks	31
Butler	30
Crawford	29
York	29
Mercer	28
Schuylkill	28

The mileage of townships in the above counties was 36,384 miles. It will be noted that Chester County has so many second class townships that it is included in both groups. Slightly more than $2,250,000 was allocated among these 1,049 townships. The allocation replaced a former expenditure under township auspices of approximately $2,200,000.

Revision of Mileage Basis of Parkinson Act - Although the defects of the mileage basis were known in advance as having no relation to the needs of a particular mile of road, this basis was regarded as a better one by township officials than leaving the entire expenditure to the discretion of the State Highway Department. It would not be difficult to apply a more equitable basis of allocation by combining the elements of population and mileage.

To illustrate this, a township may have only 25 miles of road all macadam surfaced with a suburban location involving only a 35 minute drive from a large city. Such a township would have heavy traffic, dense population, high taxable valuation and a heavy township road tax as in Easttown Township, Chester County. Another township not far distant may have 125 miles of unimproved earth road, almost no traffic or population, a low taxable valuation and a low township road tax. Here roads would have a low maintenance cost. Under the Parkinson Act, $117 per mile would be spent by the state in both townships for maintenance work. Should not the township which contributes more gasoline tax and license fees receive a larger return? If population as well as mileage were taken as a basis of allocation a fairer result would be achieved and taxpayers in both townships would benefit by proportionately equal reductions of township tax. Many townships have such a relation of mileage and population as to be unaffected in amount of allocation. Extreme cases could be met by provision for a fixed maximum and minimum allocation. Such a change should be made regardless of who performs the township road maintenance work, the state or the supervisors.

A Public Highway Program - The development of a comprehensive public highway program has received attention in several recent studies and it can be only briefly mentioned here. The equitable allocation of highway funds by means of a balanced program in which all types of road surface, the different classes of municipalities and the traffic needs of various population areas are considered may be said to be the essential purpose of such a plan. The factors to be taken into consideration in the apportionment of funds under a comprehensive program include the following:

1. Inexpensive traffic surveys.
2. Equitable adjustment of rural and urban traffic needs.
3. Degree of regularity in apportionment by types of roads, including concrete, macadam, brick, stone, construction and replacement over a period of years.
4. A planning and survey commission to work out the allocation.
5. Preference given to highway purposes in the use of highway funds.
6. Consideration given to sources of funds in determining areas for expenditure.
7. Total highway mileage by municipal divisions and by types of surfaces in each class of municipality.

It was possible in Illinois to secure agreement for a comprehensive plan of this type in which the numerous urban and rural and state interests were provided for.

Conditions of State Assistance - As in the other states which render assistance to local governments out of allocations from state revenues, so in Pennsylvania, a mixed direct and indirect system of assistance has been gradually developing. Examples of this are the following:

Return of half cent of gasoline tax to counties
Allocation of $11,352,000 for state maintenance of township roads
State Reward for township roads, 1919-1929
Allotment of beer license funds to township road districts
Allotment of liquor license monies to township school districts
The $10,000,000 Talbot Poor Relief Bill
State aid to counties for Mothers' Assistance purposes
Old age pensions relieving counties of burden of support of those eligible for pension

The haphazard enactment of such legislation does not conform to a well balanced development of local government. A coordinated program of support for local governments, including townships, should be drafted for the legislature. The time has passed when the chief burden of support for local roads can be permitted to rest solely upon general property tax revenues.

Is the Commonwealth justified in extending such a large amount to townships and other local governments without specifying the conditions which may protect its expenditure? If the state could provide a series of regulations which would insure to the taxpayer an improved financial administration by local authorities savings might be affected.

In New York State a legislative commission has been studying this subject for four years and a series of reports on state and local financial problems has resulted. These studies have been of value throughout the country because of their thoroughness on timely subjects. In reference to the conditions on which state assistance should be rendered to local governments the New York Commission recommended the following:

1. A well regulated budget procedure including a change in the fiscal year to bring estimates of expenditures, the making of appropriations, the levying of taxes and the collection of revenues into their proper sequence.
2. Creation of local government board or agency to exercise the following functions;
    Provide budget system to be uniform for different classes of municipalities
    To provide up-to-date accounting systems and assist in their use
    To supervise annual audit of local accounts at the expense of the municipality

To examine local budgets as to provisions to debt service, deficiencies of revenue and allowance for tax delinquencies
To approve of local bond issues in excess of certain
percentages of assessed value of taxable property
To approve tax rates in excess of certain percentages
of the assessed value of taxable property
To approve budgets which include more than 1/3 of
their receipts from state fund
To gather and publish statistical material on local
finance
To make efficient studies and supply information at
the request of local units of government
To withold state aid payments from any local unit as
a penalty for failure to follow the requirement of
the board ¶

It was also recommended that the cost of the services of the central board should be
assessed against the local units concerned. The Commission further suggested that con-
tributions by the state and local government purposes be limited to the amount paid in
1930. Any amount in excess of the payment in that year was to be withheld by the state.

CONCLUSIONS

Roads and Bridges

FINDING: The second class township road system contains 48,461 miles of roads,
of which 43,000 have an unimproved earth surface. Road classification would disclose many roads serving no genuine public need, but local sentiment appears adverse to their discontinuance.
RECOMMENDATION: Many miles of township roads should be vacated and a state committee of township supervisors should be asked to report on means of accomplishing
such results.

FINDING: The percentage of improved roads in the 26 Losch Act townships compares
favorably with the number of miles of improved roads in second class townships
throughout the entire state.

FINDING: The organization of the township road system in 1932 consisted of 3,633
independent road districts situated in 1,511 townships. Subdivision of authority
has been carried to extreme lengths with a result that the system is wasteful and
ineffective from an administrative viewpoint.
RECOMMENDATION: The number of road districts and supervisors should be reduced two-
thirds.

FINDING: Township purchases for materials in amounts exceeding $200 have declined
from $2,300,000 in 1930 to less than $750,000 in 1933. In seven counties, no purchases in excess of $200 were made by townships during 1933.

Contracts and Purchases

FINDING: Except for the erection or alteration of public buildings involving a cost
of more than $1,000, township contracts are not required by law to be awarded to
the lowest responsible bidder.
RECOMMENDATION: That township contracts or purchases for road or other purposes in
excess of $100 should be awarded by public bidding to the lowest responsible bid-
der. This recommendation is not regarded as inconsistent with the authority now
invested in the Department of Highways for the approval of township purchases.

¶   Fourth Report of the New York State Commission for the Revision of the Tax Laws.
"Local Government Solvency Through Tax Relief and Economy." Legislative Document (1934)
No. 56, Submitted February 15, 1934. P.43.

Awards should be made to the low bidder and the Highway Department could continue to exercise its function by passing upon the necessity for the contract or purchase.

FINDING: The township code prohibits architects and engineers in the employ of the township from being interested in any contract for public work for the township. It also prohibits supervisors and roadmasters from being interested in any purchase or contract relating to roads and bridges.
RECOMMENDATION: These provisions and their penalties should be extended to include all township officers, elective or appointive, and to all purchases and contracts as well as those for road and bridge purposes.

Parkinson Act

FINDING: The Parkinson Act has made possible substantial reductions in township tax levies.

FINDING: In 1,049 townships the Parkinson Act allocation exceeded the amount spent for maintenance purposes in 1932. In 1932 the maintenance cost per mile exceeded the amount allotted under the Parkinson Act in 466 townships.
RECOMMENDATION: The Parkinson Act if continued should be amended, to provide more equitable tax reduction for all townships, by adding population to the present mileage basis for allocation of funds.

Public Highway Program for Allocation of Motor Funds

FINDING: In the event that township road maintenance is returned to the control of township officers it is
RECOMMENDED that a public highway program should be developed, in which the various classes of municipalities, including townships, participate in the distribution of available state revenues on a basis of traffic served, population, highway mileage and types of surface.

Conditions of State Assistance

FINDING: Large sums of money are advanced to local units each year without adequate supervision.
RECOMMENDATION: State assistance should be conditioned in the future upon a substantial revision of township budget and financial procedure. Among those to be considered are:

> Change of fiscal year (Chapter 12)
> Creation of local government board or agency to
> > Provide uniform budget system
> > Establish simple accounting system
> > Supervise an annual audit
> > Examine local budgets as to provision for debt service
> > Approve local bond issue in excess of fixed percentage of assessed value
> > Approve tax rates in excess of a certain maximum
> > Pass upon budgets which include more than one third of their receipts from state funds
> > Publish statistical data from annual local financial reports
> > Supply information at the request of local officers
> > Withhold state funds from any local unit as a penalty for failure to follow the requirements of the board.

CHAPTER XV

DEBT, CONSOLIDATION, STATE SUPERVISION

I. Debt

The funded and unfunded debt of second class townships is considered in this chapter. The funded or bonded debt comprises that portion for which sinking funds have been established. A serial bond is usually issued. The figures considered are net, that is, sinking funds have been deducted. The data presented are from the Division of Taxes and Assessments, Department of Internal Affairs. The unfunded debt, also called temporary borrowing, notes or short term loans, is limited by law to a period of one year. No figures are available on unfunded debt. Figures of total debt were supplied by the Pennsylvania Department of Highways, Township Engineer's office.

Townships Having No Debt - It should be noted at the outset that 537 of the 1,515 second class townships reported no road debt in 1932. Since absence of debt is one indication of good financial management, and since this applies to 35% of the townships, due credit should be extended for this accomplishment. Figures include both bonded debt and temporary loans as reported by the township supervisors to the Department of Highways.

Causes of Township Indebtedness - The funds borrowed by second class townships are used for a wide variety of purposes connected with public roads. In one case a debt of $2,000 was incurred for the purchase of a power roller for the township roads. By 1934 payments had reduced the note to $500. When the State commenced Parkinson Act maintenance it rented the roller for maintenance purposes at the rate of $1.80 per hour. Although this was apparently a good investment it will be noted that the debt was unfunded over several years.

Another township had not reduced its debt of approximately $1,000 for some years. Money had been borrowed in order to participate in State Reward for the construction of improved stone and macadam surface roads. Subsequently, the state took over this road as part of the 20,000 mile Rural Road System. The township debt of course remained.

In the same township an additional "temporary" indebtedness of $375 was paid off during 1933. It was necessary to borrow $200 shortly after payment, in order that road work in the township might continue. In this manner debts may be liquidated and renewed each year until a so-called temporary debt becomes in effect a permanent one by renewal. The practice is in part due to the commencement of the township fiscal year on January 1, while receipts from tax collections do not come in before July or August 1. This practice results in an unnecessary interest cost among second class townships, and as pointed out in another chapter a change of the fiscal year to July 1 would relieve the dearth of tax collections during the first six months.

Debt is incurred in rare instances because a member of the Board of Supervisors encourages purchases. In a centrally located county, a supervisor after his election urged the purchase of a tractor and grader, a road drag, a gasoline engine, etc. The purchases were made and money had to be borrowed upon promissory notes. This supervisor subsequently failed to be re-elected and no machinery was purchased for some years. Upon his re-election recently he urged that the township buy a roller, but without success.

In Osceola Township, Tioga County, State Reward was the cause of a township debt of $10,000. The money was borrowed to match a state grant but repayment has not been completed although the road became part of the state system some years ago. In many sections state assistance on a basis requiring equal contributions by the local units has left its mark in the form of heavy township indebtedness.

Township Debt Per Mile - The money borrowed by second class townships is used mainly for road construction. This was indicated when the total debt per mile was studied for the highest and the lowest townships in a selected group. It is significant that of these 147 selected townships, Lower Yoder, Cambria County, had the highest per mile debt in the entire group. This township ranked high in concrete surfaced roads. Furthermore, a relatively high rating was placed upon the condition of Lower Yoder roads by Highway Department engineers. East Taylor Township in Cambria County was the next highest township. About one-fifth of its roads were improved. Logan Township in Blair County had about one-third of its mileage in improved roads, and their condition was such that they too received a high rating. A similar analysis was made comparing debt with type and condition of roads for all townships in this group. The townships with the highest and lowest debt (including temporary loans) in the group of 147 selected townships were as follows:

Township	County	1932 Debt per Mile of Township Road	Type	in Miles	Condition
Lower Yoder	Cambria	$6,833	Concrete	3.	Good
East Taylor	Cambria	5,100	Concrete	.7	Fair
Logan	Blair	3,907	Macadam	9.3	Good
Adams	Cambria	3,319	Stone	33.	Fair
Middle Taylor	Cambria	2,778	Stone	4.	Fair
Upper Providence	Delaware	2,654	Macadam	3.3	Poor
Cambria	Cambria	2,450	Concrete	.4	Fair
White	Cambria	2,362	Unimproved	6.	Poor
Gray	Greene	2,313	Unimproved	3.3	Fair
Lancaster	Lancaster	2,000	Macadam	19.	Good
Lehman	Pike	10	Unimproved	38.	Good
Chest	Cambria	13	Unimproved	13.2	Fair-Poor
Rush	Dauphin	15	Unimproved	13.3	Fair
Monaghan	York	16	Unimproved	31.3	Poor
Upper Providence	Montgomery	17	Unimproved	28.61	Fair
Hunlock	Luzerne	20	Unimproved	24.9	Fair-Poor
East Brunswick	Schuylkill	34	Unimproved	61.2	Fair-Poor
Jefferson	Dauphin	35	Unimproved	13.	Fair-Poor
Amity	Berks	36	Unimproved	25.6	Fair
District	Berks	39	Unimproved	23.4	Poor

Most of the townships in the low debt group, such as Chest Township, Cambria County, and Rush Township, Dauphin County, have unimproved roads mainly and their condition is relatively poor. It is clear that the cost of improved roads is far beyond the township's financial ability unless it resorts to heavy borrowing.

Amount of Township Bonded Debt Relative to Other Local Governments, 1932 - Bonded debt for second class township purposes amounted to only $4.10 per capita in 1932. Township school district and poor district bonded debt were not included in these figures.

A study of 22 selected second class townships indicated that the net bonded debt for all purposes (township, school, county, and poor) was $1,493,230, while the net debt of these same units for township purposes only was $254,879, or about 17% of the total.

The relationship of bonded debt to taxable valuation in second class townships is also low (.67% of the taxable valuation) it being comparable to the poor district in this respect.

Second Class Township Bonded Debt by Counties, 1932 - The per capita township bonded debt in 1932 was highest in Greene, Indiana, Erie, Cambria, and Armstrong Counties. Townships in Cameron, Clinton, Fulton, Juniata, Montour, Snyder, Union, and Wyoming counties had no bonded debt in 1932. The average bonded debt for all townships was $6,145. All of the counties in which townships had no bonded debt are in the group with lowest taxable valuation except Snyder and Union. (See Chapter IX, Chart 3).

Unfunded Debt - When temporary borrowing is resorted to no sinking fund is required because of its supposed short duration. Such debt is therefore termed unfunded. This type of borrowing is especially important in second class townships. During 1932 the total debt of such townships was $14,118,858. "Total debt" includes temporary loans, interest due on all loans, state-aid maintenance and bonded debt less sinking funds. In order to secure a fair estimate of the amount of unfunded township debt, the figures on bonded debt collected by the Department of Internal Affairs have been deducted from the aggregate of township debt reported to the Department of Highways and the remainder has been considered as the unfunded total. This is only an estimate and the possibility of error is acknowledged. It is, however, the best estimate at present available. For 1932 the difference was $4,803,706, or about 34% of the total debt. This is an exceptionally large percentage for municipal short term obligations and interest. The outstanding feature of this short term borrowing is the period for which such loans are made. The Code specifies that no certificate of indebtedness may exceed the period of one year (1933 P.L. 103, Sec. 903). In many townships the practice is to allow such debts to extend over a period of years. Interest rates on this type of borrowing are usually higher than both the current rate of interest and the rate on funded debts.

Total Debt - The total debt in 1932 was $14,118,858, and the average debt per township was $9,319. The average burden of debt is greater if the 537 townships reporting no indebtedness are omitted from consideration. In that case the average debt per township for the 978 townships having indebtedness would be about $14,437.

Debt Service Costs - A brief classification of 1932 expenditures of second class townships was presented in Chapter 13. This classification showed that debt service costs consumed 31% of all expenditures in that year. The principal item, the payment of maturing short term notes, required $2,392,879, 19% of 1932 expenditures. In Chapter 13, Table 111, the payments of townships listed by counties are shown for the following items: debt service costs, interest on notes, notes maturing, interest on bonds, sinking fund payments, and redemption of bonds.

It should be noted that duplicated expenditures occur in the consideration of repayment of loans. When an amount was spent for a particular purpose, for example for a power roller, it was listed as an expenditure under new machinery. The purchase was made with borrowed money, and upon repayment of the loan, in whole or in part, in the same year the amount was also listed as an expenditure for notes matured. Further duplication occurs because some loans may be repaid during the year and again borrowed and again repaid. The result is that the same amount will appear to be expended several times.

Defaults - Although the amount of bonded debt is relatively of less importance in townships than among other classes of municipalities, it is significant that the townships furnish far more than their share of the defaults. Of 109 municipalities listed as having defaulted at one time or another in Pennsylvania, between 1928 and 1934 (most of which have since been adjusted), 36 were second class townships. Township school districts furnished 18 of the 36 defaults. About 33% of the local government defaults during this six-year period were located in second class townships, 16.5% being township road bonds and 16.5% township school district bonds.

One of the causes of default is the short term loan. The bonded debt may not be large enough to cause default, and yet the temporary loans and interest costs may place such heavy strain on the township finances that default on the bonded debt occurs.

The principal cause of default has been the increase of tax delinquency. Many municipalities have been forced to issue deficiency bonds or to borrow money in order to meet their current obligations. In the emergency thus presented, the General Assembly during the 1933 session passed an act (1933 P.L. 813) authorizing municipalities to issue bonds bearing interest at a rate not exceeding 6% per annum, payable in 10 years up to 80% of the outstanding taxes uncollected. The proceeds from the sale of such bonds are to be used for current operating expenses. The law is to remain in effect until 1938.

About one-third of the total local government indebtedness authorized in Pennsylvania between January 1, 1933 and April 1, 1934 consisted of such emergency bonds. The wisdom of borrowing for the payment of current expenses is doubtful. Such borrowing increased the already overburdened debt charges, and merely postponed the payment of current expenditures to future generations. This type of debt has encouraged groups of taxpayers with property which depreciates rapidly in value to postpone the day of payment as much as possible. This is particularly true in industries involving the exploitation of natural resources, such as mining, lumbering, oil production, etc. In these industries the taxable value of the land depreciates at a rapid rate. Some of these taxpayers know that in the future they will be called upon to pay a smaller share of governmental costs because of depreciated value. An examination of 147 selected townships showed that second class townships located in the coal and oil regions have a proportionately larger amount of bonded debt than those located in farming districts.

The following townships with the highest and lowest per capita bonded debt in the selected group when compared as to economic characteristics lend support to this view.

Township	County	1932 Per Capita Bonded Debt	Economic Characteristics
Adams	Cambria	$120.29	¼ Farming, Bituminous
Bell	Westmoreland	30.82	Farming, Gas, Bituminous
Middle Taylor	Cambria	23.32	½ Farming, Bituminous
Lower Burrell	Westmoreland	22.64	Farming,Gas,Bituminous,Wooded
Eden	Lancaster	22.45	Farming, Wooded
East Taylor	Cambria	20.49	Bituminous
Middletown	Bucks	.42	Farming
Hunlock	Luzerne	.79	1/3 Farming, Partly Wooded
Hempfield	Westmoreland	.90	Farming, Bituminous, Gas
Mount Pleasant	Westmoreland	1.00	Farming, Bituminous, Wooded
Lower Paxton	Dauphin	1.43	Farming, Wooded
Bristol	Bucks	1.52	2/5 Farming

The establishment of greater supervision over the finances of municipalities in default is urgently needed in Pennsylvania. Several forms of state supervision are briefly considered elsewhere in this report.

II. Consolidations

Consolidation of Townships - The second class township code provided for the creation of new townships formed from second class townships in the same county as follows:

> From parts of two or more townships
> Division of a township into two townships
> Consolidation of two or more entire townships

Consolidations are now effected by petitions presented to the court of quarter sessions of the county in which the townships are located. The petitions must be signed by (a) owners of real property representing 25% of the assessed value affected, or (b) by registered voters equal to 25% of the highest total vote cast for any township office in the preceding municipal election or (c) by a majority of the supervisors of each of the townships affected. The petition must set forth the name of the proposed new township and must specify the reason for the creation, division, or consolidation. It must be accompanied by an accurate map prepared by an engineer showing the lines of the proposed township and the lines of the existing townships as affected. Maps supplied by a state agency should be authorized in lieu of those now required to be furnished by an engineer. Within thirty days of the filing of a petition for consolidation any taxpayer may file exceptions. The exceptions may question the sufficiency of the number of signers, the legality of any signatures, or the accuracy or sufficiency of the map attached. The court then fixes a date for hearing at which it may sustain the petition, permit amendments, or dismiss it (1933 P.L. 103, Sec. 208).

Having gone to the expense of furnishing an engineer, an attorney, and attending the hearing held by the Court of Quarter Sessions with witnesses, the taxpayer has reached the point where a referendum may be secured. The court orders an election on the question of consolidation in the townships affected. The election is held at the next regular election occurring at least 60 days after the order of the court confirming the petition. Such an election is attended with the extra expense of posting by the township constable not less than six printed hand bills in the townships affected. If the vote fails of a majority in any township no further proceedings may be considered for two years. The costs of the court proceeding in case of a vote adverse to the creation, division or consolidation must be paid by the petitioners.

Provision is made in the township code for the adjustment of indebtedness by the supervisors of the new township with those of the existing townships. If no agreement can be reached the supervisors may petition the court of quarter sessions to appoint three disinterested commissioners to effect a settlement (1933 P.L. 103, Sec.216).

It should be possible for a smaller number, probably 5% of the taxpayers or registered voters, to petition the county commissioners to place a referendum for consolidation upon township ballots without the necessity for court action.

<u>Grounds for Consolidation</u> - Preceding chapters have pointed out many conditions for which some constructive remedy is urgently required. The declining population and consequent impoverishment of many townships, extensive areas consisting mainly of forest, the lack of adequate taxable resources in many sections and the large number of elective offices, all militate against efficient operation at reasonable cost. In each chapter individual steps for the improvement of special conditions have also been suggested. Fundamentally these fall into two main groups, those which seek to strengthen the personnel and those which are intended to reorganize the system. We shall now consider more fully the latter.

<u>Re-Organization of Township Government</u> - The particular types of government provided by the General Assembly have been fastened too rigidly upon local communities. In some sections of the state the township form of government may be better adapted to the needs of the locality than in others. In all localities, except state forest districts there should be sufficient elasticity in the available forms of local government to permit alternatives to be chosen by the people without the necessity of filing a petition with the Court of Quarter Sessions. No plan should become effective unless approved by a majority of the electors directly affected. Where more than one plan or where two conflicting provisions are adopted at the same election, the one receiving the greatest number of affirmative votes should be adopted. In order to permit such adjustments the following recommendations are made:

1. <u>Continue township organization.</u> Permit township organization to continue in those districts in which the people desire it. In rural townships it is highly regarded, but it should not be allowed to remain in
   a. More densely populated urban second class townships if the electorate desires a different form, or in
   b. Regions covered with dense state or other forest, regardless of the wishes of the electorate of such regions. A desirable optimum would be to reduce the present number of 1,514 second class townships to not more than 500.
2. <u>Referendum on county unit</u> - In order that full opportunity may be offered the voters to secure an efficient and less costly type of local administration the legislature should provide for mandatory elections in every county on the question of retaining township government or of substituting therefor a county unit in which the jurisdiction of the county officers would be extended to former township areas. If the latter were adopted, the functions of the township tax collector and secretary-treasurer would be assumed by the county treasurer, those of the roadmasters by the county engineer or highway department and those of the township supervisors by the county commissioners.
3. <u>Alternative forms authorized</u> - Legislature should provide a general law permitting alternative plans of local government to be adopted such as special county assessment districts and township manager forms.
4. <u>Proposals for alternative forms</u> - The township supervisors should be permitted by a majority vote to submit to the electors, without necessity of court approval, a plan of township or county consolidation at any time. Such a plan should also be submitted to the township electors upon the initiative of ten percent of the registered voters of the township at the last gubernatorial election. The county commissioners should be authorized to extend road maintenance or street lighting to districts (former townships) which may so desire, by special assessments on the benefited property.
5. <u>Home rule charters to be rendered constitutional</u>- The constitutional status of the township should be abolished. If, however, it is not, the constitution should be amended to provide that a legislative enactment permitting a township or townships by adoption of a home rule charter under a general law to create, change or abolish any constitutional or statutory township offices, including justices of the peace, should not be held unconstitutional as local or special legislation. Charters adopted by township electors should

be permitted to provide for the offices, their functions, duties, powers, terms and compensation.

Proposals similar to these· in many respects have been made by the New York commission for the revision of the tax laws (1934) and the Michigan commission of inquiry into county, township and school district government (1933). The latter body recommended that in order to give relief to local property taxpayers some five hundred independent township governments were to be reduced in number by state legislation. These consolidations were to take place without local elections and were to be based upon population numbers. "We therefore recommend that the legislature by mandatory act direct the boards of supervisors (1) in counties of less than 20,000 population, to attach each township therein having less than 1,000 population to an adjoining township or townships to form a township containing not less than 1,000 inhabitants; (2) in counties having a population between 20,000 and 100,000, to attach each township therein having less than 2,000 population to an adjoining township or townships to form a township containing not less than 2,000 inhabitants; and (3) in counties of more than 100,000 population, to attach each township therein having less than 3,000 population to an adjoining township or townships to form a township containing not less than 3,000 inhabitants; provided, that townships so formed shall be bounded by existing township lines" (Report, Organization and Cost of County and Township Gov't, pp. 125 and 130).

III. Present State Financial Control

In Pennsylvania, state financial control over townships may be said to include the following items:

1. Maximum tax levy
2. Control over indebtedness
3. Approval of road purchases in excess of $200 by the State Department of Highways

4. Publication of annual audit
5. Filing of annual financial report, and annual estimate with the State Department of Highways
6. Furnishing of tax and assessment data to the State Department of Internal Affairs

Maximum Tax Levy - The maximum tax levy is sponsored by large real estate owners and has recently increased in importance. It is expressed in the form of a statutory limitation on the millage that may be levied by any municipality on real estate. The effects of tax levy limitations have been referred to in Chapter 12 and need not be repeated here. The maximum, which may run as high as 31 mills, is hardly a limitation in practice since no township was found with such a high township tax.

Control Over Indebtedness - The chief legal and constitutional limitations regarding bonded debt are the same for all municipalities. They refer to:

1. The amount of the issue
2. State supervision
3. Term of the bond
4. Tax levies for indebtedness
5. Interest rates and methods of selling bonds

1. The amount of the issue - The Pennsylvania Constitution (Article XI, Section 8) imposes two limitations on the amount of bonded debt that a municipality may issue. This amount may not exceed 2% of the total assessed valuation of taxable property, without the consent of the voters. If ratified by a majority of the voters bonds may be issued not exceeding 7% of the taxable property valuation. This does not apply to Philadelphia. 2. State Supervision - While local bonded debt is thus subject to certain constitutional safeguards, up to 1927 there was little State administrative control of such issues. In 1927, the Bureau of Municipalities of the Department of Internal Affairs was authorized to supervise the issue of indebtedness. Before any bonds may be issued by second class townships, the chairman of the board of township supervisors must certify to the Department of Internal Affairs a complete and accurate copy of the proceedings relating to the incurring or increasing of such indebtedness. This must be accompanied by a statement of the assessed valuation of taxable property, and the total amount of outstanding indebtedness and deductions allowed by law (Act No. 65, 1927). In 1933, the Legislature permitted the Department of Internal Affairs to collect a fee for examining bond proceedings. The amount of the fee is estimated approximately to cover the cost of examination, that is $10.00 plus an additional fee of one-half of one mill (.0005) on each dollar of the aggregate par value of the indebtedness to be issued (Act No. 64, 1933).

The Bureau of Municipalities attempts to examine each bond issue and the condition of the general bonded indebtedness of the municipality concerned. It determines whether or not the bond is legally drawn, and whether the municipality has not exceeded its borrowing capacity, also whether there is sufficient provision for tax levies to cover the sinking fund requirements. Such examinations apply only to municipalities issuing bonds since 1927. The Bureau therefore lacks complete information on all bonded indebtedness.

3. Term of Bond - The law also specifies that no bond may be issued for a period in excess of 30 years. In case of townships of the second class where most of the borrowed money is used for highway construction, it is questionable whether 30 years is a sufficient limitation, in view of the fact that the life of the average road is far below that figure.

4. Tax Levies for Indebtedness - A further regulation in the issuance of bonds is the provision in Article IX, Section 10, of the Constitution, which provides that a municipality incurring any indebtedness must provide a sufficient annual tax to pay the interest and principal of such indebtedness within a period of 30 years. This provision during a period of depression does not provide the proper protection to the bondholders since a sufficient tax levy does not mean that a sufficient amount of money will be deposited in the sinking fund or will be used for maturity of bonds. The various township supervisors in setting the tax rates for sinking fund purposes, do not ordinarily consider the possibility of a large delinquency.

5. Interest Rates and Methods of Selling Bonds - Other limitations include a limit for interest payment not exceeding 6% per annum, payable semi-annually (Act No. 583, 1929). The township code (section 904) also prohibits the selling of municipal bonds by private sale, except where no bids are received after a legal advertisement in a newspaper of the sale of such securities, once a week for a period of three weeks. In that case, bonds may be sold at a private sale but never below par.

As to temporary unfunded loans we have already seen that Section 903 of the code provides that townships may borrow in anticipation of taxes and issue a certificate of indebtedness not exceeding one year. This limitation is known for its violations, since most of the townships continue such indebtedness over a period of years.

Approval of Road Material Purchases in Excess of $200 - The Township Code (Sec. 802), provides that all contracts for the purchase of materials, equipments, and machinery for road purposes must be in writing and that all such contracts must be approved by the Department of Highways. The result of this legislation has been that townships are not required to award such contracts to the lowest bidders. In practice it has meant that neither the Department of Highways nor the township supervisors can be held responsible for the award. The Department cannot actually pass upon this large number of small purchases in such a manner as to exercise an effective check. The township supervisors shift responsibility to the Highway Department, since the final authority for the approval of such contracts rests with that Department.

Publication of Annual Audit - Within ten days after the completion of the township audit, a concise itemized statement of receipts and expenditures of the various officers must be published in at least one newspaper of general circulation printed in the township or county, or at least five copies of such receipts and expenditures must be posted in public places in the township (Sec. 547). This law is often violated, and in places where it is followed the reports vary in size and contents. Although the Department of Internal Affairs is empowered to "install or assist in the installation and establishment of uniform systems of accounts in the various municipalities of this state" (1929 P.L. 242), the Department has made no effort to induce municipalities to adopt a system of uniform accounting. The reports of the various townships are therefore of various types.

Filing of Annual Report of Receipts and Expenditures with the Highway Department - The only uniform type of reporting in second class townships is by means of an annual statement of receipts and expenditures and estimates of expenditures for highway purposes, filed with the Department of Highways on forms supplied by the Department. This type of report, however, in many cases is inadequate. Furthermore, until recently there has been little attempt to check these reports for accuracy. Much more could be done especially when it is considered that practically all townships file reports because state rewards are based on the expenditures so reported.

CONCLUSIONS

Debts

FINDING: More than one-third of the second class townships in Pennsylvania were without bonded or temporary debt in 1932.

FINDING: Causes of township indebtedness include
      (a) Borrowing to participate in state assistance for the construction of
          improved roads
      (b) Loans incurred for the purchase of machinery
      (c) High interest costs due to frequent renewals of temporary indebtedness.

FINDING: The amount of township bonded debt is relatively small, compared with that of other local governments. Bonded debt of second class townships amounted to $9,315,152, or $4.10 per capita in 1932.

FINDING: The trend of second class township bonded debt during the seven-year period 1926-1932 indicated a decline of approximately 4%. During the same period the debt of all local governments in Pennsylvania increased nearly 2%.

FINDING: Temporary borrowing is of particular importance in second class townships. It is estimated to be approximately 34% of the total debt.

FINDING: The cost of debt service in second class townships was about 31% of all expenditures in 1932.

FINDING: The total indebtedness, including funded and unfunded debt, was $14,118,858; and the average debt per township was $9,319 in 1932.

FINDING: Townships furnished far more than their share of defaults on bonded debt during the period 1928-1934. These defaults appear to be due to increased tax delinquency and to costly short term loans.
RECOMMENDATION: More extensive supervision over the finances of municipalities with heavy debt is urgently needed. Present forms of state control over finances are too weak and ineffective to be of real assistance to second class townships. The Department of Internal Affairs or some other state agency should be authorized to render assistance and approve budgets of at least the following:
      (a) Townships in default
      (b) Townships with a bonded or temporary debt in excess of a fixed maximum
      (c) Townships with a tax levy beyond a fixed proportion of delinquency

Reorganization - Summarized

FINDING: The diversity of township conditions though great is not of such nature as to prevent all legislative effort for the strengthening of personnel and organization. Detailed findings on the urgent need for such effort have been given in the present and preceding chapters.
RECOMMENDATION: Alternative forms of township government should be made available, to be adopted as desired by the township electors by initiative and referendum.
RECOMMENDATION: Township organization should be continued and strengthened in all sections in which it is appropriate, useful and desired by the electorate. In regions covered with state or private forest areas, township organization should be supplanted by some simpler system of supplying local needs.

RECOMMENDATION: A mandatory election should be provided in every county, giving the voters opportunity to determine by a county-wide choice as to the retention of township government or the establishment therefor of a county unit. Where the latter is chosen the possibility of establishing special assessment districts under county auspices for fire, light and other services in local areas should be afforded.

RECOMMENDATION: The reduction in number of townships to not more than 500 should be made upon the basis of a plan drafted by an authoritative legislative commission in consultation with township officials.

Present State Financial Control

FINDING: Present forms of state supervision probably do not merit the name and in some instances are not intended as such. They offer a basis upon which new controls and modernized procedure, described elsewhere, may operate for those groups of townships needing assistance from outside sources.

CHAPTER XVI

CREATION AND CORPORATE POWERS

The eight chapters of this Part III are devoted to a description of the sixty-one first class townships of the Commonwealth, including their statutory basis and practical operation. The purpose has been to indicate the governmental problems facing these units and the relationship of the Commonwealth to these questions. Some problems arise from simple defects in the statutes governing the township, others from maladjustments growing out of fundamental changes. The latter seem to show a need for greater State supervision.

The data presented were obtained from (a) documentary sources such as the Pennsylvania Constitution and statutes, compilations made by various state departments, and the studies of local government in other states; (b) organizations primarily concerned with the economical operation of government such as the Pennsylvania Economic Council and active taxpayers associations. In several instances these organizations had compilations or uncompiled factual data at hand and generously permitted their use; (c) personal visits to the townships and frequent conferences with local officials. Approximately eighty officials were interviewed in twenty-five townships†: generally the secretary, treasurer, and president of the board of commissioners. In many instances the best source of information was some other official who had seen long service. Unfortunately, complete uniformity of data was not obtainable on some phases of governmental activity, since many of the persons interviewed were unwilling to refer to their accounts or to permit access to them for specific information. At times also, conflicting statements were presented by different officials of the same township. Where verification of such data was not possible, they had to be omitted.

Very few of the officials interviewed proved to be antagonistic, and in only three townships was all co-operation refused. In many places, however, the detailed, definite material required for an accurate report existed only in such form that nothing short of a legislative inquiry could extract it.

Classification Provisions. - What is the purpose of erecting and classifying townships? Do the existence of the present township areas and the present grouping of these units fulfill the purposes of the statutes governing the creation and classification of these local areas? Do the diversities of population, area, taxable resources, and the like defeat the purposes of classification and indicate a need for consolidation, annexation, and reclassification?

The problem here is to make the classification useful. The only reason for classification is to place each group of similar local areas under a form of local government adapted to its needs. The laws providing for the erection, division, consolidation and reestablishment of local areas are intended to permit the people of these areas to roganize and to place themselves as a unit, from time to time, under that form of government peculiarly suitable to their changed conditions. These ends are not now realized. The code provides that first class townships must have a population density of 300 per square mile, but this classification is defeated by optional provisions which place the decision with the voters; moreover, consolidations and annexations leave remnants which must constitute first class townships, even though their conditions do not warrant it. The early laws made township status automatic, based on population only.

Historical Summary - The first class township, like other local govern-

---

† Upper Darby, Lower Merion, Haverford, Hanover, Mt. Lebanon, Penn, Harrison, Newport, Whitehall, Shaler, South Fayette, Elizabeth, Wilkes-Barre, Baldwin, Swatara, Spring Garden, Jefferson, Tinicum, Reserve, Annville, Nether Providence, West Norriton, Braddock, Leet, and West Lebanon.

Many additional interviews were had with non-official persons such as local newspaper editors, legislators, and local residents.

mental units, is a creature of the state legislature. Originally, townships were established by special acts of the legislature and later by order of the courts (Act of 1834, P.L. 537, Acts of 1857, P.L. 93, P.L. 304). In 1899 townships were first classified (P.L. 104) upon the basis of population density (300 inhabitants or more per square mile) into first and second class townships. All legislation prior to the first class township code provided, with one exception (Section 65, Act of 1917, P.L. 840) which was later amended (Act of 1927, P.L. 336), for the automatic determination of township status upon the basis of population density, and no provision was made for electoral action thereon. In so much as the present code provides for the determination of township status primarily upon the basis of electoral action thereon, it definitely departs from the prior legislation. In view of the fact that such departure weakens the utility of the classification, its soundness is questionable.

The provisions of the code of 1917 were repealed or reenacted in so far as they relate to first class townships by the "First Class Township Law" of June 24, 1931, P.L. 1260, and, in so far as they relate to second class townships by the "Second Class Township Law" of May 1, 1933, P.L. 103. In effect these latter acts set up a code for each of the two classes, thereby superseding the code of 1917 which included both classifications. While these acts do not attempt to include all the legislation on the subject, they are nevertheless a substantial compilation of the important statutes affecting townships.

Code of 1931 - The Act of 1931, hereinafter referred to as the code, provides four methods for the establishment of a first class township: (a) By special enrollment. At any time not less than one year prior to the decennial census of the United States the owners of 25% of the assessed valuation of the real estate may present a petition to the Quarter Sessions, averring that the township has at least 300 inhabitants per square mile; if the averment is true, the court certifies such fact to the township supervisors and county commissioners, whereupon the question is submitted to the voters at the next general election. If the action of the voters is favorable, it is then organized and established as a township of the first class. The requirement of a vote of the electors is a departure from prior legislation. The previous method was automatic, since upon the ascertainment of the requisite population a court order or proclamation established the first class status.

(b) By the United States census. After each decennial census the county commissioners shall proclaim such townships as have reached the requisite population density, and shall certify such fact to the township supervisors; whereupon the status of the township must be referred to the electors as outlined above.

(c) By division of existing townships. Division may be effectuated upon petition to the Quarter Sessions either: by the owners representing twenty-five percent of the assessed valuation of the township, or by registered voters equal in number to twenty five percent of the highest total vote cast for candidates for any office in the township, at the last election. Upon granting of the petition, the question is submitted to the voters, and if a majority are in favor of division, the court will decree the erection of a new township. It is further provided that the townships created thereby shall remain townships of the first class unless and until they are reestablished or proclaimed townships of the second class. The Act of 1933 relating to second class townships provides for such procedure "irrespective of population"; the board of commissioners may of their own initiative, or shall upon petition of twenty-five percent of the electors pass a resolution submitting the question to the voters; if a majority of the votes are in favor of reestablishment, the township shall be so constituted.

These provisions on division of townships differ in three important respects from the previous legislation: (1) None of the acts cited above (1905, 1907, 1927) had provided for a petition by any persons other than the owners of real estate. The present code, in addition to this, permits petitions by registered voters. (2) The new provision that, following the division of a first class township, both the new township and the remaining portion shall have first class status. This is not in harmony with the purpose of classification since the status is made to depend upon the procedure of petition and electoral action. (3) The general provision for the reestablishment of first class townships, irrespective of population, is likewise a questionable departure. It tends to break down the very classification which the law elsewhere seeks to establish.

(d) By consolidation of two or more first class townships within the same county. This is brought about by the filing of a petition, its confirmation, and favorable vote of the electorate of each township affected. The petition may be by either of the two methods prescribed in the petitioning for division of a township, or when signed by a majority of the commissioners of each of the townships proposed to be consolidated. This provision is substantially similar to that of the general township act of 1917.

In two important respects the present code makes the procedure of consolidation more difficult. First, a petition to the quarter sessions court is required and must be made by a larger number of petitioners in one instance or by the owners of a large amount of real estate in the other. The petition itself is more formal, complicated, and may lead to further court proceedings involving hearings and appeals on the matter. Under the prior legislation only a written request to the commissioners was necessary. or the commissioners might act of their own initiative by entering into a joint agreement to that effect and have the issue submitted to the voters. Second, the costs and expenses of the entire proceeding are thrown upon the petitioners whenever the petition is refused by the court or rejected by the voters. Under the 1917 act such costs were paid equally by each of the townships affected.

The statutes governing the creation, division, consolidation and classification of townships are responsible for many of the weaknesses of local government within these areas. Notably the basis of classification, three hundred inhabitants to the square mile, is too broad in that it permits grouping of different kinds of local areas within the same classification, despite extreme diversity in character, area, population, wealth, taxable resources and government needs. These excessive diversities within the group hinder and obstruct the solution of such problems as the institution of adequate and uniform budgetary, accounting, auditing and reporting procedures, the enactment by law of acceptable qualifications for such officers as auditors, secretaries, and the like, the character, form and extent of state supervision over the performance of specific governmental services such as health and highways, and over local fiscal operations. Not only are the statutory provisions too general, they are also inconsistent. Classification is based upon a specified density of population, yet when a first class township is divided, its parts remain first class townships even though they have not the required density; when a first class unit is to be changed to second class, the action can be taken only with the approval of the electors but if this is given the change takes place regardless of population. In short, the code provides a classification and then proceeds to remove its meaning. This renders the powers and authority of first class units unfitted to their areas in many instances and imposes needless financial burdens upon them.

Corporate Powers: The corporate powers of first class townships are similar to those of the boroughs and cities. This is due to the urban character of the majority of townships within this class.

(1) Mandatory and Optional Powers: The mandatory powers may be briefly listed: They are to administer and enforce the statutes and regulations in regard to health and to maintain township highways. The remaining functions are optional in character.

(2) Power to create offices: The first class township code specifically sets up the following offices: commissioner, treasurer, tax collector, secretary, auditor, and health officer. In the cases of auditing and health administration, the code provides optional features - it vests discretion in the commissioners as to the particular type or form such office shall assume: in the former, either a court appointed auditor or a locally elected controller instead of three elective auditors; in the latter, a board of health or health officer.

In addition to these mandatory offices the code provides for the establishment of the following, within the discretion of the commissioners: solicitor, controller, engineer, chief of police (and other police officers) township manager, committing magistrate, building inspectors, park, recreation and adjustment boards, and shade tree, zoning and planning commissions. Finally, section 1501 contains blanket authority empowering the commissioners to create any office or department deemed necessary for the interests of the township. Under this clause several townships have established the following offices: fire marshall and assistants, meat, food and sewer inspectors, road supervisor or superintendent, superintendent of sewers, sewer engineer, and a host of minor offices.

(3). Primary powers. The township is vested with the power to perform such primary and essential functions as the following:

(a) To provide for the public safety by: establishing a police force, providing lockups, regulating disorderly practices, designating committing magistrates, arresting vagrants and setting them to work, making regulations for the prevention of fires, making appropriations to volunteer fire companies, providing for the housing of fire apparatus, and by inspecting, regulating and abating fire hazards.

(b) To provide for the public health by: enforcing state and local regulations, establishing "emergency hospitals" and maintaining ambulances, regulating the construction and maintenance of house drains, cesspools, etc., inspecting places which may constitute a nuisance or menace to public health and abating such nuisances, and providing for the inspection and regulation of milk and milk products.

(c) To provide for the construction and maintenance of public highways, roads, streets, sidewalks and bridges by laying out, opening and widening the same, exercising the right of eminent domain, and assessing property owners for their improvements.

(d) To provide for the lighting of streets and highways by contract with public utilities.

(e) To provide for an adequate water supply by entering into contracts with private persons or public utilities, and by establishing and maintaining water works.

(f) To provide for the construction and maintenance of sewers and sewerage systems.

(g) To provide for public recreation by establishing and maintaining playgrounds, parks, swimming pools, parkways, community centers, playfields, gymnasiums, public baths, and the like.

(h) To exercise all such fundamental governmental powers necessary for the performance of the foregoing functions by: enacting and enforcing (by penalty) ordinances and resolutions; levying and collecting taxes for the following purposes- building and maintaining of housing for fire equipment, caring for trees, for procuring a lot and/or the erection of a townshouse, for the lighting of streets, for providing a water supply, an annual tax for the payment of interest on indebtedness and sinking fund charges, and real estate and occupational taxes for general township purposes; levying and collection of special assessments for such purposes as permanent improvements, street lighting, fire protection and the like. Entering into contracts with private persons for the construction of streets, sewers, etc.; acquiring and leasing such real and personal property as the township may require; exercising the right of eminent domain; and the right to sue and be sued in the name of the township.

(4) Regulatory and/or police powers. In addition to the powers outlined above. the township is granted the following specifically stated regulatory powers:

(a) To regulate buildings or parts thereof in respect to construction, alteration, use, sanitation, water supply, etc., and to classify them for the above purposes.

(b) To make regulations respecting foundations, party walls, and partition fences.

(c) To prohibit or regulate the manufacture, sale and use of fireworks and other explosives.

(d) To make zoning regulations.

(e) To regulate and/or license market houses, peddlers, hucksters, junk dealers, vendors, pawnbrokers, and the like.

(f) To regulate or abate slaughter houses, manure pits, dumps, cesspools, etc. or any other nuisance.

(5) <u>Power to appropriate for specific purposes.</u> The commissioners are authorized to appropriate sums for the following purposes:

(a) Memorial day services.

(b) The erection and maintenance of armories.

(c) Pension fund for police officials.

(d) Two hundred dollars, annually, for the support of any voluntary association composed of electors, formed for the study of the welfare and management of townships and for the recommendation of beneficial legislation. (Also appropriations to any forest protection association).

(6) <u>Power to erect, maintain, establish and purchase public utilities and other enterprises.</u> This power includes the following:

(a) Plants for water supply.

(b) Market houses.

(c) Incinerating plants.

(d) Comfort stations.

(e) Watering troughs and drinking fountains.

(f) Emergency hospitals.

(g) Ware houses and storage houses.

(h) Community centers, swimming pools, public bath houses, and the like.

(7) <u>Miscellaneous powers.</u> Within this group may be included the power:

(a) To purchase burial grounds for soldiers.

(b) To acquire land for armory purposes.

(c) To contract insurance for township property and workmens compensation, and group insurance for employees.

(d) To acquire forest land.

(e) To create a revolving fund for permanent improvements.

(f) To provide for registration of real estate.

(g) To display flags.

(h) To enter into agreements with other municipalities for joint construction and maintenance of highways, sewers and water works, and for the establishment of parks, playgrounds, etc.

(i) To establish a pension fund for police officials.

(8) <u>Blanket authority.</u> In addition to the power to create any office or depart-

ment deemed necessary, the commissioners are given broad authority in respect to the use and acquisition of real and personal property, and are empowered to adopt resolutions and ordinances prescribing the manner in which powers of the township shall be carried out, and for the general regulation of township affairs.

Diverse Conditions. Most of these broad powers are not exercised in the majority of townships. The larger, more populous units have their sewerage systems, police, parks, playgrounds, zoning, swimming pools, parkways, building regulations, food inspection, and the like. The smaller, less populous, less wealthy, and more rural townships such as South Fayette, Jefferson, Elizabeth and Swatara find little occasion to perform more than the two mandatory functions of highway and health administration. The units composing what might be termed a "middle" group, neither urban nor rural but semi-urban, such as Nether Providence, Baldwin, Penn and Harrison perform one or more of these optional services in whole or in part. Nether Providence has established detailed zoning and building codes; Baldwin has created a zoning code, not generally enforced, and has provided sewerage for portions of the township only.

Nature and Extent of Diversity. Table I shows the diversity in area, population and taxable resources, also the assessed valuation, and the mileage of township and state roads. It does not show an interesting fact about these units, viz., their tendency to cluster around urban centers. This tendency reflects their urban and residential character. The data on assessed valuation were taken from a variety of sources such as the annual reports filed with the Department of Highways, the annual audit statements, and the county assessment authorities. The continuous changes which are made in assessed valuation statistics and in road mileage prevent absolute accuracy in figures.

The extreme range of population is shown by the fact that in the smallest of these townships, South Versailles, the total population is only three hundred and thirty-six, while in the largest, Upper Darby, the population is forty-six thousand six hundred and twenty-six. This diversity is also shown in that there are eighteen townships with a total population less than three thousand; twelve with a population of three to five thousand; sixteen with population of five to ten thousand; and, fifteen with a population of ten thousand or more.

The same diversity is apparent in the density of population within these townships. The population per square mile varies from one hundred and ninety six in Jefferson to five thousand three hundred and twenty-two in Upper Darby. Nine of the townships have a population of two thousand or more per square mile; thirteen range from one to two thousand; twenty seven from five hundred to one thousand; and, twelve with density less than five hundred. Likewise as to taxable resources or assessed valuation: Lower Merion has the highest assessed valuation, $86,750,715, while South Versailles has the lowest, $111,033. As would be expected, the more urban townships have the greatest density of population and the largest amount of taxable resources. Where rural units such as South Fayette, Butler, Elizabeth and Jefferson have higher assessed valuations than the more urban units such as Annville, Darby and Reserve, this fact is entirely accounted for by the greater areas of the former group.

The table thus shows a group of rural townships which by nature are in reality second class units, another group which are semi-urban, semi-rural and for which the first class type was originally created, and a third group of communities which in density and number of population and taxable resources really constitute cities of the third class.

This is confirmed further by the variations in the tax rate. The more urban townships, having more governmental services, levy the higher rates while the rural townships levy the lower rates. The apparent exceptions to this rule are explained by one or more of the following: disparity in area; greater proportion of improved road miles; a high proportion of miles of state roads to the total road mileage; and either a decreasing density of population or a slower rate of population growth.

More specifically those townships having a tax rate of six mills or less are as follows: Hanover, South Fayette, Ross, Swatara, Collier, O'Hara, Plymouth, Jefferson, Nether Providence, West Norriton, Springdale, Neville, and South Versailles. None of these units are either urban or predominantly urban. South Fayette, and Jefferson are exclusively rural townships while Swatara, Collier, Plymouth, West Norriton, and Springdale are predominantly rural. The remaining units, Hanover, Ross O'Hara, Nether Providence, Neville, and South Versailles are urban communities in part. The low mill-age in Hanover may be accounted for by the fact that it is a Losch Act township and by the existence of a strong, vigilant taxpayers association. In Ross, O'Hara, Nether Providence, Neville and South Versailles the millage is explained by one or more of the following: smallness in area, proportion of unimproved roads, large proportion of state road miles, decrease in population, and the slower rate of population growth. Generally, it may be stated that those townships of large area, high proportion of un-improved roads, high ratio of state road miles, low tax rate and least density of popu-lation, are rural in character.

From the data here presented it would appear that some state agency should re-draft the present classification. An enlarged basis is needed which will not-only in-clude density of population but also the amount of taxable resources. Consideration should be given to the transfer of the large urban townships into the third class city group where they belong, as well as to the reclassification of the rural townships as second class units. The remaining townships would then be homogeneous in their semi-urban, semi-rural character and in the governmental problems which they must face.

The alternative should also be considered of abolishing the present division into two classes and setting up instead a single class with subdivisions according to density of population and property valuation. Further provision should be made for annexation to one or more of the larger adjoining governmental units for those townships, such as West Lebanon, South Versailles, Darby, Reserve, and Braddock, which cannot justify an independent existence of their own. In the first two, rapidly decreasing population and taxable resources, due to industrial conditions, combined with the restricted area of each would seem to call for annexation or consolidation with stronger adjacent units. While the population in Braddock is not actually decreasing, the rate of in-crease is rapidly declining, and this, in conjunction with lack of adequate taxable resources, smallness of area, and comparative poverty of the population indicates the same need for annexation. The president of the board of commissioners of this town-ship stated that the township did not possess adequate resources to support itself and that annexation was highly desirable. Reserve and Darby should be similarly dealt with because of small area and proximity to urban centers.

That some steps will have to be taken to define the character of townships of the first class more clearly is evidenced by population trends within these units. In 1910 the range of population among these units was small. From 1910 to 1920 the differences increased by the rapid growth of those areas bordering upon large urban centers. Be-tween 1920 and 1930 this change was accelerated and numerous boroughs were erected from parts of townships. This latter step usually impaired the taxable resources of the township affected and left the remaining portion predominantly rural in character. In order to meet this situation the Act of May 18, 1933, P.L. 818 amending the Act of 1927 was passed. It provides that where a borough is to be erected from a portion of a township of the first class a petition to that effect must be signed by two-thirds of the resident freeholders and by the owners of three-fourths of the assessed valua-tion of real estate in the township affected. This petition is presented to the court which in turn lays it before the grand jury. If approved it is certified to the court, and if the court confirms the action a borough is erected. The new act is a step in the right direction. Meanwhile the steady increase in township population calls for even more optional code features to meet the needs of the large urban areas. This means that the value of township classification disappears and is additional evidence that a new classification is needed.

Another indication of the future trend toward greater diversity is found in the tendency of most townships of this class to cluster around large urban or metropolitan centers. Eighteen are either adjacent to or directly bordering upon the City of Philadelphia. Twenty-four are similarly situated in the Pittsburgh metropolitan center including Pittsburgh, Clairton and McKeesport. The remaining townships are either ad-jacent to or within close proximity of such cities as Harrisburg, Scranton, Erie, Allentown and Lebanon, or such large urban centers as the boroughs of Norristown, Ambridge and Tarentum.

The population trends within such townships reflect the expansion of the metro-politan center to which they are adjacent. Consequently provision should be made for their automatic absorption within such expanding urban area, or for their erection as independent boroughs or cities where the size of the area and other relevant conditions justify an independent existence.

Finally, consideration should be given to the possibility of consolidating bor-oughs and first class townships. The code is intended to provide a form of government suitable to residential districts which are semi-urban and semi-rural. If such form of government can be successfully established, many of the separate units now existing could be merged in such manner as to constitute political communities of this charac-ter. This consolidation would strengthen the taxable resources of these sections and at the same time eliminate the overhead costs arising from separate organizations for small areas. Such a change is especially needed in those townships where one or more boroughs have been erected within the boundaries of the existing units. Examples are seen in Hanover township which completely envelops the boroughs of Warrior Run, Sugar Notch, and Ashly; Coal township which completely surrounds the borough of Shamokin; and Lower Merion within the confines of which the borough of Narberth has been erected. Consideration should also be given to similar action where the erection of boroughs has completely separated one or more portions of the township from the remaining por-tions as in the case of Upper Darby and Darby townships, and in those cases where a small borough and a first class township are directly adjacent as in the case of Nether Providence, Whitehall, and others. The experience of the townships visited leaves no doubt that in the present financial stress each local unit be so located as to include within its territory adequate taxable property resources to finance its governmental functions or it must be abolished, by merger, as incapable of performing those same functions. These conditions show the need for a state-wide examination by some offic-ial body in order to redraft the present classification.

CONCLUSIONS

1. FINDING - Many first class townships cannot be distinguished from third class cities, boroughs and second class townships respectively in their number and density of population, taxable resources, area and other conditions. This is due to the statutory provisions on classification, erection, reestablishment, consolidation, corporate powers and organization. These provisions are so broad, loose and inconsist-ent as to nullify completely the purposes of the classification.

Sections 207, 218, 225, 226 of the code permit townships with a population density of the first class to remain in the second class or to become second class townships. These sections provide that remnants from the consolidation or division of first class townships retain first class status regardless of density of population. They fail to provide any requirement which will assure that a first class township shall have suffi-cient taxable property valuation to finance its functions. The present classification of townships has therefore lost its meaning and purpose and has become a cause of con-fusion and heavy tax costs.

RECOMMENDATION - That a legislative commission in consultation with officers of both first and second class townships and other persons, should prepare a new basis of class-ification with either:

a) Two or more classes, or
b) A single class with options suited to township conditions

RECOMMENDATION - That such legislative commission examine and include in its report proposals relative to

a) The inclusion of assessed valuation of taxable resources in the basis of classification
b) A more homogeneous grouping of local areas in each class
c) A maximum and minimum requirement as to density of population and assessed valuation, in order to provide for a more automatic classification
d) The feasibility of annexing small, paritally defunct townships or those which

are so closely integrated with a large municipality as to be indistinguishable therefrom. Such townships as West Lebanon, South Versailles, Wilkes-Barre, Braddock, Lackawanna, Darby, Reserve, and Spring Garden, should be examined with this end in view.

2. FINDING - The lack of taxable resources, the possibility of cutting overhead costs and other factors make the consolidation or annexation of many of the townships desirable. The present statutory provisions seriously handicap either possibility.

RECOMMENDATION - That in view of these statutory obstacles, the procedure for consolidation be facilitated by relieving the petitioners of the burden of paying the costs of such procedure where the proposal for consolidation is rejected by the electorate.

TABLE I

DIVERSITY OF CONDITIONS

Township	Population			Percent of Increase or Decrease		Population per Square Mile	Area in Square Miles	Total Assessed Valuation 1933	Mileage of Township Maintained Roads			Mileage of State Roads
	1910	1920	1930	1920 over 1910	1930 over 1920				Improved	Unimpr.	Total	
Upper Darby	5,385	8,956	46,626	66.31%	420.61%	5,322.60	8.76	$58,698,433	86.32	6.15	92.47	21.59
Lower Merion	7,671	23,866	35,166	211.11	47.34	1,501.53	23.42	86,750,715	108.02	13.20	121.22	31.68
Haverford	3,989	6,631	21,362	66.23	222.15	2,232.18	9.57	31,211,142	58.11	9.08	67.19	29.43
Coal	6,105	17,574	12,929	187.86	13.40	1,729.95	11.52	10,623,862	7.23	33.60	40.83	17.34
Abington	5,896	8,684	18,648	47.28	114.73	1,230.89	15.15	23,877,571	65.48	40.00	105.48	34.55
Hanover	6,965	11,139	17,770	59.92	59.52	396.56	44.81	49,056,022	23.34	4.80	28.14	27.06
Plains	10,529	13,986	16,044	32.83	14.71	988.54	16.23	8,335,519	5.25	16.75	22.00	18.12
Cheltenham	9,434	11,015	15,731	30.60	42.81	1,793.73	8.77	25,456,015	60.52	1.85	62.37	20.88
Mount Lebanon	----	2,258	13,403	----	493.57	2,084.45	6.43	34,575,435	42.31	3.80	46.11	4.02
Stowe	8,145	10,665	13,368	30.93	25.34	2,370.21	5.64	11,533,590	6.50	16.00	22.50	4.35
Penn	6,207	8,342	13,337	34.39	59.87	680.11	19.61	13,408,880	2.85	43.60	46.45	32.31
Harrison	6,687	8,389	12,387	40.40	31.93	1,722.80	7.19	10,223,610	36.00	6.00	42.00	11.22
Radnor	7,094	8,181	12,263	15.32	49.89	857.55	14.30	22,050,100	31.00	5.66	36.66	31.23
Newport	10,277	10,992	12,087	6.95	9.96	740.63	16.32	17,456,152	4.90	4.00	8.90	12.70
Whitehall	9,350	9,886	10,707	5.73	8.30	851.79	12.57	6,263,529	27.90	42.40	70.30	19.24
Shaler	5,154	6,306	9,273	22.35	51.80	844.92	11.38	10,098,489	----	25.00	25.00	11.40
South Fayette	8,761	9,221	9,147	5.25	-.80	430.85	21.23	4,033,725	----	30.00	30.00	25.98
Butler	6,235	7,692	8,641	23.36	12.33	391.88	22.05	8,532,636	9.20	49.80	59.00	23.00
Ross	3,812	4,984	8,581	29.82	73.38	2,189.03	3.92	10,068,930	3.80	41.50	45.30	8.77
Ridley	2,945	5,342	8,326	81.39	55.85	1,607.34	5.18	8,947,306	12.75	35.00	47.75	12.44
Mifflin	8,076	11,267	8,167	39.51	-27.51	383.60	21.29	4,645,355	1.50	72.00	73.50	16.72
Elizabeth	7,410	6,563	7,479	-11.43	13.95	339.34	22.04	5,625,970	----	23.40	23.40	25.74
Wilkes-Barre	5,703	6,608	7,176	15.86	8.59	2,677.78	2.70	5,287,485	1.90	7.80	9.70	2.00
Baldwin	8,230	4,928	6,371	-40.12	29.28	364.47	17.48	5,596,280	18.00	17.00	35.00	12.84
Swatara	5,630	5,847	6,333	3.85	8.31	508.68	12.45	2,423,500	12.30	15.65	27.95	20.19
Scott	5,737	4,927	6,203	-14.11	25.89	1,466.43	4.23	5,247,065	2.00	34.00	36.00	4.40
Collier	4,211	4,651	6,091	10.44	30.96	460.04	13.24	3,691,910	----	23.00	23.00	16.66
East Bethlehem	1,929	4,969	5,765	157.59	16.01	977.12	5.90	2,036,135	2.50	14.50	17.00	9.22
North Versailles	4,435	4,844	5,668	9.22	17.01	641.17	8.84	3,841,460	----	22.00	22.00	12.19
Springfield (Montg. Co.)	2,994	3,374	5,541	12.69	64.22	837.00	6.62	7,850,375	20.92	6.15	27.07	11.76

O'Hara	3,917	4,672	5,127	19.27	9.73	376.98	13.60	$6,991,580	----	17.80	17.80	5.10
Spring Garden	2,209	2,766	4,675	25.21	69.01	678.52	6.89	4,231,575	25.00	6.50	31.50	11.85
Springfield (Del.Co.)	1,132	1,298	4,589	14.66	235.54	677.84	6.77	6,642,380	11.60	4.50	16.10	12.81
Plymouth	2,706	4,201	4,369	18.29	36.48	512.19	8.53	3,818,660	14.50	----	14.50	15.31
Jefferson	6,771	5,009	4,138	-26.02	-17.38	196.49	21.06	4,406,640	----	24.00	24.00	26.52
Wilkins	3,626	3,455	4,127	-4.71%	19.45%	603.36	6.84	2,947,100	----	14.00	14.00	9.19
Upper Moreland	----	2,195	3,989	----	81.73	519.40	7.68	3,966,330	33.00	7.00	40.00	18.37
East Deer	3,702	3,506	3,928	-5.29	12.03	2,146.44	1.83	7,099,456	.50	2.00	2.50	3.59
Tinicum	1,135	2,500	3,630	120.26	45.20	870.50	4.17	4,873,161	9.00	15.00	24.00	5.27
Lower Chichester	1,250	2,581	3,473	106.48	34.56	1,490.65	2.33	2,292,741	----	2.00	2.00	3.34
Lawrence Park	----	----	3,241	----	----	1,988.34	1.63	4,526,935	2.55	3.04	5.59	3.54
Reserve	2,024	2,605	3,108	28.70	19.30	1,079.16	2.88	2,145,095	----	5.30	5.30	1.70
Lackawanna	2,756	3,050	3,077	10.66	3.88	596.32	5.16	900,022	----	8.00	8.00	1.51
Annville	2,482	2,651	2,997	6.80	13.05	1,998.00	1.50	1,996,880	5.50	----	5.50	3.04
Nether Providence	1,941	2,314	2,833	20.76	20.86	529.53	5.35	4,525,100	16.06	4.00	20.06	12.97
Darby	6,305	7,922	2,773	25.64	-64.99	909.18	3.05	1,299,870	4.20	5.30	9.50	2.00
Aston	2,135	2,107	2,659	-1.31	26.19	359.81	7.39	1,606,970	7.50	6.50	14.00	14.96
West Norriton	812	1,775	2,588	69.33	88.21	508.44	5.09	3,097,565	19.00	.25	19.25	10.35
West Pottsgrove	1,507	1,709	2,400	13.40	40.43	895.52	2.68	1,219,110	4.29	5.20	9.49	1.71
Rochester	1,227	1,688	2,380	37.57	40.99	648.50	3.67	1,418,013	11.50	3.50	15.00	5.54
Stonycreek	2,293	843	1,860	-63.23	120.64	461.54	4.03	1,507,440	----	7.37	7.37	3.32
Patterson	434	875	1,846	101.61	110.97	1,109.16	1.31	979,960	----	7.00	7.00	4.79
Braddock	739	1,215	1,794	36.41	47.65	949.20	1.89	979,610	----	15.00	15.00	2.69
Springdale	667	910	1,565	----	71.97	660.33	2.77	822,220	----	3.50	3.50	3.89
Neville	634	1,272	1,532	101.00	20.44	1,021.33	1.50	8,265,225	----	3.50	3.50	3.00
Leet	569	627	1,272	10.19	102.87	300.00	4.24	1,114,970	2.00	.50	2.50	.53
Crescent	893	980	1,107	9.74	12.95	540.00	2.05	775,720	----	2.00	2.00	1.78
White	633	1,221	1,031	92.89	-15.56	2,713.15	.38	339,472	----	2.50	2.50	2.04
West Lebanon	1,098	1,030	921	-6.19	10.58	2,361.54	.39	484,360	4.00	----	2.00	1.11
Borough	639	670	885	4.85	32.08	811.93	1.09	613,230	----	5.00	5.00	2.76
South Versailles	417	303	336	-27.33	10.89	685.71	.49	111,033	----	4.00	4.00	1.40

CHAPTER XVII

INTERNAL ORGANIZATION

The problem of internal organization is one of adapting the structure of the township to the conditions of those local areas having substantially the same governmental needs and desires. As such this problem is a part of, and supplementary to, the problem of establishing and maintaining an adequate classification of local governmental areas.

The structure required to meet the needs of a highly urbanized community will of necessity be more complex, broader in its scope of powers and more costly in operation than the structural forms adapted to rural communities. Obviously it is impossible to erect any one form which will at the same time adequately serve the needs of both rural and urban areas. Nevertheless, this is exactly the situation created by our present system of classification which groups into one class many heterogenous communities with political needs and problems that are often entirely dissimilar.

The innumerable variations of governmental organization among the townships of the first class, as we have seen, arise from the permissive character of the present code in regard to corporate powers and offices, and from the wide diversities, already indicated, among these units as to such matters as population, area, taxable resources, and governmental needs. On the one hand such highly urbanized communities as Upper Darby, Lower Merion, Mt. Lebanon, and the like have established forms which, in their complexity and scope of administration, excel many of those set up by third class cities under the code relating to that class of governments. On the other hand, such predominantly rural townships as Jefferson, Elizabeth, South Fayette, Plymouth, and the like have created governmental forms that, in their simplicity and restricted scope of administration, are indistinguishable from those found in the second class townships.

Between these two extremes are to be found many variations of form adapted to the particular needs, taxable resources, and desires of the semi-urban, semi-rural townships which compose the greater number of townships within this first class. The optional features of the code are such that a second class township, a borough, or a third class city can readily find a place within its folds and can readily establish a form of organization adapted to its needs by a process of careful selection and rejection of the code provisions. This is no theoretical possibility; it has been done. Among the sixty-one first class townships are several that are indistinguishable from a third class city, a borough, and a second class township respectively. It might well be said "you take your petition to the court and the electorate, and you take your choice of form of government". The result is "home rule" with a vengeance.

It might be contended that such elasticity is particularly desirable since it permits a community to adopt the form of government best suited to its needs. The difficulty with such a viewpoint is that the very diversity thereby created in turn creates governmental problems which cannot be adequately met as long as such diversity exists. A modern accounting or budgetary procedure adaptable to townships expending several hundred thousand dollars per year for varied governmental services will be neither adequate, scientific, nor logical for a unit expending only several thousand dollars per year for highways and health. It is one thing to establish by law a concentrated administrative responsibility in one office such as the controller, manager, or secretary in a township which can afford to employ competent full time personnel but it is another matter to make such centralization mandatory in a small unit with restricted resources and inability to finance any but a part time staff. In brief, the question resolves itself into one of the value and purposes of classification. If such classification is of value and serves a useful purpose, then it must be so established and maintained as to perform its function. If it serves no useful purpose, which no one would hold, then it should be entirely discarded.

The permissive features of the code in regard to governmental offices allow the board of commissioners, in practice, to determine the structure of government in these townships.

A brief review of the township offices as to terms, qualifications, optional or mandatory and appointive or elective character indicates more particularly the extent to which the organization may be so determined. The mere reading of the statutory provisions concerning township offices in itself affords no picture of the actual townships' existence which can only be obtained by the separate study of the organizations operating in the communities within this classification. Yet in spite of this fact the legislature is compelled to enact legislation which is supposed to apply to these communities as a group. The problem which this situation creates for the legislature is amply demonstrated by the contrast between the possibilities of organization as set forth by the code and the actual structure now in existence among these varied communities.

Governmental organization by offices - elected:

1. The commissioners. Where a township is divided into five or more wards, a commissioner is elected from each ward. In townships having less than five wards, the number of commissioners shall be five, one such commissioner being elected from each ward, and the remaining number to be elected at large. Where a township is not divided into wards, five commissioners are elected at large. The commissioners, whether elected at large or by wards, are elected for a four year term, the terms being so arranged by a "stagger" system that only a part of the board is elected every two years. In townships where the commissioners are elected from wards, the terms of the commissioners from the even numbered wards expire at one time and the terms of those from the odd numbered wards expire two years earlier or later. Where a township is not divided into wards the code provides that either two or three of the five commissioners shall be elected at each municipal election (which occurs at two year intervals). The only qualification for the office is the general one found in section 501 which provides,"no person shall be eligible to any elective office in any township unless he is an elector of the township for which he is chosen." When a vacancy occurs the court of quarter sessions, "upon the petition of twenty qualified voters of the ward or township, as the case may be, in which the vacancy occurs" may appoint a qualified elector as commissioner. If such petition is not presented to the court within thirty days after the vacancy occurs, the board of commissioners may fill the vacancy by a similar type of appointment. In either event the term of the appointee is limited to the next municipal election at which time a commissioner is to be elected to fill the unexpired term.

2. The treasurer-tax collector. Unlike the provisions of the second class township code which establish two distinct offices, treasurer and tax collector, the first class code combines these two offices in the township treasurer. The term of office is four years and expires along with the terms of the commissioners elected from the even numbered wards. The treasurer may also hold the office of treasurer of the school district and such is the universal practice. The only qualification in addition to that specified in section 501 is the requirement that the treasurer must give bond "in a sum to be prescribed by ordinance or resolution and at least equal to fifty percentum of the probable amount of the annual township tax. Neither the treasurer nor township auditors may hold any additional office in the township. When a vacancy occurs the commissioners are empowered to fill the same, the term of the appointee being limited to the next municipal election at which time a treasurer is elected to fill the unexpired term.

3. The auditors. Where the elective auditor system prevails each township elects one auditor for a six year term at each municipal election. No qualification for office is stipulated other than that in section 501. Vacancies are filled by the court of quarter sessions, upon the petition of electors, by an appointment for the unexpired term.

4. The controller. When a township establishes the office of controller by ordinance the office of elective auditor, above described, is abolished. The controller is elected, where established, for a four year term. Such controller "shall be a competent accountant and an elector of the township for at least four years prior to his election." In addition to these qualifications he is required to give bond "in the sum of twenty thousand dollars, conditioned for the faithful discharge of his duties." Vacancies occuring in this office are filled in the same manner as described in the case of such vacancies occurring in the office of township treasurer.

5. Miscellaneous elective officials: In addition to the elective officials cited above, there are others who are not included in this description because they discharge county rather than township duties. Among these are: (a) two locally elected justices of the peace and two constables, each serving for four year terms: (b) assessor, assistant assessor, and two assistant triennial assessors locally elected for a term of four years in all counties except those of the first, second, and third classes, where county assessors are appointed by the county board of assessment. Where the office of township assessor has been retained an assistant assessor is also sometimes elected. (c) Election officials who although locally elected are considered as performing a county function under the county commissioners.

Governmental organization by offices - appointive: In addition to the elective offices, the township commissioners are vested with the authority of appointment in regard to the following mandatory and optional offices.

1. The township secretary. The commissioners appoint a township secretary "who must be a qualified voter of the township, and not a member of the board". Such secretary serves at the pleasure of the board.

2. The Board of health. The commissioners must provide for the administration of health either by the appointment of a board of health or a health officer. Where the former method is adopted the board is composed of five members chosen by the commissioners for a term of five years, a member of the board being appointed annually for a five year term by the "stagger" system, thereby providing for a continuing personnel. No qualifications are laid down for the members of such board other than that at least one of the members "shall be a reputable physician of not less than two years' experience in the practice of his profession, if one resides within the township".

Where the latter procedure is adopted and a health officer appointed such officer serves at the pleasure of the board unless otherwise provided by ordinance. Such health officers "shall have had some experience or training in public health work in accordance with rules and regulations established by the advisory health board of the State Department of Health (attendance at designated colleges for a six week summer course prescribed by the Health Department is sufficient to meet the certification requirements), and "shall not enter upon the performance of their duties until.......... certified so to do by the Secretary of Health of the State Department of Health." To these qualifications is added that of giving a bond for the faithful discharge of duties in such sum as the commissioners shall, by ordinance, stipulate.

3. Township solicitor and township engineer. The commissioners may create such offices and appoint persons thereto for a term of two years. Each such official gives bond in such sum as the commissioners by ordinance direct. It is further stipulated that the solicitor must be "a person learned in the law" and that the engineer "shall be a registered engineer." Vacancies in both cases are to be filled by appointment by the commissioners for the unexpired term.

4. Court appointed auditor. Where the commissioners so elect they may by ordinance vest authority in the court of common pleas to appoint an auditor to make the annual audit of the township accounts. Where this procedure is adopted by the commissioners the elective office of township auditor is abolished. Such auditor is appointed for one year only and "shall be either a person or firm licensed as a certified public accountant, or a person skilled in auditing work."

5. Park, recreation and adjustment boards; shade tree and planning commissions. The commissioners are vested with the authority to create any of the above offices and make appointments thereto. Where any or all of such offices are established, the terms of such appointees are specified in the code. Thus in each case a "stagger" system is employed: (a) The park board is to consist of five members one of whom shall be appointed annually for a term of five years. A similar provision exists in the case of the recreation board. (b) The board of adjustment (zoning) is to consist of three members one of whom shall be annually appointed for a term of three years. (c) The shade tree commission is to consist of three members one of whom shall be appointed for a term of five years, another for a term of four years, and the third for three years. (d) The

planning commission to consist of five members one of whom shall be annually appointed
for a term of five years. The successors to these appointees shall each be appointed
for a five year term. No specific qualifications are stated except in the case of the
planning commission where the code provides that "all members shall reside within the
zone of jurisdiction of said commission, as hereinafter defined." The planning commission's jurisdiction extends over a zone covering the township and "a distance of
three miles" beyond the township boundaries unless such zone would conflict with that
of another unit. No township official, except commissioners, "shall be ineligible to
serve as a member of the Planning Commission." In all cases, wherever a vacancy occurs
in any of these offices, the commissioners are empowered to make an appointment for
the unexpired term. The code further stipulates that members of the board of adjustment "shall be removable for cause, by the appointing authority, upon written charges
and after public hearing." Aside from these qualifications none other are stated, nor
does section 501 of the code apply since that qualification, cited in the preceding
chapter, only affects the elective offices.

6. Offices created under the broad section 1501 of the code. All of the larger
townships have utilized this blanket authority to establish a governmental organization which approximates that of the larger boroughs and the third class cities. The
terms, qualifications, compensation, powers, and duties of such offices are wholly
determined by the commissioners in the ordinance establishing and regulating such
offices. This broad authority is well illustrated in the case of Mt. Lebanon township
which has established the office of township manager under section 1501 of the code.
While the office is specifically mentioned, the nature of the office is entirely
determinable by the commissioners.

The office of township manager was established in this township by Ordinance No.
696, February 27, 1928 and later amended in part by Ordinance No. 761, December 17,
1928. The commissioners have been fortunate in the selection of a capable man who
has served continuously since the creation of the office. The term of office is two
years, the manager to be appointed by a majority vote of the board and to be "subject
......to removal at any time by a vote of the majority of the Commissioners of said
Township." The compensation was at first fixed at thirty-six hundred dollars per
annum, payable in monthly installments, and no qualifications for such office were
required other than the posting of a bond ($1,000) by such official "for the faithful
performance of the duties of his office."

The ordinance outlined the manager's duties and powers as follows: (a) To supervise the construction, maintenance and repair of the sewer system, and, streets and
highways; (b) To supervise the maintenance, repair and sanitary condition of the township buildings; (c) To perform the duties of "Building Inspector" in accordance with
ordinances relating to the same; (d) To supervise the collection and removal of garbage,
and to investigate and act on complaints concerning the same; (e) To inspect all fire
hydrants, and equipment and to maintain the same in serviceable order; (f) To keep a
record of all faulty or dangerous street lights or wires and to report on their condition
to the commissioners; (g) To inspect and report upon all public work being done by the
township, and, to stop work upon such projects where it is not being done in accordance
with contractual stipulations until such time as action may be taken by the Highways
Committee and the township engineer; (h) To direct the care and custody of township
tools and machinery and to report the need of repair to the commissioners; (i) To purchase all equipment and supplies pursuant to action of the commissioners, and to purchase minor supplies without such action; (j) To receive complaints and requests for
service and to investigate and report upon the same to the commissioners; (k) To supervise all township employees in the various departments placed under his control; and
(l) To perform "such other duties as may be from time to time delegated to him by the
Township Commissioners". In addition to these specific duties, the manager "shall
further undertake the general management of all other business of the Township not
specially given by statute or by ordinance of the Township to other officers thereof".

In accordance with this ordinance the township manager is to perform the above
duties "subject to the supervision and control of the proper committee of said Board
of Commissioners having in charge the respective matters to be done and performed by
said Township Manager". Thus, while the source of his authority proceeds directly from

the board of commissioners, he is made responsible directly to the' committees of the board which are vested with the control of administration over specified township functions.

This ordinance placed the general supervision over township administration in the township manager. In one instance, however, it provided that he himself was actually to perform an administrative function, that of inspecting buildings within the township to assure compliance of private property owners with the provisions of the building code. The ordinance of December 17, 1928 provided that he was to supervise such building inspection through the employment of building inspectors who were to do the actual inspection. This same ordinance also raised the compensation to forty-two hundred dollars per annum and added the supervision of police to his duties.

This particular office has been described in detail, not only because it is an interesting innovation but also as an illustration of the broad authority with which the commissioners are vested in the creation of township offices not mentioned in the code. While no other townships have established this office, the position of the secretary in the following townships, among others has been so modified as to closely approximate that of the manager above outlined: Lower Merion, Haverford, Shaler and Upper Darby. As can readily be seen the commissioners' authority is so broad that they can practically determine township structure and the distribution of administrative powers. While in theory their powers are limited by the specific provisions of the code, yet in practice this limitation is insignificant because of the extraordinary number of optional features in the code. The free use made of these options has led to a remarkable diversity in various townships. In all cases there is a board of commissioners, though the number of such commissioners varies from five to thirteen; there is a township secretary, township treasurer, auditors and a health officer or a health board. These offices are mandatory and must be established in every unit. The remaining offices are purely optional or within the creative authority of the commissioners and are therefore not found in all of the townships.

The results of an intensive survey of twenty-five townships show that the office of solicitor had been established in twenty-five, that of township engineer in twenty-one, fire chief or marshall, in seven, chief of police in nine, building inspector in eleven, plumbing inspector in eight, health officer (instead of a board of health in two, health board in twenty three, milk control officer or inspector in five, road superintendent in eleven, and sewer superintendent in three. Three had established a park board, three had established a shade tree commission, six a board of adjustment, one a zoning commission, and none had instituted either a recreation board or planning commission. Many offices, not specifically mentioned in the code, were found in some townships such as meat inspector, sewer inspector, community nurse, water supply superintendent and the like. In some cases these offices are similar in character to offices mentioned in the code, e.g. the sewer inspector being similar to the code "plumbing inspector."

There is no necessary correlation between such offices and the performance of a specific function. While only eleven townships are mentioned as having a road superintendent this does not mean that there is no road construction or repair in the remaining fourteen units. It merely indicates that in those units no separate office has been created or established in which control and supervision over highway construction and maintenance has been vested. In such townships the commissioners either directly supervise such work or they hire a road foreman who usually carries out the detailed orders of one or more of the commissioners. In the following chapter we shall discuss the extent to which the administration is vested in some administrative office such as a township manager, secretary, engineer, and the like, and the extent to which such administrative control proceeds directly from the commissioners or a committee of the board.

It is interesting to note that in eight of these twenty-five units there were no full time positions, and that in five others all positions were part-time with the exception of the treasurer and one or two other officials. In the semi-urban townships practically every position is a full time one and additional employees such as assistant secretaries, bookkeepers, accountants and the like are employed. Other natural diversities in organization occur from the combination under one office of functions which in other townships are vested either in different or separate offices. One township will combine the offices of plumbing and sewer inspector, while another will have a separate office for each. Among other combinations may be noted: chief of police and fire marshall, plumbing and building inspector, superintendent of water supply and sewage disposal, superintendent of roads and sewers, health officer and police, and health officer and plumbing inspector.

Further variations caused by more complex needs may be seen in the accompanying charts showing the internal organization of township government in Lower Merion, Baldwin, and West Lebanon. These three townships have been selected because they are typical of the different governmental structures found in the highly urbanized, the semi-rural or semi-urban and the rural townships.

The Lower Merion chart shows the demand for a larger more complex form of organization to administer such urban services as police and fire protection, parkways, zoning, and food inspection. In townships of this character each office also employs numerous additional personnel not indicated in the chart.

The Baldwin township organization is typical of such units as Nether Providence, Spring Garden, Harrison, and Neville which communities have become sufficiently urbanized, in part, to demand the performance of services alien to rural communities. It will be noted that a large number of these offices, as contrasted with the former group, are filled by part-time personnel. Thus, the building inspector, solicitor, township secretary, engineer and health officer will receive a compensation of less than one thousand dollars annually whereas in Lower Merion these officials have full-time positions. Furthermore, many of these officials are provided with assistants in the larger townships, typified by Lower Merion, while in Baldwin and similar units the entire burden falls upon the lone appointee. The secretary of Lower Merion township, for example, is provided with the services of three full-time assistants and numerous other clerical help whereas the secretary of Baldwin township must attend to all bookkeeping, clerical, and other duties pertaining to the office of township secretary himself.

The West Lebanon chart is descriptive of both rural and small urban communities which are gradually losing their urban character by decreasing population and lower taxable resources. It will be noted that the officials are all compensated upon a part-time basis and that their remuneration is considerably less than that received by similar part-time officers in townships of which Baldwin is typical. This indicated the restricted scope of the services which they perform. Obviously such officers have no assistant personnel as there is no occasion for any more extended service nor are the taxable resources available to support additional services. With the exception of the water department, this form of organization is duplicated in the predominantly rural townships such as South Fayette, Jefferson, Plymouth, and others.

CONCLUSIONS

1. FINDING - The township board of commissioners are vested with such broad authority in the creation of offices and the distribution of powers between such offices, that they are in effect able to determine the particular form of the local governmental structure.

2. FINDING - The internal organization within these local areas varies so greatly that no concept of the local township government can be obtained by a mere study of the code itself.

3. FINDING - The complexity of the governmental structure within this class of townships varies in direct proportion to the urban or rural character of the local community. The more urban areas, the peoples of which demand the performance of urban governmental services, have set up an extensive governmental organization while the rural areas have been content with more simplified forms.

4. FINDING - The more urban townships governed by a complex political organization employ the more competent personnel in full-time positions. The rural units are governed by less competent personnel filling part-time positions.

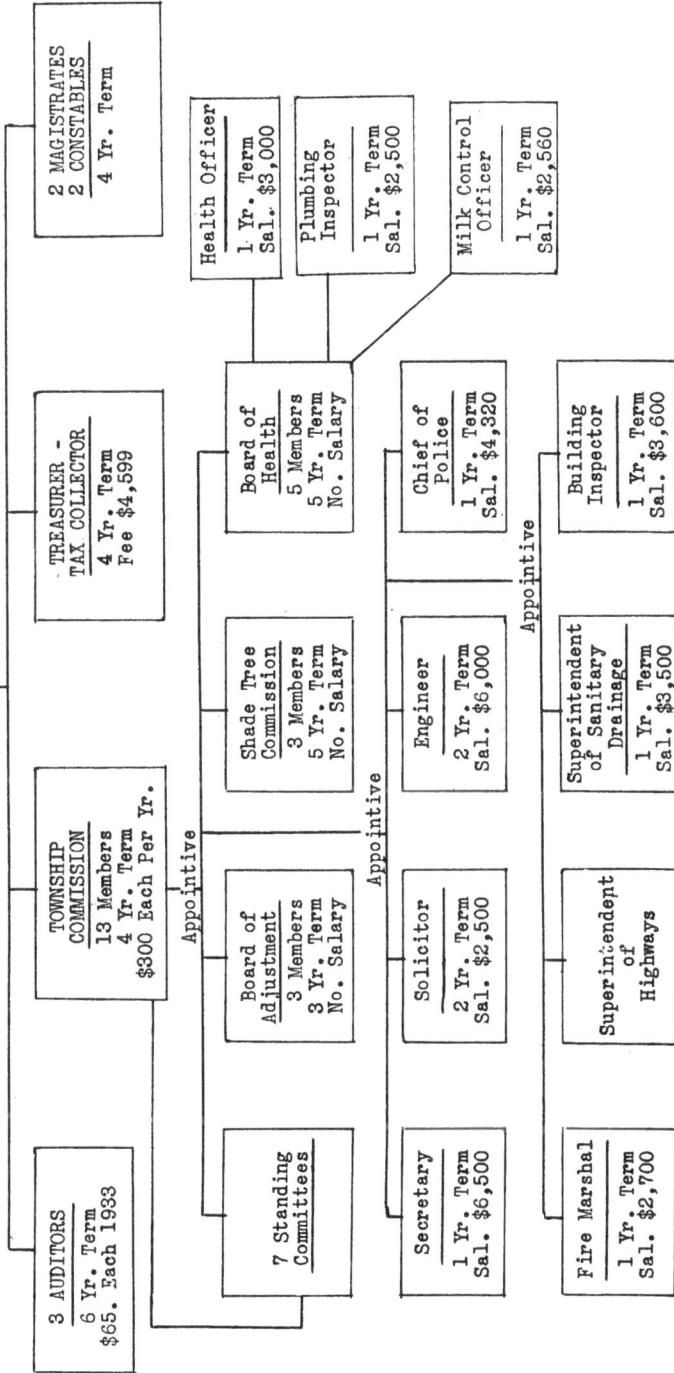

LOWER MERION TOWNSHIP
Montgomery County, Pa.

GOVERNMENTAL ORGANIZATION

TOWNSHIP
ELECTORATE

Elective

3 AUDITORS
6 Yr. Term
$65. Each 1933

TOWNSHIP
COMMISSION
13 Members
4 Yr. Term
$300 Each Per Yr.

TREASURER -
TAX COLLECTOR
4 Yr. Term
Fee $4,599

2 MAGISTRATES
2 CONSTABLES
4 Yr. Term

Appointive

7 Standing
Committees

Board of
Adjustment
3 Members
3 Yr. Term
No. Salary

Shade Tree
Commission
3 Members
5 Yr. Term
No. Salary

Board of
Health
5 Members
5 Yr. Term
No. Salary

Health Officer
1 Yr. Term
Sal. $3,000

Plumbing
Inspector
1 Yr. Term
Sal. $2,500

Milk Control
Officer
1 Yr. Term
Sal. $2,560

Appointive

Secretary
1 Yr. Term
Sal. $6,500

Solicitor
2 Yr. Term
Sal. $2,500

Engineer
2 Yr. Term
Sal. $6,000

Chief of
Police
1 Yr. Term
Sal. $4,320

Appointive

Fire Marshal
1 Yr. Term
Sal. $2,700

Superintendent
of
Highways

Superintendent
of Sanitary
Drainage
1 Yr. Term
Sal. $3,500

Building
Inspector
1 Yr. Term
Sal. $3,600

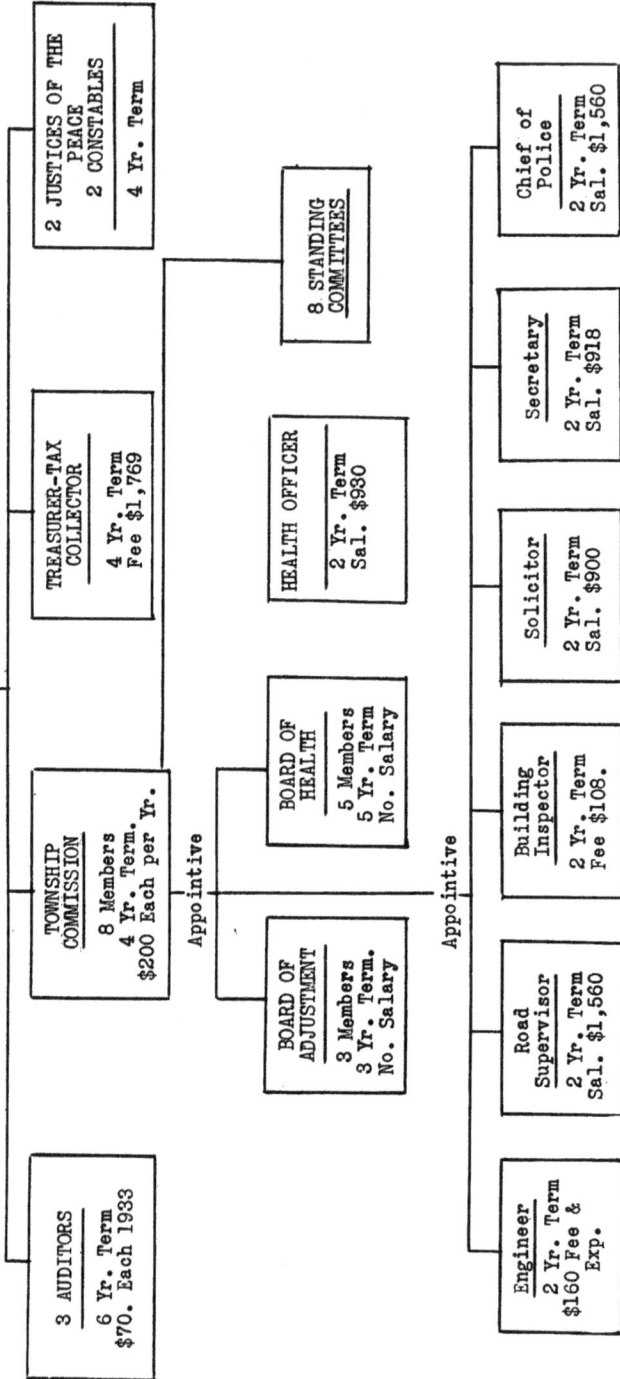

BALDWIN TOWNSHIP
Allegheny County
Penna.

G O V E R N M E N T A L   O R G A N I Z A T I O N

T O W N S H I P
E L E C T O R A T E

Elective

2 JUSTICES OF THE PEACE
2 CONSTABLES
4 Yr. Term

TREASURER-TAX COLLECTOR
4 Yr. Term
Fee $1,769

8 STANDING COMMITTEES

HEALTH OFFICER
2 Yr. Term
Sal. $930

TOWNSHIP COMMISSION
8 Members
4 Yr. Term.
$200 Each per Yr.

Appointive

BOARD OF HEALTH
5 Members
5 Yr. Term
No. Salary

BOARD OF ADJUSTMENT
3 Members
3 Yr. Term.
No. Salary

Appointive

3 AUDITORS
6 Yr. Term
$70. Each 1933

Chief of Police
2 Yr. Term $1,560
Sal. $1,560

Secretary
2 Yr. Term
Sal. $918

Solicitor
2 Yr. Term
Sal. $900

Building Inspector
2 Yr. Term
Fee $108.

Road Supervisor
2 Yr. Term
Sal. $1,560

Engineer
2 Yr. Term
$160 Fee &
Exp.

WEST LEBANON TOWNSHIP
Lebanon County, Pa.

GOVERNMENTAL ORGANIZATION

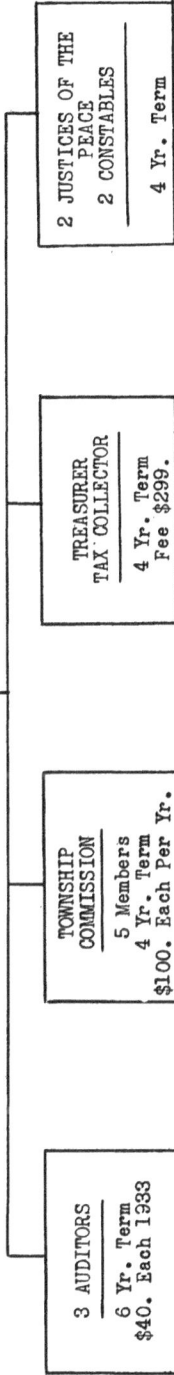

TOWNSHIP ELECTORATE

Elective

Appointive

2 JUSTICES OF THE PEACE
2 CONSTABLES
4 Yr. Term

6 Standing Committees

Solicitor
2 Yr. Term
Sal. $80.

TREASURER
TAX COLLECTOR
4 Yr. Term
Fee $299.

Health Officer
Sal. $90.

TOWNSHIP COMMISSION
5 Members
4 Yr. Term
$100. Each Per Yr.

Board of Health
5 Members
5 Yr. Term
No. Salary

3 AUDITORS
6 Yr. Term
$40. Each 1933

Water Superintendent and Road Supervisor
1 Yr. Term
Salary $960.

CHAPTER XVIII

ADMINISTRATIVE METHODS
AND PROCEDURE

It is generally agreed that greater efficiency and definiteness of responsibility are promoted by vesting the care of administrative detail and the policy-determining functions in distinct offices and that the official directing administration should be immediately responsible to those charged with policy determination. It is similarly agreed that the control and direction of administration should be centralized in a single office. One important purpose of the Survey was to determine to what extent a distinction was made between administrative and legislative functions not only by statutory provisions but also in the actual operation of these local units, also to what extent direct administrative responsibility was provided for by statute or in practice. Where these principles have not been observed it was sought to discover what shortcomings, if any, of local administration could be attributed to this failure. Finally an attempt was made to ascertain to what extent procedural defects could be partially or wholly remedied by a greater degree of state supervision.

Responsibility for legislation and administration - The code places all legislative, policy-determining powers in the township board of commissioners. It provides that the board shall have the power to enact ordinances and resolutions by means of which it may perform such governmental functions as are prescribed or permitted by statute. Ordinances must be published and do not take effect until ten days after the date of publication. No such requirement exists in regard to resolutions. This sometimes proves unfortunate in that some boards enact what are really ordinances, but put them in the form of "resolutions" on the ground that they thereby save publication expense. Many of these resolutions are on matters which should be brought to public attention. Particularly is this true where the board violates the provisions of the code and amends a budget ordinance by means of "resolution".

The code also provides that in part the responsibility for administration shall vest in the board. This is accomplished in several ways: (a) Certain administrative duties are placed directly upon the board. It is required to prepare and enact a budget, to file an annual statement of road administration with the State Highway Department; to construct, maintain, and repair highways; to make purchases of supplies; to enter contracts; to hire and/or appoint personnel, etc. (b) Definite administrative duties are attached to specific offices. Most of the officials filling such offices are in turn made responsible to the board. Thus, the board of health or health officer are required to perform specific matters in regard to the regulation of public health, and the secretary is required to maintain an account ledger and to charge expenditures against the proper appropriations. (c) The board has the option of establishing certain offices such as a zoning commission, board of adjustment, shade tree commission, controller, solicitor, engineer, and planning commission, to which offices definite administrative functions have been attached. (d) The board may also establish a township manager and such other offices as are necessary to the good government of the township. Where the board exercises this discretion it is similarly empowered to determine the nature and scope of the administrative duties attached to such offices. (e) With the exception of the treasurer and auditors, the board may vest duties in such offices additional to the ones specifically set forth in the code.

The commissioners are required to organize themselves as a board, biennially, for the purpose of carrying out the above legislative and administrative powers. The code stipulates that such organization shall take place "on the first Monday of January of each even-numbered year", and that if the commissioners fail so to organize themselves "within ten days from the time prescribed by this section, the court of quarter sessions, upon the petition of ten taxable inhabitants,.... shall issue a rule upon the delinquent (commissioner) or delinquents to show cause why their seats should not be declared vacant." After a hearing the court may declare such seats vacant and appoint successors until others are elected at the next municipal election to fill the unexpired terms of those so removed. Organization at biennial intervals is required in order to permit the newly elected commissioners to take their places on the board.

The present statutes permit each township to be divided into wards upon the granting of a petition. Where such division results in the creation of a number of wards totaling to an even number, political difficulties hinder prompt organization and efficient government. To illustrate: Baldwin township is divided into eight wards from each of which one commissioner is elected. In 1933 the four commissioners from the even-numbered wards were elected while those who had been elected from the odd-numbered wards remained in office. As a result of a deadlock between the old commissioners and the newly elected ones the organization of the board was delayed until April, 1934. Other township business is similarly delayed. This situation provides an excellent opportunity for those officials who are politically minded to press their own advantage at the expense of the township since the compromise usually reached is a political one. If the ward system is to be retained, provision should be made for the election of one commissioner at large in all instances where a township is divided into an even number of wards.

At the time of organization the commissioners are instructed to select a president and vice-president from among their number and to appoint a township secretary who shall not be a member of the board. The general practice of the commissioners is to appoint a number of standing committees at this time. Of twenty-five townships upon which data were obtained: four, Hanover, Newport, Whitehall, and Swatara, have no standing committees; two, Reserve and Wilkes-Barre, appoint standing committees which take the form of committees of the whole, and therefore, can hardly be said to apportion the work of the board, and one, Spring Garden, in lieu of committees apportions the work of the board among the commissioners, making each the head of a department. All the remaining units appointed standing committees having in charge such subjects as highways, police, fire, street lighting, sewers, finance, printing, supplies, ordinance, water supply, sidewalk, public property, public safety, administration, zoning, etc.

The number of the committees and the number of the personnel on each varies considerably. Braddock township sets up three such committees while Haverford creates eleven. The membership ranges from three to seven. In many units one or more of these committees are committees of the whole. This is generally true of the finance committee.

In many townships the evidence shows that committees adopt a formal or routine attitude towards their work. It will be noticed that these duties are both legislative and administrative in character, since they include both preparation of ordinances or resolutions and supervision of administrative services.

Following their organization the commissioners are required to meet monthly, at least this appears to be the general practice. In most townships there are few meetings other than the regular monthly ones, with the exception of the budgeting period. Generally the public does not attend, for two reasons: (a) Lack of interest except in such matters as budgeting and the like. (b) The tendency of some boards to discourage public attendance.

Integrated versus disconnected administration. The township law makes no provision for centralizing responsibility as is clear from the following features: (1) The township treasurer, an elected officer, is empowered to collect revenues and pay out township moneys. (2) Administration of health is placed in the hands of a board of health or a health officer. (3) Certain duties are by law given to expressly designated township officers and agencies, such as the solicitor, and zoning commission, which may be established at the option of the commissioners. The centralization of administrative power and responsibility can be accomplished only by legislative changes vesting such authority in a manager, secretary or controller. This would be desirable.

To what extent have the township boards attempted to centralize control in actual practice? (1) By appointing a manager as in Mt. Lebanon. (2) By vesting power in the hands of the secretary, as has been done in Lower Merion and other townships. (3) By placing one commissioner in charge of a certain function such as highways, as in Spring Garden township. (4) By placing all administrative power in each ward in the hands of its commissioners, as in Whitehall. (5) By placing all road supervision in a ward in the hands of the ward representative on the commission as in Pennsylvania. (6) By distributing functions to different committees. (7) By retaining control of all matters in the control of the board itself. Many townships use a combination of several methods. The general practice has been to decentralize control. Progress towards centralization must come by statutory changes.

Control of General Administration. This includes keeping of accounts, collections of revenues, care of public property, payment of moneys, publication of ordinances, printing, postage, and purchase of supplies. Such power reposes by statute in the hands of the commissioners, secretary and treasurer. The collection and disbursement of moneys belongs in the province of the treasurer and the keeping of accounts in the province of the secretary. Other powers are in the hands of the commissioners. They may exercise these functions themselves or delegate them to township officers. As a rule, commissioners in the larger urban townships tend to give this work to the secretary. In the smaller, rural areas such duties are handled by the commissioners themselves.

The average length of service of the secretary in twenty-three townships was nine years. One secretary served twenty-three consecutive years. Compensation ranged from $100 per year in West Lebanon to $6,500 in Lower Merion. Most officials are on part-time only. In only four of the twenty-three secretaries' offices were there assistants. The secretary, except for the treasurer, is usually the highest paid township officer.

That the commissioners should be divested of any immediate supervision of administrative detail will perhaps be clear from the following considerations: (a) They are all-part time officials receiving a very small compensation. (b) They usually tend to be influenced by the district or ward which they represent and consequently tend to neglect administrative matters except in so far as their own particular section of the township is affected. This is particularly true where the township is divided into wards. (c) Their length of service is considerably shorter than that of the secretaries. Data were obtained on the length of service of ninety-two out of one-hundred and seven commissioners in sixteen townships. The length of service ranged from one to twenty years and the average number of years served was 5.08 or approximately four years less than the average of the secretaries. Of the remaining fifteen commissioners, thirteen were serving their first term and two their second. (d) Efficient government profits from a separation between legislative and administrative authority, and, between the appropriating and spending authority. (e) Administration calls for a knowledge of, and attention to, numerous factual and detailed matters; these can only be mastered by some official who can devote all or the greater portion of his time to the work.

Several of the commissioners who were opposed to divesting the office of administrative powers suggested that the salary should be so increased as to permit the commissioners to devote more time to township affairs. This solution was generally opposed by everyone except the commissioners themselves and many of these expressed themselves against such proposal. It is felt that this would afford no adequate remedy since such provision would thrust too great a financial burden on the smaller townships, would cost more than the employment of a full time secretary, would not separate administrative and legislative authority, and would generally make the office of commissioner more political in character.

Both the township ordinances and their administration are adversely affected by the practice of dividing townships into wards and electing one commissioner from each of these districts. All branches of administration, general, highways, police and fire protection, sewers and the like are impaired.

The general practice is to permit the individual commissioners, the secretary, and other township officials to purchase supplies and have the bills therefor sent to the secretary. Near the end of the month these bills are presented to the board for approval after which, if approved, warrants are issued for their payment. In many of the townships the bills are first approved by the finance or other committees and then recommended for payment to the board. In any event the purchases are generally not made through a single office such as that of the secretary or manager. Obviously such a purchasing method is uneconomical and open to serious political implications.

The practice among the better managed townships is to have all purchases made through the office of the secretary by means of requisitions presented by the commissioners and other local officials. While this system is most productive of large savings in such townships as Mt. Lebanon, Lower Merion, and Haverford there is no sound reason why similar practices should not be instituted in the smaller townships.

Highway Administration - The control and supervision of the construction, maintenance, and repair of highways is vested in one of the following authorities: (a) The board of commissioners as a group. (b) The individual ward commissioners. (c) The township engineer or the superintendent of roads. (d) The township manager or secretary. And (e), the commissioner selected to head the highway department.

In those townships such as West Lebanon, Annville, Braddock, Wilkes-Barre, West Norriton, and Swatara, where the board directly controls and supervises the construction and maintenance of highways, the practice is to hire a road foreman or supervisor. These men are paid a small compensation either in the form of an hourly rate slightly in advance of that paid to the laborers, as in Annville where the rate is fixed at forty cents per hour, or a small salary stipend, as in West Lebanon where the pay is eighty dollars per month. The road foremen are in effect, mere "gang" foremen who are directly controlled and supervised by the board. The methods of road maintenance are not of the best and this defect is not remedied by the supervision of commissioners who generally are no better informed upon matters of highway administration than is the foreman whom they are directing.

One of the most unfortunate methods of administration is that of placing the township roads in each ward under the control of the ward commissioner. The general practice is for the board to make a specific appropriation to the ward commissioners upon the basis of the total number of township-maintained road miles in each ward. The ward commissioner is then given full discretion in the spending of this sum.

The fact that the road conditions within a single township tend to be similar and that the boards universally determine the sums allotted on the total road mileage of each ward indicates that a cost per mile tabulation derived from a comparison of the sums expended in each ward with the total road mileage of such ward should provide a fair index of the character of this form of highway administration. The results obtained in Penn township are typical:

Ward	1933 Expend.	Total Twp. Miles	Cost per Mile
1st	$1,099.76	2.62	$ 419.75
2nd	1,559.99	5.26	296.58
3rd	1,449.94	6.12	236.92
4th	2,799.41	13.28	210.79
5th	2,001.78	4.78	418.78
6th	1,997.15	6.23	320.57
7th	249.42	0.87	286.69
8th	1,864.31	5.09	366.27
9th	2,298.95	6.68	344.12
10th	485.12	1.70	285.36
11th	1,699.92	4.18	406.68

There is a variation of approximately two hundred and twenty-dollars between the highest and lowest costs per mile. Since the variation exists within the bounds of a single township which has a total mileage of township maintained highways of slightly over forty-six miles (46.45) and of which total less than three miles are "improved" highway, it would seem that the difference is primarily one of differences in administrative efficiency. It is significant that in this township each ward commissioner takes direct charge of road maintenance without the assistance of a road foreman.

A number of townships such as Nether Providence, Baldwin, Harrison, and Shaler, place the control and supervision of highway administration in a road supervisor or superintendent who is appointed by the board and is directly responsible to it. The exact powers of such officials are difficult of determination but generally they are given the power of hiring and firing, purchasing directly or through the secretary, determining the character and extent of repairs and the like. No comparison of costs per mile can be made between these townships and those in which other methods of administration are in vogue, because of the variations as to miles of improved roads, differences in types of improved roads, and differences of other conditions such as subsoil, wage rates, and drainage. It is evident that much more efficient methods can be utilized by an experienced supervisor than by commissioners and that the opportunities for political influences are smaller.

Those townships which vest control of highway administration in the office of secretary or manager generally combine such authority upon the part of the latter with a direct responsibility either to the board or to the highway committee of the board. In none of these cases does the secretary actually supervise highway administration. He delegates this duty to a highway director, township engineer or the like who is made responsible to the secretary. In several cases the office of the secretary has very little control, such control being actually exercised by a highway committee. All of the highly urbanized and wealthier townships provide for some such form of highway administration.

Of the twenty-five townships upon which data were obtained, Spring Garden alone provided for the administration of highways by a so-called highway department headed by a commissioner. In fact this was the only unit where administrative functions were combined in five departments under the control of one of the five commissioners. The commissioner of the highway department also acts as supervisor since no road foreman is designated.

As this summary indicates, there is no uniform control of highway administration. Each township has employed the particular procedure that suited its own tastes.

The total miles of first class township-maintained highways in the Commonwealth are approximately one thousand six hundred and sixty-eight. Of this total approximately eight hundred and twenty miles are "improved" while slightly more (847.95 miles) are unimproved. If the road mileage of the larger townships such as Lower Merion, Upper Darby, and Mt. Lebanon were to be subtracted from this total, the remainder would be under one thousand miles. This fact combined with the number of highway miles within these townships now maintained by the state indicates the desirability of divesting the numerous local units of this function and centralizing it in either the county or state governments, except in the larger semi-urban units.

This conclusion is further supported by the character of the administration by local officials. The following townships among others do not employ a regular road force but hire a large number of men for a few days labor off and on during the year: White-hall, Newport, West Lebanon, Braddock, Baldwin, and Plains. The regularly employed force of those townships is generally small; in Annville, the road foreman and one laborer; Reserve and Tinicum, two laborers regularly employed; Harrison, five; Spring Garden, three; Nether Providence, road foreman and two laborers; Swatara, four; Shaler and Mt. Lebanon, fourteen. One township, Wilkes-Barre, has no road force and makes only such expenditures for roads as are absolutely unavoidable.

Several units, Braddock, Baldwin, Swatara, and others, permit local taxpayers to work out their taxes upon the roads. This is a relief measure and while not many of the townships permit it, the practice of employing a large number of men for a short period during the year amounts to the same. West Lebanon township paid out three hundred and sixteen dollars in wages for road labor in 1933. This sum was distributed among approximately fifty individuals, each being hired for a day or two at the rate of three dollars and fifty cents per day. In Whitehall township approximately one hundred and ninety-five individuals were employed on the road during the same period. The number so employed in Newport was even greater than that in Whitehall. It is difficult to see how the acknowledged low efficiency of road work can be improved by a personnel recruited in this way.

TABLE I

COMPARISON OF HIGHWAY AND TOTAL EXPENDITURES, 1933
for 50 Townships

Township	Highway Expenditures	Total Expenditures	Ratio Highway to Total Exp.
Upper Darby	$ 65,540	$1,210,195	5.41%
Lower Merion	255,107	876,570	29.10
Coal	42,037	190,800	22.03
Abington	57,920	432,196	13.40
Hanover	11,314	229,718	4.93
Cheltenham	70,800	513,833	13.78
Mt. Lebanon	24,562	317,102	7.75
Stowe	11,025	95,785	11.51
Penn	34,065	108,328	31.45
Harrison	16,441	91,089	18.05
Newport	19,810	168,632	11.75
Whitehall	47,297	125,837	37.59
Shaler	18,084	160,586	11.26
South Fayette	8,334	13,324	62.55
Mifflin	8,806	32,358	27.06
Elizabeth	7,357	17,245	42.66
Baldwin	7,003	59,083	11.85
Swatara	4,384	12,570	34.88
Scott	8,078	42,632	18.95
Collier	6,255	14,550	42.99
East Bethlehem	6,030	15,420	39.11
North Versailles	9,154	27,279	33.56
Springfield (Montg. Co.)	33,807	141,269	27.47
O'Hara	14,310	32,831	43.59
Spring Garden	11,424	40,414	28.27
Springfield (Del. Co.)	8,702	78,640	11.07
Plymouth	13,006	22,591	57.57
Jefferson	6,984	11,296	61.83
Upper Moreland	11,855	58,300	20.33
East Deer	6,139	47,533	12.92
Tinicum	9,986	62,947	15.86
Lower Chichester	2,731	14,670	18.62
Lawrence Park	9,491	36,992	25.66
Reserve	2,854	21,555	13.24
Lackawanna	2,546	5,879	43.31
Annville	4,461	14,796	30.15
Nether Providence	6,619	21,256	31.14
Darby	1,260	16,579	7.60
West Norriton	8,842	19,508	45.32
Rochester	2,469	11,446	21.57
Patterson	3,975	10,553	37.67
Braddock	834	17,778	4.69
Springdale	630	2,484	25.36
Neville	2,464	57,666	4.27
Leet	1,101	8,668	12.70
Crescent	2,487	5,806	58.06
White	774	3,052	30.52
West Lebanon	1,446	6,793	21.29
Borough	109	6,298	1.73
South Versailles	312	453	68.87

Some indication of the relative importance of highway administration is shown by the above table. It shows the great differences that exist among the townships of the first class. Those which spend most of their money on roads are the rural ones. None of such rural units should be first class townships since they are indistinguishable from those

of the second class which are also rural and which spend the greater amount of their revenues for road administration. The urban communities, on the other hand, spend a relatively much smaller amount on highways since they make large expenditures for such matters as street lighting, sewers, zoning, fire and police protection, building regulations, parks, recreation and the like.

Finally, it may be noted that the townships which spend the larger proportion of
their funds for highways are also the ones in which highway administration is most defective. In most of these units no regular road force is employed, and control or supervision is retained in the board or the individual commissioners thereof. The townships where the proportion of total road expenditures is smaller, on the other hand,
have generally centralized control and supervision of highway administration in one
official, and employ a regular road force.

Sewer Administration.- Most of the townships visited had no sewerage system, this
being taken care of by the installation of cesspools. In the eleven which had sewers
only a portion of the community was served. Of these eleven only five had provided for
a separate administration of sewers, the remainder taking care of this matter through
their highway forces. Six of the eleven townships had entered into agreements with other
local units for the construction, maintenance or operation of their systems.

The regional character of this function requires that it be placed under the control
and supervision of regional bodies in those cases where a township is located within a
large metropolitan area.

Health Administration - This function is vested in the board of health or, where the
commissioners by ordinance so provide, in a single health officer. In only two units,
Annville and Braddock, has the township board so provided. In Reserve the Health Board,
contrary to the code, is comprised of only four members each of whom is elected from one
of the four wards. Where a board exists it is to be composed of five members one of
whom must be a physician. This requirement was not met in at least four of the twenty-
five units, Wilkes-Barre, South Fayette, Leet, and West Lebanon. The board is instructed to appoint a health officer whose duties are generally the same as those of the
health officer appointed by the township commissioners. These duties may be briefly
described: To enforce the state health laws, to enforce local health regulations, to
inspect restaurants and other public places, to quarantine those afflicted with contagious diseases, to abate nuisances, to inspect house drains, soil pipes, cesspools, and
the like. The board of health or health officer is further empowered to: enact and
publish local health regulations, create emergency hospitals, etc.

These broad powers are rarely used except in the more thickly populated townships.
Only nine of the twenty-five townships had established the office of plumbing inspector. Several townships provide for some form of food inspection in addition to that of
the health officer. Three townships employed or contributed to the salary of a community nurse. Several units have entered an agreement with nearby hospitals for the
placing of those afflicted with contagious diseases in hospital wards maintained for
that purpose at a stipulated weekly or monthly rate.

In the great majority of cases the township provides no other health services than
those performed by the health officer. This latter official is a part-time officer in
twenty-one of the twenty-five townships studied, the salary range being from sixty
dollars, in Leet, to two thousand five hundred and sixty dollars per annum in Lower
Merion. The restricted character of the duties performed, the part-time nature of the
position, and the lack of training on the part of the personnel all indicate that in
the greater number of townships of this class, the administration of health should be
taken over by the Commonwealth as has been the case in the second class townships.

Police Administration - The code permits the commissioners to establish a police
force and to set up a pension system for their retirement. Seven of the twenty-five
townships make no provision for police protection other than that which is provided by
the regular township constables. Of the remaining eighteen, only ten employ one or more
full-time officers the salary range being from four thousand three hundred and twenty,
in Lower Merion for the chief of police, to sixty dollars per annum in Braddock. The
larger townships such as Lower Merion, Upper Darby, Haverford, Mt. Lebanon, and others

employ a chief and a number of subordinates. The smaller townships employ only one or two part-time policemen.

Fire Protection - The code provides the commissioners with the power to: create fire and water districts and to levy a special tax upon such districts for purposes of fire protection; to make fire regulations; to purchase equipment; to donate to volunteer companies; and to establish suitable places for the housing of fire apparatus.

None of the townships have established their own fire houses. The general practice is to either donate to the volunteer companies or to purchase equipment for them and in a few cases the township commissioners do both. This aid is supplemented by the amounts which the local units receive from the state on foreign insurance premiums.

The administration of these companies is generally retained in the volunteer organizations which designate an individual member to serve as chief of the company. In some townships, the commissioners annually appoint the chief of one of these companies, on the basis of rotation, as the township fire chief who serves without pay. In only seven of the twenty-five townships did the commissioners appoint a fire chief or marshall who received some compensation, and in one of these the offices of fire and police chief were combined in one person. Two townships, South Fayette and Wilkes-Barre, were entirely dependent upon adjoining municipalities for their fire protection as they had neither volunteer nor township companies.

None of the governmental services are more in need of some form of centralized administration than that of fire protection. The practice of making donations to the volunteer companies with no supervision of the manner in which the money is spent, of permitting these companies to generally purchase and designate their own type of equipment, of permitting them to select their own chiefs, and of allowing each of the volunteer companies within the township to run its own affairs without any supervision upon the part of the township officials tends toward a minimum of fire protection at a maximum cost.

Most of the units cannot afford to appoint a full-time chief or even a part-time one who will be sufficiently compensated for actually supervising this service. However, no sound reason appears why such control and supervision of fire administration should not be placed in the township secretary along with his other administrative functions. If no provision is made for the centralization of all administration in the secretary's office or a similar one, the legislature should, nevertheless, provide for such centralization in the case of fire protection. This provision should be so worded as to place supervision in the secretary, in those townships where the commissioners do not appoint a fire chief with similar authority.

Zoning Administration - The commissioners are authorized to make regulations concerning the foundations, drainage, size, etc., of buildings whether or not they establish a zoning system. Where the latter is created, the location, use, height, etc. of all buildings may be regulated in relation to zoning districts. The actual administration of the former is usually vested in a building inspector who may be directly responsible either to the township secretary, the board, or some other official. Zoning administration is vested in a zoning commission or a board of adjustment, whichever the board of commissioners establishes. A building and zoning code or ordinance is usually adopted by the board setting out the regulations in detail and designating the enforcing officials.

Only eleven of the twenty-five townships have provided for any form of building regulation. In several of these - Braddock, Baldwin, and Hanover - little, if any real regulation and enforcement is attempted. Eight of the twenty-five have adopted a zoning system and in one of these eight, Baldwin, the provisions are not generally enforced. All of these units are predominantly urban in character.

Miscellaneous Services - Several townships provide other services such as playgrounds, parks, libraries, and water supply. One township, Mt. Lebanon, has established a public library to which it contributes, while several others contribute an annual sum to an established public library within the township. A number of the units

have also established their own water systems; and a few purchase their water supply
from some adjoining municipality or a public utility and then in turn sell it to their
residents. None of the twenty-five townships had entered any public utility enterprise
other than that of water supply.

Inter-Municipal Agreements - The code is extremely liberal in permitting townships
to enter into agreements with other municipalities for the joint performance of governmental services. Thus, the commissioners may enter such agreements for the joint construction and maintenance of highways, sewers, and water works; the establishment and
operation of "parks, playgrounds, playfields, gymnasiums, public baths, swimming pools,
or indoor recreation centers", the administration and enforcement of health laws, and
the like. A number of townships have utilized this authority.

Thirteen agreements relating to the construction, maintenance, and repair of sewer
lines and disposal plants were in operation among those units visited. These agreements
are all of the same general type providing that each of the municipalities to the contract pay a specified proportion of the construction and maintenance expense.

The problem of sewage disposal is not a local problem. Those townships situated
within metropolitan or large urban centers such as Philadelphia and Pittsburgh should
have their systems planned and constructed as a portion of a unified network of disposal
plants and trunk lines. In actual practice one or more of these local communities establish a sewer system that is adaptable to its local watersheds. Where this is done
it is usually preceded by endless bickering as to division of costs, type of system,
and the like, often ending in the refusal of one or more of the municipalities to go
through with the agreement. As other adjoining communities develop and require the ex-
tablishment of some form of sewage disposal they are compelled, where topographical conditions make it necessary, to make some form of arrangement with another municipality.
The advantage which one community has topographically over another permits it to impose
its conditions in any agreement that is effected.

The Borough of Narberth and Lower Merion have made an agreement whereby the latter
township provides police protection for the former upon a purely cost basis. No other
formal agreement concerning police services was found although several other units have
understandings whereby they may call upon either the state police or those of an adjoining municipality whenever police protection is needed.

Several informal arrangements exist as to fire protection. South Fayette township
which has no local fire companies has an understanding with the adjoining boroughs of
Oakdale and Bridgeville whereby the latter provide such protection and are paid by the
township on the basis of service rendered. An exchange of services has been arranged
between Spring Garden and the city of York, and between Mt. Lebanon and adjacent municipalities.

Two agreements as to health administration have been set up by the townships of
Lower Merion and Leet respectively. The former township has entered into an arrangement with adjoining municipalities for the maintenance of an office to analyse milk and
milk products. The latter unit has joined with adjacent municipalities for the maintenance of a community nurse.

The existence of many of these agreements indicates the inability of townships to
individually bear the cost of certain services. Where such services are not purely
local matters such as highways and sewage disposal, the control of the service should
be placed in an authority that is capable of coping with the problem presented. Where
the agreements concern the more efficient or economical performance of local services
they should be encouraged by some form of initiative upon the part of a state agency.

CONCLUSIONS

(1) FINDING: Administrative and policy-determining functions are indiscriminately combined in the board of township commissioners, both by statute and in the practical
operation of government. This has resulted in uneconomical and inefficient administration in the performance of such functions as highway construction and mainten-

ance, the purchasing of supplies, fire and police protection and the like.
RECOMMENDATION: That the township board of commissioners be divested of all immediate supervision of administrative matters.

(2) FINDING: Immediate control and supervision of administration is differently divided
both by statute and in actual practice among numerous township offices some of which
are responsible to the commissioners and some of which are not.
RECOMMENDATION: That authority for the direct control and supervision of administrative matters be vested in a single official such as the township manager or secretary who shall be made directly responsible to the board of township commissioners.

(3) FINDING: The present system of dividing control over matters of general administration viz., the keeping of accounts, the purchasing of supplies, the care of public
property, the payment of moneys, and the like among the commissioners, township secretary treasurer and other local officials is uneconomical and inefficient.
RECOMMENDATION: That in the event recommendation number two is not adopted, all control and supervision of general administration be vested in the office of the township secretary.

(4) FINDING: The small mileage of township maintained roads within the townships of the
first class does not justify the needless duplication of personnel and equipment
that is necessitated by the retention of local control over this function. The problem of street construction and maintenance is primarily a local problem and as such
should remain within the control of the urban townships.
RECOMMENDATION: That all township-maintained roads or highways be taken over by the
Commonwealth. That all street construction and maintenance be continued as a local
function. That whenever a township opens a street or whenever any highway or portion
thereof becomes a street, the construction and maintenance of the same shall come
within the control of the local unit.

(5) FINDING: The practice of dividing the control and supervision of highway administration among the ward commissioners is attended by high costs and general ineffective administration.
RECOMMENDATION: That the practice of placing each ward commissioner in charge of
the township highways or streets within his ward be prohibited by statute.

(6) FINDING: The administration of health laws is at present vested in part-time, untrained local health officers. Their duties, in all but the predominantly urban
communities, are very restricted in scope and perfunctorily performed. The small
sums expended by the individual townships do not permit of more efficient health
service and yet such sums taken in the aggregate constitute a considerable amount.
RECOMMENDATION: That the responsibility for health administration be vested in the
Commonwealth for all townships which are unable to employ full-time personnel.

(7) FINDING: The township authorities either donate money to, or purchase equipment and
supplies for, the volunteer fire companies within the township. In most of the townships the management of the equipment and the expenditure of the money donated is
neither controlled or supervised by the local board or officials. The general practice is for the board to permit each company to elect its own chief to direct its
activities. Occasionally townships appoint one of these chiefs, in rotation, township fire chief. Little or no authority and small remuneration accompany the office.
RECOMMENDATION: That in the event recommendation number two is not adopted, the
commissioners be required to either hire a full-time fire chief where taxable resources and need would justify such procedure, or where this is not necessary, to
vest the control and supervision of fire administration in the township secretary.
That the township commissioners be prohibited from purchasing equipment for, or making donations to, volunteer fire companies where these conditions are not met.

(8) FINDING: Those townships situated within large urban centers are compelled to enter

numerous agreements for the performance of such services as sewage disposal, police protection, and water supply. The agreements so reached are not sufficiently integrated with regional needs and planning.

RECOMMENDATION: That consideration be given to the establishment of regional bodies for the purpose of centralizing regional functions such as sewage disposal, zoning, and the like.

CHAPTER XIX

ACCOUNTING PROCEDURE

In the fiscal operations of a township, an accounting system offers the mechanism by which, and by which only, the financial operations of the unit may be organized and planned. (1) It provides the basis for the estimation of revenues. No accurate estimate of anticipated receipts from taxes can be reached unless the accounting system shows prior years' tax collections, tax delinquencies, penalties, exonerations, and collections from past due and liened taxes. No accurate estimate of expected total revenues can be attained unless the ledgers show collections from other sources such as licenses, departmental earnings, fines, grants, special assessments and the like in previous years. (2) It provides the basis for the estimation of expenditures. An accurate estimate of probable needs or expenditures requires that the bookkeeping system shall depict the amounts expended in prior years for the specific purposes of each major function, viz.; highways, highway materials, labor, machinery, repairs, construction; total amount spent for police, personnel, supplies, capital outlays, and the like. (3) It provides the basis for the adjustment of estimated needs to anticipated revenues. It is impossible to operate any governmental unit within the limits of its current income without an approximately accurate knowledge of the expectable income and the anticipated needs of the unit governed. It is likewise impossible to plan indebtedness where there is no distinction available as to indebtedness incurred for the current operations of government and for that incurred to meet capital outlays. (4) As a corollary to (3), it provides the basis for a knowledge of functional costs. Unless the budgeting authority knows exactly how much has been expended upon such matters as highways, sewers, health, administration, etc., in previous years, such authority faces an impossible task in the face of declining revenues in attempting to restrict estimated expenditures within the limits of anticipated revenues. Shall expenditures be decreased for highways or health, or administration. Obviously, no answer can be given to such questions until it is known what was previously expended for these functions and how much of such functional expenditure is in the nature of fixed or absolutely essential costs. (5) It provides the only basis for budgetary control. How can a unit continue to operate fiscally within its budget plan unless it can at any given time compare its operations to date with the limits set to such operations in its budget. The answer is that it cannot, unless its operations are so portrayed in its accounting system that a comparison of its actual and proposed operations is readily available.

It will be seen that these five purposes or ends of an accounting system provide the foundation stones for budgeting. But the most complete ledger accounts are of little value unless the responsible personnel can utilize them. This raises another problem i.e. to formulate an accounting system that will meet the above purposes and yet be sufficiently simple for operation by the type of local personnel usually charged with fiscal affairs.

There is still another basic function of an adequate accounting system. It should be designed not only to aid officials in planning fiscal affairs of the unit but also to make these affairs readily understandable by the average layman-taxpayer. This latter purpose is too often ignored not only by the officials themselves but also by those seeking improvements in fiscal procedures. Accounting must either be so designed as to carry out its basic fiscal purposes and be readily understandable to the layman, or it must be so designed that it may readily be translated by officials into such understandable form, and it must be made mandatory upon such officials so to translate township accounts and present them to the public. If local self government is to survive, two elements are necessary: (a) Knowledge of governmental operations upon the part of the citizen-taxpayer; and (b) participation by such informed citizens in the budgeting and other policy-determining processes of government. Under our present practices of accounting, budgeting, auditing and reporting the officials themselves are usually ignorant of the true state of affairs.

Accounting practices - To what extent do existing accounting practices perform the above functions? If they do not, is this ascribable to faulty and inadequate statutory provisions? To the township personnel invested with authority over fiscal operations? To the lack of any guidance by some outside source such as a state agency?

1. _The keeping of township accounts_ - Who shall be responsible for the maintaining of all township accounts? The code and relevant statutory provisions give no answer. The code provides as follows:

The secretary "shall provide suitable books, the cost of which shall be paid out of the township funds, wherein he shall enter all matters of which he is required to keep a record" (Section 901).

"The secretary of a township shall keep a record of the appropriations made by the township commissioners and the amounts chargeable thereto. He shall furnish to any person so requesting, a statement showing the amount available for future charges against any appropriated fund" (Section 902).

"Every township treasurer shall take charge of all township moneys from all sources, and promptly deposit the same in a bank..............and keep distinct accounts of all sums received from taxes and other sources, which accounts shall at all times be open to the inspection of the commissioners and township auditor or controller" (Section 803).

"The treasurer shall, on the first day of each month, make a true statement, in writing, to the secretary of the Board of Township Commissioners, of all taxes collected during the previous month, giving names of taxables and the amount collected from each and the total amount received" (Section 1716).

The statute would seem to indicate that the keeping of revenue accounts is to vest in the treasurer, the keeping of expenditure accounts, with such other accounts as the commissioners might authorize the secretary to keep, is to vest in the secretary, and that by means of cooperation between the secretary and treasurer combined with such monthly statements of revenues, the secretary would be sufficiently informed of revenue receipts. In any event such is the general practice. The effect and probable intent of the statute is to divide responsibility for the keeping of the township accounts between the treasurer and the secretary. The results have been wholly bad.

Data on thirteen townships reveal that: (a) In seven, the treasurer made complete monthly reports to the secretary. (b) In two, no reports were made, the secretary being compelled to depend on the bank balance statements. (c) In two, no such reports were made until recently, when the treasurer was finally induced, through pressure by the secretary, to make occasional reports. (d) In one township monthly statements were exchanged, but they were of such defective character that they have not been balanced for years.

The first requisite of any accounting system, or for the performance of any duty for that matter, is the designation of some one to be responsible for the administration of the function. All township accounts should be kept by the secretary and if the present legislature abolishes the local tax collector this result will be automatically achieved in the first class townships. In addition the statute should specifically charge the secretary with the duty of maintaining a record of the revenues received, nature of sources, and the like. The general practice at the present time is for the secretary to transcribe such monthly reports of the treasurer in the township minutes book where it is rapidly forgotten; in a few cases such statements are merely filed without any transcription.

Naturally, there are many townships where complete cooperation exists between the offices of the secretary and treasurer. This group, however, comprises a relatively small portion of the first class townships in the State.

2. _Types of township accounts_ - In what form or forms shall the accounts of fiscal operations be kept? In effect the statutory provisions permit the local units to prescribe whatever accounting system they deem necessary. Consequently, each township has set up its own system, each of which differs from all the others. The well-nigh universal practice in accounting, among these units, is to set up a general fund to which all receipts are allocated and from which expenditures are made in accordance with

the specific appropriations set forth in the budget ordinance. While this is the basic procedure there are so many variations that it is impossible to arrive at any definite classification of the accounting systems. Broadly, however, the following types may be described:

(a) Those townships in which only a general fund exists. This is the simplest type of bookkeeping. All receipts are entered in this fund as debit items and the expenditures or warrants are paid out of this fund and listed as credit items. At the end of the year the balance between these two is struck and the cash balances indicated.

(b) Those townships in which all revenues and expenditures are set up as credit and debit items of a general fund and in which are also maintained separate accounts for such matters as bonded indebtedness, sewer, highway, etc., construction, savings accounts, and the like. This accounting procedure goes one step further than the first in that the appropriation made by the commissioners is followed by an actual cash transaction in which either the sum of the appropriation or a lesser amount is actually set aside for debt retirement purposes and a separate account of such fund is made and retained.

(c) Those townships which maintain two or more distinct funds such as general fund, sinking fund, highway fund and the like. The general practice is that of setting up a general fund and in addition thereto some form of indebtedness fund and one or more capital outlay funds such as sewer extension fund, paving construction fund, etc. These funds are maintained as distinct in that: (1) Revenues are allocated directly to each separate fund instead of by way of an appropriation from the general fund. (2) Expenditures are paid directly from each separate fund and listed as credit items against the debit items, (revenues) of each fund separately. (3) Cash balances are maintained for each fund separately. And (4), transfers within these distinct funds are restricted to such transfers as in effect amount to a temporary loan from one fund to another or such transfers as are based on the need of charging against one fund an expenditure, properly assignable to it, that has been made out of another fund for reasons of convenience or necessity. Where these four elements are missing there can in reality be no separate funds since such matters as are set up separately from the general fund merely amount to the setting up of an account of one or more of the several general fund transactions or operations.

(d) Those townships in which a general and one or more additional funds are maintained distinct from each other and in which one or more transactions or operations of the general fund are set up separately into a specific account. A list of the funds which are set up independently of the general fund will prove instructive. In one township such funds are created for: health (milk control phase only), sewers (lateral connections and minor incidentals only), highways (curbing and sidewalks primarily), sewer construction; in a second for: electric lighting, sewers, and water supply; in a third for: fire; in a fourth for: a general savings account. These do not include the indebtedness funds such as special assessment funds, paving (assessment) funds, bond and sinking funds. A list of accounts of the general fund operations would prove just as varied.

3. _Revenue and expenditure accounting_ - First as to revenues, how are they accounted for? The division of responsiblity between the township secretary and the treasurer as to revenue accounting has been previously mentioned. There is no uniformity of practice. Among the smaller units the general custom appears to be for the secretary to copy the statement received from the treasurer into the minutes book or if such statement is not available to compile some statement from the receipts that go through the secretary's hands and attempt to obtain data on the remainder either by inquiry of the treasurer or from the bank statements where these are sufficiently detailed. In at least four townships where access was had to the minutes book these revenues were itemized and totalled for the month. However, the sources of revenue were not separated by columns, all amounts received being placed in one column and totaled, with a few words placed before each amount stating its source. This was the sole record of receipts in such townships aside from the accounts maintained by the treasurer. If the statements of the officials may be accepted this is the bookkeeping process of the smaller units, there being no record of such receipts other than that in the minutes

books, the monthly statements of the treasurer to the secretary where such statements were kept, and the accounts of the treasurer himself. These minutes are generally in long hand, and since calculating machines are not available, many errors creep in. Thus in one township the receipts in the minutes book totaled $10,012.82 while the auditors report showed a total of $9,401.89 though the difference in the two figures is partially accounted for by the failure of the auditors to include a temporary loan of $650 in their audit. In a second unit the receipts for the fiscal year 1933 totaled $15,559 in the secretary's book (though his total was not checked) whereas the auditors report showed only a total of $15,163.92.

Among the smaller group of larger and wealthier townships revenue receipts are entered into a separate ledger at monthly intervals. These ledger books are so arranged as to permit the allocation of the various items of revenues to their proper sources. Consequently a brief glance at such ledgers indicates not only the total revenue receipts for the month but also the amounts received from each source such as current and past due tax collections, special assessments, licenses, permits and the like. In several cases these ledgers also contain a statement of the deposit of these revenues upon the basis of the monthly bank and treasurers' statement.

More specifically the following methods were observed: (1) The keeping of separate ledger accounts in which all receipts are specifically itemized and allocated, monthly and yearly totals being inserted. (2) The recording of itemized but not allocated revenues in the minutes books with totals struck in some cases and not in others. (3) The recording of itemized and allocated expenditures and revenues in a single loose leaf book showing balances remaining for various appropriations and cash balance in bank. (4) The filing of the treasurer's report without transcription upon any ledger or minutes book.

Expenditure accounting. Where all fiscal operations are based upon a general fund all sums expended are listed as credit items. Where independent funds exist in addition to the general fund, expenditures will be classified on the basis of the various funds and the sums expended entered as credit items in the appropriate one viz.: where there is an independent sinking fund such expenditures as bond redemption, interest on debt, state bond tax, and other debt charges will be entered as credit items in the sinking fund. It is instructive that even in regard to such a restricted subject matter as just exemplified there is no uniformity of practice. Thus one township will charge all items but redemption of principal against the general fund while in another, interest, state tax, and principal payments will be charged against the sinking fund and in still others these debt charges will be credited in each fund indiscriminately as it seems convenient.

Regardless of the number or nature of funds set up, the general practice is to enter all expenditures in the minutes book and also in what is usually termed an "appropriation book". In this latter book the secretary will enter the appropriations made by the commissioners and then proceed to allocate the sums expended to these various appropriations. A few townships have established an accounting system in which expenditures are not only entered against particular appropriations but also allocated to specific purposes. To illustrate: In 1933 the Spring Garden township commissioners appropriated $10,600 for highways as follows; $7,500 for stone and material, $100 for team hire, and $3,000 for salaries. As bills are paid for road purposes these same payments will be allocated to one of these three purposes.

A small number have established an accounting system which provides for the specific allocation of all sums expended, according to a detailed classification, and in addition includes the "double entry" method of bookkeeping. Under this system all bills that are approved for payment are immediately entered in the ledger, against the sum appropriated, as debit items and upon actual payment are again entered as credit items. This system is especially needed as a form of budgetary practice since it shows immediately the actual amount of obligations incurred against a specific appropriation and the amount of bills payable.

The general practice in all townships is to enter bills approved and bills paid in the minutes book. In some townships the warrant number and the purpose for which the expenditure was made are given. In others only the warrant number and amount are given.

Incredible as it may seem there are several townships.where the only record of expenditures to be found is in this book.

Types of books kept in various townships. Perhaps no better illustration of the lack of uniformity and the absence of any adequate accounting system can be given than a brief description of such books.

Annville:
    Minutes book.
    Loose leaf typewritten ordinance book.
    Appropriation book - an ordinary inexpensive note book such as children use
        for school purposes. Bills charged against appropriate appropriation
        with amount and purpose listed. Totals at irregular intervals and
        annually. No allocation. No statement of bills payable.

West Lebanon:
    Minutes book - the sole book. All ordinances kept here. Receipts and ex-
        penditures listed with amounts and sources and/or purposes. Monthly
        and yearly totals given. No attempt at either allocation of receipts
        or expenditures. Balances between actual expenditures and appropria-
        tions shown at irregular intervals. No statement of bills payable.

South Fayette:
    Minutes book - Ordinances receipts and expenditures entered.
    Appropriation book - Monthly disbursements listed by warrant for each
        appropriation and without specifying purpose of sums expended. Monthly
        totals of sums expended and balances between appropriations and actual
        expenditure.
    Road District book - itemized account of all road expenditures for each
        district. No allocation of expenditures. Monthly totals and balances.

Penn:
    Minutes book - Ordinances, receipts and expenditures entered. Receipts
        itemized. Expenditures itemized showing warrant number, name of payee,
        and amounts but not the purposes.
    Ordinance book - in long hand.
    Appropriation book - Expenditures listed against each appropriation. War-
        rant number, name of payee, and amounts given but not purposes. No
        allocation of expenditures. No monthly totals or balances.
    Account book - Monthly receipts entered but not totalled. Totals of monthly
        expenditures entered but not itemized. Yearly balances and totals-
        struck.
    Ward book - Road expenditures itemized but not allocated for each ward.

Elizabeth:
    Minutes book - Ordinances, receipts and expenditures entered.
    Balance book - Balances struck continuously as bills are paid, showing dif-
        ference between appropriations and actual expenditures.
    Distribution book - Disbursements itemized but not allocated. Yearly
        recapitulation and balances.

Shaler:
    Minutes book - Total of monthly warrants authorized and itemized, though not
        allocated, receipts entered.
    Ordinance book - Ordinances pasted in the book.
    Appropriation book - Expenditures entered against proper appropriations. No
        allocation. Balances and totals carried continuously as bills are paid.
    Analysis book - Expenditures of general and road accounts itemized and
        allocated.
    Water Ledger Account book - operations of water department entered and
        analysed.
    Bills Payable book - Bills payable entered and classified according to
        appropriations.

Sewer Assessment and Lien book - Record of receipts and accounts payable
    maintained.
Information book - Analysis of police car expenditures, indebtedness main-
    tained. Monthly statements of treasurer entered.

Other illustrations might be given but these suffice to indicate the lack of uniformity and the inadequacy of township accounts. Many divergencies in practice are
apparent: (1) Expenditures are not itemized as to purpose in some townships making
resort to either the bills or the minutes books (where adequate records are maintained
therein) necessary. (2) Generally expenditures are itemized as to purpose but no
attempt is made to allocate such sums by segregating the amounts paid for specific purposes. (3) In a small group of units expenditures are both itemized as to purpose and
allocated by means of the segregation of the amounts paid for specific purposes.
(4) Where expenditures are so allocated the type or classification of purposes differs.
(5) A few townships have established the "double entry" type of bookkeeping automatically showing the amount of bills payable. (6) Among the remaining units where the
double entry system is not utilized bills payable are either recorded monthly or annually in a separate ledger or such sum cannot be determined without resort to the minutes
book.

Enough has been shown to indicate that there is little attention paid to functional
cost accounting. Such cost accounting can only be had where a careful classification
is made of expenditures, based on the functions performed, and a careful allocation of
sums expended to their proper classification. Even such accounting is of small value
unless costs within each function are in turn grouped as fixed and flexible costs,
personnel and other costs, and similar classes.

The almost universal practice is to set up an accounting system in which expenditures are listed according to appropriations. This means that where the appropriations
are not properly classified, the accounts will not be properly classified. Practically
all appropriation ordinances contain an appropriation to what is variously termed
"general", "general road", "road", "contingent", "miscellaneous" fund. Some units
appropriate to two or more such broad purposes such as Harrison township which appropriates to a "contingent fund" (which is in effect an appropriation for administration)
and which expends certain sums from a so-called "unappropriated fund" which is in effect
the unexpended balance of the operations of the "general fund" in the prior year. This
latter fund is at once the basic fund for all expenditures except certain sinking fund
payments and the account against which miscellaneous administrative expenditures are
charged. While the accounts are here fairly well classified and itemized, the distribution of different types of expenditures within these three so-called funds, contingent,
unappropriated, and general, makes a complete statement of functional costs impossible
without a reallocation of items between these accounts and without a further classification within each account. Even in these townships where only a "general" appropriation
is set up under one name or another the same difficulty appears. The tendency is to
include in this broad appropriation many sums expended for items properly coming within
some one of the several other appropriations.

Functional costs are as difficult to obtain where several independent funds are
maintained. This is true even where such funds are created on a purely functional basis
such as funds set up for water departments or sinking funds and where a distinction is
made within a function such as capital outlays and operating expenditures, with an independent fund set up for each phase within the function.

The evidence indicates, then, that in most of the townships a true picture of
functional costs is unobtainable. This is particularly true of the smaller units which
compose the greater number of the townships of this class. The incompleteness and in-
adequacies of the type of ledger accounts, voucher systems, method of billing, and the
recording of minutes by the secretary make it impossible to secure such costs. In the
larger townships it was found that these data were not at hand but were only obtainable
by resort to various ledger accounts and other sources. Several of the units have established an audit form by means of which such costs are fairly well presented to the local
taxpayers and officials in the annual audit statement although such data are not readily
available during the year.

That an accounting system which will provide a continuous picture of functional costs during the fiscal year is feasible even for the smaller units is amply demonstrated by the example of several of the smaller townships among which may be mentioned Swatara and Spring Garden. The accounting systems of these two townships present an approximately accurate statement of functional costs which are immediately available without resort to vouchers, bills, minutes, or independent ledger accounts.

Contact with local officials in charge of the maintenance of township accounts revealed that these men are overwhelmingly in favor of the institution of some uniform accounting system. It is significant that not a single official expressed himself against the introduction of uniform accounting although several of them described what they considered to be obstacles in the setting up of such systems. These may be grouped under two headings: (1) The tendency of some officials to resist the introduction of any system that was not substantially like their own because they had either originated their own system, in which case the personal element enters, or because they have by long service become habituated to the present methods. (2) The difficulty of properly classifying the units for such purposes. There is such a great diversity in population, assessed valuation, revenues, expenditures, fiscal operations generally, and governmental problems that many officials felt that it would prove exceedingly difficult to devise a system that would fit all the units in this class. These objections, especially the latter, are of much importance and must be given full consideration in drafting any new system. To them may be added a third, the necessity of devising a simplified plan which can be operated by the local personnel.

4. <u>Uniform reporting</u> - Closely related to the problem of adequate accounting is that of reporting to a central agency the fiscal operations as they are recorded in the local accounts. The objectives of such reporting are: to determine whether the required accounting has been properly undertaken; to ascertain whether, in appropriations and actual sums expended the local officials have complied with the legal requirements as to budgeting; to give the state supervisory authorities all necessary information on local finances; to make possible a comparison of functional costs in different townships (a comparison which would be gladly received by township officers themselves) and to give local taxpayers and officials an accurate and timely picture of the local fiscal operations.

The problem here is two-fold: (a) The need for adequate legislation; (b) The establishment of such uniform accounting, budgetary, and auditing forms as will permit of accurate reporting without the needless duplication of effort now involved in transferring fiscal data from the ledger accounts to divergent auditing and reporting forms.

(a) Present reporting legislation - The revelant statutory provisions are as follows:

"The township commissioners shall annually, on or before the first day of February in each year, make a sworn statement to the State Department of Highways on blanks furnished to them by the said Department, of the whole amount of tax levied during the preceding year for road and bridge purposes. They shall specify in such report, the amount expended for maintenance or repairs of roads and bridges, for opening and building new roads and bridges, and for macadamizing or otherwise permanently improving roads, the number of miles of roads thus made, and the total number of miles of township roads in the township. Such report shall also contain such other matters and things pertaining to roads and bridges as the department may require" (Section 704).

The statute clearly does not contemplate an annual report of the fiscal operations of the reporting units. It merely provides for what might be termed a report of highway operations. The commissioners are requested to submit data only upon the road tax duplicate and road expenditures. The department is not authorized to request date upon matters not pertaining to highways such as sewers, police, fire, administration, health, and the like.

The statutory provisions are also defective in that they do not authorize a report upon the entire fiscal operations of the reporting unit. No sound reason appears for limiting such reporting to highway affairs. This limitation was primarily a matter of accident since the Highway Department in cooperating closely with the local units, particularly the second class townships, found a need for more complete information upon local highway administration and sought to provide itself with such data by requesting such authority as the statute now provides. In fact some form of reporting is authorized wherever there is a close cooperation and association between local and state officials in the administration of a particular function or where administrative supervision over local administration is vested in some state agency. Thus the local units are compelled to report either regularly or upon stated occasions to each of the following state authorities: (1) Department of Highways (2) Department of Health (3) Department of Revenue (4) Department of Internal Affairs.

It seems clear that any system of uniform accounting, budgetary and auditing procedure requires the close cooperation, supervision, and study of some central agency. This means either that some state office should be created in which these functions could be centered or some existing agency designated for their exercise. Such uniform reporting as is here proposed would confer important and lasting advantages in many other respects, as we have seen.

Much of the value of reporting lies in the opportunity it provides for aiding officials and the public in determining the relative standing of their local units by means of accurate comparative data on such matters as functional costs, revenue collections and delinquencies, and indebtedness. The present legislative mandate does not even provide for the compilation and publication of such data as is reported. The Bureau of Statistics in the Department of Internal Affairs compiles a record of assessment and tax statistics for local governments through field men working in cooperation with county officials. The Department of Highways makes a similar compilation of such data from the reports filed in that department by the local units reporting. Each compilation duplicates the other and the two compilations arrive at different conclusions as to figures and totals. In neither case is the material, so gathered, published nor is a summary of the same required to be made public.

Finally, the present provisions fail to include any adequate enforcement of penal clauses. What if a local unit fails or refuses to report as requested? In the case of the second class townships such failure can be penalized by the withdrawing of state aid for local highway administration but this is not true in the case of the first class units where state supervision and cooperation in highway administration is practically non-existent and not desired by the local officials.

(b) The present reporting form - The form issued to the local units by the State Highway Department might be improved as a report of local fiscal operations in several respects: (1) Too much stress is placed upon highway operations and not sufficient space allotted to other functions. (2) Simplified as the form is, there are several items which are not clear to the officials who fill out such forms. To take two instances: Among the sources of receipts appears the item "fines and penalties". Does this mean tax penalties only, or both tax penalties and fines recovered for the violation of local ordinances and the like, or only fines? Each of these three interpretations was held to be the only correct one by local officials compiling such forms. Among the items of expenditure listed is one of "notes maturing". Does this mean that a note which has matured should be included here though not actually paid? Should a note negotiated and paid before the close of the fiscal year be listed? Reporting officials again held varying opinions on the matter. (3) The form, while suited to the character of the fiscal operations of some units, is not readily adaptable to the operations of the larger units. Thus it provides no functional classification for such matters as parks, playground, library, public utility enterprises such as water departments, or for tax collection or auditing. Several of these functions assume large proportions in the larger units and one or two of them are important even in many of the smaller townships. In short, the form should provide an adequate basis for the determination of functional costs. (4)Finally, the form should provide for a sufficient distinction between capital and operating outlays for functions other than highways.

(c) The Reports - The statute provides that local officials shall file such reports on or before February 1st of each year. This mandate is as generally ignored as that requiring auditors to file a copy of the annual statement within ten days of its completion with the county clerk of the quarters session court. On July 25, 1934 fifty-five of the sixty-one townships had filed their report with the department. Of these fifty-five only thirteen had filed their reports on or before February 2nd while seventeen had filed sometime after May 1st, 1934. Generally all reports are eventually filed after the department has sent out several requests and reminders to the local officials.

The reports themselves are unreliable sources of information and, unless used in conjunction with the annual audit statement or supplemented by contact with the local accounts or officials, are almost worthless. Even where the state forms may be accurately filled in, they provide an incomplete and inadequate picture of the fiscal operations. Thus, the Braddock township report lists $8,183.40 as "other receipts" out of a total of only $20,738.97 revenues for 1933; the Harrison report lists $9,033.69 "other receipts" out of a total of $105,118.96; Tinicum lists $17,091.25 out of a total of $72,963.40. The absurdity of the situation is further shown by a comparison of these large "other receipts" totals with the amounts itemized for other purposes. Thus, the Whitehall state report for 1933 lists a total of $39,155.43 "other receipts" and $66. for revenues derived from the "rental of township property"; the Lower Merion report lists $151,799.64 "other receipts" and $690 revenues derived from rental of township property; Spring Garden lists $2,473.05 "other receipts" and $45.46 revenue derived from "interest on bank balances"; the Coal report lists $63,378.74 derived from "other receipts" and $379.15 revenues from "fines and penalties". Many other examples of a similar character may be cited.

The same difficulty occurs in regard to expenditures reported. One or two illustrations will suffice. The Mt. Lebanon report shows $18,706.69 miscellaneous administration expense, $1,470.97 miscellaneous highway expense, $4,223.74 miscellaneous police expense, and $11,031.24 "other expense" not even broadly allocated to any general function making a total of $33,961.67 miscellaneous expenditures out of a total of $363,988.95. Lower Merion lists the following "miscellaneous" sums, $25,342.43, $9,747.05, $21,578.56, $34,948.37, for administration, highways, sewers, and police respectively. To these sums must be included a third, non-allocated "other expenditures" amounting to $31,854.92 making in all a total "miscellaneous" expenditure of $123,471.33 out of a total of $767,961.35. It might be said as some officials maintained that the state report is so unsuited to the present accounting system of the local units and to the reporting of the large fiscal operations of the large units that the fault is due to the state form alone.

In completing the preliminary work of the present Survey, these state reports were studied, duplicated and partially compiled and analzed before officials in the local units were interviewed. This was done with the hope of obtaining a fairly accurate knowledge of the area before approaching a local official for further information. It is significant that these officers without exception, could no longer interpret the state reports which they had compiled but a month or two previous to the interview. When inconsistencies were pointed out to them between the data presented in the state report and the data set forth in their accounts or in the annual audit statement, they invariably stated, with some show of feeling, that the state report form did not conform to their accounting system and that such great difficulties were occasioned in transferring the data from the local accounts to the state form that the latter report was subject to considerable error.

This situation contrasts strongly with that in the second-class townships. The latter are almost exclusively rural and the primary governmental service is that of highway administration. Under the Parkinson act and other legislation the state highway department is interested in obtaining accurate and detailed data as to the highway operations of these units. Consequently it has taken steps to secure more complete and accurate data from them. Among the other means used was that of preparing a budget ordinance, account book, a warrant book, a payroll sheet, and a state report in each of which expenditures and receipts are identically itemized and allocated. This in effect provides a uniform accounting and reporting system for the second class townships. It is a simple operation for the township secretary to transfer figures from warrants to the account book or from the payroll sheet to the account book and then from the account book to the state report. No addition, subtraction, or multiplication is necessary. Consequently, the data on file in the state highway department are accurate and provide an excellent picture of the fiscal operations of the reporting units.

This uniform accounting and reporting for second class townships was obtained without any legislative compulsion. The state highway department drew up forms for budgeting, accounting, and reporting in such a manner as to constitute a uniform system. It printed such forms and distributes them free of charge to the township officials. By this means and by close contact with these local officials the department has succeeded in getting the great majority of the second class townships to adopt this uniform system. Many of the units adopting the system have carried it one step further by using the state report form as their annual audit form thereby completing the uniformity of the system to cover the four fundamentals of budgetary, accounting, reporting, and auditing procedure. Why is there such a great contrast in procedure and results in regard to the two classes of townships? The answer is clear. In the one case the reporting unit is rural in character and its primary governmental service is that of highway construction and maintenance in which the state highway department is primarily interested in several ways. On the other hand the first class townships are primarily urban in character providing many important governmental services in addition to that of highway administration and in which the highway department has little, if any, interest. In the one instance an interested state agency has taken the initative and succeeded in producing most commendable results. In the other instance there has been no state agency sufficiently interested or authorized to take the initiative.

What, then, should be done to obtain these same results in the case of the first class units? Legislation should be enacted, providing some state agency with the definite responsibility of instituting, maintaining, and supervising a system of uniform budgetary, accounting, auditing, and reporting forms and procedure. Cannot these ends be obtained without legislation and by an indirect process as in the case of second class township? The answer is likewise clear and definite. The governmental services of these units are so varied that no one state agency has, at present, such a broad interest in their local administration as to institute such procedures and forms. Can the first class townships be expected to provide such a system for themselves on their own initiative? In the first place these units are scattered over the state and primarily interested only in their own local problems. In the second place the institution of such a system is not an overnight task. The diversity existing among these units, the breadth of the governmental services that they perform, and the lack of any informational or experimental data make it absolutely essential that some central agency undertake the task. The problem is one of magnitude, requiring competent personnel, adequate resources, sufficient authority, and application to the task over a continuing period of time. Only a state agency can meet these requisites.

CONCLUSIONS

1. <u>FINDING</u> - The present accounting systems are neither uniform nor adequate. They fail to show functional costs, they make only a haphazard distinction between fixed and other charges, and they frequently give no totals or balances other than the annual ones.
   <u>RECOMMENDATION</u> - That, in view of the chaotic state of present accounting practices, a uniform accounting procedure and standard ledger forms be adopted for the first class townships.

2. <u>FINDING</u> - Lack of training and the part-time character of the local accounting personnel largely explain the present bookkeeping methods. The officials themselves are generally capable, needing only some form of outside initiative and aid in the perfecting of accounting practices.
   <u>RECOMMENDATION</u> - That the institution and administrative supervision of a uniform accounting system be vested in a state agency, alternatively designated.
   <u>RECOMMENDATION</u> - That the primary consideration, in the institution of standard accounting forms and methods be that of the utmost simplicity compatible with their essential purpose.
   <u>RECOMMENDATION</u> - That the state authority be instructed to publish and distribute among the local accounting personnel, handbooks describing methods to be followed, types of balances to be kept and the like.

RECOMMENDATION - That the state agency similarly furnish such standard forms, ledgers, and books, as may be required for the operation of the uniform system either free of charge or at cost.

RECOMMENDATION - That an adequate enforcement and penalty clause be attached to such uniform accounting act.

3. FINDING - There is no adequate system of reporting local fiscal operations. This is due to the lack of sufficient statutory provision for such reporting, to the defective character of the present state highway department forms, and to the lack of coordination between the accounting and the reporting systems.

RECOMMENDATION - That the state agency authorized to institute uniform accounting, budgeting, and auditing procedures and forms be similarly authorized to institute uniform reporting forms.

RECOMMENDATION - That the standard reporting forms adopted be closely correlated with the standard budgeting, accounting, and auditing forms.

RECOMMENDATION - That in view of the need for more information on such matters as comparative functional costs, indebtedness, revenues, and other local fiscal operations, the data obtained through the institution of a uniform reporting system be published by the state authority in such manner as to permit of the comparison of the fiscal operations of the local reporting units.

4. FINDING - The local reporting officials have shown a general disregard for the statutory provision requesting them to file their annual reports, upon such forms as the highway department designates, by February 1st. The reports that are filed are generally incomplete, unreliable and often late.

RECOMMENDATION - That adequate penalty provisions be provided by law and that the enforcement of the same be placed within the powers of the above recommended state authority.

CHAPTER XX
BUDGETARY PROCEDURE

The problem of budgeting is one of seeking an acceptable adjustment between available revenue, anticipated expenditures, and the satisfaction of the political needs of a particular community. Estimates of revenue and expenditures can be obtained only after the compilation of a large amount of factual data. Even then accurate estimates are not easily made and are necessarily subject to change as unforeseen contingencies arise. No one governmental official or group of officials should be responsible for both the preparation of the factual data and for the determination of fiscal policy. Many defects in the budgeting process have their foundation in this procedural basis. Questions to be treated in this chapter include the following: What are the elements of an adequate budgetary system? To what extent are such elements observed in the existing budgetary practices of the first-class townships? How may existing defects be remedied?

Elements of an Adequate Budget System - A budget must include estimates of all expenditures and of all revenues. Such estimates must be based upon data drawn from past experiences and upon prospective future conditions within the budget period. It is essential that these estimates and data be in sufficient detail to provide a financial plan for the fiscal year in which the budget is to operate. The process should include the following steps:

(1) Collection and preparation of estimates for all expenditures and revenues.
(2) Comparison of: (a) appropriations and expenditures of prior years for each governmental function, viz., highways, health, police, etc; (b) tax collections and tax delinquency or estimated receipts and actual receipts for each category of income, viz., taxes, licenses, special assessments, etc.
(3) Revision of estimated receipts and expenditures upon the basis of the above comparisons.
(4) Assembling of the budget document including the appropriation and tax levying ordinances.
(5) Holding a public hearing which tax payers may attend to express their views upon the budget.
(6) Formal adoption of the budget.
(7) Control of the transactions authorized by the budget.
(8) Checking actual operations - periodic statements to commissioners of actual expenditures and revenues compared with estimated expenditures and revenues.

(1) Collection and preparation of estimates - Who shall collect and prepare such estimates? The First Class Township Code and relevant statutes are silent. Section 1701 of the Code provides, "The board of township commissioners of townships shall each year, within sixty days after the first Monday of January, estimate the several amounts of money which will be required for the several specific purposes of township government and expenses during the current fiscal year, and by ordinance appropriate, out of the revenues available for the year, the specific sums required. The tax levied by the township authorities shall be fixed at such figure, within the limit fixed by law, as, with all other sources of revenue, will meet and cover said appropriations."

This provision makes no mention of the collection and preparation of revenue and expenditure estimates. It goes no further than to compel the commissioners to make an estimate, which estimate is, in practice, a mere guess unsupported by any factual material. The statute makes no distinction between budget preparation and budget enactment; between the assembling of the budget data and the determination of fiscal policy. The former function is an administrative matter to be performed by a qualified administrative official such as the township secretary; the latter is a policy-determining function of the board of commissioners. The commissioners have neither the time nor the requisite information for assembling the budget document and the task lies outside their proper functions.

The legislature should (a) distinguish between budget preparation and budget enactment, and (b) vest responsibility for budget preparation in the township secretary, together with adequate powers to obtain the information needed for its preparation.

No distinction between budget preparation and budgetary policy has been established in actual practice. Nor is there any one form of procedure; in the absence of any statutory provision each township has evolved its own methods. These may be grouped as follows: (a) Preparation and enactment of the budget by the board of commissioners; (b) preparation by the secretary and enactment by the commissioners; (c) preparation by the finance committee and enactment by the commissioners; and (d) preparation by the solicitor, a commissioner, the ordinance committee or an unofficial person or body such as a taxpayers association or civic committee.

(a) <u>Preparation and enactment of budget by the commissioners.</u> - This is the prevailing practice. Examples are seen in West Norriton, Braddock, Penn, Leet, Annville, West Lebanon, Whitehall, Newport, and Hanover. The commissioners have an informal meeting, at which there is a general discussion in regard to the tax rate to be levied for the coming fiscal year and the amount of money to be appropriated for the various township functions. There is no prepared budgetary document available to aid in reaching conclusions. There are no prepared comparisons of prior years' tax collections, tax delinquencies, revenue collections from sources other than taxation, etc. Reliance is placed upon three general sources of information: (1) The budget ordinance of the prior year. The discussion centers upon the question whether or not the amounts appropriated for the prior year are to be increased, decreased, or retained.

A superficial survey of the budget ordinances of successive years will reveal the strong tendency of the commissioners merely to reenact or readjust, with slight variations, the sums appropriated in prior years. In 1933 the Swatara commissioners appropriated $5,800 for highways exclusive of bridges, culverts, and cross drains and of this sum only $3,400.37 was actually expended. Despite the fact that there was a difference of $2,399.63 between the estimated expenditure and the actual outlay, the commissioners appropriated the same amount in the budget ordinance of 1934. This is but one of many examples; it is illustrative of a general tendency.

(2) The second source of information is the township secretary. He is present at all meetings of the commissioners and knows more about the township affairs, fiscally, and otherwise, than any other township official. In a few cases the secretary prepares a brief statement for this informal budgetary meeting such as the statement of last year's appropriations, actual expenditures, and amount of difference between the two for each function. This is done in Baldwin, Elizabeth, South Fayette, and Harrison. He also acts as a general source of information on the results of the past year's operations. Many of these secretaries indicated a desire to enter more fully into the process of budget making but were held back because they considered they had no right to participate beyond the direct requests of the commissioners.

Most of these capable men would be able to contribute a great deal toward the establishment of an adequate budgetary system if a statutory provision invested them with the requisite authority. Such a provision, combined with the educational guidance and advice of some central state agency supplying handbooks and field supervisors, would remove one of the chief obstacles to an adequate budgetary procedure.

(3) Finally, many commissioners rely heavily upon the amounts actually expended for each item in the prior year. If such outlay has varied greatly from the appropriation the board will fix a sum slightly in advance of the actual expenditure. Where the variance is small the board will merely reenact the prior appropriation.

The procedure outlined is clearly not one of budget making. It is merely an effort to enact an appropriation ordinance upon the basis of meagre data. In some cases the reliance is placed upon prior year appropriation figures, in others upon prior years' outlay for general functions, and in a few upon a haphazard combination of both prior year appropriations and actual expenditures. In short, the commissioners determine budgetary policy without any budget preparation. The same procedure exists as to revenue estimates where heavy reliance is placed upon the prior year's tax rate and the total expenditures.

(b) - Preparation of budget by secretary. - The most adequate budget procedure is found, without exception, in those units where responsibility is placed upon this official for the preparation of a budget document. Examples are lower Merion, Haverford, Mt. Lebanon, Upper Darby and Shaler. In all such cases a typewritten statement is prepared giving detailed estimates of income and expenditure upon the basis of comparisons of estimated and actual receipts and outlay of several preceding years. Allowance is made, upon a scientific basis, for tax delinquency, collection of past due taxes, abatements and the like. When the board meets, the prepared document is before it, and it can proceed to the determination of fiscal policy upon the basis of reasonably comprehensive data. There is no occasion for needless haranguing over facts nor is time consumed in a last minute effort to obtain them.

(c) Preparation of the budget by the Finance Committee. - Where the finance committee is a committee of the whole, the budget procedure rests with the board of commissioners. Examples are seen in Baldwin (a committee of the whole), Nether Providence, Tinicum, and Neville. The budget document is nothing more than the proposed appropriation ordinance. In a few cases the secretary prepares the document and the finance committee then irons out the major difficulties and makes a preliminary revision after which the revised fiscal statement is placed before the commissioners for formal action. Where this committee makes a preliminary report, which is considered budget preparation, it merely reports the amounts which it thinks should be appropriated to the specific township functions together with a proposed tax rate. In no case does it prepare a document estimating detailed receipts and expenditures upon the basis of the factual experience of prior years. Since this committee, though not always a committee of the whole, is composed of the more influential members of the board, its recommendations are largely conclusive.

(d) Preparation by Other Groups. - The budget ordinance is in some places prepared by the ordinance committee, in others by the township solicitor, in still others by a taxpayers' association. The latter is allowed to do the work because of its control of expert services.

(2) Revenue Estimates - What is the objective of the estimate? This may be briefly stated - to permit the township to operate within the limits of its annual income, prepare for unforeseen contingencies, plan its capital outlays and its indebtedness. The estimate should provide an approximately accurate knowledge of the amount of moneys to be received by the governmental unit in the coming fiscal year. Consequently such an estimate must include all possible sources of income, and allow for such matters as abatements, delinquencies, and collections on prior years' duplicates.

To what extent do present practices in revenue estimation meet these elementary requisites? The general practice is for the commissioners to set a tax rate which when multiplied by the estimated assessed valuation will equal the amount of estimated expenditures. The resultant sum is accepted as the total revenue available for the coming fiscal year. To take an actual example - Penn township in 1933 with an assessed valuation of $13,408,880 at 7 mills tax rate has a tax duplicate of $93,862,16. This latter sum was accepted as the estimate of total revenues available for the year and in accordance therewith an appropriation of $91,627 was made. The commissioners gave no consideration to such matters as abatements, delinquencies, collections from prior years' duplicates, and the like. A brief investigation would have shown the officials that the amount of tax delinquency had been rapidly increasing (in excess of 20% for 1930, of 23% for 1931, and of 34% for 1932) and that a similar trend existed in the amounts by which current delinquency exceeded collections of taxes from prior years (current delinquency exceed collections from previous years taxes by $6,767 in 1930, by $11,701 in 1931, and by $21,441 in 1932). Investigation would also have revealed that approximately five or six thousand dollars could be anticipated from sources other than taxation.

Applying a simple procedure, the commissioners could have made allowance for a delinquency of thirty-five per cent or more on the current duplicate or approximately thirty-three thousand dollars. On the basis of past experience the commissioners would then anticipate that approximately thirteen thousand dollars (40% of the estimated current delinquency) would be offset by collections on taxes of previous years and by

five thousand dollars from revenue sources other than taxation. This would have provided a roughly accurate estimate of $78,000 as the total amount of revenues anticipated
as available for the year 1933. Instead of then appropriating a total of $91,627, the
commissioners should have revised their expenditure estimates to conform to anticipated
revenues. The result of failing to base their revenue estimates upon some such procedure as outlined above is amply demonstrated in Table I which shows that their total
expenditures exceeded total receipts.

A comparison of total appropriations with the total amount of outstanding taxes
(current duplicate included) at the beginning of the fiscal year will show immediately
whether or not the commissioners have arrived at their revenue estimate by a procedure
of making allowances for delinquencies and the like. Where these two sums are nearly
equal, there has been obviously no attempt to make an accurate estimate. Spring Garden
township affords a typical illustration. The total amount of outstanding taxes at the
beginning of the fiscal year 1933 was approximately $48,000 while the amount appropri-
ated for the year was $44,675. The commissioners, thus, made allowance for less than
four thousand dollars of tax delinquency despite the fact that in the prior year nearly
twelve percent of the current duplicate had remained unpaid. The 1934 revenue estimate
is even more defective. The total appropriation was $44,500 although the total of the
tax duplicates was only $43,541.33 or $958.67 less than the total amount appropriated.
A brief investigation reveals the following pertinent facts: Three thousand dollars
represents a generous estimate of revenues derivable from sources other than taxation
($2,588.11 in 1933). The percentage of tax delinquency had been steadily increasing
and amounted to over sixteen percent in 1933. The collections on taxes of previous
years had been steadily decreasing and amounted to less than thirty-four percent of the
current delinquency in 1933. Ignoring such matters as allowances for abatements and
exonerations, the consideration of these facts reveal that $40,000 would have been a
generous estimate of total revenues to be anticipated or $4,500 less than actual appro-
priations for the year. This township provides a typical illustration of the failure
to base revenue estimates upon any adequate procedure of investigation and adjustment.
The following table provides additional examples.

Township	1933 Appropriation	1933 Duplicate	1933 Collection	Total Tax Duplicate	Total Tax Collection
Annville.....	14,821	12,986	11,228	14,888	11,841
West Lebanon.	7,200	5,571	3,704	9,475	5,736
		4,359	2,886	7,858	3,758
Nether Provid.	24,820	24,909	19,375	31,428	19,692
Penn	91,627	93,957	55,683	185,825	65,494
Jefferson	21,000	13,220	7,884	23,771	11,793
Braddock......	20,900	9,793	4,420	17,061	6,515
Elizabeth.....	23,000	20,846	14,049	35,770	15,905
Harrison	40,850	45,023	31,460	63,628	39,074
Leet	9,161	7,247	4,713	9,219	6,325

It will be noted that the above data present a comparison between current and total
duplicates outstanding at the beginning of the year and total appropriations. This
latter sum is used as a contrast to the amount of taxes outstanding because total appropriations represent the only tangible evidence of the amount of revenue estimated by
the commissioners in the absence of any budget document other than the appropriation
ordinance. When local officials are queried as to whether or not allowances are made
for such matters as tax delinquency, abatement, exoneration, penalty, and the like they
invariably reply in the affirmative. Hence it is necessary to check such statements
against the facts. Since it is to be assumed that officials will confine their appropriations within the sums which they consider will be available as revenue, it is safe
to take the appropriation figure as indicative of their revenue estimate. A comparison,
then, of appropriations and of outstanding taxes will indicate the extent to which such
matters as delinquencies and the like have entered into the revenue estimate. Where
the appropriation equals or exceeds the amount of outstanding taxes, it is obvious that
no allowances have been made. Where it falls below the latter figure, the procedure
outlined in the illustrations of revenue estimation in Penn and Spring Garden townships
must be applied to discover if sufficient allowance for these factors has been made.

Proceeding upon this basis, the data reveal that six of the townships listed appropriated more than the full duplicate for the current year while the remainder appropriated practically the full amount of the duplicate. This alone shows the failure to arrive at a revenue estimate by a process of adjusting delinquency, exonerations, collections on previous years taxes and the like. If additional evidence is needed it is supplied by the fact that seven of the nine townships were compelled to resort to temporary loans during the year while Braddock received revenue from the sale of bonds. In one township Leet, a temporary loan of $1,100 was still outstanding at the end of the year as we have seen and the bills payable amounted to over two thousand dollars.

These illustrations clearly show that there is no systematic effort at revenue estimation. The commissioners begin the budget process by observing the previous years expenditures for each function. If the appropriation of the prior year exceeds the actual expenditure, the tendency is to reenact the same appropriation. Where the reverse occurs the tendency is to raise the appropriation slightly above the prior year's expenditure. Spring Garden township will serve as an illustration of this process. The 1933 budget appropriated $2,000 for miscellaneous expenditures. Actual requirements were $517.31 and the 1934 budget appropriated $1,700 for this purpose. The 1933 budget appropriated $2,500 to the contingent fund. Actual requirements were $1,232.83 and the 1934 budget set aside $2,000 for this purpose. Other examples could be readily provided. The commissioners, then, having reached an appropriation figure "safely" in excess of actual requirements, center their attention upon revenues. This process naturally requires that revenue estimates be subordinated or adjusted to proposed appropriations. The tendency to over-appropriate forces the local officials into a position where little, if any, allowance can be made for such matters as delinquency since to do so would result in the anomaly of having appropriations exceed the revenue estimate - a situation prohibited by the code.

The fruits of this form of revenue estimation are shown in one of three ways; by the necessity for temporary loans; by the retirement of unfunded debt through the issuance of bonds; or by either holding expenditures far below the appropriations or by piling up excessive amounts of unpaid bills. Each of these methods is quite common although the number of units that definitely attempt to hold their expenditures below appropriations is small. Only a small number of townships, unfortunately, attempt to draw up some document upon which an approximately accurate estimate of revenues may be based.

(3) _Expenditure Estimation:_ Expenditures should be determined upon the basis of: (a) The revenues estimated as available during the coming year. (b) Necessary outlays for fixed charges. (c) Necessary outlays for functional requirements. And (d) Expenditures for contingent requirements. The objective is the same as in the case of revenue estimation - to permit the township government to operate within its current income.

To what extent do present practices fulfill these requirements? No attempt is made to distinguish between fixed and flexible charges, except in the case of debt service, or to itemize actual needs. The commissioners base their estimate upon three things: (a) Prior year expenditures. (b) Prior year appropriations. And (c) their own untabulated knowledge. In most instances the commissioners set appropriations sufficiently in advance of prior year expenditures to assure themselves that these will not be exceeded. Obviously this is not budget making in the proper sense of the word.

In many cases the commissioners make no adjustments in expenditure estimates, even though they have sufficient facts indicating the necessity for such adjustment. For example, in 1933 and 1934 the commissioners of Whitehall township estimated expenses for engineering services at $1,000, in spite of the fact that in the three previous years the expense for such service amounted to twice as much. A similar situation existed in the estimate for auditing services. Braddock township is a typical example of many other units:

TABLE I

COMPARISON OF APPROPRIATIONS, RECEIPTS AND EXPENDITURES

Township	1933 Appropr.	Total Rev. Receipts 1933	Total Receipts 1933	Excess 1933 Approp. over Rev. Receipts
Upper Darby	$763,975	$680,743	$1,187,480	$ 83,232
Lower Merion	810,459	871,363	992,439	-60,904
Coal	175,500	163,912	213,092	11,588
Abington	320,350	281,280	383,474	39,070
Hanover	274,647	231,762	232,584	42,885
Cheltenham	356,750	428,213	479,284	-71,463
Mt. Lebanon	291,415	226,195	227,678	66,220
Stowe	197,960	115,485	125,771	82,475
Penn	91,627	79,968	86,342	11,659
Harrison	40,850	70,614	89,814	-29,764
Newport	175,990	167,890	167,890	8,100
Whitehall	118,930	66,331	124,771	52,599
Shaler	90,197	129,094	144,535	-38,897
South Fayette	15,600	14,098	14,567	1,502
Mifflin	-------	33,816	33,862	------
Elizabeth	23,000	16,681	18,691	6,319
Baldwin	79,073	45,681	50,691	33,392
Swatara	9,075	12,836	13,336	-3,761
Scott	45,000	41,123	43,307	3,877
N. Versailles	-------	21,519	21,679	------
Springfield (Montg. Co.)	80,763	70,503	157,880	10,260
Spring Garden	44,675	39,569	47,319	5,106
Springfield (Del. Co.)	61,942	73,617	79,311	-11,675
Plymouth	22,676	18,799	33,105	3,877
Jefferson	21,000	12,923	12,923	8,077
East Deer	27,790	62,429	64,656	-34,639
Tinicum	35,470	53,997	69,165	-18,527
Lower Chichester	--------	22,792	22,792	-----
Lawrence Park	53,000	35,893	35,911	17,107
Reserve	15,315	18,231	20,241	- 2,916
Annville	14,821	13,802	14,714	1,019
Nether Providence	24,820	23,245	25,360	1,975
Darby	15,335	10,057	22,953	5,278
West Norriton	-------	20,652	20,888	------
Rochester	13,850	8,908	10,908	4,942
Braddock	20,900	8,194	20,123	12,706
Neville	-------	62,007	62,033	------
Leet	9,161	5,173	8,526	3,988
Crescent	-------	5,700	5,700	------
West Lebanon	7,200	6,874	7,874	326

TABLE I

COMPARISON OF APPROPRIATIONS, RECEIPTS AND EXPENDITURES

Total Gov't'l Exp. 1933	Total Expenditures 1933	1934 Appropr.	Excess 1934 Approp. over 1933 Exp.	Excess '33 Exp. over '33 Appropr.	Percent over and under '33 Approp.
$616,126	$1,210,195	$788,650	$-421,545	$446,220	36.9
656,185	876,570	986,556	19,986	66,111	7.5
115,138	190,801	165,536	- 25,265	15,301	8.0
276,745	432,196	482,000	49,804	111,846	25.9
132,675	229,718	385,593	155,875	- 44,929	-19.6
314,334	513,833	349,200	-164,633	157,083	30.6
238,819	317,102	302,800	- 14,302	25,687	8.1
87,078	95,784	275,764	179,980	-102,176	106.7
79,172	108,328	91,026	- 17,302	16,701	15.4
65,228	91,089	-------	-------	50,239	55.2
91,410	168,632	171,374	2,742	- 7,358	4.4
73,550	125,837	134,281	8,444	6,907	5.5
75,721	161,586	90,054	- 71,532	71,389	44.2
13,324	13,324	14,800	1,476	- 2,276	17.1
32,172	32,358	61,721	29,363	-------	-----
15,275	17,245	20,475	3,230	- 5,755	33.4
26,043	59,083	54,632	- 4,451	- 19,990	33.8
12,070	12,570	10,105	- 2,465	3,495	27.8
30,942	42,632	50,330	7,698	- 2,368	5.6
19,079	27,279	32,734	5,455	-------	-----
50,632	141,269	88,836	- 52,433	60,506	35.8
37,414	40,414	44,500	4,086	- 4,261	10.5
55,975	78,646	44,950	- 33,696	16,704	21.2
14,344	22,591	35,878	12,287	- 85	0.4
11,287	11,296	20,000	8,704	- 9,704	85.9
41,093	47,532	28,400	19,132	19,742	41.5
30,657	62,947	42,750	- 20,197	27,477	43.7
14,650	14,670	19,669	4,999	-------	----
25,212	36,992	49,000	12,008	- 16,008	43.3
18,844	21,555	14,765	- 6,790	6,240	28.9
12,225	14,796	13,399	- 1,397	- 25	0.2
17,911	21,256	-------	-------	- 3,564	16.8
6,054	16,579	13,480	- 3,099	1,244	7.5
16,069	19,508	22,003	2,495	-------	----
6,762	11,446	-------	-------	- 2,404	21.0
7,378	17,778	15,650	- 2,128	- 3,122	17.6
52,911	57,666	55,498	- 2,168	-------	-----
6,447	8,668	9,971	1,303	- 493	5.7
5,806	5,806	9,463	3,657	-------	-----
5,793	6,793	6,850	57	- 407	6.0

Appropriation 1933		Expenditure 1933	Appropriation 1934
Administration	- $2,500	$2,227.58	$2,500
Roads & Highway	- 3,000	834.84	3,000
Health	- 500	204.98	400
Fire & Water	3,000	98.61	2,500
Lock-up	- 400	1,429.82	250
Light	- 2,500	1,650.00	2,000
Debt service	- 9,000	11,331.95	5,000
Totals	$20,900	$17,777.78	$15,650

It is apparent from this table that the 1934 estimates were based on the 1933 appropriations with practically no consideration for actual expenditures of the year 1933. This type of so-called budgeting is accompanied by the following significant facts in the above township: (1) No amending or transferring ordinances are enacted in regard to budgeting. (2) No budget or any other statement is presented to the commissioners at the budget meetings. (3) The budget ordinance was not printed prior to 1934. (4) Over fifty percent of the receipts for 1933 were in the form of bond sales. (5) The proceeds from the bond issue of 1933 were devoted to the payment of temporary loans accumulated over a period of years and negotiated to meet operating costs. (6) A delinquency of over fifty percent upon the 1933 duplicate.

Jefferson township expended $16,161.98 in 1932, - appropriated a total of $21,000 for 1933. Expenditures for 1933 totalled $11,295.07 and $20,000 was appropriated for 1934. There were no unforeseen contingencies which prevented greater expenditures nor did any contingencies fail to materialize. The commissioners merely appropriated a sum that approximated the current and past duplicates.

Table I provides some indication of the nature of budgeting practices in these townships. The first column shows the appropriation at the beginning of the year. As such it represents the adjustment of needs to estimated revenues. The second column includes only revenue receipts such as taxes, special assessments, licenses, interest etc. The total receipts in the third column include "non-revenue" income derived from such sources as temporary loans, bond sales, refunds, sale of materials, and the like, as well as revenue receipts. Governmental expenditures include only functional costs while total expenditures include sums expended for debt service, permanent improvements, and other non-functional expenditures.

Over-appropriation: (a) - In relation to receipts. Twenty-five of the forty townships are shown to have appropriated amounts in excess of revenue receipts, and fifteen (Hanover, Mt. Lebanon, Stowe, Penn, Newport, South Fayette, Elizabeth, Baldwin, Scott, Jefferson, Lawrence Park, Annville, Rochester, Braddock and Leet) amounts that exceeded the total of all receipts. Appropriation in excess of revenues may not in itself indicate poor budgeting since bond sales etc., may have been anticipated. However, Table I in the chapter on Functional Costs shows that only four (Upper Darby, Springfield (Mtg.), Whitehall and Braddock) of these townships floated bonds in 1933 and only three received gifts or grants in excess of $1,000. On the other hand sixteen negotiated temporary loans. A definite correlation exists between the tendency to over-appropriate and the negotiation of temporary loans as shown by the fact that of the thirty-one townships which negotiated temporary loans in 1933 (Table on Receipts, Chap. XXII) over fifty percent appropriated more than revenue receipts justified. Furthermore the tendency to over-appropriate is also accompanied by either a decrease in assets, an increase in liabilities, or both. (Table IV, Chap. XXIII.)

(b) - In relation to expenditures. Seventeen of the twenty-five townships which appropriated in excess of revenue receipts also appropriated in excess of total expenditures. Of these only six* could be considered as having come within a reasonable

---

* This may be seen by referring to the last column which indicates the approximate accuracy of the budget estimates.
    The six townships are: Newport (4.4%), Scott (5.6%), Leet (5.7%), Plymouth (.4%), Annville (.2%), and West Lebanon (6.0%).

degree of accuracy in estimating such appropriations. However, this does not mean that the budgeting authorities made a proper adjustment between appropriations and expenditures. The adjustment was brought about in Leet by temporary loans which were not paid, indicating that the appropriations exceeded the revenues available. The adjustment in Newport and Scott was effected by large amounts of bills payable, also indicating an excessive appropriation. It is significant that of these seventeen townships, thirteen appropriated in excess of both revenue and total receipts.

Summarizing the tendency to over-appropriate it may be noted that: (a) Twenty five of the forty townships appropriated in excess of revenue receipts. (b) Fifteen appropriated in excess of total receipts. (c) Seventeen appropriated in excess of total expenditures. (d) Fourteen appropriated amounts which exceeded total receipts and total expenditures. (e) A close correlation between appropriations and expenditures was found to exist in only six townships. But this was not the result of good budgeting in most cases since the amounts appropriated and expended exceeded total receipts in five of these townships thereby necessitating a decrease in cash balances, an increase in liabilities, or both. The strong tendency toward excessive appropriation is accompanied by: (a) The necessity of borrowing. (b) The failure to meet short term notes out of revenues. (c) A decrease in cash balances. (d) An increase in liabilities in the form of bills payable.

Under-appropriation. Twenty three of the forty townships show a distinct tendency to under-appropriate. All but four appropriated more than their governmental expenditures but less than their total expenditures. However only four, namely Coal, Mt. Lebanon, Whitehall and Darby can be considered as having established a reasonable balance between appropriations and expenditures. In the latter two this was only achieved by temporary loans which remained unpaid at the close of the year. Coal township succeeded only by means of the accumulation of a large item of bills payable, while in Mt. Lebanon a heavy decrease in cash balance and a large accumulation of bills payable took place. The tendency to under-appropriate is caused primarily by the practice of using the prior year appropriation as a basis. The appropriations for 1933 and 1934 vary only slightly in East Deer, Swatara, Shaler, Reserve, and Tinicum. It is significant that most of these townships repaid all temporary loans negotiated and that four of them paid off more than they borrowed.

Adjustment to available data. Generally little or no adjustment is made between the prior year expenditures and the appropriations for the coming year. Those townships which tend to over-appropriate such as Penn, Stowe, Elizabeth, Spring Garden, Jefferson, Lawrence Park and Leet continue to over-appropriate in 1934 in face of the lower expenditures in 1933. Those which show the opposite tendency such as Upper Darby, Cheltenham, Swatara, East Deer and Reserve continue to make lower appropriations even though in 1933 the expenditures had greatly exceeded appropriations.

A few illustrations will indicate the failure to make adjustments. The Stowe township commissioners appropriated $102,176 in excess of actual requirements in 1933. The 1934 appropriation was seventy-seven thousand dollars more than that of 1933! The 1933 appropriation in Jefferson township exceeded expenditures by $9,704. The 1934 figure is then decreased by one thousand dollars and placed at twenty thousand dollars although the total receipts for 1933 were only $12,923! Leet offers another example. It appropriated $9,161 for the year 1933, which sum exceeded the current and total receipts of the township. Despite the fact that the revenue receipts for 1933 were only $5,173, that the year had closed with bills payable totalling $2,332, and that $1,100 were still outstanding on a temporary loan negotiated in that year - with all these considerations before them, the Leet commissioners increased the 1934 appropriation to $9,971!

A comparison of the 1934 appropriations with the revenue receipts, total receipts, and total expenditures for the year 1933 will further show the tendency toward over-appropriation and the failure to base the appropriation upon that of the previous year. Of the forty townships, seventeen appropriated in excess of total receipts and total expenditures of the year 1933! There can be no justification for such procedure. This was done in the face of high delinquency in tax collection, an unfavorable situation in outstanding loans and bills payable, and the depletion of cash balances. Three units, Baldwin, Mt. Lebanon and Penn, appropriated in excess of total receipts for 1933, though not in excess of total 1933 expenditures. The significance of this is that the excess of expenditures over receipts in 1933 had led to severe decreases in the cash balances, and in two units such decreases had been accompanied by the cumulation of a large item of bills payable. Yet in the face of these facts, the commissioners appropriated an amount in excess of total receipts for 1933. While in Baldwin there was a substantial decrease in the 1934 appropriation, yet an amount in excess of the prior year's receipts during a period of large tax delinquency seems unjustifiable.

Finally, seven townships appropriated sums in excess of their 1933 revenue receipts. This policy may not be bad where the township is free of encumbrances (Lower Merion for example), and where the percentage of tax delinquency may be expected to remain stationary. However, this is not true of all these units. Thus Darby ended the year with outstanding temporary loans amounting to approximately $7,000. In spite of this and with revenue receipts totalling only $10,000 the commissioners appropriated over $13,000 for 1934.

Summary. A careful scrutiny of the comparisons afforded by this table indicates that there is very little of what might be termed adequate budgeting among these townships. While it is true that the strained financial conditions of the period, and the growing urban character of many of these units make it difficult to establish and maintain a sound budget, yet the failure of the commissioners to take into account the fiscal data of prior years, when making their budgets, indicates the absence of any effort on their part to establish sound budgeting practices.

(4) The Budget Ordinance: The code provides that the commissioners shall estimate the revenues for the coming year and by ordinance appropriate such revenues to the specific requirements of the township government within sixty days after the first Monday in January. Section 1501 requires that such appropriation ordinances be published "in one newspaper of general circulation in the township" and that the "ordinance shall not become effective until ten days after the publication aforesaid." In addition, sections 1702 and 1703 stipulate that such appropriations are not to be exceeded and "no work shall be hired to be done, no materials purchased, no contracts made and no orders issued for the payment of any moneys" which would cause such appropriations to be exceeded. Where this provision is violated or where "contracts, hirings, or purchasing made, or orders or warrants issued, (are) not provided for by an appropriation by the township commissioners as is required by law" such contracts, hirings, purchases made, etc., are invalid unless the commissioners are ordered to make payment therefor by either the court of quarter sessions or common pleas.

The budget ordinance constitutes the sole budget document in most of the townships. In many, the statutory requirements of enactment within the stipulated sixty day period and of publication are ignored. The following units, among others, failed to enact their 1934 budget ordinances within the sixty day period: Baldwin, Braddock, Scott, Lower Chichester, West Lebanon, Radnor, Crescent, North Versailles, and Neville. The following townships among others, do not usually publish their budget ordinances: Springdale, South Versailles, Braddock, South Fayette, and West Pottsgrove.

The code makes no specific provision for a budget hearing. In the case of Hiscox v. Township of Hanover et al. (Luzerne County, In Equity No. 6, July term, 1934) the court held that sections 1701 and 1502, "predicate the powers of the commissioners to levy a tax, within the limit fixed by law, upon the prior enactment of an ordinance appropriating the specific sums required out of the revenues available and the enactment of this ordinance became a condition precedent to the levy" and it was further stated that "through the medium of publication the taxpayers may know of proposed appropriations and may have opportunity to appeal to the court for redress against such appropriations as are unlawful or actuated by motives other than the public welfare."

According to the opinion of the court the budget ordinance must be enacted and published at least ten days prior to the tax levying ordinance. This opinion is based upon section 1502 requiring that all ordinances must be published and that they do not become effective until ten days after such publication, and upon the interpretation of section 1701 in such manner as to provide a period of time for the local taxpayers to express themselves upon such budget ordinance. The interpretation of the court has the effect of providing a partial substitute for a budget hearing.

Many of the townships are further violating the code in enacting and publishing their budget and tax-levying ordinances simultaneously. The following townships, among others, published the 1933 ordinances simultaneously: Spring Garden, Penn, Leet, Whitehall, Hanover, Upper Darby, East Deer, Ross, Springfield (Del. County), and Ridley. The following townships, among others, did likewise in 1934: Coal, Penn, West Lebanon, Annville, Leet, Whitehall, Newport, Hanover, Upper Darby, Elizabeth, East Deer, Crescent, Mifflin, Neville, North Versailles, Ross, Darby, Springfield (Del. County), Lawrence Park, Plains, and Ridley.

There can be little opportunity for the participation of the local taxpaying public in the budgeting process where the budget ordinance is not published. Even where it is, there is little public participation for the following reasons: the tendency of the commissioners either to enact such ordinances at meetings from which the public are excluded or at meetings the date and place of which are unknown to the public; the tendency to reach an agreement upon such documents prior to the public meeting and to force through the ordinance so agreed upon with little discussion; and the absence of any specific and mandatory provision in the code calling for such budget hearing meetings. Consequently, the taxpayers are compelled to resort to the courts in order to gain an effective hearing. Such appeals are costly and complicated, requiring the services of legal counsel. The result is that the local taxpayers have practically no voice in the budgeting process except in those restricted localities where, due to peculiar conditions, strong taxpayer associations are in operation and constantly checking the local fiscal operations of these units. For these reasons the opinion of the court, in the case just cited, provides a sound interpretation upon the enactment of such ordinances since it affords the public an opportunity of studying the appropriations and of endeavouring to make their wishes effective by attacking the tax levying ordinance. This, however, is a poor substitute for a mandatory budget hearing. The present statutory provisions should be so revised as to require the commissioners to hold such a hearing, at which the public would be invited to express its views, following the adoption of a tentative budget and prior to its enactment in final form. This would aid in the establishment of an adequate budgetary control. It would also obviate the numerous budget appeals that are continually being filed in equity by the larger taxpayers.

The code makes no provision as to the particular form in which appropriations shall be made other than to state that the commissioners shall by ordinance appropriate "the specific sums required". For this and other reasons the budget ordinances display the same lack of uniformity that characterizes the local accounts. The forms utilized may be broadly classified as follows: (a) Those which appropriate a lump sum for each general township function such as highways, health, administration, and sewers, without any further description of purpose. This is the most common form. These lump sum appropriations do not meet the code provision for the appropriation of "specific sums". (b) Those which appropriate a lump sum for each general township function and in addition thereto describe in a loose, general way the purposes for which each lump sum appropriation is to be expended. (c) Those which, in appropriating a lump sum for each general function, allocate such sum to specific purposes such as personnel service, supplies or materials, salaries, and stationery. This form fully meets the requirement of the code and provides the public with a more complete picture of proposed fiscal operations. The budget ordinance form utilized by Upper Darby is the best of this type. Few of the townships follow this practice though the following do so to some extent: Spring Garden, Mifflin, Penn, and Ridley. (d) Those which, while coming within one of the above classifications, also make appropriations to the individual wards within the township. The outstanding illustration of this type is Whitehall township which appropriates to each of the five wards sums for police, road, fire, and other purposes.

A small number of townships do not publish their entire budget document but only a summary thereof. Among such units may be included Lower Merion, Shaler, Haverford,

Plains, Hanover, and Mt. Lebanon. The reason advanced for this practice is that the cost of such publication would be prohibitive since the detailed appropriations generally cover several typewritten sheets. While this is a sound argument it must be remembered that the purpose of the code in requiring the publication of such ordinances is to provide the local taxpayers with information upon proposed fiscal operations. There appears to be no sound reason why an acceptable compromise cannot be reached by some such publication as that of Upper Darby which in addition to making lump sum appropriations for each general function also allocates such sums specifically to each of the following items: personnel service, services other than personnel, supplies, materials, equipment, and fixed charges.

The budget ordinances include, with few exceptions, one or both of the following: (a) An appropriation variously known as "miscellaneous", "general", "general fund", "contingent" or "contingent fund", and the like. The following townships, among others, include such items in their ordinances: Abington, Cheltenham, Plymouth, Rochester, North Versailles, Spring Garden, Whitehall, Ridley, Springfield (Montgomery County), Darby, Swatara, and Elizabeth. (b) A phrase such as "and all other incidental expenses", or "and incidental and administrative expenses not contained in other appropriations", following the stated purposes of a general appropriation to such functions as highways and health. Lawrence Park, Jefferson, Neville, Baldwin, and Mt. Lebanon townships afford illustrations.

Such vague titles furnish very little information for the taxpayers and provide the commissioners with full discretion as to expenditures even after a so-called budget has been adopted. There can be no real budgetary control where these practices are prevalent. This is especially true where appropriations are made to two or more such vague purposes as "general fund", "miscellaneous", or "contingencies" as in Spring Garden and several other townships. Nor can there be such control where each appropriation includes a phrase authorizing expenditures for any and all purposes "not contained in other appropriations".

The State Highway Department has attempted to eliminate these defects in the budget ordinance forms by the creation of a model budget ordinance which it has succeeded in getting many of the townships of the second class to adopt. No attempt however has been made to establish such uniform ordinance for the first class units and consequently each unit has adopted its own form. Only two of the first class townships have attempted to adopt budget forms that would conform with the state highway report. In one of these townships, Rochester, the form adopted follows the state highway form in detail while in the other, Elizabeth, the form merely groups the items of the state highway report into five lump sum appropriations.

(5) Budgetary control: The primary purpose in the adoption of a budget has been described. The effect of making unrestricted appropriations may best be shown by a few brief illustrations. The Abington Township appropriation ordinance includes fourteen items one of which is entitled "miscellaneous". This item amounts to $20,000 and $20,275 in the 1933 and 1934 ordinances respectively and in both cases is the fifth largest item in the fourteen. The Rochester ordinance appropriates $4,655 to the "general fund". This item is the largest of the twelve appropriations although the others include such items as the compensation of the treasurer, secretary, and commissioners. The commissioners of Springfield Township (Montgomery County) appropriated the sum of $3,379.53 in 1933 "for all other expenditures".

In other townships an appropriation to highways, for example, will include an item labelled "miscellaneous". Thus, the Baldwin budget ordinance for 1933 appropriates $12,247.72 to highways and allocates $5,500 of such sum to "miscellaneous expenditures not otherwise provided for". One might wonder how these commissioners were able to estimate highway needs so closely as $12,247.72 and yet were compelled to allocate such an extremely large portion of the estimate to "miscellaneous".

The importance of all these practices is in their bearing upon budgetary control. The usual method of maintaining such control is that of setting up the appropriation figure in the books, where such books are maintained, and then placing all items of expenditure under one or the other of the various appropriation figures, as described in the chapter on Accounting. The secretary then adds at monthly or irregular intervals the various sums expended under each appropriation, subtracting such sums expended from the latter figure to show the unexpended appropriation. In most townships the unexpended appropriation balances are shown each month and presented at the monthly meetings of the commissioners by the secretary and, in a few, such monthly statement of the secretary not only shows the unexpended appropriation balances but also the balances of revenue or funds available over actual expenditures of the prior month. Among the townships in which monthly unexpended appropriation balances are shown may be included Baldwin, Wilkins, Elizabeth, Spring Garden, and Whitehall. Among those townships where balances are shown between monthly receipts and expenditures may be included Mt. Lebanon, Shaler, and Spring Garden.

When the commissioners appropriate to such accounts or subject as "contingencies", "general fund", "reserve fund", "miscellaneous", "all other expenditures" and the like-what sums expended shall be charged by the secretary against such appropriations? Obviously any expenditures that he deems fit to so charge or such sums as he is directed to so charge by the commissioners. What is to prevent the secretary from including expenditures for fire, police, highways, and the like, even where appropriations are made for these matters, as credit items in the "contingent" or other general appropriation? Nothing. He may enter any kind of an expenditure against such general appropriation. Furthermore, such sums are so charged against these general appropriations in actual practice.

There is but one conclusion from these facts - the present budget ordinances and documents do not present a complete and definite plan by which the fiscal operations of the township may be directed. When the plan permits any type of expenditure to be charged against a general appropriation there cannot be said to be a plan.

To what extent is the budget plan adhered to? The natural reply would seem to be that, with all the flexibility which the present budgets generally provide, the commissioners should experience little difficulty in meeting this condition. Yet such is not the case. In spite of statutory provision above cited, expenditures above appropriations are frequent.

The following townships among others exceeded one or more appropriations in the fiscal year 1933: Whitehall, Penn, Hanover, Newport, Plains, Coal, and Mt. Lebanon. Because of the difficulty of correlating the annual audit statement on expenditures with the budget appropriation, it is impossible in most cases to determine whether or not appropriations have been exceeded. Many of the officials in the above townships maintained that their budgets had not been exceeded upon the ground that the annual audit statement or the secretary's book showed unexpended balances for each appropriation. The following typical illustrations, taken from Newport in 1933, describe the nature of these unexpended appropriation balances:

Function	Appropriation	Expenditure	Unpaid bills	Amount of excess
Lighting	$ 8,600	$ 2,926.52	$ 5,597.58	----------
Health	5,020	4,339.70	1,090.20	409.90
Water	1,600	1,206.53	778.64	385.17
Fire	16,830	5,153.67	4,044.65	----------
Police	12,430	9,922.97	2,847.52	340.59
Road	16,745	17,907.24	4,164.34	5,326.58
(Coal Twp.)				
Road	53,000	41,215.57	14,769.75	10,685.32
Fire	18,000	8,730.07	14,277.90	5,907.97

The officials point to the fact that expenditures are within the appropriations ignoring the unpaid bills created and also ignoring section 1702, above quoted, which prohibits the incurring of any obligations which would have the effect of causing appropriations to be exceeded. The actual state of affairs is worse than the above figures indicate for in almost every case the expenditures listed are only the ones charged against the appropriation stated and do not include expenditures made for these purposes and charged against the general fund. Thus in the case of Coal Township the sum $4,833 (fire expenditures charged against other appropriations) must be added to $8,730.07 which would make the amount by which the fire appropriation was exceeded even greater than here indicated. Many other townships follow the same practice.

Many likewise attempt to give the appearance of operating within their budget by making payments from a general appropriation. It is surprising that this is done when section 1702 expressly permits amending ordinances, with the stipulation that such ordinance "shall set out the reasons for and the character of such change". Few townships enact such ordinances, although many of them enact transferring ordinances or resolutions at the end of the fiscal year. Some inkling of the accuracy of budget estimation is given by the amount of the change made in the original budget by such transfer ordinances. In Baldwin township the sum of $7,900 was appropriated in 1933 "for the payment of principal, interest, and state taxes of the bonded indebtedness of the Township". Despite the supposedly fixed nature of such expenditure, the amending ordinance adopted near the close of the fiscal year increased this sum to $12,900.

The practice of making specific appropriations for each ward in the township has shown poor results. In most cases the sums so appropriated are solely for road purposes although in at least one township, Whitehall, the appropriation is for such matters as fire and police protection as well. The amount appropriated is based upon the miles of township highway in each ward and this sum is spent at the sole discretion of the commissioner from that ward. A few illustrations will show the effect upon budgetary control in the townships which adhere to this practice.

Township	Ward	Appropriation	Expended	Balance	Date
Whitehall (1934)	No.1	$ 3,248.17	$ 3,248.17	none	July 6, 1934
	No.2	1,158.01	1,158.01	none	" " "
Elizabeth (1933)	No.1	687.50	687.50	none	Dec. 31, 1933
	No.2	632.50	632.32	.18	" " "
	No.3	687.50	687.48	.02	" " "
	No.4	770.00	770.33	-.33	" " "
	No.5	797.50	797.41	.09	" " "
	No.6	770.00	770.00	none	" " "
	No.7	687.50	687.15	.35	" " "
	No.8	687.50	687.45	.05	" " "
	No.9	715.00	716.33	-1.33	" " "
S. Fayette (1933)	No.1	788.00	714.98	74.02	" " "
	No.2	1,080.00	1,085.05	-5.05	" " "
	No.3	630.00	784.34	-154.34	" " "
	No.4	1,170.00	1,112.95	57.05	" " "
	No.5	1,710.00	1,795.42	-85.42	" " "
	No.6	1,012.50	855.12	157.38	" " "
	No.7	461.50	554.08	-92.58	" " "
Penn (1933)	No.1	1,100.00	1,099.76	.24	" " "
	No.5	1,600.00	2,001.78	-401.78	" " "
	No.8	1,000.00	1,864.31	-864.31	" " "
Penn (1934)	No.4	2,500.00	1,589.64	910.36	July 1, 1934
	No.5	1,500.00	1,362.79	137.21	" " "
	No.6	1,900.00	1,172.15	727.85	" " "
	No.7	275.00	229.00	46.00	" " "
	No.9	1,900.00	1,401.97	498.03	" " "
	No.10	750.00	679.86	70.14	" " "
	No.11	1,550.00	1,060.35	489.65	" " "

The above figures indicate three tendencies where the ward system of appropriation is in effect: (a) for the commissioners to spend the entire sum appropriated; (b) to

do this in the first portion of the fiscal year instead of planning their outlays; (c)·
to exceed their appropriations.

That it is impossible to maintain budgetary control when a specified sum is set
aside for the sole use of a commissioner at his own discretion is attested to not only
by the data presented but also by the universal testimony of officials in these townships. Several stated, however, that they saw no way of eliminating the system because
of the fear of various portions of the township that they would not receive their fair
portion of expenditures and improvements under any other system.

Finally, the problem of budgetary control is further complicated by the lack of relationship between the tax collection and the fiscal year. The commissioners are instructed to enact their budget within sixty days of the first Monday in January. This
means that the budget ordinance should be passed and effective by the early days of
March. In most townships the local fiscal year begins on January first, or within a
few days thereafter. This means that where the budget is not adopted until March first
or thereafter, under the practice in most of the townships as above shown, the local
officials have been operating for three or more months without a budget. Exceptions are
seen in the better managed townships like Lower Merion and Mt. Lebanon. Such a situation does not improve budgetary control and the code should unquestionably be so revised
as to require the adoption and enactment of the budget prior to the beginning of the
fiscal year instead of some sixty days thereafter.

Summary - Our examination of both the statutory provisions and the existing practices
in budgeting discloses a wide-spread absence of scientific or adequate procedure either
in making up the budget or in exercising a budgetary control. The problem is therefore
not so much that of strengthening a defective procedure as of creating a new one.

CONCLUSIONS

1. FINDING - At present the budget forms used by first class townships are not uniform
nor are they sufficiently complete to make comparisons between different townships.
Since the forms do not itemize expenditures according to township function it is impossible to ascertain with reasonable accuracy the township functional costs.
RECOMMENDATION - That in view of the urgent need for comparative data on functional
costs, a uniform budgetary procedure and set of forms be adopted for first class townships.

2. FINDING - The weaknesses disclosed in accounting systems and in budget procedure
are in part due to the part-time character and lack of training of the budgeting personnel.
RECOMMENDATION - That administrative supervision and guidance coupled with assistance
in supplying the requisite handbooks of instruction and forms be vested in a state a-
gency. That in view of occasional past failures of state agencies to exercise their
powers and duties, an alternative state authority be also designated.

FINDING - The statutory requirements that the budget be enacted within a sixty day period, that it contain specific appropriations, that it be published and that no appropriations shall be made in excess of the budget ordinance have all been ignored on so
many occasions and in so many townships, without remedial action having been taken,
that the value of these legislative provisions has all but disappeared.

RECOMMENDATION - In view of this general disregard of the statutory mandates, more adequate penalty provisions should be provided by law and their enforcement should be entrusted to the state authority above recommended.

4. FINDING - The estimation of revenues and expenditures is so haphazard and inaccurate
that the local units are compelled to negotiate frequent temporary loans, to accumulate
large amounts of bills payable or to reduce their assets by drawing down their cash
balances. This is due to inaccurate estimation in part, also to the total lack of correlation between the fiscal year and the assessment and tax collection periods.
RECOMMENDATION - That the state agency above recommended, be authorized to examine and
report to the state legislature a new arrangement of the assessment and collection periods in relation to the fiscal year and that the same agency be empowered to prescribe

necessary rules of procedure for the estimation of revenues and expenditures.

5. FINDING - The most acceptable budgetary practices were found in those townships where the secretary or manager was made responsible for the preparation of a detailed budget document. These men were the most informed of all local officials upon local fiscal matters.

RECOMMENDATION - That in view of the undoubted superiority of these officials, the responsibility for the preparation of a detailed budget document and its submission to the commissioners by the secretary be vested by statute in the office of secretary or manager.

6. FINDING - Budgets are not adopted until several months after the beginning of the fiscal year. This defeats the objectives of budgeting and of budgetary control.

RECOMMENDATION - That the enactment of a budget for the next succeeding fiscal year be required, by statute, prior to the beginning of such fiscal year.

7. FINDING - The public is not given sufficient opportunity to participate in the process of budget making. The policy of having taxpayers appeal to courts upon budgetary matters is undesirable as placing too great a burden upon the taxpayer and as subjecting judicial process to political influences.

RECOMMENDATION - That a mandatory budget hearing, for the benefit and participation of the public be prescribed, to be held after the adoption of a tentative budget by the commissioners and prior to its enactment in final form by the commissioners.

8. FINDING - A number of local boards amend their budgets by means of unpublished resolutions. This practice is undesirable in that it permits the board secretly to destroy the essence of the budget.

RECOMMENDATION - That all changes in the budget be ineffective until ten days after their publication in ordinance form or until ten days after the filing of the same with the state authority.

9. FINDING - The forms and procedures of budgeting, accounting, auditing, and reporting are closely interrelated. Where defectives exist in one or more of these practices they invariably affect the others. Good budgeting depends upon the completeness and accuracy of data available in the accounts. Good accounting is dependent upon the form of budgeting.

RECOMMENDATION - That in view of this close correlation, the state authority designated with the responsibility for instituting and maintaining a uniform budgeting process be likewise designated as to accounting, auditing, and reporting; and that to this end all local units be compelled to file their budgets as well as annual audit statements and reports with such state agency within fifteen days of the enactment or certification thereof.

CHAPTER XXI

AUDITING PROCEDURE

We shall consider briefly the following features of township auditing:

(1) The auditing authority
(2) The elective auditing personnel
(3) Auditing practices under the elective system
(4) Publication and filing of audit statements and reports
(5) Audit forms.

(1) The Auditing Authority - Who shall audit local accounts and fiscal operations? The provisions of the code invest the commissioners with the option of having this function performed either by: (a) Three elective auditors, each of whom serves for a six year term, one being elected every two years, (b) An auditor appointed by the county court of common pleas to make the annual audit, or (c), a controller, elected every four years. The general practice is to have the auditing performed by three elected auditors. Fifty-four of the sixty-one first class townships employed the latter system in 1933, while only two employed the controller, and only three the court appointed auditor.

An examination of the audits made by controllers and court appointed auditors shows that these audits are generally superior to those made by the elected auditors. They are superior in that: (a) Such audits are more complete, more detailed, and better compiled. They present such matters as bills payable at the close of the fiscal year, analyses of functional costs, debt service, outstanding debt, assets and liabilities, and the like. (b) They contain recommendations to the commissioners concerning inefficient bookkeeping practices, illegal transactions and those of doubtful legality, and warnings of the dangers inherent in specified practices. Thus, the Haverford township audit (1933) pointed out that the code provided, section 1801, that all contracts or purchases in excess of five hundred dollars should be based upon an advertisement for bids and that the commissioners had, in one instance, made such a purchase without complying with this provision. It was also suggested that certain minor changes be made in bookkeeping methods and that the commissioners take steps, in lieu of the small amount of fines turned over to the township by the local justices of peace, to have such revenues turned into the township treasury. The auditor of Hanover township in addition to calling the attention of the commissioners to this latter problem, in his 1933 audit, also described the tendency of the board to expend moneys in excess of appropriations, indicating that this was a violation of sections 1702-1703 of the code. (c) These audits are also superior in that they include a consolidated statement of actual rather than nominal revenues and expenditures.

The superiority of the court appointed auditor would seem to be due to two facts: (a) In each case where such an auditor has been appointed the court has selected either a certified public accountant or an accounting firm. (b) Such auditors have made it a point to acquaint themselves with the financial provisions of the First Class Township Code and have not hesitated to set forth the discrepancies between actual practices and the legal stipulations. In addition their training and experience has provided the basis for suggestions as to improvements in the keeping of township accounts.

In view of these facts it is surprising that only two of the twenty-four townships visited have provided for this type of annual audit. Inquiry as to the reasons why their units had not given this method a trial brought forth the following arguments from local officials: (a) The system is too costly. This was the general and primary argument advanced by commissioners and other local officials. They pointed out that according to section 520 of the code, the court is to appoint an auditor "who shall be either a person or firm licensed as a certified public accountant, or a person skilled in auditing work", and that while it was an advantage to employ persons trained in accounting yet such persons demanded more remuneration than the township was compelled to pay for the audit under the elective system. Furthermore, as these officials stated, this same section provides that "the compensation of any such appointed auditor shall be fixed by the court "which subjects the commissioners to the payment of an uncertain sum.

(b) A second objection to a court-appointed auditor was that such appointments would tend to be political. No evidence was advanced upon this issue though invariably reference would be made to the results of this system some years ago when all local audits were made by court appointed auditors. Obviously this argument is not applicable to the employment of certified accountants by authority of the commissioners alone. Officials were generally agreed that if the commissioners were vested with such authority the employment of an accountant would be most desirable providing that the costs of such audits could be placed within the financial means of the unit.

(c) A few officials expressed themselves as opposed to any audit by what they termed an "outsider". These men were generally opposed to court appointment on the ground that the court might as well select someone not resident in the township to make the annual audit. They also objected to the employment of an accountant by the commissioners on the ground that in many cases it would be difficult, if not impossible, to obtain such person within the limits of the township. The general opposition to an "outsider" was placed on a somewhat vague basis, the intimation being that only a resident of the township would be acquainted with the local situation and competent to deal with local accounts and also that the local taxpayers' money should be retained within the community.

(d) Finally, many opposed any system of auditing other than that by three elected auditors upon purely political grounds. As one president of the board of commissioners stated, "the elective auditor system is a source of political strength, it provides three jobs with little work and additional remuneration". Politically and publicly he stated that he was opposed to any other system. Privately he had no hesitation in stating that from an "efficiency" standpoint the auditing of the local accounts by a controller, a court-appointed auditor, or an outside certified public account hired by the commissioners were all preferable to the three elective auditors.

(2) Auditing Personnel under the elective system - What is the type of auditing personnel obtained under the elective system? In twenty townships where information relating to the occupations of the sixty auditors, responsible for local auditing, was obtained, the following occupations were included: Clerks- store, railway, unemployed, factory, and ledger clerks were among the types described - (12); Coal miners- (6) Banking - teller, cashier and messenger were among those described - (7); Auditors for private firms - (3); Chemists - (3); Mill foreman (3); Carpenters - (2); Painters - (2); Printers - (2); Housewife - (2); Laborer - (2); Accountants - (2); and one each of the following occupations, - retail baker, shipper, manager of specialty shoppe, farmer, bricklayer, cost-estimator, bookkeeper, boilermaker, manufacturer, fruit grower, salesmanager, treasurer of private concern, and trucker.

In these same units, the length of service of these sixty men is indicated by the fact that: eleven were serving their first year of a six year term, six were serving their second year, seven their third year, six their fourth year, four their fifth, four their sixth, one the seventh, five their eighth, three their tenth, and one the fifteenth year. In several other cases it was impossible to obtain an accurate statement of the number of years in service but seven were serving their first term, two were serving a second term, and three were in their third term of office. Otherwise stated this means that out of the sixty elective auditors, seventy-five percent of them were serving their first terms, eighteen and two-thirds percent were serving their second terms, and, six and one-third percent of them only were serving for a third term. These facts indicate a rather high rate of turnover among these officials though the undesirable effects usually attended by such turnover are somewhat counteracted by the employment of the stagger system whereby one auditor is elected every two years for a six year term. The length of the term itself, six years, should provide ample opportunity for the elected auditor to acquaint himself with his duties in time to render efficient service to the township within his first term alone.

The present auditing personnel, under the elective system, is not capable of rendering an adequate auditing service. This conclusion arises from a study of the present auditing practices, hereinafter described, from contact with the auditors themselves, and from the general opinions expressed by other local officials. It is further attested by the fact that their occupations indicate that only fifty percent of their number have had any prior accounting, bookkeeping or auditing experience. It is significant that those who have had some prior bookkeeping experience, are distributed among only twelve of the twenty units, the books of the remaining townships being audited by men who, as far as their occupations indicate, are unqualified to perform such duty. Whether a simplified accounting system combined with state supervision would permit of an acceptable audit by such men is a question that only actual practice could verify. The annual auditing of township accounts by trained accountants is the only solution that can be unreservedly recommended.

(3) Auditing practices under the elective system - What procedure is followed in the auditing of township accounts? Relevant portions of the code provide as follows:

"The auditors of townships shall meet annually, on the day following the day (first Monday of January) which is fixed by this act for the organization of the township commissioners, and shall audit, settle, and adjust the accounts of the township commissioners, township treasurer, tax collector, secretary, and other officers and persons receiving and disbursing or authorizing the disbursement of the moneys of the township during the preceeding fiscal year" (Section 1001).

"The auditors shall cancel all orders and vouchers presented to them, which they find have been paid, by writing the word "audited" on the face thereof" (Section 1004).

"The accounts of the officer or officers in question may be investigated de novo. the figures and facts found and stated by the auditors in their report of audit shall be taken as prima facie correct, as against any such officer, and the burden shall be upon each officer whose accounts are in question to establish the validity of the credits which he claims" (Section 1015).

"Two auditors shall constitute a quorum" (Section 1001).

In addition to these procedural stipulations the auditors are given certain specified powers such as to subpoena records, to compel attendance, to administer oaths, to employ an attorney, to surcharge local officials for balances remaining against them, and to enforce the collection of the same by filing and prosecuting actions in the courts.

It is extremely difficult to obtain any detailed statement as to the auditing procedure. While much difficulty was encountered in securing information on exact details of auditing prodecure, the following features deserve comment: (a) The burden of actual auditing usually falls upon one or two of the three auditors; (b) The auditors place a heavy reliance upon the secretary and treasurer in making their audit; (c) They rarely exercise the powers, cited above, which are granted to them by the code; (d) They exercise no controller powers, that is they do not go beyond a specific bill to investigate such matters as price, quantity, and quality; and (e), they do not attempt to enforce such portions of the code as sections 1702-1703 which stipulate that all payments in excess of appropriations are to be considered void. Nor do they even go so far, generally, as to point out such matters and recommend some form of action to the commissioners.

The auditors rarely exercise their power of surcharging township officers for illegal financial acts. Such surcharging would quickly put an end to such measures as the payment of warrants in excess of appropriations, or for purposes for which no appropriation has been made, or payments made by virtue of an account described as an "unappropriated fund," or payments of warrants out of improper accounts as where an appropriation is made for such matters as highways, health, and the like and a temporary loan from the sinking fund totalling $500 is repaid by taking one hundred dollars from each of several funds, as was the case found in one township.

It is significant that the rare instances of surcharging and appeals from the annual audit statement are usually found in those units and regions where a strong taxpayers' association or civic organization exists and carefully scrutinizes local administration.

The auditors generally do not go beyond the specific bills for which payment has been made in the auditing of the accounts. In short verification of the mere existence of a bill of specified amount, a voucher for such payment, and a proper entry in the ledger is held a sufficient performance of their duty, unless of course, a lack of balance is shown in these items. No attempt is made to verify the bill or to investigate the validity of the charge.

Whether the present systems of auditing are retained or not, it would be highly desirable that the legislature clearly and in unequivocal language empower the auditing authority to exercise a restricted power of passing upon the excessive amount and the propriety of bills. Such a power might be limited to investigation of those which the auditor or auditors hold to be unreasonable upon their face. An unlimited authority over this matter would not be necessary or advisable.

The auditors rarely make any recommendations to the commissioners as to such matters as illegal practices, bad bookkeeping methods and the like. Not only would such recommendations and warnings be an aid to efficient budgetary control but they could be the means of securing and maintaining better accounting methods. Where the auditing is performed by an accountant appointed by the court such recommendations and warnings are placed before the commissioners. No instance was found of a similar practice among the elective auditors except in the rare instances where one of the elective auditors was versed in methods of accounting.

Why do auditors fail to exercise these powers? The reasons are varied:

(a) Ignorance of the law upon the part of the auditors.
(b) The difficulty which these officials have in understanding and interpreting the law relating to their office.
(c) The lack of knowledge and general lethargy of many such officials. The latter of these causes can only be eliminated by the elimination of personnel of this type. However, it is felt that the former causes are in their nature remediable. No sound reason exists for placing the burden upon these officials to obtain copies of the law relating to their office and to interpret the same. An idealist might argue that no one should assume an office without acquainting himself with its duties. In a practical world, however, it is too much to ask that a man otherwise occupied, obtaining small remuneration, active officially only for a short time each year, lacking knowledge as to possible sources of information, and not versed in legal phraseology, should acquaint himself with the legal basis of his office and make interpretations of the same. If the elective system is retained handbooks should be provided for the auditing personnel describing the nature, duties, and powers of the office in concise, simple language. They should likewise be informed that the state agency invested with general supervision over local financial administration is at all times available to them in an advisory capacity.

(4) <u>The publication and filing of audit statements or reports</u> - The relevant statutory provisions are as follows:

"The auditors shall complete their audit, settlement and adjustment within as short a time as possible.....They shall, within ten days thereafter, publish, by advertisement in at least one newspaper of general circulation printed in the township or county, a concise itemized statement of the receipts and expenditures of the several officers for the preceeding fiscal year, they shall also, within ten days thereafter, file a copy of such statement with the secretary of the township and another copy of such statement with the clerk of the court of quarter sessions" (Section 1003).

Both the letter and the spirit of this section are generally ignored.

Audits are not made within the time limit set by law. An even greater laxity exists in the filing of audit statement with the clerk of the quarter sessions court. This filing is of utmost importance since it is the only permanent record maintained of such statements for most units, as local papers and records are continually being lost,

destroyed either wilfully for political reasons or by fire, and discarded for want of
space. Three townships publish neither an audit statement or even a summary thereof.

In Allegheny county the following townships had not as yet filed their 1933 audit
on August 20th, 1934: East Deer (last report to be filed was that of 1930 which was
filed on March 16, 1931), Neville, Penn, Ross, Scott, and Springdale. In Delaware county the following townships had still not filed their 1933 audit reports on June 25, 1934:
Upper Darby, Springfield, Tinicum, Lower Chichester, Darby and Aston. Investigation further revealed that the annual audit statements of Annville and West Lebanon had not been
filed since 1925 and that Lackawanna auditors had not filed their annual report since
1928.

Township	Date Certified	Date Filed
Collier	Jan. 5, 1934	Feb. 7, 1934.
Jefferson	Mar. 7, 1934	Mar. 26, 1934.
Neville	Apr. 3, 1934	Not filed- Aug. 20, 1934.
Scott	Feb. 12, 1934	Not filed- Aug. 20, 1934.
Wilkins	Feb. 16, 1934	Mar. 31, 1934.

In the above group are listed those units whose audit statements for the year 1933
were certified as completed and approved by affidavit and the date of the receipt of
such audit statements by the clerk of courts. The auditors, in each case, failed to file
their report within the stipulated ten day period. The illustrations here set forth are
not exceptional, they are typical of a general disregard of the statutory mandate.

Section 1005 of the Act provides for the failure of any auditor to comply with the
above provisions "a penalty of one hundred dollars, to be recovered by suit, instituted
in the name of the township, upon the complaint of any taxpayer, in the same manner as
debts of like amount are recoverable." This penalty provision is ineffective since it
depends upon the investigation, knowledge and initative of a tax payer. What is needed
is an enforcing stipulation which will confer not only on the tax payers but on the
clerk of courts or some other interested party the same right of instituting an action,
with the further proviso that costs of such suit shall be defrayed either by the township or the auditors upon the conviction of the latter.

A similar laxity exists in regard to the publication of the annual audit statement.
The following townships do not publish their annual statements: Annville, West Lebanon,
Ross, South Versailles, Braddock, West Pottsgrove, and Wilkes-Barre. Although statements
were made by local officials in other townships to the effect that their annual audit report was published, an investigation failed to uncover such publication. The reason
generally advanced for failure to publish such reports is that it would involve an unnecessarily large printing charge. While many units do publish either an audit report
or a portion thereof, the publication is either so condensed or so defective that it can
by no stretch of the imagination be considered as fulfilling the requirement of the law
calling for the setting forth of itemized expenditures and receipts.

(5) Auditing forms - What purposes should an audit form or statement serve? Two
primary purposes may be listed, among others: (a) To provide a concise picture of the
fiscal operations of the specific unit during the fiscal year just closed and of the fiscal status of the township as of the date of the audit. (b) To place such concise fiscal summary before the local officials and local taxpayers in order that they may pass
judgment upon past operations and form judgments as to future fiscal policies. This
appears to be the intent of the statute, section 1003 above cited, which provides that
the auditors shall publish a "concise itemized statement of the receipts and expenditures
of the several officers for the preceeding fiscal year".

A distinction must first be made between the published audit statement and the one
actually compiled. Some townships publish the full and complete audit statement compiled by the auditors. Among these may be listed: Lower Merion, Nether Providence,
Swatara, Jefferson, Newport, Harrison, Spring Garden, Shaler, and Abington. Others pub-

lish only a summary of the audit statement. Among these may be listed: Haverford, Whitehall, Hanover, Baldwin, Penn, and Radnor.

The audit forms are adapted to the accounting systems of the units audited. Consequently the diversities of the accounting system are repeated in the audit forms. The defects of the former are displayed in the latter. Accordingly audit statements may be classified or grouped upon the basis of the accounting system. (1) Those which show merely the transactions of the general fund. Eleven out of forty audit forms are set up on this basis. Among these units may be listed: Crescent, Tinicum, Swatara, Hanover, and West Norriton. (2) Audit forms which set forth fiscal operations (receipts and expenditures) within two or more independent funds. Seven out of forty audit statements were so set up including the following units: East Deer, Newport, and Springdale. (3) Forms which show all fiscal operations upon the basis of a general fund and in addition thereto include one or more separate accounts of one of the general fund transactions such as debt service. Twelve audit statements were of this type including the following units among others: South Fayette, Coal, Elizabeth, Annville, Cheltenham, Nether Providence, and Leet. (4) Audit statements describing fiscal operations within two or more independent funds and in addition thereto including one or more separate accounts of a single transaction within one of the independent funds. The remaining nine units utilize this method. They include among others, Lower Merion, Radnor, Rochester, Harrison, and West Lebanon.

The audit statement may be further classified upon the basis of its treatment of expenditure items. (1) Audit forms in which expenditures are not only itemized but also classified according to function. Twenty-three out of forty units may generally be included in this group although there is no uniformity either in the classification of expenditure or in the extent to which expenditures are itemized. (2) Audit forms in which expenditures are itemized but not classified. Eleven townships come within this group. (3) Those in which the sums expended are classified by totals but not itemized. (4) Those which classify expenditures by total primarily but also set up itemized expenditures for one or more classes.

Difference of practice exists as to such important matters as the amount of bills payable and/or warrants outstanding at the close of the year, and assets and liabilities. In the case of bills payable a study of the annual statements of forty townships for the year 1933 revealed that only seven out of the group included the amount of bills payable, twelve included the amounts of warrants outstanding, and two included both the amounts of warrants and bills outstanding. Only twenty-one of the units, therefore, included any statement as to these matters while nineteen made no mention of either of them.

Twenty-seven of this same group of townships included a statement of assets and liabilities in some form in their audit statements while thirteen did not. Of these twenty-seven, however, one included a statement of assets only while two others merely included an inventory. Of the remaining twenty-four, five limited the statement of their assets to accounts receivable such as outstanding and liened taxes. Eleven audits included somewhat condensed statements of property values in their assets while only seven presented such property assets in sufficient detail. In no case did the audit include an item showing an estimate of depreciation of property value though this fact does not preclude an allowance for such amounts in the estimation.

The evidence points to the definite conclusion that the present audit statements of these units do not, except in rare cases, perform the function of presenting a concise picture of the fiscal operations during the previous year and of the fiscal status of the township as of the date of the audit. Do these audit statments, then, provide the local officials and the local taxpayers with sufficient understandable, informative data from which judgments may be formed as to future fiscal policy? The data presented are incomplete as already shown. They are compiled in such manner as to render an intelligent interpretation impossible without the aid of the township accounts and the explanations of the local officials. Aside from incompleteness the following difficulties confront the taxpayer and local official in understanding and interpreting these audit statements: (a) The practice of setting up such statements upon a basis of nominal receipts and/or expenditures. (b) The practice of setting up independent funds without accompaning the same with consolidated balance sheets. (c) The practice of making numerous transactions within funds. (d) The practice of including in such statements the number of each warrant, the payee thereof, and the purpose, without prividing any consolidated balance sheet.

Several townships make a definite attempt to inform the local taxpayers as to the fiscal status and operations of their government at the close of each fiscal year by the issuance of an annual report. The report of Lower Merion is the most elaborate and complete. It is of course not adapted to the smaller units with restricted operations and resources. Attention is called, however, to the Springfield report as being within the means of many of the first class townships and as being worthy of imitation with the modifications necessary to the peculiar conditions of each unit. The Springfield report is a small pamphlet of ten mimeographed pages containing, among other things, a consolidated balance sheet of receipts and expenditures, itemized expenditures for each specific function and for debt service, itemized statement of tax and other receipts, the annual reports of the building inspector, board of health, board of adjustment, and other administrative agencies, and a graphic statement of the utilization of the tax dollar and per capita costs for the main functions of government. The only criticism lies in the fact that several nominal transactions are included in the consolidated statement, and, therefore, the taxpayer is compelled to sift these transactions before he can obtain a picture of actual receipts and expenditures. The report could also be improved by the insertion of budget appropriations and more detail upon bonded indebtedness.

Summary - Present practices in the annual auditing of township accounts are characterized by the same lack of uniformity and adequacy that characterize local budgetary and accounting procedures. The primary purpose of an audit is to check the regularity of local accounts and fiscal operations and to inform the taxpayers of the township of its fiscal status at the end of each year. Obviously the limitations of auditing procedure are largely determined by the limitations of the accounting system. Where the latter segregates and allocates properly the audit may show such segregation and allocation. Where the accounts are defective, the audit will also be defective unless, as is not usually the case, the auditing personnel are especially versed and trained in accounting. The basic problem is one of establishing an adequate and uniform accounting system. The supplementary problem is one of obtaining auditing personnel capable of handling such accounts by experience and training. The latter problem should not be under-estimated but it is not too much to suggest that the establishment of an adequate and simplified accounting system would go far toward the solution of this difficulty. The present deficiencies of the local auditing procedures are not solely ascribable to the incompetency of the auditing personnel. They are inherent in the diversities and confusion now existent in the bookkeeping methods employed by these units. With much truth it can be said that no auditing procedure or form is better than the accounting system or forms which it presumes to audit.

CONCLUSIONS

(1) FINDING - The annual audit statements do not provide an adequate, readily understandable description of the fiscal operations of the township. Nor do they indicate the fiscal status of the unit at the close of the year. This is largely due to the incompleteness of the data presented, the inclusion of purely nominal items, and the complexity and frequency of inter-fund transactions.

RECOMMENDATION - That in view of the need for more complete and uniform audit reports, a uniform auditing procedure be adopted for the first class townships.
RECOMMENDATION - That the institution and maintenance of uniform auditing procedures and forms be vested in a state agency similarly empowered in regard to accounting, budgeting, and reporting.

(2) FINDING - The defects of the audit practices are due to the part-time and untrained character of local auditing personnel.
RECOMMENDATION - That the commissioners shall be permitted to appoint a certified public accountant to make the annual audit.
RECOMMENDATION - That the commissioners be empowered to request the state authority,

above recommended, to make the annual audit. Such audit to be made by certified accountant employees of the state agency at cost.

RECOMMENDATION - That, in view of occasional past failures of state agencies to exercise their powers, an alternative state authority be designated for this purpose.

(3) FINDING - The local auditors rarely exercise their powers of surcharging local officials for violations of the code provisions in regard to fiscal matters. This is partly due to a lack of familiarity with the law, partly to a wilful disregard of its provisions.

RECOMMENDATION - That in view of this situation, the state authority vested with supervision of local auditing be empowered to inspect and audit the books and accounts of any township at any time.

RECOMMENDATION - That this audit be made either upon the initiative of the state authority or upon the petition of a specified percentage of the local taxpayers.

RECOMMENDATION - That the state agency be empowered to prosecute any township official revealed by such inspection and audit to be guilty of malfeasance in office, in the name of the Commonwealth.

(4) FINDING - The range of diversity within this group of townships is so great that a single, uniform auditing procedure and form is not feasible without further classification.

RECOMMENDATION - That in view of this diversity, the state agency should be empowered to further classify these units in the instituting of uniform auditing forms and procedures.

(5) FINDING - Auditing costs must be kept within the financial means of the various townships. The present auditing costs are proportionately higher than is necessary due to the faulty nature of their budgeting and accounting systems.

RECOMMENDATION - That in view of these facts, the auditing procedures and forms adopted in the institution of a uniform system should be of utmost simplicity and should be closely correlated to the uniform budgeting and accounting systems.

(6) FINDING - The local auditing personnel has shown a general disregard of the statutory provisions concerning the period in which the audit must be completed and filed.

RECOMMENDATION - That all audits be completed within thirty days of the close of the fiscal year.

RECOMMENDATION - That a copy of the annual audit of every township be filed with the designated state agency within ten days of its completion.

RECOMMENDATION - That no auditor be considered as discharged of his duties until the audit has been so filed and received the approval of said agency. Provided, that if the state agency fails to act within thirty days the audit shall stand approved. Provided further, that such approval or disapproval shall not affect the right of local taxpayers to file an appeal from the annual audit.

RECOMMENDATION - That adequate penalty provisions to be enforced by the state agency, be provided by law.

(7) Alternative Recommendations - That in the event recommendations included under number three are not adopted; the following modifications be made in the existing legislation -

    (a) That no persons be qualified to audit local accounts unless they shall have had at least one year of accounting training or experience.

    (b) That the court of common pleas shall appoint a non-resident certified accountant to make the annual audit upon the petition of the owners of twenty five percent of the assessed real estate valuation of said township.

    (c) That all audits be completed within thirty days of the close of the fiscal year.

    (d) That the annual audit statement be filed with a designated state agency within ten days of the completion of the audit and that such state authority be properly empowered to enforce such provisions under adequate penalty stipulations.

(e) That all auditors be compelled, as a condition precedent to the assumption of the duties of their office, to register with the said state agency. That upon such registration the auditors shall be provided, without cost, with a handbook describing the nature of their powers and duties and recommending the more acceptable and efficient procedures to be followed.

(f) That the state authority employ field agents to aid and advise the local auditing personnel.

(g) That auditors be authorized to investigate the propriety or excessiveness of any bill or warrant which they deem to be unreasonable on its face.

CHAPTER XXII
REVENUES

The first class townships derive their income from two general sources: revenue and non-revenue receipts. Revenue receipts include all income which increases the aggregate assets of the township without increasing liabilities or decreasing existing assets, e. g. taxes on real estate and personal property, special assessments against real estate for improvements, licenses and fees required of certain trades and businesses, fines and forfeits, interest received on cash balances, gifts and grants from the State or other units of government and occasionally from individuals, and the earnings of departments from services such as sewage disposal and water supply. Non-revenue receipts, which increase assets by creating corresponding liabilities, include proceeds from the sale of bonds, from temporary loans, refunds and the sale of property and supplies. Receipts from non-revenue sources are sporadic and uncertain depending upon the necessity for borrowing, the opportunity or necessity of selling surplus or unsuitable supplies, and the extent to which errors or joint construction lead to refunds. The two groups are distinguished to permit a clearer analysis of current operations and to determine the extent to which townships must borrow in order to carry on established services.

As is shown in Table IV, some 68% of the total income of all first class townships is derived from taxation. The code specifies that the tax levy be made upon "all property and upon all occupations within the township made taxable for township purposes, as ascertained by the valuation for county purposes made by the assessors of the several counties of this Commonwealth for the year for which the township taxes are levied." This makes the assessment of property a county function. It has been so treated in the preceding chapters on the county. The subject need be considered here only as it affects the period of tax collection and the budgeting process. (See Chapter XX on Budgeting.) Delay by the county authorities in providing township officials with the assessment duplicate until well after the beginning of the fiscal year causes a discrepancy between the periods of incoming revenues and expenditures, which makes budgetary control difficult.

The Levy of Taxes - Since the township commissioners usually do not receive the assessment duplicate from the county authorities until after the sixty day period within which they are required to adopt a budget and its accompanying appropriation and tax-levying ordinances, they must either base their tax rates upon estimates of the assessed valuation or delay the process of budgeting beyond the statutory period. Both procedures are followed in different townships.

The scientific method of fixing a tax rate in normal times would be to estimate the anticipated needs of the coming fiscal year and then divide the amount so determined into the assessed valuation. The result would be the number of mills that would have to be levied. In practice, however, the commissioners first turn their attention to the millage rate and determine whether the public will be satisfied with the retention of the rate for the preceding year or will demand a decrease.

The purposes and rates of taxation have been delimited in part by the State legislature. Under the present code the commissioners may levy any or all of the following taxes:

1. "An annual tax for general township purposes not exceeding ten mills." This rate may be increased 5 mills upon petition of the commissioners to the court of quarter sessions, but if the owners of 50 per cent of the assessed valuation of real estate object at the court hearing the petition is to be denied. Due to the scant limitations upon levies for numerous special purposes such petitions are extremely rare.

2. "A tax for the purpose of building and maintaining suitable places for the housing of fire apparatus." Such a tax is levied by Spring Garden, Hanover and West Norriton townships. In 1933 Springfield township (Delaware County) also levied such a tax.

3. "A... tax, not exceeding one-tenth of one mill, for the purpose of caring for

trees."

4. "A.... tax, not exceeding fifty percentum of the rate of assessment for township purposes, for the purpose of procuring a lot and/or erecting a building thereon for a town house, and for the payment of indebtedness incurred in connection therewith."

5. "A ... tax sufficient to pay interest on indebtedness and sinking fund charges." Approxiamately two-thirds of the townships levy such a tax.

6. "A tax, not to exceed two mills on the dollar of the assessed valuation of taxable property in such township, for the purpose of maintaining and operating parks, playgrounds, playfields, gymnasiums, public baths, swimming pools and recreation centers."

7. "A tax, not exceeding two mills on the dollar of the assessed valuation of taxable property in such township, for the purpose of establishing a public library" (Act of July 20, 1917, P. L. 1143). Upper Darby was the only township of those visited where such a tax was levied. Other townships, however, make annual donations to libraries from their general funds.

In addition the commissioners may levy other taxes in the nature of special assessments, e. g., for street lighting, water supply, and fire protection. Such taxes may be assessed, like other special assessments, against the abutting property owners or the commissioners may create districts within the township for these purposes and levy the special taxes against the property in those districts. Several townships have established street lighting and fire districts.

At present there is no definite limit to the total township tax rate because maximum limits have not been placed upon taxes for such purposes as housing fire apparatus or debt service. Local taxpayers' associations frequently demand reductions of particular rates but even when they succeed in reducing some rates they may meet with higher rates for other purposes. The total rate cannot well be fixed by statute because of the diversity in such matters as taxable valuation and local needs. If the total rate is to be regulated at all it must be by administrative action of a safeguarding nature invoked upon application of taxpayers.

Table I gives a general view of the rates levied in first class townships in 1930 and 1932. It shows the extreme range of millage rates, of total tax levies, and of per capita levies. The data on total taxes levied and millages were taken from the assessment and tax collection statistics of the Bureau of Statistics in the Department of Internal Affairs. All other data were computed.

In 1932 the township levy comprised less than thirty per cent of the total levy in 52 of the 61 first class townships. This was in decided contrast to the school levy which accounted for more than 50 per cent of the total in 39 townships. The ratio of the township tax to the total was only slightly different in 1930, in that year being less than 30 per cent of the total in 48 townships. Thus between 1930 and 1932 township officials were slightly more successful than school boards in adjusting their tax rates to the depression.

Between different townships there is a notably wide range in the tax rate, in the per capita tax burden and in the ratio of the township tax to the total levy. The lowest millage in 1930 was four (Ross) and the highest sixteen (Patterson) while the average for all 61 townships was slightly under 10 mills (9.85). In 1932 the millage ranged from 3.25 in Ross to 16 in Patterson and the average was slightly over nine (9.27) mills. Patterson retained its high position in 1933 with a 16 mill levy as contrasted with a 4 mill levy in South Fayette, Ross and Springdale. The average millage in 1933 was 9 mills, or approximately 9 per cent less than in 1930.

Similar variations are found in comparing per capita tax burdens. The largest per capita township levy in 1930 was that of Neville ($38.50) and the lowest that of Leet ($1.90). In 1932 the highest and lowest rates were, respectively, those of Neville ($37.30 and South Versailles ($2.00). There is a close correlation between the size of

T A B L E I
DISTRIBUTION OF TAX LEVY AND MILLAGE

Township	1930			1932			Tax Rate in Mills	
	Total Taxes Levied	Per Capita	Percent of Total	Total Taxes Levied	Per Capita	Percent of Total	1930	1932
Upper Darby								
Township	$ 704,193	$ 15.10	32.2 %	$ 760,389	$ 16.31	30.9 %	15.50	13.25
School	1,213,614	26.03	55.5	1,272,120	27.28	51.7	23.00	23.00
County	222,134	4.76	10.2	287,594	6.17	11.7	3.25	4.25
Poor	47,745	1.02	2.1	138,951	2.98	5.7	.90	2.50
Total	2,187,686	46.92		2,459,054	52.74		42.65	43.00
Lower Merion								
Township	97,829	27.72	33.9	782,424	22.25	32.0	12.00	9.00
School	1,044,806	29.71	36.4	1,035,099	29.43	42.3	13.00	12.00
County	852,542	24.24	29.7	627,652	17.85	25.7	4.00	3.00
Poor	-------	-----	----	-------	-----	----	-----	-----
Total	2,872,177	81.67		2,445,175	69.53		29.00	24.00
Haverford								
Township	328,617	15.38	29.7	357,739	16.75	27.8	12.00	11.50
School	607,946	28.46	54.9	671,402	31.43	52.3	22.00	23.00
County	146,547	6.86	13.2	181,959	8.52	14.1	3.25	4.25
Poor	24,575	1.15	2.2	72,974	3.42	5.8	.90	2.50
Total	1,107,685	51.85		1,284,074	60.11		38.15	41.25
Coal								
Township	114,013	5.72	15.9	141,921	7.12	16.5	7.50	9.50
School	493,506	24.76	68.9	448,173	22.49	52.2	32.00	30.00
County	108,682	5.45	15.2	120,125	6.03	13.9	7.00	8.00
Poor	-------	-----	----	149,391	7.50	17.4	----	10.00
Total	716,201	35.94		859,610	43.13		46.50	57.50
Abington								
Township	309,023	16.57	30.0	298,476	16.01	29.7	10.50	10.50
School	585,545	31.40	56.8	599,732	32.16	59.7	25.00	25.00
County	135,899	7.29	13.2	106,126	5.69	10.6	4.00	4.00
Poor	-------	----	----	-------	----	----	----	----
Total	1,030,467	55.26		1,004,334	53.86		39.50	39.50
Hanover								
Township	665,274	37.44	27.9	272,160	15.32	19.0	9.10	5.60
School	805,791	45.35	33.8	510,013	28.70	35.5	11.00	10.30
County	652,617	36.73	27.4	385,928	21.72	26.9	8.90	7.80
Poor	255,875	14.30	10.9	267,300	15.04	18.6	3.50	5.50
Total	2,379,557	133.91		1,435,401	80.78		32.50	29.20
Plains								
Township	113,336	7.06	14.4	92,620	5.77	16.5	10.00	11.00
School	533,624	33.26	67.9	355,899	22.18	63.3	49.00	43.00
County	99,737	6.22	12.7	67,309	4.20	12.0	8.90	7.80
Poor	38,379	2.39	5.0	46,355	2.89	8.2	3.50	5.50
Total	785,076	48.93		562,183	35.04		71.40	67.30
Cheltenham								
Township	375,135	23.85	28.3	383,635	24.39	32.9	15.00	15.00
School	574,101	36.49	43.4	565,141	35.93	48.4	23.00	22.00
County	374,381	23.79	28.3	217,992	13.86	18.7	4.00	3.00
Poor	-------	-----	----	-------	-----	----	----	----
Total	1,323,617	84.14		1,166,768	74.17		42.00	40.00
Mt.Lebanon								
Township	267,990	19.99	29.9	261,278	19.49	27.0	10.00	9.00
School	368,806	27.52	41.1	377,501	28.17	39.1	15.00	13.00
County	222,621	16.61	24.7	246,555	18.40	25.5	8.38	8.38
Poor	37,234	2.78	4.3	81,485	6.08	8.4	1.50	3.00
Total	896,651	66.90		966,819	72.13		34.88	33.38

T A B L E   I
DISTRIBUTION OF TAX LEVY AND MILLAGE

Township	1 9 3 0			1 9 3 2			Tax Rate In Mills	
	Total Taxes Levied	Per Capita	Percent of Total	Total Taxes Levied	Per Capita	Percent of Total	1930	1932
Stowe								
Township	$ 144,395	$ 10.80	26.8 %	$ 141,170	$10.56	27.9 %	13.50	13.50
School	238,074	17.81	44.0	244,051	18.26	48.3	23.00	23.00
County	91,797	6.87	17.0	89,708	6.71	17.8	8.38	8.38
Poor	66,046	4.94	12.2	30,500	2.28	6.0	1.50	3.00
Total	540,312	40.42		505,429	37.81		46.38	47.88
Penn								
Township	114,363	8.57	23.8	103,788	7.78	20.5	9.00	13.50
School	231,313	17.34	48.1	257,937	19.34	50.7	18.00	23.00
County	114,948	8.62	24.1	110,233	8.27	21.7	8.33	8.33
Poor	19,238	1.44	4.0	35,867	2.69	7.1	1.50	3.00
Total	479,862	35.97		507,825	38.08		36.88	47.88
Harrison								
Township	94,391	7.62	24.6	97,924	7.91	23.6	10.00	10.00
School	189,986	15.34	49.5	206,897	16.70	49.9	19.00	20.00
County	85,078	6.87	22.2	83,108	6.71	20.0	8.38	8.38
Poor	14,110	1.14	3.7	27,037	2.18	6.5	1.50	3.00
Total	383,565	30.97		414,966	33.50		38.88	41.38
Radnor								
Township	195,110	15.91	28.4	215,736	17.59	27.7	10.00	10.00
School	335,899	27.39	48.9	327,312	26.69	42.0	17.00	15.50
County	138,223	11.27	20.1	182,973	14.92	23.5	3.25	4.25
Poor	17,607	1.44	2.6	52,888	4.31	6.8	.90	2.50
Total	686,839	56.00		778,909	63.52		31.15	32.25
Newport								
Township	237,085	19.61	18.2	176,844	14.63	21.6	9.50	10.00
School	752,431	62.25	57.9	405,755	33.56	49.7	30.00	22.60
County	222,730	18.43	17.2	138,679	11.47	17.0	8.90	7.80
Poor	87,347	7.23	6.7	95,278	7.88	11.7	3.50	5.50
Total	1,299,593	107.52		816,556	67.56		51.90	45.90
Whitehall								
Township	84,449	7.89	33.0	62,965	5.88	34.9	12.50	12.50
School	141,137	13.18	55.1	97,963	9.15	54.4	19.00	20.00
County	30,312	2.83	11.9	19,305	1.80	10.7	4.00	4.00
Poor	------	----	----	------	----	----	-----	-----
Total	255,898	23.90		180,233	16.83		35.50	36.50
Shaler								
Township	90,304	9.43	28.1	95,720	10.00	27.5	10.00	10.00
School	134,382	14.04	41.9	143,075	14.95	41.3	15.00	15.00
County	83,123	8.68	25.8	82,159	8.58	23.7	8.38	8.38
Poor	13,553	1.42	4.2	26,119	2.73	7.5	1.50	3.00
Total	321,362	33.57		347,073	36.26		34.88	36.38
South Fayette								
Township	29,935	3.27	15.9	23,688	2.59	13.7	7.50	6.00
School	117,223	12.82	62.2	105,106	11.49	60.7	29.00	26.00
County	35,389	3.87	18.7	33,363	3.65	19.3	8.38	8.38
Poor	6,004	.66	3.2	10,969	1.20	6.3	1.50	3.00
Total	188,551	20.61		173,126	18.93		46.38	43.38
Butler								
Township	85,685	9.92	28.1	76,794	8.89	26.1	10.00	9.00
School	145,428	16.83	47.7	144,235	16.69	49.0	16.00	16.00
County	73,635	8.52	24.2	73,413	8.50	24.9	8.50	8.50
Poor	-------	-----	----	-------	-----	----	-----	-----
Total	304,748	35.27		294,442	34.07		34.50	33.50

T A B L E   I
DISTRIBUTION OF TAX LEVY AND MILLAGE

Township	1 9 3 0			1 9 3 2			Tax Rate In Mills	
	Total Taxes Levied	Per Capita	Percent of Total	Total Taxes Levied	Per Capita	Percent of Total	1930	1932
Ross								
Township	$ 37,368	$ 4.35	14.8 %	$ 31,845	$ 3.71	11.7 %	4.00	3.25
School	119,052	13.87	47.2	127,802	14.89	47.1	13.25	13.25
County	82,653	9.63	32.7	84,758	9.88	31.3	8.38	8.38
Poor	13,361	1.56	5.3	26,905	3.14	9.9	1.50	3.00
Total	252,434	29.42		271,310	31.62		27.13	27.88
Ridley								
Township	85,223	10.24	33.0	92,064	11.06	32.3	10.00	10.00
School	134,839	16.19	52.2	128,852	15.48	45.3	16.00	14.00
County	30,587	3.67	7.8	41,968	5.04	14.7	3.25	4.25
Poor	7,670	.92	3.0	21,853	2.62	7.7	.90	2.50
Total	258,319	31.02		284,737	34.20		30.15	30.75
Mifflin								
Township	59,061	7.23	24.3	52,920	6.48	23.7	12.00	12.00
School	133,072	16.29	55.0	120,869	14.80	54.2	30.00	30.00
County	43,267	5.30	17.7	37,058	4.54	16.7	8.38	8.38
Poor	7,343	.90	3.0	12,083	1.48	5.4	1.50	3.00
Total	242,743	29.72		222,930	27.30		51.88	53.38
Elizabeth								
Township	33,351	4.46	18.2	25,846	3.46	14.7	7.00	5.50
School	98,170	13.13	53.5	96,938	12.96	55.3	21.00	21.00
County	44,701	5.98	24.4	40,431	5.41	23.0	8.38	8.38
Poor	7,152	.96	3.9	12,232	1.64	7.0	1.50	3.00
Total	183,374	24.52		175,447	23.46		37.88	37.88
Wilkes-Barre								
Township	78,053	10.88	16.4	13,197	1.84	4.7	10.00	2.50
School	300,497	41.88	63.3	193,375	26.95	69.3	39.00	36.00
County	69,410	9.67	14.6	43,368	6.04	15.5	8.90	7.80
Poor	27,119	3.78	5.7	29,033	4.05	10.4	3.50	5.50
Total	475,079	66.20		278,973	38.88		61.40	51.80
Baldwin								
Township	55,013	8.63	29.2	51,194	8.04	28.3	9.00	9.00
School	69,925	10.98	37.0	66,192	10.39	36.5	12.00	12.00
County	54,817	8.60	29.0	48,288	7.58	26.7	8.38	8.38
Poor	8,985	1.41	4.8	15,466	2.43	8.5	1.50	3.00
Total	188,740	29.62		181,140	28.43		30.88	32.38
Swatara								
Township	21,635	3.42	21.6	14,512	2.29	15.7	9.00	6.00
School	62,968	9.94	62.9	62,228	9.83	67.4	24.00	23.00
County	15,458	2.44	15.5	15,593	2.46	16.9	6.00	6.00
Poor	------	----	----	------	----	----	----	----
Total	100,061	15.80		92,333	14.58		39.00	35.00
Scott								
Township	49,352	7.96	30.6	47,300	7.63	27.4	10.00	10.00
School	61,446	9.91	38.0	72,204	11.64	41.9	12.00	15.00
County	43,251	6.97	26.8	39,585	6.38	23.0	8.38	8.38
Poor	7,410	1.19	4.6	13,351	2.15	7.7	1.50	3.00
Total	161,459	26.03		172,440	27.80		31.88	36.38
Collier								
Township	23,982	3.98	20.7	21,049	3.46	18.7	10.00	6.00
School	54,885	9.01	47.3	53,518	8.79	47.5	26.00	15.00
County	31,527	5.18	27.2	28,426	4.67	25.2	8.38	8.38
Poor	5,516	.91	4.8	9,673	1.59	8.6	1.50	3.00
Total	115,910	19.03		112,666	18.50		45.88	32.38

T A B L E   I

DISTRIBUTION OF TAX LEVY AND MILLAGE

Township	1 9 3 0			1 9 3 2			Tax Rate In Mills	
	Total Taxes Levied	Per Capita	Percent of Total	Total Taxes Levied	Per Capita	Percent of Total	1930	1932
East Bethlehem								
Township	$ 22,035	$ 3.82	18.2 %	$ 14,359	$ 2.49	13.6 %	10.00	7.00
School	80,047	13.88	66.1	73,622	12.77	69.6	34.00	34.00
County	19,013	3.30	15.7	17,809	3.09	16.8	8.50	8.50
Poor	------	-----	----	------	-----	----	-----	-----
Total	121,095	21.00		105,790	18.35		52.50	49.50
North Versailles								
Township	36,953	6.52	20.8	30,287	5.34	17.1	10.00	8.00
School	99,643	17.58	55.9	102,337	18.05	57.8	27.00	27.00
County	35,863	6.33	20.1	34,108	6.02	19.3	8.38	8.38
Poor	5,732	1.01	3.2	10,236	1.81	5.8	1.50	3.00
Total	178,191	31.43		176,968	31.22		46.88	46.38
Springfield (Montg.Co.)								
Township	77,657	14.01	24.3	78,970	14.25	29.4	10.00	10.00
School	146,888	26.51	45.9	126,654	22.86	47.2	19.00	16.00
County	95,329	17.20	29.8	62,823	11.34	23.4	4.00	3.00
Poor	------	-----	----	-------	-----	----	-----	-----
Total	319,874	57.73		268,447	48.45		33.00	29.00
O'Hara								
Township	35,495	6.92	22.5	41,121	8.02	21.1	6.00	6.00
School	58,342	11.38	37.1	73,881	14.41	37.9	9.50	10.50
County	54,902	10.71	34.9	60,509	11.80	31.0	8.38	8.38
Poor	8,873	1.73	5.5	19,511	3.81	10.0	1.50	3.00
Total	157,612	30.74		195,022	38.04		25.38	27.88
Spring Garden								
Township	40,106	8.58	26.8	42,118	9.01	27.8	10.50	10.00
School	63,392	13.56	42.5	69,905	14.95	46.3	16.00	16.00
County	45,962	9.38	30.7	39,150	8.37	25.9	10.00	8.00
Poor	-------	-----	----	-------	-----	----	-----	-----
Total	149,460	31.97		151,173	32.33		36.50	34.00
Springfield (Del. Co.)								
Township	76,579	16.69	31.4	79,258	17.27	31.3	12.00	12.00
School	135,483	29.52	55.7	125,028	27.25	49.2	21.00	20.00
County	25,667	5.59	10.5	33,657	7.33	13.3	3.25	4.25
Poor	5,805	1.26	2.4	15,645	3.41	6.2	.90	2.50
Total	243,534	53.07		253,588	55.26		37.15	38.75
Plymouth								
Township	28,158	6.44	24.8	28,640	6.56	23.8	7.50	7.50
School	67,216	15.38	59.3	77,411	17.72	64.3	16.00	18.50
County	17,983	4.12	15.9	14,320	3.28	11.9	4.00	3.00
Poor	------	-----	----	------	-----	----	-----	-----
Total	113,357	25.95		120,371	27.55		27.50	28.00
Jefferson								
Township	31,447	7.60	27.5	15,956	3.86	13.3	8.00	4.00
School	39,586	9.57	34.7	58,796	14.21	48.8	10.00	15.00
County	37,182	8.99	32.6	34,850	8.42	29.0	8.38	8.38
Poor	5,900	1.43	5.2	10,720	2.59	8.9	1.50	3.00
Total	114,115	27.58		120,322	29.08		27.88	34.38
Wilkins								
Township	29,328	7.11	21.4	30,323	7.35	21.1	11.00	11.00
School	80,260	19.45	58.7	83,146	20.15	57.8	30.00	30.00
County	23,221	5.63	17.0	22,815	5.53	15.9	8.38	8.38
Poor	4,007	.97	2.9	7,421	1.80	5.2	1.50	3.00
Total	136,816	33.15		143,705	34.82		50.88	52.38

T A B L E   I
DISTRIBUTION OF TAX LEVY AND MILLAGE

Township	1 9 3 0			1 9 3 2			Tax Rate In Mills	
	Total Taxes Levied	Per Capita	Percent of Total	Total Taxes Levied	Per Capita	Percent of Total	1930	1932
Upper Moreland								
Township	$ 35,269	$ 8.84	24.6 %	$ 31,912	$ 8.00	27.5 %	9.00	8.00
School	86,791	21.76	60.4	67,530	16.93	58.3	20.00	16.00
County	21,581	5.41	15.0	16,420	4.12	14.2	4.00	3.00
Poor	------	-----	----	------	-----	----	-----	-----
Total	143,641	36.01		115,862	29.05		33.00	27.00
East Deer								
Township	53,819	13.70	22.4	52,614	13.39	25.5	7.50	7.50
School	87,373	22.24	36.4	75,765	19.29	26.7	12.00	11.00
County	83,859	21.35	34.9	57,661	14.68	28.0	8.38	8.38
Poor	14,943	3.80	6.2	20,259	5.16	9.8	1.50	3.00
Total	239,994	61.10		206,299	52.52		29.38	29.88
Tinicum								
Township	47,464	13.08	26.3	48,303	13.30	27.3	10.00	10.00
School	94,870	26.13	52.6	96,267	26.52	54.3	20.00	20.00
County	15,738	4.34	8.7	21,105	5.81	11.9	3.25	4.25
Poor	4,271	1.18	2.4	11,509	3.17	6.5	.90	2.50
Total	162,343	44.73		177,184	48.81		34.15	36.75
Lower Chichester								
Township	20,745	5.97	28.5	21,402	6.16	25.0	10.00	10.00
School	42,847	12.34	58.8	49,332	14.20	57.6	20.00	22.00
County	7,463	2.15	10.2	9,954	2.87	11.6	3.25	4.25
Poor	1,868	.54	2.6	4,943	1.42	5.8	.90	2.50
Total	72,923	21.00		85,631	24.66		34.15	38.75
Lawrence Park								
Township	44,134	13.62	28.2	45,277	13.97	25.5	10.00	10.00
School	80,542	24.85	51.4	91,195	28.14	51.3	18.00	20.00
County	31,981	9.87	20.4	41,332	12.75	23.3	7.00	9.00
Poor	------	-----	----	------	-----	----	-----	-----
Total	156,657	48.34		177,804	54.86		35.00	39.00
Reserve								
Township	19,084	6.14	26.0	19,690	6.34	23.7	10.00	10.00
School	34,191	11.00	46.6	42,027	13.52	50.6	17.00	20.00
County	17,266	5.56	23.5	16,137	5.19	19.4	8.38	8.38
Poor	2,880	.93	3.9	5,171	1.66	6.2	1.50	3.00
Total	73,421	23.62		83,025	26.71		36.88	41.38
Lackawanna								
Township	12,942	4.20	22.4	11,354	3.69	23.2	15.00	14.00
School	33,618	10.93	58.3	28,812	9.36	58.9	35.00	35.00
County	7,645	2.48	13.3	5,515	1.79	11.3	7.75	6.75
Poor	3,451	1.12	6.0	3,244	1.05	6.6	4.00	4.00
Total	57,656	18.73		48,925	15.90		61.75	59.75
Annville								
Township	14,768	4.93	21.0	14,010	4.67	21.3	7.50	7.00
School	43,548	14.53	62.0	40,414	13.48	61.5	21.00	19.00
County	10,914	3.64	15.5	10,145	3.39	15.4	3.00	3.00
Poor	986	.33	1.4	1,200	.40	1.8	.50	.60
Total	70,216	23.43		65,769	21.94		32.00	29.60
Nether Providence								
Township	33,706	11.90	22.3	33,536	11.84	22.0	8.00	7.50
School	90,984	32.12	60.2	76,740	27.09	50.4	20.00	16.00
County	22,664	8.00	15.0	31,198	11.01	20.5	3.25	4.25
Poor	3,853	1.36	2.6	10,657	3.76	7.0	.90	2.50
Total	151,207	53.37		152,131	53.70		32.15	30.25

T A B L E   I
DISTRIBUTION OF TAX LEVY AND MILLAGE

Township	1930			1932			Tax Rate In Mills	
	Total Taxes Levied	Per Capita	Percent of Total	Total Taxes Levied	Per Capita	Percent of Total	1930	1932
Darby								
Township	$ 14,276	$ 5.15	22.5 %	$ 13,108	$ 4.73	21.2 %	10.00	10.00
School	42,925	15.48	67.7	39,326	14.18	63.7	30.00	30.00
County	4,904	1.77	7.7	6,021	2.17	9.8	3.25	4.25
Poor	1,342	.48	2.1	3,276	1.18	5.3	.90	2.50
Total	63,447	22.88		61,731	22.26		44.15	46.75
Aston								
Township	15,805	5.94	30.4	15,698	5.90	26.7	10.00	10.00
School	27,929	10.50	53.7	29,926	11.25	50.9	18.00	18.00
County	6,971	2.62	13.4	9,536	3.59	16.2	3.25	4.25
Poor	1,327	.50	2.6	3,605	1.36	6.1	.90	2.50
Total	52,032	19.57		58,765	22.10		32.15	34.75
West Norriton								
Township	30,857	11.92	34.6	26,519	10.25	30.3	10.50	8.50
School	44,229	17.09	49.6	47,840	18.49	54.6	13.00	13.00
County	14,175	5.48	15.9	13,222	5.11	15.1	4.00	3.00
Poor	------	-----	---	------	-----	----	-----	-----
Total	89,261	34.49		87,581	33.84		27.50	24.50
West Pottsgrove								
Township	12,086	5.04	33.3	12,186	5.08	30.7	10.00	10.00
School	19,170	7.99	52.7	23,396	9.75	58.9	15.00	16.00
County	5,091	2.12	14.0	4,131	1.72	10.4	4.00	3.00
Poor	------	----	----	------	----	----	-----	-----
Total	36,347	15.14		39,713	16.55		29.00	29.00
Rochester								
Township	14,708	6.18	20.9	14,130	5.94	23.0	10.00	10.00
School	37,701	15.84	53.5	32,803	13.78	53.5	23.00	23.00
County	15,096	6.34	21.4	11,588	4.87	18.9	10.00	8.00
Poor	2,958	1.24	4.2	2,826	1.19	4.6	2.00	2.00
Total	70,463	29.61		61,347	25.78		45.00	43.00
Stonycreek								
Township	17,177	9.23	25.5	15,255	8.20	24.1	10.00	10.00
School	37,899	20.38	56.2	35,200	18.92	55.6	21.00	20.00
County	12,318	6.62	18.3	12,875	6.92	20.3	6.75	8.00
Poor	------	-----	----	------	-----	----	-----	-----
Total	67,394	36.23		63,330	34.05		37.75	38.00
Patterson								
Township	14,276	7.73	25.6	14,714	7.97	28.3	16.00	16.00
School	26,484	14.35	47.6	27,526	14.91	52.9	28.00	28.00
County	13,153	7.13	23.6	7,997	4.33	15.4	10.00	8.00
Poor	1,785	.97	3.2	1,839	1.00	3.5	2.00	2.00
Total	55,698	30.17		52,076	28.21		56.00	54.00
Braddock								
Township	9,503	5.30	21.5	9,776	5.45	21.2	10.00	10.00
School	24,845	13.85	56.2	25,731	14.34	55.9	26.00	25.00
County	8,458	4.71	19.1	7,958	4.44	17.3	8.38	8.38
Poor	1,433	.80	3.2	2,599	1.45	5.6	1.50	3.00
Total	44,239	24.66		46,064	25.68		45.88	46.38
Springdale								
Township	4,197	2.68	10.9	3,211	2.05	8.6	5.00	4.00
School	26,026	16.63	67.3	25,561	16.33	68.5	31.00	31.00
County	7,196	4.59	18.6	6,421	4.10	17.2	8.38	8.38
Poor	1,265	.81	3.3	2,126	1.36	5.7	1.50	3.00
Total	38,674	24.71		37,319	23.85		45.88	46.38

T A B L E   I
DISTRIBUTION OF TAX LEVY AND MILLAGE

Township	1930			1932			Tax Rate In Mills	
	Total Taxes Levied	Per Capita	Percent of Total	Total Taxes Levied	Per Capita	Percent of Total	1930	1932
Neville								
Township	$ 58,978	$ 38.50	81.9 %	$ 57,142	$ 37.30	25.7 %	8.85	7.00
School	59,580	38.89	32.2	73,110	47.72	32.9	9.00	9.00
County	56,513	36.89	30.5	68,227	44.52	30.7	8.38	8.38
Poor	10,001	6.53	5.4	24,086	15.72	10.8	1.50	3.00
Total	185,072	120.80		222,565	145.27		27.73	27.38
Leet								
Township	2,435	1.90	16.5	7,241	5.69	17.3	7.00	6.50
School	7,872	6.19	53.3	22,726	17.87	54.4	22.00	20.00
County	3,843	3.02	26.0	8,803	6.92	21.1	8.38	8.38
Poor	609	.48	4.1	2,980	2.34	7.1	1.50	3.00
Total	14,759	11.60		41,750	32.82		38.88	37.88
Crescent								
Township	6,370	5.75	20.4	6,452	5.83	20.8	8.50	8.50
School	17,078	15.43	54.7	16,293	14.72	52.5	22.00	20.00
County	6,646	6.00	21.3	6,306	5.70	20.3	8.38	8.38
Poor	1,125	1.01	3.6	1,997	1.80	6.4	1.50	3.00
Total	31,219	28.20		31,048	25.05		40.38	39.88
White								
Township	4,969	4.82	24.6	5,008	4.26	23.5	15.00	15.00
School	11,081	9.75	54.9	12,894	12.51	60.6	29.00	35.00
County	3,452	3.35	17.1	2,722	2.64	12.8	10.00	8.00
Poor	675	.65	3.4	669	.65	3.1	2.00	2.00
Total	20,177	19.57		21,293	20.65		56.00	60.00
West Lebanon								
Township	6,037	6.55	43.2	5,535	6.01	44.9	11.50	11.50
School	5,917	6.42	42.3	5,879	6.38	47.7	11.00	11.00
County	1,764	1.92	12.6	627	.68	5.1	3.00	3.00
Poor	263	.29	1.9	291	.32	2.4	.50	.60
Total	13,981	15.18		12,332	13.39		26.00	26.10
Borough								
Township	5,990	6.77	23.0	6,004	6.78	23.7	10.00	10.00
School	12,602	14.24	48.4	12,878	14.55	50.8	20.00	20.00
County	6,275	7.09	24.1	5,225	5.90	20.6	10.00	8.00
Poor	1,198	1.35	4.6	1,223	1.38	4.8	2.00	2.00
Total	26,065	29.45		25,330	28.62		42.00	40.00
South Versailles								
Township	737	2.19	16.5	674	2.00	16.3	6.00	6.00
School	2,474	7.36	55.4	2,308	6.87	55.9	20.00	20.00
County	1,075	3.20	24.1	874	2.60	21.2	8.38	8.38
Poor	182	.54	4.1	271	.81	6.6	1.50	3.00
Total	4,468	13.30		4,127	12.28		35.88	37.38

the per capita tax levy and the urban or rural character of the township. The predominantly urban townships all have high per capita levies while the rural ones invariably have low per capita tax rates. In the intermediate group a clear tendency can be observed for the per capita levy to increase in proportion to the urban nature of the unit. This reflects the diversity in conditions in these first class townships and reflects the demand of urban communities for a more complex and expensive form of government. Examples of this are as follows: (a) Strictly urban townships - Upper Darby ($16.31), Lower Merion ($22.25), Haverford ($16.75), Abington ($16.01), Cheltenham ($24.39), Mt. Lebanon ($19.49), Radnor ($17.59), Lawrence Park ($13.97), Neville ($37.30). (b) Predominantly urban - Harrison ($7.91), Shaler ($10.00), Ridley ($11.06), Spring Garden ($9.01), East Deer ($13.39). (c) Predominantly rural - Plymouth ($6.56), West Pottsgrove ($5.08), Mifflin ($6.48). (d) Strictly rural - Ross ($3.71), Elizabeth ($3.46), Swatara ($2.29), Collier ($3.46), Jefferson ($3.86), Springdale ($2.05), Crescent ($5.83).

Collection of Taxes - The first class township code charges the township treasurer with the responsibility for collecting township, school, county and poor taxes in addition to his duty of receiving and paying out all township moneys. The code further requires him to make regular deposits of money received and to present a monthly report to the township secretary showing the sources and amounts of money received during the month. The commissioners name the depository, which in turn must furnish a surety bond. This has recently become impossible because of the failure of surety companies. Instead the depositories (banks and trust companies) have deposited collateral as surety. The code should be amended to legalize this practice. It should also require depositories to furnish the township with an itemized monthly statement. As a guarantee of the faithful discharge of his duties the treasurer is compelled to give bond in an amount "equal to fifty percentum of the probable amount of the annual township tax."

The treasurer's compensation for the collection of township taxes is "a sum equal to five percentum of all township taxes received or collected by him, and, in addition thereto, a sum equal to one per centum on all other moneys received or collected by him for the township, unless a different rate or annual compensation shall be fixed by ordinance . . . . finally adopted thirty days prior to his election." In no event shall his total compensation exceed ten thousand dollars annually. For collecting other local taxes the treasurer's compensation is fixed by the taxing authorities.

The treasurer is the highest paid of all the local officials, as is shown in the table on functional costs in the following chapter. Many townships have taken steps to reduce his compensation either by placing him on a smaller fee basis or upon a straight salary. Some have also reduced or abolished the treasurer's fee of one per cent on all "other moneys." Where the compensation has been reduced by either of these two methods the township generally assumes the cost of the treasurer's bond and the expenses incidental to his office.

The code does not provide the treasurer with a fee upon the amount of moneys paid out by him. Despite this fact the treasurer of West Lebanon township received a five per cent fee on the amount of State tax on indebtedness which this township was compelled to pay. Presumably this mistake arose from a misapprehension as to the nature of this tax which was looked upon as a receipt rather than as an expenditure.

In a previous chapter the strong grounds for placing all local tax collections in the office of the county treasurer have been set forth. If this is not done and the present system is retained it would seem advisable to place the office on a salary basis. Maximum and minimum salaries should be stipulated according to township population or, preferably, according to assessed valuation.

Tax Collection Procedure - The county commissioners prepare the county duplicate and deliver it to the local authorities. The code provides that "As soon as possible after the receipt of the duplicate from the county commissioners, the board of township commissioners shall deliver a duplicate of the assessment of township taxes to the township treasurer, together with their warrant for the collection of the same." Upon the delivery of the township duplicate to the treasurer the latter is to give public notice of its receipt by posting printed notices and by publication in a local newspaper. Within "thirty days after receiving the tax duplicate, (he) shall notify every taxable whose name appears on such duplicate. Such notice shall contain the rate of taxation, the occupation of such taxable, and the full amount of taxes for which said taxable is liable for the current year."

The taxpayer is given a period of sixty days from the date of such notification to pay his taxes, after which he is charged a penalty of two per cent of the amount of the tax, and thereafter an additional penalty of one per cent for each month or fractional part thereof during the remainder of the fiscal year. Any tax remaining unpaid at the end of the fiscal year in which it was levied carries a penalty of six per cent per annum until paid.

We next encounter an inexcusable defect in the law, which has caused loose, slipshod financial management and waste in many first class townships. The code makes no statement as to when the county commissioners are to furnish local officials with the county duplicate. Consequently they often fail to furnish the duplicate until several months after the beginning of the fiscal year. In numerous instances this was not done until May or June, some five or six months after the beginning of the townships' fiscal year. Allowing a reasonable time for the transcription of the duplicate and the mailing of notices by the treasurer, the date for the imposition of the penalty upon late tax payments will be postponed until the latter part of August or early in September. Since most people defer payment until shortly before the penalty period this means that well over half the fiscal year will have passed before any considerable portion of the current tax duplicate is collected. The commissioners will have been operating for the first half of the fiscal year either upon the basis of the collections of the prior year or by means of temporary loans in anticipation of taxes or by the accumulation of bills payable. That this negligent and wasteful method is entirely unnecessary is shown by a contrast with the school code in which sections 501, et seq., provide a procedure that should be prescribed in the township code.

The first class township code should be amended, as suggested in the chapter on budgeting, to establish a close relation between the fiscal year and the tax collection period. This might be done by changing the fiscal year to commence in June or July, instead of January the first or shortly thereafter as at present, or by requiring the assessing authorities, by statute, to deliver the duplicate within thirty days of the beginning of the present fiscal year. This latter provision would still permit a spread of three months or more between the beginning of the year and the period of tax collection and therefore would not offer a complete solution to the problem. Furthermore, it is desirable to have the duplicate placed in the hands of the tax collector by the beginning of the fiscal year rather than thereafter. Consequently, unless the county authorities can be made to change the period during which assessments are made, the only acceptable solution would seem to be that of changing the first class township fiscal year.

Due to growing tax delinquencies the State legislature has authorized all local taxing districts to provide by ordinance or resolution for the payment of taxes by installments. A township adopting the installment plan may fix the number and period of such installments. If a taxpayer elects to pay by this method a five per cent penalty on each overdue installment is required. Furthermore, if one-half or more of the total number of installments become delinquent an additional penalty of three per cent is added to each delinquent installment. Of the twenty townships where data was obtained on this matter, only three, Leet, Shaler, and Baldwin, took advantage of the above act. Several townships also permit the taxpayer to work out his taxes in road labor.

Tax Exonerations. Section 1718 of the code provides that:

"The township commissioners shall at all times make abatements or exonerations for occupation taxes, mistakes, indigent persons, unseated lands, deaths, removals, et cetera, as to them shall appear just and reasonable. The township secretary shall enter in a book, or books, to be kept for that purpose, the names of all persons abated or exonerated, together with the reason why, the amount of the tax, and the date when levied, and give to the treasurer a certificate stating the nature of the tax and the amount exonerated." †

---

† Exoneration is also used elsewhere to describe the action of the commissioners in exonerating the tax collector (not the taxpayer) from the collection of taxes turned over to the county treasurer after such taxes have become delinquent.

The general practice in exonerations is for the treasurer to submit a list of individuals to the commissioners for their action. Usually the commissioners appear to ratify this list with little or no investigation. When one commissioner was queried as to the adequacy of this procedure he replied that the power of exoneration was abused and that the commissioners placed too much discretion in the treasurer. However, he also stated that no commissioner was familiar enough with the list of taxables to pass an intelligent judgment on the question of exonerations without an investigation which they did not have time to make. Consequently he was able to suggest no method of improving the present practice.

Several secretaries indicated that some improvement might be made if the treasurer were to give each commissioner a list of the persons submitted for exoneration, living in the commissioner's ward or district. If such list were given to the commissioner in sufficient time to make a few investigations he would be able to pass intelligently upon most of the proposed exonerations. As the matter now stands the treasurer submits his list, the commissioners give it a brief glance and then proceed to ratify the list on the same evening it is submitted.

The code specifies that the secretary shall maintain a book or books in which a list of such exonerations shall be maintained. This stipulation is generally ignored since in most cases no such list is maintained by the secretary and where maintained does not generally specify the reasons for the exoneration in sufficient detail. The failure to keep such a list is especially to be deplored since it is the only means of publicity in regard to tax exonerations. The school code provides for publicity by permitting the directors to publish a list of such exonerations in a newspaper of general circulation in the local community. In the township provision of some form of publicity combined with a more careful scrutiny of proposed exonerations by the commissioners would go far toward the elimination of the abuses which the facts indicate are inherent in the exoneration procedure.

Tax Delinquency - In Table II on tax delinquency in first class townships, the figures on taxes levied, amount of delinquency and collections from prior years' levies were obtained from the Bureau of Statistics, Department of Internal Affairs. The other portions of the table were computed. This table shows the extent of delinquency by percentage and by amount for the years 1930 and 1932, the trend toward greater delinquency in the two year period, and the degree to which current delinquency was offset by current collections from previous years' levies. The percentage of township tax delinquency in 1932 was either double or more than double the 1930 percentage of delinquency in nineteen of the sixty-one townships. In five townships where this was not true, the amount of delinquency in 1932 was more than fifty percent of the township tax levy. In only seven of the townships was the percentage of delinquency for 1932 less than that of 1930. Finally, the table shows that in less than one-fourth of the townships (15) was the 1932 tax delinquency less than twenty percent of the township tax levy.

That the township officials were not suddenly confronted by this problem is indicated by the growth of delinquency in the following sixteen townships:

Township	Per Cent of Current Taxes Remaining Unpaid at the End of Each Year.			
	1930	1931	1932	1933
Upper Darby	23.6	29.4	37.5	38.5
Haverford	1.3	19.6	22.1	26.2
Radnor	4.8	----	12.0	13.5
Ridley	39.7	29.6	39.5	42.95
Swatara	16.1	11.7	20.2	22.7
East Bethlehem	15.7	18.7	27.8	35.9
Spring Garden	6.8	10.8	11.5	16.01
Springfield (Del.Co.)	18.8	23.0	32.6	36.4
Tinicum	27.0	7.7	22.7	3.1
Lower Chichester	----	23.4	40.0	33.4
Annville	9.3	10.6	13.6	13.91
Nether Providence	8.1	11.7	17.7	20.4
Darby	41.2	52.8	64.0	64.1
Aston	17.6	----	42.1	100.0
Stony Creek	22.0	21.4	32.8	32.8
West Lebanon	16.0	17.7	27.5	30.0

T A B L E   II

T A X   D E L I N Q U E N C Y

TOWNSHIP	Total Township Taxes Levied		Amount Delinquent at End of Year		Per Cent Delinquent at End of Year		Taxes Collected in Current Year, from Previous Years' Levies		Amount by Which Current Delinquency Exceeds Taxes Collected from Previous Years	
	1930	1932	1930	1932	1930	1932	1930	1932	1930	1932
Upper Darby	$704,193	$760,389	$166,443	$285,187	23.6%	37.5%	$112,160	$307,329	$54,283	$-22,142
Lower Merion	974,829	782,424	117,279	156,371	12.0	20.0	80,596	41,308	36,683	115,063
Haverford	328,617	357,739	41,331	79,440	12.6	22.1	11,682	47,740	29,649	31,700
Coal	114,013	141,921	31,281	24,001	27.4	16.1	-----	-----	31,281	24,001
Abington	309,023	298,476	41,017	71,812	13.3	24.1	83,712	24,786	42,695	47,026
Hanover	665,274	272,160	80,667	10,357	12.1	3.8	-----	-----	80,667	10,357
Plains	113,336	92,620	2,205	18,313	1.9	20.0	-----	2,799	2,205	15,514
Cheltenham	375,135	383,635	43,522	67,403	11.6	17.5	25,910	13,241	17,612	54,162
Mt. Lebanon	267,990	261,278	41,606	89,836	15.5	34.4	18,913	34,748	22,693	55,088
Stowe	144,395	141,170	21,886	45,531	15.4	32.2	7,381	10,371	14,505	35,160
Penn	114,363	103,788	23,040	35,506	20.2	34.1	16,273	14,065	6,767	21,441
Harrison	94,391	97,924	11,446	20,298	12.1	20.4	9,244	9,160	2,202	11,138
Radnor	195,110	215,736	9,364	25,983	4.8	12.0	715	11,104	8,649	14,879
Newport	237,085	176,844	71,126	4,571	30.0	2.6	-----	-----	71,126	4,571
Whitehall	84,449	62,965	9,677	15,210	11.5	24.1	5,389	3,555	4,288	11,655
Shaler	90,304	95,720	14,715	27,989	16.3	29.2	6,506	8,344	8,209	19,645
South Fayette	29,935	23,688	5,212	6,055	17.4	25.6	1,283	2,164	3,929	3,891
Butler	85,685	76,794	11,516	18,046	13.4	23.5	3,979	6,880	7,573	11,166
Ross	37,368	31,845	7,805	8,679	20.9	27.3	2,879	2,689	4,926	5,990
Ridley	85,223	92,064	33,783	36,304	39.7	39.5	5,817	16,845	27,966	19,459
Mifflin	59,061	52,920	18,313	29,816	31.0	56.4	10,982	3,375	7,331	26,441
Elizabeth	33,351	25,846	5,285	7,195	15.8	27.8	3,992	2,013	1,293	5,182
Wilkes-Barre	78,053	713,197	1,132	4,288	1.5	.5	-----	-----	1,132	4,288
Baldwin	55,013	51,194	12,299	18,407	22.1	35.9	5,854	4,731	6,445	13,676
Swatara	21,635	14,512	3,490	2,929	16.1	20.2	-----	-----	3,490	2,929
Scott	49,352	47,300	10,840	15,157	21.9	32.1	6,460	3,924	4,380	11,233
Collier	23,982	21,049	6,362	6,348	26.5	30.2	1,221	1,104	5,141	5,244
East Bethlehem	22,035	14,359	3,451	3,991	15.7	27.8	1,479	1,449	1,972	2,542
North Versailles	36,953	30,287	15,542	16,506	42.0	54.5	7,087	3,415	8,455	13,091
Springfield (a)	77,657	78,970	-----	9,699		12.3	3,424	4,925	- 3,424	4,774
O'Hara	35,495	41,121	2,784	7,643	7.8	18.6	1,611	1,805	1,173	5,838
Spring Garden	40,106	42,118	2,785	4,832	6.8	11.5	1,971	3,159	814	1,673
Springfield (b)	76,579	79,258	14,226	25,916	18.8	32.6	7,426	7,296	6,800	18,620
Plymouth	28,158	28,940	6,506	9,706	23.1	33.8	8,915	2,117	- 2,409	7,589
Jefferson	31,447	15,956	4,536	4,342	14.4	27.2	1,749	4,207	2,787	135
Wilkins	29,328	30,323	6,969	15,568	23.8	51.3	4,716	3,267	2,253	12,301
Upper Moreland	35,269	31,912	3,964	9,763	11.2	30.6	5,775	5,218	- 1,811	4,545

TABLE II

TAX DELINQUENCY

TOWNSHIP	Total Township Taxes Levied		Amount Delinquent at End of Year		Per Cent Delinquent at End of Year		Taxes Collected in Current Year, from Previous Years' Levies		Amount by Which Current Delinquency Exceeds Taxes Collected from Previous Years	
	1930	1932	1930	1932	1930	1932	1930	1932	1930	1932
East Deer	$ 53,819	$ 52,614	$ 1,897	$ 5,975	3.5%	11.4%	$ 2,049	$ 341	$ -152	$ 5,634
Tinicum	47,464	48,303	12,866	10,988	27.0	22.7	833	843	12,033	10,145
Lower Chichester	20,745	21,402	----	8,554	----	40.0	----	3,586	----	4,968
Lawrence Park	44,134	45,277	2,879	2,501	6.5	5.5	299	715	2,580	1,786
Reserve	19,084	19,690	3,955	6,683	20.8	33.9	2,256	1,708	1,699	4,975
Lackawanna	12,942	11,354	4,996	5,486	38.6	48.3	5,010	1,764	-14	3,722
Annville	14,768	14,010	1,371	1,904	9.3	13.6	1,268	1,067	103	837
Nether Providence	33,706	33,536	2,717	5,924	8.1	17.7	845	----	1,872	5,924
Darby	14,276	13,108	5,881	8,410	41.2	64.0	1,517	1,475	4,364	6,935
Ashton	15,805	15,698	2,780	6,618	17.6	42.1	610	1,372	2,170	5,246
West Norriton	30,857	26,519	1,753	4,599	5.7	17.7	2,756	1,998	-1,003	2,601
West Pottsgrove	12,086	12,186	849	2,744	7.0	22.5	866	748	-17	1,996
Rochester	14,708	14,130	4,780	7,438	32.4	57.8	1,483	89	3,297	7,349
Stonycreek	17,177	15,255	3,777	5,003	22.0	32.8	2,346	1,673	1,431	3,330
Patterson	14,276	14,714	3,800	7,919	26.6	53.7	2,133	1,439	1,667	6,480
Braddock	9,503	9,776	2,977	6,359	31.3	65.0	937	418	2,040	5,941
Springdale	4,197	3,211	888	691	21.2	21.5	709	567	179	124
Neville	58,978	57,142	2,461	2,064	4.2	3.6	786	773	1,675	1,291
Leet	2,435	7,241	823	1,798	36.9	25.2	352	209	471	1,589
Crescent	6,370	6,452	1,402	2,489	22.0	38.6	735	815	667	1,674
White	4,969	5,008	1,628	2,690	32.8	53.8	689	732	939	1,958
West Lebanon	6,037	5,535	968	1,520	16.0	27.5	1,068	539	-100	981
Borough	5,990	6,004	719	1,900	12.0	31.6	964	262	-245	1,638
South Versailles	737	674	254	524	34.5	76.6	386	146	-132	378

(a) Montgomery Co.
(b) Delaware Co.

Since none of the townships with exceptionally high delinquencies have been included in this group, the trend toward greater delinquency may be taken as typical. The total amount of taxes unpaid at the end of the year on the 1930 tax levies of the sixty-one townships comprised over 15% of the total levies; in 1931 the amount delinquent comprised over 19% of the total levies; and in 1933 slightly over 23%. In the face of this alarming growth it is surprising to note the general failure to make adjustment in budget appropriations as found in Table I of the budget chapter.

The extent to which delinquency on the current levy has been offset by collection of taxes of previous years is shown in column four. This column shows conclusively that with the exception of one township (Upper Darby) the collections from unpaid taxes of prior years do not even approach the amount of delinquency on the current tax duplicate. This means that each year the total amount of outstanding or unpaid taxes increases. In the townships with high delinquency rates the total amount of such outstanding taxes approximates the current levy. Finally, it may be noted that in only four townships, Coal, Hanover, Reserve, and Tinicum, did the excess of current delinquency over collections on the unpaid taxes of prior years decrease in 1932 as compared with 1930.

In summary then the table clearly shows: a growing amount of delinquency in the collection of township and other local taxes; an increasing excess of current delinquency over collections on unpaid taxes of prior years; and a gradual increase in the net amount of outstanding unpaid taxes.

The data for Table III were taken from the official audit statements, from the records of the Pennsylvania Economic Council, and from materials collected in the field. An examination of this table summarizing the receipts of fifty townships, reveals the relative importance of their sources of revenue. Approximately 68% of all receipts in first-class townships are derived from taxes and 32% from all other sources. Temporary loans appear as the next greatest source of revenue, receipts from such loans accounting for nearly 11½% of the total. It must be remembered, however, that money received from temporary loans, as recorded in official reports, is often greater than the actual receipts for the year. Consequently, the money received from the sale of bonds would in reality constitute the second largest item in the total receipts. Departmental earnings (3.83%), licenses and fines (3.07%), and special assessments (2.79%) respectively constitute the next most productive sources of revenue.

Table IV, summarizing receipts of fifty townships for 1933 by population groups, shows that with a few minor exceptions the proportionate amount received from the various revenue sources is about the same for each group. Thus, the percentage accounted for by tax collections in the first group (68.0%), second group (62.1%), third group (69.0%), and the last group (70.0%) shows little variation. The low proportion occurring in the second group is entirely accounted for by the fact that Springfield township (Montg.Co.) received over eighty thousand dollars in 1933 from the proceeds of bond sales. Likewise, the large percentage of money derived from special assessments in the second group as contrasted with the other three is accounted for by the fact that growing urban townships such as Shaler and Baldwin tend to place the burden of permanent improvements upon abutting property owners. Consequently, these special assessments bulk larger in this group where the presence of rural townships such as South Fayette, Elizabeth Collier, and Swatara, which provide few governmental services other than highway maintenance, makes for a much smaller total of receipts and expenditures.

A distinct correlation may be observed between the urban character of the townships and the per capita tax burden. The first group which contains the largest number of the more urbanized communities has a per capita tax of over twelve ($12.80) dollars while the remaining groups show a decreasing per capita burden. The unusually low per capita burden ($4.34) in the second group is accounted for by the presence of a large number of rural or predominantly rural townships. The same correlation exists as to per capita non-revenue receipts. The larger urban communities have been generally unwilling to sacrifice or curtail governmental services and consequently as tax collections have decreased they have endeavoured to meet the difference by temporary and long term borrowing. The rural townships have not been faced with the problem of such curtailment of services and consequently they have had less occasion for short or long term borrowing. The borrowing in groups two, three, and four is almost entirely restricted to the urban communities within these groups, as may be seen by reference to Table III which analyzes the receipts of the fifty townships individually.

T A B L E   III

SUMMARY OF RECEIPTS - 1933

FOR 50 TOWNSHIPS, TOTAL POPULATION 378,702

	RECEIPTS	Amount	Per Capita	Percent of Total Receipts
REVENUE	Taxes	$ 3,753,915	$ 9.91	67.58 %
	Special Assessments	155,202	.41	2.79
	Licenses and Fines	170,674	.45	3.07
	Departmental Earnings	212,718	.56	3.83
	Interest	19,406	.05	.35
	Water Rents	48,460	.13	.87
	Gifts and Grants	32,434	.09	.58
	Miscellaneous	9,352	.02	.17
	Total Revenue Receipts	4,402,161	11.62	79.25
NON-REVENUE	Sale of Bonds	480,449	1.27	8.65
	Temporary Loans	632,312	1.67	11.38
	Sales	1,381	.003	.02
	Refunds & Joint Construc.	13,778	.04	.25
	Miscellaneous	24,976	.07	.45
	Total Non-Rev. Receipts	1,152,896	3.05	20.75
	GRAND TOTAL RECEIPTS	$ 5,555,057	$14.67	100.00 %

TABLE IV

SUMMARY OF RECEIPTS - 1933

FOR 50 TOWNSHIPS, BY POPULATION GROUPS

12 Townships, Population over 10,000

	RECEIPTS	Amount	Per Capita	Percent of Total Receipts
REVENUE	Taxes	$ 2,932,732	$12.80	68.0 %
	Special Assessments	99,439	.43	2.3
	Licenses and Fines	117,119	.51	2.7
	Departmental Earnings	188,922	.82	4.4
	Interest	13,501	.06	.3
	Water Rents	-------	----	----
	Gifts and Grants	28,999	.13	.7
	Miscellaneous	3,044	.01	.1
	Total Revenue Receipts	$ 3,383,756	14.76	78.5
NON-REV.	Sale of Bonds	381,500	1.66	8.9
	Temporary Loans	532,653	2.32	12.4
	Sales	269	.00	.006
	Refunds & Joint Construction	10,966	.06	.3
	Miscellaneous	1,475	.01	.03
	Total Non-Rev. Receipts	926,863	4.04	21.5
	TOTAL RECEIPTS	$ 4,310,619	$18.81	100.0 %

12 Townships, Population 5,000 to 10,000

	RECEIPTS	Amount	Per Capita	Percent of Total Receipts
REVENUE	Taxes	$ 353,595	$ 4.34	62.1 %
	Special Assessments	45,628	.56	8.0
	Licenses and Fines	26,271	.32	4.6
	Departmental Earnings	1,239	.01	.2
	Interest	1,337	.02	.2
	Water Rents	14,775	.18	2.6
	Gifts and Grants	1,441	.02	.3
	Miscellaneous	3,433	.04	.6
	Total Revenue Receipts	447,719	5.50	78.5
NON-REV.	Sale of Bonds	87,377	1.07	15.3
	Temporary Loans	33,245	.41	5.8
	Sales	489	.006	.09
	Refunds & Joint Construction	1,474	.02	.3
	Miscellaneous	------	-----	-----
	Total Non-Rev. Receipts	122,585	1.50	21.5
	TOTAL RECEIPTS	$ 570,304	$ 7.00	100.0 %

TABLE IV

SUMMARY OF RECEIPTS - 1933

FOR 50 TOWNSHIPS, BY POPULATION GROUPS

11 Townships, Population 3,000 to 5,000			Percent of Total Receipts
RECEIPTS	Amount	Per Capita	
REVENUE			
Taxes	$ 311,957	$ 7.39	69.0 %
Special Assessments	10,002	.24	2.2
Licenses and Fines	21,994	.52	4.9
Departmental Earnings	19,926	.47	4.4
Interest	1,393	.03	.3
Water Rents	18,881	.45	4.2
Gifts and Grants	260	.006	.1
Miscellaneous	2,467	.06	.5
Total Revenue Receipts	386,880	9.16	85.6
NON-REV.			
Sale of Bonds	-------	----	----
Temporary Loans	47,702	1.13	10.6
Sales	20	.0004	.004
Refunds & Joint Construction	262	.006	.1
Miscellaneous	17,201	.41	3.8
Total Non-Rev. Receipts	65,185	1.54	14.4
TOTAL RECEIPTS	$ 452,065	$10.71	100.0 %

15 Townships, Population under 3,000			
RECEIPTS	Amount	Per Capita	Percent of Total Receipts
REVENUE			
Taxes	$ 155,631	$ 6.02	70.0 %
Special Assessments	133	.005	.1
Licenses and Fines	5,290	.20	2.4
Departmental Earnings	2,631	.10	1.2
Interest	3,175	.12	1.4
Water Rents	14,804	.57	6.7
Gifts and Grants	1,734	.07	.8
Miscellaneous	408	.02	.2
Total Revenue Receipts	183,806	7.11	82.8
NON-REV.			
Sale of Bonds	11,572	.45	5.2
Temporary Loans	18,712	.72	8.4
Sales	603	.02	.3
Refunds & Joint Construction	1,076	.04	.5
Miscellaneous	6,300	.24	2.8
Total Non-Rev. Receipts	38,263	1.48	17.2
TOTAL RECEIPTS	$ 222,069	$ 8.59	100.0 %

CONCLUSIONS

(1) <u>FINDING:</u> The present statutory provisions fail to limit the total township tax levy although such limitations are imposed in the case of a number of the individual taxes that may be levied at the discretion of the commissioners.
<u>RECOMMENDATION:</u> A maximum millage should be prescribed which would limit the total tax rate.

(2) <u>FINDING:</u> There is an extreme variation in the total township tax levy among these townships.

(3) <u>FINDING:</u> The present practice of tax exoneration is accompanied by abuses which may be remedied, in part, by some form of publicity requirement.

(4) <u>FINDING:</u> There is a definite and accentuated trend toward greater tax delinquency and an increase in the total amount of outstanding taxes.

(5) <u>FINDING:</u> Tax collections comprise more than two-thirds of the total receipts of the first-class townships.

(6) <u>FINDING:</u> There is a distinct correlation between the per capita tax burden and the urban character of the township, the more urban townships having a greater per capita burden than the more rural ones.

CHAPTER XXIII

FUNCTIONAL COSTS

The cost of township government has been tabulated by functions and by offices for 50 of the 61 first class townships. The data were obtained mainly from audit statements and state reports and, as to 25 townships, such material was supplemented by interviews with local officials. In some cases the defective nature of the local accounts, statements, and reports barred the possibility of absolute accuracy except where access was had to vouchers or ledger accounts. Where such access was not possible the reliability of the data is largely dependent upon the accuracy of the audit and state reports, though in many cases further checking through detailed statements on file at the court house or the records of an active taxpayers' association was possible.

Expenditures of Individual Townships - Table I shows the functional costs of these 50 townships in 1933 and, for comparison, the expenditures of 25 of the same townships in 1930. Data for 1933 were obtained in part by field visits and in part from annual audit statements, state reports, and the records of the Pennsylvania Economic Council. All of the 1930 data were obtained from the Pennsylvania Economic Council.

Most of the column headings are self-explanatory. "Miscellaneous governmental expenditures" include donations for Memorial Day services, care of cemeteries, premiums on officers' bonds, compensation insurance, etc. Under "non-governmental expenditures" are listed those expenses which are not operating costs, including permanent improvements, bond retirements and sinking fund payments, and such miscellaneous items as refunds, damage claims, etc. Permanent improvements are properly governmental costs but are not included under that heading because they are not regularly recurring charges.

Relative Cost of Township Functions - Table II shows the relative cost of the various governmental services. Highway maintenance (including sewer maintenance) and police protection are the two most expensive functions of the first class townships. The relatively low proportion of expenditures for highway maintenance and the comparatively high costs of police and fire protection are in sharp contrast with the like governmental costs in second class townships. Highway maintenance consumes well over 50 per cent of of all the expenditures of the chiefly rural second class townships. In the first class townships total highway expenditures, including salaries, maintenance costs, street lighting, and engineering, are only approximately 19 per cent of the total, although highway maintenance is the most costly single function. The second largest expenditure (18 per cent) is for protection of life and property. General government accounts for 8 per cent of the total; health and sanitation, 4½ per cent; recreation, 1 per cent; and water supply, approximately 1 per cent.

Non-governmental expenditures amount to 36.44 per cent of township expenditures. Repayment of debts is the largest item under this classification, taking 26.61 per cent of the total. Total debt service, including interest payments, take more than 38 per cent of all township expenditures. Thus when all the objects of township expenditures are ranked together according to their cost they are as follows: debt service, 38 per cent; highways, 19 per cent; protection of life and property; 18 per cent; permanent improvements, 8.5 per cent; general government, 8 per cent; health and sanitation, 4.5 per cent; recreation, 1 per cent; and water system, .8 per cent. Other miscellaneous expenditures account for slightly more than 2 per cent of total expenditures.

Mandatory Township Expenditures - It is significant that mandatory expenditures account for only a relatively small proportion of the total governmental costs of townships of the first class. Even if the broadest definition of the term "mandatory" is adopted, such expenditures would include only the following items: the salaries and expense of the commissioners, secretaries, auditors, and treasurer; board of health (but not the costs of rubbish collection), highway salaries and maintenance, bond maturity and sinking fund payments, and interest on funded debt. The repayment of temporary loans and the interest charges on such loans are not properly classified as mandatory items since such costs are voluntarily contracted by the board. However, since no distinction has been made as to interest charges for temporary loans and for funded indebtedness, the entire interest item will be included even though only a portion of it may be considered of a mandatory nature. If all these items are added they will constitute only forty-five per-

T A B L E I.
FUNCTIONAL COSTS

GOVERNMENTAL EXPENDITURES

GENERAL GOVERNMENT

	1	2	3	4	5	6	7
TOWNSHIP	Commissioner's Salaries	Secretary's Salaries	Secretary's Expense	Auditor's Salary & Exp.	Treasurer's Salary & Exp.	Solicitor's Salary & Exp.	Public Property
Population of 10,000 and Over							
1933 Upper Darby	1500	17066	5164	1952	14149	5084	14401
1930	1500	7959	2166	1970	16107	6736	18612
1933 Lower Merion	3900	15396	4299	264	5534	6084	13197
1930	1245	12885	5734		5998	3747	19945
1933 Coal	2600	2340	1118	1145	7279	4330	92
1933 Abington	2400	7141	1718	1982	7213	3441	9549
1930	1380	5629	1260	1053	6242	5122	5098
1933 Hanover	987	1882	111	1015	11083	2393	360
1933 Cheltenham	1200	9641	1387	100	7250	2250	3684
1930 Mt.Lebanon	1200	7955	1035	-	8991	2464	3382
1933 Stowe	1000	11904 ⑦	2228 ⑧	465	5456	4277	5201
1933 Penn	5185	-	5146	-	2959	-	-
1933 Harrison	2200	1200	705	125	2173	1000	-
1930	783	1650	237	909	2335	896	1426
1933 Newport	900	1710	1039	1741	8049	2293	1056
1930	600	1440	2341	951	7468	100	36
1933 Whitehall	1000	960	500	270	2098	500	1158
1930	500	960	-	-	2297	600	400
Population 5,000 to 10,000							
1933 Shaler	1400	2161	558	461	4323	600	614
1930	692	1659	1546	206	4808	4100	924
1933 South Fayette	1200	500	79	114	606	486	55
1933 Mifflin	1600	600	223	225	2177	4000	1762
1930	1500	1200	857	525	3841	2310	100
1933 Elizabeth	1800	360	301	101	843	300	404
1930	879	360.	373	123	1705	300	839
1933 Baldwin	1600	918	-	213	1769	1220	517
1930	775	1020	-	-	2733	1500	1163
1933 Swatara	783	150	229	15	437	100	-

Year		1	2	3	4	5	6	7
1930	Scott	500	150	335	-	874	100	-
1933	Scott	1350	760	604	109	3202	950	936
1933	Collier	900	200	433	178	770	518	24
1933	East Bethlehem	1000	295	310	-	553	250	69
1933	North Versailles	1400	512	147	311	1142	248	405
1933	Springfield (Montgomery County)	1000	1800	215	-	3105	500	1103
1930	O'Hara	500	900	820	-	2981	500	2162
1933	O'Hara	1000	660	184	31	1701	728	60
	Population 3,000 to 5,000							
1933	Spring Garden	1000	350	120	80	725	500	50
1933	Springfield (Delaware County)	500	2776	409	63	2833	735	241
1930	Springfield (Delaware County)	500	3004	193	146	3441	800	254
1933	Plymouth	500	400	-	-	999	100	-
1930	Plymouth	500	400	-	-	1600	50	-
1933	Jefferson	892	250	213	81	548	704	269
1933	Upper Moreland	500	2475	892	-	1634	750	614
1930	Upper Moreland	500	1150	512	-	1432	-	17
1933	East Deer	692	600	136	233	2617	478	3532
1930	East Deer	681	750	784	137	3966	1608	3631
1933	Tinicum	500	500	3397*	90	239	1237	24
1930	Lower Chichester	500	500	1743	75	2446	350	-
1933	Lower Chichester	900	17	155	108	1024	295	-
1933	Lawrence Park	500	250	76	91	1500	206	-
1930	Lawrence Park	653	215	688	101	1659	453	111
1933	Reserve	450	270	245	111	889	366	989
1933	Lackawanna	500	125	27	-	327	350	-
	Population Under 3,000							
1933	Annville	500	120	144	15	270	100	72
1933	Nether Providence	500	592	65	-	1157	225	-
1933	Darby	500	200	46	45	466	305	192
1933	West Norriton	500	420	697	30	880	100	843
1930	Rochester	500	420	473	30	1593	128	300
1933	Rochester	500	325	-	-	360	233	71
1930		163	339	44	-	672	-	-
1933	Patterson	613	125	152	-	388	225	331
1930	Patterson	500	150	98	-	1185	150	-
1933	Braddock	500	150	293	105	366	202	117
1933	Springdale	525	100	75	35	127	103	251
1930	Neville	260	60	-	-	220	125	175.
1933	Neville	492	500	480	375	2351	787	5973

	1	2	3	4	5	6	7
1933 Leet	-	140	43	10	300	121	-
1933 Crescent	500	200	33	16	256	200	46
1933 White	500	100	2	-	211	50	-
1930 West Lebanon	144	40	21	30	234	50	-
1933 West Lebanon	500	100	25	125	299	80	-
1930 Borough	500	100	-	125	-	80	-
1933 Borough	500	100	33	53	222	200	-
1930 South Versailles	500	100	-	60	324	200	49
1933 South Versailles	48	22	-	-	31	-	-
Total 1933 Expenditures for 50 Townships	52300	91013	34693	13146	117225	51100	69688
Total 1933 Expenditures for 25 Townships	25913	66174	21527	7798	69256	30869	53352
Total 1930 Expenditures for 25 Townships	17172	49345	20973	5778	82817	31573	57198

FOOT NOTES:

(1) Includes library expenditures and assessors' salaries.
(2) Includes Planning Commission expenditures and assessors' salaries.
(3) Includes Planning Commission expenditures.
(4) Includes assessors' salaries.
(5) Includes refund of taxes: Newport $23,524; Hanover $29,674.
(6) Includes salaries and other expenses.
(7) Includes salaries of Township Manager and clerks.
(8) Includes expenses of Township Manager, Clerks, and Planning Commission.
(9) Includes library expenditures.
† Includes personnel service and other expense.
‡ Includes election expenditures.
* Includes administrative expenditures.

TABLE I.

FUNCTIONAL COSTS

GOVERNMENTAL EXPENDITURES

	PROTECTION OF LIFE AND PROPERTY			HEALTH AND SANITATION		RECREATION	HIGHWAYS			
	8	9	10	11	12	13	14	15	16	17
TOWNSHIP	Police	Fire	Bldg. & Plumbing Inspection	Board of Health	Rubbish Collection		Highway Maintenance Salaries	Highway Maintenance Expense	Street Lighting	Engineering
Population of 10,000 and over										
1933 Upper Darby	144234	31661	3814	8236	37797	2203	36017	27792	68107	17255
1930	148246	36365	–	14983	46868	–	81995	60569	69892	–
1933 Lower Merion	209325	53019	9629	27014	50351	14404	58566	92334	54759	–
1930	200346	43403	12028	25839	41655	4370	65232	149742	39767	–
1933 Coal	135	13563	–	2709	3215	–	17935	24102	5000	2250
1933 Abington	59550	20181	4366	6951	20092	21161	27696	14728	28000	1098
1930	63806	26095	6993	8677	24185	10280	9020	32323	23004	–
1933 Hanover	32277	24039	–	1718	–	–	10142	1172	–	96
1933 Cheltenham	64729	19616	66	5590	14833	10331	21454	43767	29498	–
1930 Mt.Lebanon	67947	20334	193	8016	15833	14843	32523	88526	22703	1380
1933 Mt.Lebanon	28692	22143	–	1265	15300	10114	13259	6353	15728	211
1933 Stowe	19454	1519	–	1238	8167	–	8515	1510	–	2359
1933 Penn	2160	13498	–	1239	–	–	13295	4613	17776	4547
1933 Harrison	6705	3794	–	550	1354	–	14708	1733	4670	–
1933 Newport	14227	16926	–	5108	–	–	13064	6746	2927	785
1930	9494	12309	–	3479	–	–	16228	6627	4817	1050
1933 Whitehall	4131	6350	–	854	162	–	6000	25745	11600	1867
1930	3481	4139	–	993	–	–	–	22457	10866	–
Population 5,000 to 10,000										
1933 Shaler	2762	354	–	1157	–	–	10971	7113	240	1078
1930	4408	1998	–	1377	–	–	23407	20384	243	4503
1933 South Fayette	252	828	–	500	–	–	4729	3605	–	–
1933 Mifflin	2267	201	–	597	–	–	2003	6803	–	–
1930	9010	4392	–	1171	–	–	522	1821	2335	–

	8	9	10	11	12	13	14	15	16	17
1933 Elizabeth	-	2197	-	829	-	-	4945	2372	-	-
1930	-	1500	-	996	-	-	22151	638	-	-
1933 Baldwin	1420	113	-	987	-	-	4177	2826[6]	511	-
1930	2170	4071	-	1012	492	-	-	5432[6]	5465	-
1933 Swatara	173	1875	-	416	1279	-	1851	2533	2849	37
1930	130	1190	-	384	-	-	4426	12407	2909	210
1933 Scott	67	3186	-	120	-	-	3790	2788	5195	610
1933 Collier	-	1626	-	465	-	-	3989	2266	2166	-
1933 East Bethlehem	1439	414	-	432	-	-	3863	2167	-	-
1933 North Versailles	-	642	-	807	-	-	6525	2479	-	-
1933 Springfield (Montgomery County)	11841	7522	-	1605	-	-	1269	3724	8704	3600
1930 O'Hara	16039	7020	-	1222	-	-	15424	22578	8601	3687
1933	3464	2098	237	496	-	-	7793	6517	-	394
Population 3,000 to 5,000										
1933 Spring Garden	3428	10734	-	200	3232	-	3340	8084	3154	946
1933 Springfield (Delaware County)	10359	7285	241	801	1076	-	3090	6612	6021	-
1930	11532	9276	367	943	1400	-	5806	19317	5168	12
1933 Plymouth	-	2125	-	457	-	-	2105	3094	3627	-
1930	-	2030	-	296	-	-	4323	4971	3278	-
1933 Jefferson	-	-	-	973	-	-	4743	2241	-	60
1933 Upper Moreland	5838	1715	-	2733	739	-	-	9080	2877	-
1930	4531	500	-	334	-	-	5772	14056	-	-
1933 East Deer	3974	1105	-	239	1582	-	3235	2424	4126	432
1930	5266	780	-	505	3120	1045	4580	2879	4260	113
1933 Tinicum	4673	2058	-	1135	-	3191	3134	6852[6]	3563	-
1930	5694	3100	-	1613	1394	-	6665	5578	2869	-
1933 Lower Chichester	565	2220	19	139	790	-	-	2731[6]	1654	-
1933 Lawrence Park	1838	4070	35	878	960	-	977	4108	5424	-
1930 Reserve	3067	6772	-	993	2568	-	900	4170	8493	250
1933 Reserve	-	590	-	202	-	-	-	2854[6]	3032	-
1933 Lackawanna	966	320	-	110	-	-	1759	787	200	-
Population Under 3,000										
1933 Annville	9	1992	-	136	-	-	742	3719	2876	-
1933 Nether Providence	-	1374	-	533	-	-	-	6619	4723	50
1933 Darby	2001	-	-	647	-	-	920	340	-	-
1933 West Norriton	-	3066	-	446	-	-	2256	3147	2684	-

	8	9	10	11	12	13	14	15	16	17
1930 Rochester	–	1479	–	529	–	–	3534	11263	2597	–
1933	239	673	–	100	–	–	–	2469†	594	–
1930 Patterson	–	–	–	–	–	–	600	992	–	–
1933	–	1038	–	103	–	–	2332	1643	–	–
1930	–	2372	–	146	–	–	1389	2218	1610	62
1933 Braddock	147	1464	47	205	–	–	641	193	1652	–
1933 Springdale	–	50	–	182	–	–	543	87	–	–
1930	–	–	–	200	–	–	–	1313⑥	–	–
1933 Neville	6316	4268	–	639	3462	–	9	690	2361	300
1933 Leet	235	672	–	70	1110	–	–	1101⑥	920	–
1933 Crescent	100	–	–	–	–	–	1826	661	1748	–
1933 White	–	75	–	–	–	–	–	774⑥	1017	–
1930	–	–	–	–	–	–	716	1759	1684	–
1933 West Lebanon	–	243	–	100	–	–	316	1134	1105	–
1930 Borough	–	324	–	181	–	–	–	4403⑥	1218	–
1933 Borough	–	66	–	65	–	–	–	109⑥	1146	–
1930	–	–	–	100	–	–	–	940⑥	1090	–
1933 South Versailles	–	–	–	–	–	–	–	3126⑥	–	–
Total 1933 Expenditures for 50 Townships	649992	294568	18419	81976	164714	59258	328564	366649	312234	39270
Total 1933 Expenditures for 25 Townships	541580	183584	18135	66583	128084	49144	206041	277012	239379	26149
Total 1930 Expenditures for 25 Townships	555167	189449	19616	73989	138302	32684	305213	497363	222869	11267

T A B L E   I.

FUNCTIONAL COSTS — Continued

	TOWNSHIP	18 GOVERNMENTAL EXPENDITURES Water Works	19 Interest	20 Miscellaneous	21 Total Governm't'l. Expend.	22 Per Capita Governm't'l. Expend.
	Population of 10,000 and Over					
1933	Upper Darby	-	177369	4277	616126	13.21
1930		-	150016	6372	670288	14.38
1933	Lower Merion	-	16363	20041 [1]	656185	18.66
1930		-	20411	15135 [1]	667746	18.99
1933	Coal	-	26115	1210	115138	5.78
1933	Abington	-	35237	4241 [2][8]	276745	14.84
1930		-	8933	3093 [3][8]	242193	12.99
1933	Hanover	-	43445	1955	132675	7.47
1933	Cheltenham	-	74775	4163	314334	19.98
1930	Mt. Lebanon		81307	383625	383625	24.39
1933			92781	2442 [9]	238819	17.82
1933	Stowe		31026	-	87078	6.51
1933	Penn		11080	3561	79172	5.94
1933	Harrison		22360	1118	65228	5.27
1933	Newport		13608	1231	91410	7.56
1930			32079	3105	102124	8.45
1933	Whitehall		6193	4162	73550	6.87
1930			5768	2545	55006	5.14
	Population 5,000 to 10,000					
1933	Shaler	14599	21690	5640 [4]	75721	7.91
1930		10016	9504	543	90318	9.43
1933	South Fayette	-	-	370	13324	1.46
1933	Mifflin	-	9029	685	32172	3.94
1930		-	9811	683	39578	4.85
1933	Elizabeth	-	30	753	15275	2.04
1930		-	-	949	30813	4.12
1933	Baldwin	-	6934	2346	26043	4.09
1930		-	5830	261	32711	5.13
1933	Swatara	-	6	616	12070	1.91

Year	Place	18	19	20	21	22
1930	Scott	-	12	700	24327	3.41
1933	Collier	-	7125	150	30942	4.99
1933	East Bethlehem	-	1	214	13750	2.26
1933	North Versailles	-	140	423	11355	1.97
1933		-	2658	405	19079	3.36
1933	Springfield (Montgomery County)	-	3538	1106	50632	9.14
1930	O'Hara	-	794	450	83678	15.10
1933		-	80	1383	26826	5.23
	Population 3,000 to 5,000					
1933	Spring Garden	-	455	1116	37414	8.00
1933	Springfield (Delaware County)	-	13040	893	55975	12.20
1930	Plymouth	-	15401	536	78096	17.01
1933		-	-	937	14344	3.28
1930	Jefferson	-	-	1198	18646	4.27
1933		-	-	313	11287	2.73
1933	Upper Moreland	-	720	2558	33125	8.30
1930		8459	-	1140	29944	7.51
1933	East Deer	21172	6429	800	41093	10.46
1930	Tinicum	-	3778	585 ‡	56595	14.92
1933		-	3255	-	30657	8.45
1930	Lower Chichester	-	4110	-	36637	10.09
1933	Lawrence Park	-	3872	180	14650	4.22
1933		-	2860	410	25212	7.78
1930	Reserve	5542	3184	446	37949	11.71
1933		-	3256	48	18844	6.06
1933	Lackawanna	-	77	331	5879	1.91
	Population Under 3,000					
1933	Annville	-	1311	219	12225	4.08
1933	Nether Providence	-	1269	804	17911	6.32
1933	Darby	-	102	290	6054	2.18
1933	West Norriton	-	-	1000 ⑨	16069	6.21
1930	Rochester	-	584	516	23946	9.25
1933		-	648	550	6762	2.84
1930	Patterson	-	3455	1327	7592	3.19
1933		-	1629	402	8981	4.87
1930		-	1416	154	11450	6.20

	18	19	20	21	22
1933 Braddock	-	1114	182	7378	4.11
1933 Springdale	-	7	99	2184	1.40
1930 Neville	-	-	337	2690	1.72
1933 Neville	14598	8657	653	52911	34.54
1933 Leet	-	1536	189	6447	5.07
1933 Crescent	-	90	130	5806	5.24
1933 White	-	145	178	3052	2.96
1930 West Lebanon	-	169	103	4950	4.80
1933 West Lebanon	1436	174	160	5793	6.29
1930 Borough	1820	212	583	9546	10.36
1933 Borough	-	804	-	3298	3.73
1930 South Versailles	-	850	181	4394	4.96
1933 South Versailles	-	-	40	453	1.35
Total 1933 Expenditures for 50 Townships	44634	653033	74977	3,517453	
Total 1933 Expenditures for 25 Townships	24494	394483	57251	2,486808	
Total 1930 Expenditures for 25 Townships	33008	357124	45935	2,746842	

T A B L E I. - Continued

FUNCTIONAL COSTS - Continued

		NON-GOVERNMENTAL EXPENDITURES				Total Non-Gov't'l. Exp.	Per Capita Non-G. Exp.	TOTAL EXPEND-ITURES	PER CAPITA TOTAL EXPENDITURES
	TOWNSHIP	Permanent Improv'ts.	Temporary Loans Repaid	Bond Maturity and Sink. F. Paym'ts.	Miscel-laneous				
		23	24	25	26	27	28	29	30
	Population of 10,000 and Over								
1933	Upper Darby	73220	380827	138822	1200	594069	12.74	1210195	25.95
1930		591351	233000	55321	60	879732	19.87	1550020	33.24
1933	Lower Merion	78021	120000	22364	-	220385	6.27	875570	24.93
1930		672932	345000	32740	3238	1053910	29.97	1721656	48.96
1933	Coal	3230	47433	25000	-	75663	3.80	190801	9.57
1933	Abington	37366	90000	27265	820	155451	8.34	432196	23.18
1930		118387	110000	48259	4455	281101	15.07	523294	28.06
1933	Hanover	-	-	56713	40330 ⑤	97043	5.46	229718	12.92
1933	Cheltenham	68083	53000	78095	321	199499	12.68	513833	32.66
1930	Mt. Lebanon	182598	-	75128	-	257726	16.38	641351	40.77
1933	Stowe	6588	-	71695	-	78283	5.84	317102	23.66
1933	Penn	1000	1000	6706	-	8706	.65	95784	7.17
1933	Harrison	16156	6000	7000	-	29156	2.19	108328	8.12
1930		-	18569	7256	36	25861	2.09	91089	7.35
1933	Newport	-	-	53615	23607 ⑤	77222	6.39	168632	13.95
1930		68420	50000	33900	-	152320	12.60	254444	21.05
1933	Whitehall	15552	20929	15806	-	52287	4.88	125837	11.75
1930		41645	18587	33397	-	93629	8.74	148635	13.88
	Population 5,000 to 10,000								
1933	Shaler	46383	19925	16201	3356	85865	8.97	161586	16.87
1930		177954	7900	-	196	186050	19.43	276368	28.86
1933	South Fayette	-	-	-	-	-	-	13324	1.46
1933	Mifflin	-	-	-	186	186	.02	32358	3.96
1930		115496	-	8500	353	124349	15.23	163927	20.07
1933	Elizabeth	-	1970	-	-	1970	.26	17245	2.31
1930		-	-	-	10	10	.001	30823	4.12
1933	Baldwin	-	25000	8040	-	33040	5.18	59083	9.27
1930		18494	35795	-	-	54289	8.52	87000	13.65
1933	Swatara	-	500	-	-	500	.08	12570	1.98

	23	24	25	26	27	28	29	30
1930 Scott		1000		410	1410	.22	25737	4.06
1933 Collier	1500	2000	8003	187	11690	1.88	42632	6.87
1933 East Bethlehem		800			800	.13	14550	2.39
1933 North Versailles			4065		4065	.71	15420	2.67
1933	150	2000	6000	50	8200	1.45	27279	4.81
1933 Springfield(Montgomery County)	87637		3000		90637	16.36	141269	25.50
1930 O'Hara	43015	29000			72015	13.00	155693	28.10
1933		6000		5	6005	1.17	32831	6.40
Population 3,000 to 5,000								
1933 Spring Garden		1000	2000		3000	.64	40414	8.64
1933 Springfield(Delaware County)		5000	15401	2270	22671	4.94	78646	17.14
1930 Plymouth	6712		9356		16086	3.50	94164	20.51
1933	8247				8247	1.89	22591	5.17
1930 Jefferson	14136				14136	3.24	32782	7.50
1933				9	9	.002	11296	2.73
1933 Upper Moreland	2775	13000	9400		25175	6.31	58300	14.61
1930 East Deer	498				498	.12	30442	7.63
1933	864	25536	5000	575	6439	1.64	47532	12.10
1930 Tinicum	108534	19040	10000		144070	36.68	202665	51.59
1933			13250		32290	8.90	62947	17.34
1930 Lower Chichester		25000	3820		28820	7.94	65457	18.03
1933 Lawrence Park	10179			20	20	.01	14670	4.22
1930 Reserve	6769	8790	1601		11780	3.63	36992	11.41
1933		2000	1438	10	17007	5.25	54956	16.96
1933			690	21	2711	.87	21555	6.94
1933 Lackawanna							5879	1.91
Population Under 3,000								
1933 Annville		600	1916	55	2571	.86	14796	4.94
1933 Nether Providence		2000	1345		3345	1.18	21256	7.50
1933 Darby	6817		3708		10525	3.80	16579	5.98
1933 West Norriton	3439				3439	1.33	19508	7.54
1930 Rochester	1243	22700			22700	8.77	46646	18.02
1933	7948	798	2643		4684	1.97	11446	4.81
1930 Patterson	1000	1500			9448	3.97	17040	7.16
1933	2335		572		1572	.85	10553	5.72
1930		1000	1022		4357	2.36	15807	8.56

	23	24	25	26	27	28	29	30
1933 Braddock	-	8500	1890	10	10400	5.80	17778	9.91
1933 Springdale	-	300	-	-	300	.19	2484	1.59
1930 Neville	1550	-	-	-	1550	.99	4240	2.71
1933 Neville	1765	-	2990	-	4755	3.10	57666	37.64
1933 Leet	-	1400	821	-	2221	1.75	8668	6.81
1933 Crescent	-	-	-	-	-	-	5806	5.24
1933 White	-	-	-	-	-	-	3052	2.96
1930 West Lebanon	-	100	-	-	100	.10	5050	4.90
1933 West Lebanon	-	-	1000	-	1000	1.08	6793	7.38
1930 Borough	-	-	1515	-	1515	1.64	11061	12.00
1933 Borough	3039	2000	1000	-	3000	3.39	6298	7.12
1930 South Versailles	-	400	1000	-	4439	5.02	8833	9.98
1933 South Versailles	-	-	-	-	-	-	453	1.35
Total 1933 Expenditures for 50 Townships	471215	851591	620873	73058	2016737		5534190	
Total 1933 Expenditures for 25 Townships	434009	752289	413075	32335	1631708		4118516	
Total 1930 Expenditures for 25 Townships	2181813	915308	315396	8732	3421249		6168091	

TABLE II.

SUMMARY OF FUNCTIONAL COSTS - 1933

FOR 50 TOWNSHIPS, TOTAL POPULATION 378,702

	FUNCTIONAL COSTS	Amount	Per Capita	Percent of Total Costs
GOVERNMENTAL	Commissioners' Salaries	$ 52,300	$ .14	.94 %
	Secretaries' Salaries	91,013	.24	1.64
	Secretaries' Expenses	34,693	.09	.63
	Auditors' Sal. & Exp.	13,146	.03	.24
	Treasurers' Sal. & Exp.	117,225	.31	2.12
	Solicitors' Sal. & Exp.	51,100	.13	.92
	Public Property	69,688	.18	1.26
	Police	649,992	1.72	11.75
	Fire	294,568	.78	5.32
	Bldg. & Plumbing Inspection	18,419	.05	.33
	Board of Health	81,976	.22	1.48
	Rubbish Collection	164,714	.43	2.98
	Recreation	59,258	.16	1.07
	Highway Salaries	328,564	.87	5.94
	Maintenance Expense	366,649	.97	6.63
	Street Lighting	312,234	.82	5.64
	Engineering	39,270	.10	.71
	Water Works	44,634	.12	.81
	Interest	653,033	1.72	11.80
	Miscellaneous	74,977	.20	1.35
	Total Gov. Expen.	3,517,453	9.29	63.56
NON-GOV'TAL	Permanent Improvements	471,215	1.24	8.51
	Temporary Loans Repaid	851,591	2.25	15.39
	Bond Matur. & Sink. F. Pay'ts	620,873	1.64	11.22
	Miscellaneous	73,058	.19	1.32
	Total Non-Gov. Expen.	2,016,737	5.33	36.44
	GRAND TOTAL COSTS	$ 5,534,190	$ 14.61	100.00 %

cent of the total expenditures.† Furthermore, these figures include the cost items of many townships of a rural character where the amount of money expended for highways is disproportionately large due to the restricted nature of the functions performed. If only the urban and semi-urban units, which render many services in addition to highway maintenance, were taken into consideration* the proportion of mandatory expenditures would be considerably decreased.

Variations According to Population - Table III shows how the different items of expenditure and governmental services vary with changes in population. The most striking trend is the tendency of overhead (general government) costs to increase proportionately as the population decreases. Thus general government costs constitute approximately 5½ per cent of total expenditures in the first group of townships (population - 10,000 or over); approximately 9½ per cent in the second group (population - 5,000 to 10 000); approximately 10 per cent in the third group (population - 3,000 to 5,000); and nearly 14 per cent in the fourth group (population - under 3,000). If the solicitor's salary and expenses are included the difference is even more striking: group one, 6.5 per cent; group two, 11.4 per cent; group three, 11.7 per cent; and group four, 14.3 per cent. This trend shows a need for consolidations among the smaller townships.

A second observable trend is the decrease in the proportion of expenditures for protection of life and property and for health and sanitation, which accompanies a decrease in population and size of the townships. Thus protection of life and property constitutes 19 per cent of the total expenditures in the first group, only about 8 per cent in group two, 16 per cent in group three, and approximately 10 per cent in group four. The failure of an absolute and continuing decrease in the proportion of such expenditures is readily explainable by reference to Table II showing the functional costs of all these townships. In the third group 4 townships, Springfield (Delaware County), Lawrence Park, Spring Garden, and Tinicum, are predominantly urban in character, and hence demand services of this nature. The same situation occurs in group four. These communities, because of their urban character and their close proximity to large urban centers, demand governmental services which are alien to the more rural areas.

It will be observed that in the first group, highway maintenance constitutes approximately 11 per cent of the total expenditures, in the second nearly 18 per cent, in the third 17.5 per cent, and in the fourth 16 per cent. The tendency for highway expenditures to increase sharply in the second group and then to fall off slightly in the remaining two groups indicates the presence of townships in the second and third groups of a predominantly rural character. The functional differences between urban and rural townships are shown in that in the smaller units the importance of such services as police, fire, health, etc., is much less while highway maintenance is relatively higher in the list, while in the larger units the reverse is true.

Analysis of Current Operations - Table IV shows the extent to which the townships are operating within the limitations of their current income. Total revenue receipts are derived from such sources as taxes, special assessments, etc. The income from these sources is of a recurring nature while that from the non-revenue group (bond sales, temporary loans, sale of material, etc.) is sporadic and can in no sense be considered as recurrent income. For this reason the amount received from non-revenue items has been excluded from this table.

Current expenditures exceeded total revenue receipts in 20 of the 50 townships listed. Failure of a township to operate within the limitations of its revenue receipts is usually accompanied either by a decrease in its assets (indicated in Table IV by a decrease in its cash balance), or by an increase in its liabilities (shown by the amount of warrants and bills outstanding), or by a combination of both. This situation, if permitted to continue, must ultimately result in one or more of the following: (a) a higher tax rate; (b) an accumulation of unpaid temporary loans or bills payable; or (c) the flotation of bonds.

---

† Several small items of mandatory character such as compensation insurance appear in the miscellaneous group. This item is so small that it would not affect the total percentage of mandatory costs by more than one percent at the outside.
* Application of the same procedure to Tables III A and III B will show that the mandatory expenditures of these townships are much smaller than the figures given here since these townships expend much larger sums for such matters as police, fire, recreation, and the like.

T A B.L E   III. A

SUMMARY OF FUNCTIONAL COSTS - 1933

FOR 50 TOWNSHIPS, BY POPULATION GROUPS

12 Townships, Population over 10,000

	FUNCTIONAL COSTS	Amount	Per Capita	Percent of Total Costs
GOVERNMENTAL	Commissioners' Salaries	$ 23,655	$ .10	.5 %
	Secretaries' Salaries	70,890	.31	1.6
	Secretaries' Expenses	23,652	.10	.5
	Auditors' Sal. & Exp.	9,722	.04	.2
	Treasurers' Sal. & Exp.	75,578	.38	1.7
	Solicitors' Sal. & Exp.	32,548	.14	.8
	Public Property	50,124	.22	1.2
	Police	585,619	2.56	13.4
	Fire	226,309	.99	5.2
	Bldg. & Plumbing Inspection	17,875	.08	.4
	Board of Health	62,472	.27	1.4
	Rubbish Collection	151,271	.66	3.5
	Recreation	58,213	.25	1.3
	Highway Salaries	240,651	1.05	5.5
	Maintenance Expense	250,595	1.09	5.8
	Street Lighting	238,065	1.04	5.5
	Engineering	30,468	.13	.7
	Water Works	------	----	---
	Interest	550,352	2.40	12.6
	Miscellaneous	48,401	.21	1.1
	Total Gov. Expen.	2,746,460	11.98	63.0
NON-GOV'TAL	Permanent Improvements	299,216	1.31	6.9
	Temporary Loans Repaid	737,758	3.22	16.9
	Bond Matur. & Sink. F. Pay'ts	510,337	2.23	11.7
	Miscellaneous	66,314	.29	1.5
	Total Non-Gov. Expen.	1,613,625	7.04	37.0
	TOTAL COSTS	$ 4,360,085	$ 19.03	100.0 %

T A B L E   III. B

SUMMARY OF FUNCTIONAL COSTS - 1933

FOR 50 TOWNSHIPS, BY POPULATION GROUPS

12 Townships, Population 5,000 to 10,000

	FUNCTIONAL COSTS	Amount	Per Capita	Percent of Total Costs
GOVERNMENTAL	Commissioners' Salaries	$ 15,033	$ .19	2.6 %
	Secretaries' Salaries	8,916	.11	1.6
	Secretaries' Expenses	3,283	.04	.6
	Auditors' Sal. & Exp.	1,758	.02	.3
	Treasurers' Sal. & Exp.	20,628	.25	3.6
	Solicitors' Sal. & Exp.	9,900	.12	1.7
	Public Property	5,949	.08	1.0
	Police	23,685	.29	4.2
	Fire	21,056	.26	3.7
	Bldg. & Plumbing Inspection	237	.00	.04
	Board of Health	8,411	.10	1.5
	Rubbish Collection	492	.01	.1
	Recreation	-----	---	----
	Highway Salaries	55,945	.69	9.8
	Maintenance Expense	45,193	.55	7.9
	Street Lighting	19,665	.24	3.5
	Engineering	7,117	.09	1.3
	Water Works	14,599	.18	2.6
	Interest	51,231	.63	9.0
	Miscellaneous	14,091	.17	2.5
	Total Gov. Expen.	327,189	4.02	57.4
NON-GOV'TAL	Permanent Improvements	135,670	1.66	23.8
	Temporary Loans Repaid	58,195	.71	10.2
	Bond Matur. & Sink. F. Pay'ts	45,309	.56	8.0
	Miscellaneous	3,784	.05	.7
	Total Non-Gov. Expen.	242,958	2.98	42.6
	TOTAL COSTS	$ 570,147	$ 7.00	100.0 %

T A B L E   III. C

SUMMARY OF FUNCTIONAL COSTS - 1933

FOR 50 TOWNSHIPS, BY POPULATION GROUPS

11 Townships, Population 3,000 to 5,000

	FUNCTIONAL COSTS	Amount	Per Capita	Percent of Total Costs
GOVERNMENTAL	Commissioners' Salaries	$ 6,934	$ .16	1.7 %
	Secretaries' Salaries	8,013	.19	2.0
	Secretaries' Expenses	5,670	.13	1.4
	Auditors' Sal. & Exp.	857	.02	.2
	Treasurers' Sal. & Exp.	13,335	.32	3.4
	Solicitors' Sal. & Exp.	5,721	.14	1.5
	Public Property	5,719	.14	1.5
	Police	31,641	.75	7.9
	Fire	32,222	.76	8.0
	Bldg. & Plumbing Inspection	260	.01	.1
	Board of Health	7,867	.19	2.0
	Rubbish Collection	8,379	.20	2.1
	Recreation	1,045	.02	.2
	Highway Salaries	22,383	.53	5.6
	Maintenance Expense	47,867	1.13	11.9
	Street Lighting	33,678	.80	8.4
	Engineering	1,338	.03	.3
	Water Works	14,001	.33	3.5
	Interest	33,964	.80	8.4
	Miscellaneous	7,586	.18	1.9
	Total Gov. Expen.	288,480	6.83	72.0
NON-GOV'TAL	Permanent Improvements	22,065	.52	5.5
	Temporary Loans Repaid	40,040	.95	10.0
	Bond Matur. & Sink. F. Pay'ts	47,342	1.12	11.8
	Miscellaneous	2,875	.07	.7
	Total Non-Gov. Expen.	112,342	2.66	28.0
	TOTAL COSTS	$ 400,822	$ 9.49	100.0 %

T A B L E III. D

SUMMARY OF FUNCTIONAL COSTS - 1933

FOR 50 TOWNSHIPS, BY POPULATION GROUPS

15 Townships, Population under 3,000

	FUNCTIONAL COSTS	Amount	Per Capita	Percent of Total Costs
GOVERNMENTAL	Commissioners' Salaries	$ 6,678	$ .26	3.3
	Secretaries' Salaries	3,194	.12	1.5
	Secretaries' Expenses	2,088	.08	1.0
	Auditors' Sal. & Exp.	809	.03	.4
	Treasurers' Sal. & Exp.	7,684	.30	3.8
	Solicitors' Sal. & Exp.	2,931	.11	1.4
	Public Property	7,896	.31	3.9
	Police	9,047	`5	4.5
	Fire	14,981	.58	7.4
	Bldg. & Plumbing Inspection	47	---	---
	Board of Health	3,226	.12	1.5
	Rubbish Collection	4,572	.18	2.3
	Recreation	-----	---	---
	Highway Salaries	9,585	.37	4.7
	Maintenance Expense	22,994	.89	11.3
	Street Lighting	20,826	.81	10.3
	Engineering	350	.01	.1
	Water Works	16,034	.62	7.9
	Interest	17,486	.68	8.7
	Miscellaneous	4,896	.19	2.4
	Total Gov. Expen.	155,324	6.01	76.5
NON-GOV'TAL	Permanent Improvements	14,264	.55	7.0
	Temporary Loans Repaid	15,598	.60	7.6
	Bond Matur. & Sink. F. Pay'ts	17,885	.70	8.9
	Miscellaneous	65	---	---
	Total Non-Gov. Expen.	47,812	1.85	23.5
	TOTAL COSTS	$ 203,136	$ 7.86	100.0 %

T A B L E IV.

ANALYSIS OF CURRENT OPERATIONS

Township	Total Revenue Receipts	Current Expenditures	Cash Balance Beginning of Year	Cash Balance End of Year	Bills and Warrants Outstanding
Upper Darby	$ 680,743	$ 756,148	$ 189,878	$ 167,158	$ 33,456
Lower Merion	871,363	578,549	142,602	258,462	-----
Coal	163,912	140,138	185,493	75,540	46,809
Abington	281,280	304,830	90,693	41,972	89
Hanover	281,762	229,718	18,834	21,697	-----
Cheltenham	428,213	392,750	80,297	46,512	-----
Mount Lebanon	226,195	310,514	134,726	45,303	11,228
Stowe	115,485	93,784	2,488	5,211	1,379
Penn	79,968	86,172	44,074	22,086	4,553
Harrison	70,614	72,520	1,938	665	21,562
Newport	167,890	168,632	1,871	1,131	32,735
Whitehall	66,331	89,356	-----	-----	-----
Shaler	129,094	94,278	53,631	37,579	96
South Fayette	14,098	13,324	2,339	3,583	210
Mifflin	33,816	32,358	4,428	5,930	-----
Elizabeth	16,681	15,275	3,696	5,199	67
Baldwin	45,681	34,083	12,306	3,904	86
Swatara	12,836	12,070	-----	768	-----
Scott	41,123	39,132	2,661	3,337	6,517
Collier	15,705	13,750	646	2,634	-----
East Bethlehem	14,810	15,420	717	106	242
North Versailles	21,519	25,129	8,964	3,365	6,940
Springfield (Montg. Co.)	70,503	53,632	15,674	32,285	-----
O'Hara	34,419	26,831	2,946	10,537	1,002
Spring Garden	39,569	39,414	3,238	10,145	-----
Springfield (Del. Co.)	73,617	73,640	12,802	13,472	-----
Plymouth	18,799	14,344	5,303	15,817	531
Jefferson	12,923	11,296	2,080	3,707	-----
Upper Moreland	42,430	42,525	71,102	73,247	-----
East Deer	62,429	46,668	4,461	21,588	303

Tinicum	$ 53,997	$ 43,907	$ 3,799	$ 10,015	$ -----
Lower Chichester	22,792	14,670	2,479	11,808	8,974
Lawrence Park	35,893	26,813	14,145	13,066	11,162
Reserve	18,231	19,555	9,438	8,113	5
Lackawanna	6,200	5,879	-----	319	1,700
Annville	13,802	14,196	450	367	95
Nether Providence	23,245	19,256	882	5,065	79
Darby	10,057	9,762	728	7,103	-----
West Norriton	20,652	16,069	5,542	6,923	-----
Rochester	8,908	9,405	1,012	471	133
Patterson	9,563	9,553	40	86	10,397
Braddock	8,194	9,124	1,613	3,960	-----
Springdale	2,875	2,184	351	1,264	55
Neville	62,007	55,901	43,718	48,087	42
Leet	5,173	7,268	250	114	2,332
Crescent	5,700	5,806	1,674	1,571	-----
White	2,205	3,052	1,244	532	405
West Lebanon	6,874	6,793	1,528	1,610	-----
Borough	4,104	4,298	172	78	-----
South Versailles	447	453	12	5	-----
TOTALS:	$ 4,404,727	$ 4,210,224	$ 1,188,965	$ 1,053,497	$ 203,133

Many of the townships, such as Lower Merion, Cheltenham, and Shaler, show a decrease in their cash balances even though their total revenue receipts exceeded their current expenditures. Where this situation is found it indicated that the township is either paying for permanent improvements with current revenues or repaying temporary loans accumulated in prior years, or both.

CONCLUSIONS

1. FINDING - Highway maintenance and police protection comprise the two largest items of functional expenditure in the townships of the first class.

2. FINDING - Expenditures for debt service account for more than one-third of the total expenditures.

3. FINDING - Mandatory expenditures constitute less than forty-five percent of the total outlay of these townships.

4. FINDING - General governmental costs increase proportionately as population decreases.

5. FINDING - Decrease in population is accompanied by a decrease in the amount of expenditures for the protection of life and property, and health and sanitation. It is also accompanied by an increase in the relative importance of expenditures for highway maintenance.

6. FINDING - There is a strong trend toward decreasing the assets and increasing the liabilities of the townships resulting from the tendency of current expenditures to exceed total revenue receipts.
   RECOMMENDATION - In view of the serious and continuing financial difficulties which now confront many of the first class townships, the consolidation and annexation of the weaker units should be undertaken under some general plan. It is recommended that a legislative commission prepare such a plan, in consultation with township officers and others.

PART IV—THE BOROUGH

CHAPTER XXIV

THE BOROUGH

Purpose and Scope of Study - It is the purpose of this part of the survey to show briefly the form of government created for boroughs and how the affairs of such municipalities are managed. To do this it has been necessary to examine the Constitution and statutes of the Commonwealth insofar as they pertain to this type of government and, in addition, the official acts, records and reports of a number of boroughs.

Number of Boroughs Studied and Sources of Materials Used - At present there are 936 boroughs whose populations, according to the census of 1930, range from 17 to 35,853. The number of boroughs in each of six population groups is as follows:

Population Range	Number of Boroughs
10,000 or more	49
5,000 to 10,000	97
2,500 to 5,000	148
1,500 to 2,500	123
500 to 1,500	238
Less than 500	239
Population unknown (created since 1930)	6

Three boroughs had more than 25,000 inhabitants each in 1930 while 32 had less than 150 each.

To obtain first-hand information trips were made during the summer of 1934 to 50 boroughs selected as being fairly typical of the boroughs in each population group, although in comparison with the entire number of boroughs in the state the sample contained a relatively high proportion of the larger boroughs, of the smallest boroughs, and of those newly created since 1930. In making the selection care was taken to include boroughs in the anthracite and bituminous coal regions, farming regions in different parts of the state, the metropolitan areas surrounding Philadelphia and Pittsburgh, and the sparsely populated mountain and timber regions. The list includes a number of boroughs which employ borough managers as well as boroughs of approximately the same size which have no such officer. Also included are approximately equal numbers of boroughs operating water or electric plants and those which do not.

In these boroughs numerous officers were interviewed, official records and reports were examined and when possible official forms were obtained for further comparison and analysis. In many boroughs merchants, former officials, and representatives of taxpayers' associations were also interviewed. As a rule most of the information desired was obtainable from the borough secretary or from the manager, in boroughs having managers. Material obtained from other sources was usually merely corroborative. Extremely valuable information was obtained from the active officers of the State Association of Boroughs.

A large part of the statistical data with reference to the levy and collection of taxes was taken from the official figures compiled by the Bureau of Statistics in the State Department of Internal Affairs. Valuable material was also obtained from the Bureau of Municipalities in the same department. The basic data for tables dealing with borough receipts, expenditures and debts were obtained from the analyses of borough auditors' and controllers' reports prepared by the Pennsylvania Economic Council. The computations based upon this material were made under the direction of the Survey staff.

Laws Governing Boroughs - Article III, section 34, of the Constitution of Pennsylvania, as amended in 1923, provides that: "The Legislature shall have power to classify.... boroughs .... according to population, and all laws .... with reference to, any class,

---

This is a condensed summary of a larger collection of data and conclusions which because of limitations of space, could not be included.

shall be deemed general legislation within the meaning of the Constitution; but ....
boroughs (shall be divided) into not more than three classes."

In spite of this grant of power the Legislature has not seen fit to classify boroughs. Several of the older boroughs still operate under special charters or are subject to special laws but for the great majority of boroughs the law is uniform. Most of the basic law relating to boroughs is incorporated in the General Borough Act, 1927 P. L. 519, as amended, which superseded a similar Act of 1915, P. L. 312. This act provides (section 107) that "Any borough or incorporated town, incorporated or acting under any local or special act of assembly, may surrender the provisions of its special acts in their entirety, or so far as they are inconsistent with this act, and be governed by the provisions of this act." Omitted from the General Borough Act are provisions relating to a number of subjects affecting boroughs but on which the law is uniform for several forms of local government. Among these subjects are the assessment of taxes, the procedure for collecting municipal claims, the amount and method of incurring or increasing indebtedness, the election of officers, the giving of municipal consent to public service corporations, state roads, state-aid roads and private roads, validations of elections, bonds, ordinances, and acts of corporate officers, and crimes and offenses by borough officers. Poor districts, boards of health, schools, borough and ward constables, and justices of the peace are also dealt with in separate laws. So also are zoning and public libraries.

Creation of Boroughs - Article II of the General Borough Act provides that the courts of quarter sessions may incorporate any town or village within their jurisdiction into a borough upon application by a majority of the freeholders residing within the limits of the proposed borough. No limitations are prescribed as to the size or density of population. If there is little local opposition to the proposed incorporation it is customarily granted. Discretion, however, is in the court. In the case of the proposed incorporation of Linfield Borough (Q. S. Montgomery County, 1929, No. 352) a charter was denied although the freeholders of the proposed borough were unanimously in favor of it. In his decision Judge Knight pointed out that incorporation would materially and unfairly increase taxation in the remainder of the township, that the reasons for incorporation, namely, lack of curbing and paving and water for fire protection, may be taken care of by the township supervisors upon public application, and that the expediency of incorporation depends upon the advantages and disadvantages to the community as a whole. He quoted with approval from the decision in re Prospect Park Borough, 166 Pa. 502, as follows: "The question of necessity or expediency of incorporating a village and adjacent territory into a borough does not depend so much upon the will of a majority of the freeholders residing outside the limits of the proposed borough, or upon the unanimous consent of those residing within the proposed lines, as it does upon the fact that the advantages to the whole people, as a community will overbalance the disadvantages."

Such is the law. In fact, boroughs are frequently incorporated for reasons of temporary or local self-interest without regard to the general interest of entire communities. Cherry Valley, in Butler County, had a population of 80 persons in 1930; it is strictly a farming community, and performs no functions of a municipal nature. Its only expenditures are for roads and the necessary overhead expenses which result from its separate existence. Persons interviewed in this borough were agreed that the reason for incorporation was the belief that while it was still part of the township a disproportionately small part of the township taxes were spent within its limits. Similar considerations led to the incorporation of Churchill Borough, in Allegheny County, in 1934. In this case, however, the borough is a growing suburb of Pittsburgh and presumably will perform some municipal services in the course of time. The Borough of Allenport, in Washington County, was created for a somewhat similar reason. In many other cases excessively small boroughs are operating because at one time their inhabitants either wished to escape some part of the township tax burden or desired the performance of some functions which the township could not or would not render.

Changes of Borough Limits - Several methods are provided in the General Borough Act by which boroughs may extend their limits to include contiguous territory which is not necessarily in the same county. The annexed territory may be another borough, a township, or a part thereof. Likewise territory may be detached from a borough and annexed to a contiguous township or townships, or made into a separate borough. Boroughs which include parts of two counties are governed for borough and school purposes as one borough and, for county and poor purposes, as parts of the counties and poor districts where the territory is actually situated.

Annulment of Borough Charters - By Article III of the General Borough Act the courts of quarter sessions, upon petition of two-thirds of the taxable inhabitants of a borough, are authorized to annul its charter. The petition in such cases must state that the territory embraced within the limits of the borough shall revert to the township from which it was taken or that it shall be made a new township of the second class. It is noteworthy that while a petition to create a borough need be signed by a mere majority of the freeholders within the proposed limits of the borough, a petition for annulment of a charter requires the signatures of two-thirds of the taxable inhabitants. Presumably this provision was designed to prevent undesirably frequent changes in the form of government. In view, however, of the excess of governmental units in the state it might be well to reverse the requirements and provide for the annulment of charters upon petition by a majority and the granting of charters only upon petition by two-thirds of the taxable inhabitants. The case of the Borough of Narberth is in point. There the residents of the borough would undoubtedly profit from a merger with Lower Merion township but the two-thirds rule has prevented action to date.

Unfortunately the boroughs which have had their charters annulled within the last fifteen years have become new townships without enlarging their limits, so that the possible advantages of consolidation with existing townships have not materialized. The reverse process whereby boroughs extend their limits to include parts of townships occurs frequently, however.

Wastefulness Caused by Multiplicity of Governmental Units - As will be seen from Tables X to XIV in this chapter direct expenditures by small boroughs are of negligible size and are not unduly burdensome. Officers' salaries in such boroughs are at a minimum in most cases. Per capita expenditures for all purposes except street maintenance and street lighting are far below the average for all boroughs chiefly because few small boroughs are active in other ways than these. Per capita expenditures for street maintenance are naturally about as great in large or small places. Similarly it is to be expected that the per capita cost of street lighting would be greater in a small and sparsely settled community than in a more densely populated region. Also, it is probably true that in a very small borough the residents are in much closer contact with their local officers, are better informed concerning the purpose of expenditures than are the taxpayers of large communities, and therefore exercise more direct control over such expenditures. Hence small boroughs spend much less per capita for borough government than do large boroughs.

Nevertheless a large amount of indirect waste results from the separate existence of several hundred needlessly small independent boroughs many of which perform no governmental functions which could not also be performed were the boroughs combined with neighboring townships. This waste takes many forms and imposes needless tax burdens upon the taxpayers of the borough itself, the adjacent townships, the county, state, and even the national government. It is offset only by the fact that the voters of these ultra-small boroughs are given a greater amount of local self-determination in respect to a limited number of governmental policies, but this involves incurring additional expenses for their neighboring communities, with the voters of those communities having no say in the matter.

The most direct additional expense which results from creating a new borough from a township is the cost of electing the borough officers. This expense is borne by the taxpayers of the county as a whole. In one extreme case (New Washington Borough) the county paid more for the rent of an election hall in 1933 than the borough raised by taxation and all other miscellaneous receipts combined. As is shown in Chapter V above, maintenance of unnecessary election districts costs thousands of dollars annually.

The greatest waste results because there is ordinarily a separate school district corresponding to each separate borough and township. This results in the existence of many uneconomically small school districts. This situation does not necessarily follow and could be corrected by amending the school laws to bring about the consolidation of school districts without eliminating any of the primary governmental units. The present extravagantly large number of school districts is, however, the direct result of the long established rule that each borough should be a separate school district.

The existence of several hundred unnecessary units of local government complicates and adds to the cost of many state departments, notably those of Internal Affairs, Health, Welfare, Highways and, above all, Public Instruction. Even the federal government is put to some extra expense. Thus, for example, both Stroudsburg and East Stroudsburg have or are about to have handsome new postoffices, either of which would be adequate for the combined population.

Although such wastefulness sometimes occurs where large boroughs are adjacent, it is most conspicuous in the case of small boroughs. Four of the boroughs visited during the summer of 1934, New Buffalo, Cherry Valley, Gouldsboro, and New Washington, had populations of less than 100 each. New Buffalo provides street lights and owns a public dump, the latter cared for without cost by persons living nearby. Otherwise street maintenance is the only function performed by these boroughs at present and that is of the most rudimentary character except for the state roads. The legislature might well consider taking steps toward revoking the charters of incorporated places which perform no municipal services. In other cases where boroughs, though small, perform some municipal functions savings would result from enlarging the territory served. The limits of many existing school districts should certainly be enlarged.

Relative Advantages of the Borough Form of Government - Forty-nine boroughs are now large enough to become cities of the third class if their inhabitants preferred that form of government. Many other boroughs are larger than the smallest existing city. Although 9 cities were created between 1910 and 1920, 7 between 1920 and 1930, and 2 since 1930, there still remain more boroughs large enough to be cities than there are cities. Persons interviewed in a number of these large boroughs attributed their preference for the borough form of government to its greater flexibility and inexpensiveness and the fact that boroughs are granted authority to exercise all of the important powers of cities. Unlike cities, boroughs may at their option adopt the efficient and economical manager form of government. Few officers need be employed by boroughs which wish to dispense with them and salaries are seldom prescribed by statute. The members of borough councils serve without pay whereas councilmen in cities receive statutory salaries. (For other comparisons of borough and city costs see the tables at the end of this and the following chapter.)

On the other hand there are not so many advantages as formerly in a small community being incorporated, for townships are now permitted to render many of the services which once led to the incorporation of boroughs. In four of the boroughs created since 1930 three closely related factors were chiefly responsible for incorporation. These were a desire for a greater degree of local control over their local government, a desire to escape payment of greater amounts of township taxes than were returned to the territory comprising the borough, and a desire to provide locally for better or more numerous services than the township government chose to offer.

Corporate Powers - (1) General Powers. Each borough is a body corporate and politic and possesses the general powers of a corporate body.

(2) Power to Create Offices. The General Borough Act provides that there shall be elected in each borough a council, burgess, high constable, assessor, tax collector, three auditors or one controller, and that a secretary and a treasurer shall be appointed by the council. Other offices which may be created by council include the following: Street commissioner, building inspector, solicitor, police, borough manager, engineer, health officer, borough planning commission, bureau of mine inspection and surface support (in the anthracite region), commission of water works, recreation board, shade tree commission, park commission, board of health, and zoning commission. In addition council may "appoint such other officers as it shall deem necessary."

(3) Power to Provide Services. In addition to the essential powers of taxation

and eminent domain, boroughs are given extensive authority to exercise municipal powers in providing streets, sidewalks, curbs, gutters, drains, bridges and viaducts, sewers, water supply, electricity, airports, wharves and docks and related facilities, parks, playgrounds, swimming pools, indoor recreation centers and gymnasiums, shade trees and forest reserves, comfort and waiting stations, drinking fountains and watering troughs, garbage and sewage disposal plants, fire apparatus and fire houses, market houses and market places, lockups, armories, rooms for veterans' associations, and community buildings. They may also contract with street railways for the removal of street railway tracks, confine and regulate watercourses, require electric wires to be placed underground, operate gas wells, acquire, dispose of, and regulate burial grounds, and provide a real estate registry.

(4) Power to Regulate. Boroughs have wide powers to make and enforce regulations in the interest of health and cleanliness, beauty, convenience, comfort, and safety. Important enumerated powers under this heading have to do with the extinction and prevention of fires, the construction and repair of buildings, the care and removal of garbage and other refuse, regulation of markets, peddling and the inspection of milk, etc.

(5) Power to Prohibit. They may also prohibit entirely any obstructions or nuisances in the highways, nuisances or dangerous structures on public or private grounds, noxious or offensive businesses, the manufacture, sale or exposure of inflammable or dangerous articles, keeping hogs within the borough, accumulations of garbage on private property, the running at large of dogs and other animals, and the erection of wooden structures.

(6) Power to Appropriate Public Funds. Boroughs may pay the expenses incidental to the exercise of any of their corporate powers. In addition certain appropriations are specifically authorized for such purposes as municipal music, Memorial Day services, burial ground maintenance, fire companies, etc.

(7) Miscellaneous Powers. A great number of miscellaneous powers are vested in boroughs. Among the more important of these are:

Power to collect municipal claims.
Power to create and invest special funds.
Power to select, etc., depositories for borough funds.
Power to license auctioneers, foreign dealers in merchandise, transient retail dealers, amusements, and vehicles.

Power to define disorderly conduct; to impose fines and penalties, and to provide for the enforcement of ordinances.

(8) Elastic Clause. Finally, the council of each borough may "enact, revise, repeal, and amend such laws, rules, regulations, and ordinances, not inconsistent with the laws of the Commonwealth, as it shall deem beneficial to the borough and to provide for the enforcement of the same."

Governmental Organization - Charts I and II show, respectively, the organization of a typical small borough not divided into wards and of a large borough divided into wards. Both have the same elective officers as shown, the only difference being in the number of councilmen and assessors. It is optional with each borough whether it will have three auditors or one controller. The minimum organization required in every borough is complete when council elects a president, a secretary and a treasurer, only the first of whom may be a member of council. As shown in Chart II the number of appointive officers may be increased as the needs of the borough dictate. In large and small boroughs alike it is customary for committees of council to supervise the administration of the borough's affairs.

Qualifications of Officers - Elective officers must be electors of the borough and ward officers must reside in the wards from which they are elected. Assessors and tax collectors must be "properly qualified persons" and a borough controller must be a "competent accountant and an elector of the borough for at least four years prior to his

CHART 1. ORGANIZATION OF TYPICAL SMALL BOROUGH NOT DIVIDED INTO WARDS

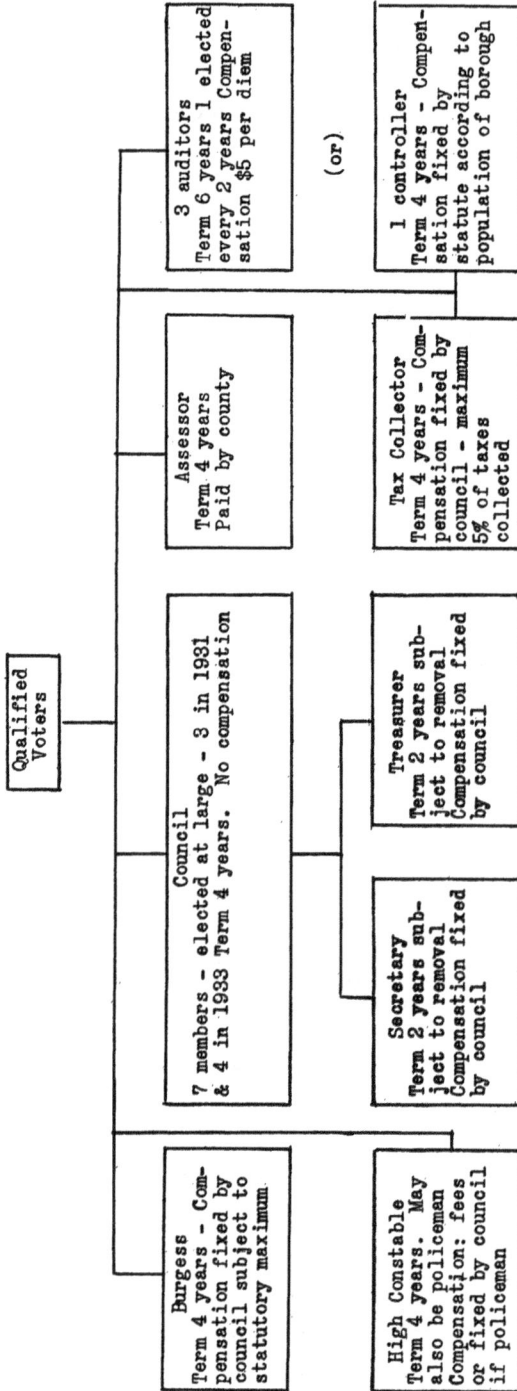

Qualified Voters

Burgess
Term 4 years - Compensation fixed by council subject to statutory maximum

High Constable
Term 4 years. May also be policeman Compensation: fees or fixed by council if policeman

Council
7 members - elected at large - 3 in 1931 & 4 in 1933 Term 4 years. No compensation

Secretary
Term 2 years subject to removal Compensation fixed by council

Treasurer
Term 2 years subject to removal Compensation fixed by council

Assessor
Term 4 years Paid by county

Tax Collector
Term 4 years - Compensation fixed by council - maximum 5% of taxes collected

3 auditors
Term 6 years 1 elected every 2 years Compensation $5 per diem

(or)

1 controller
Term 4 years - Compensation fixed by statute according to population of borough

CHART II. ORGANIZATION OF LARGE BOROUGH DIVIDED INTO WARDS

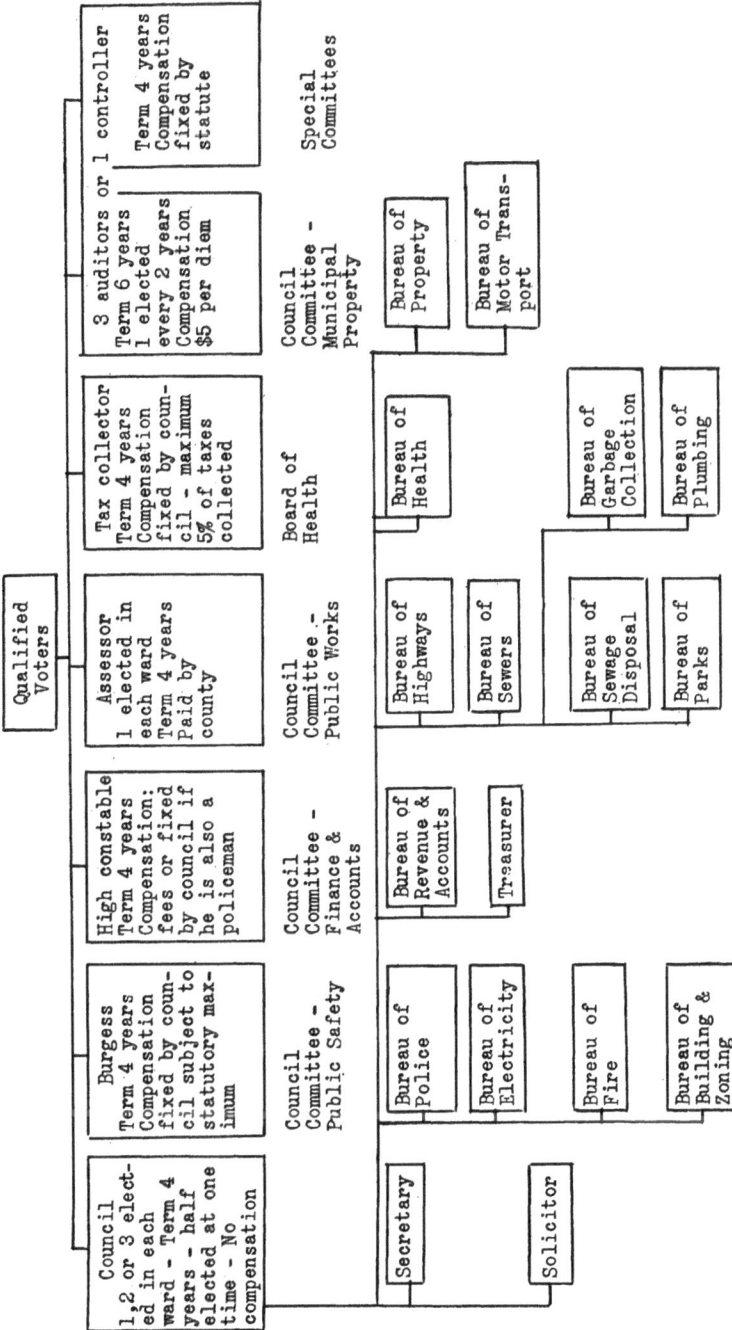

Qualified Voters

Council
1, 2 or 3 elected in each ward - Term 4 years - half elected at one time - No compensation

Burgess
Term 4 years Compensation fixed by council subject to statutory maximum

High constable
Term 4 years Compensation: fees or fixed by council if he is also a policeman

Assessor
1 elected in each ward Term 4 years Paid by county

Tax collector
Term 4 years Compensation fixed by council - maximum 5% of taxes collected

3 auditors or 1 controller
Term 6 years 1 elected every 2 years Compensation $5 per diem

Term 4 years Compensation fixed by statute

Council Committee - Public Safety

Council Committee - Finance & Accounts

Council Committee - Public Works

Board of Health

Council Committee - Municipal Property

Special Committees

Secretary

Solicitor

Bureau of Police

Bureau of Electricity

Bureau of Fire

Bureau of Building & Zoning

Bureau of Revenue & Accounts

Treasurer

Bureau of Highways

Bureau of Sewers

Bureau of Sewage Disposal

Bureau of Parks

Bureau of Health

Bureau of Garbage Collection

Bureau of Plumbing

Bureau of Property

Bureau of Motor Transport

election." These are the only requirements which the General Borough Act requires of elective borough officers. Certain offices are incompatible, however. A member of council may not be a school director, nor may a member of Congress or any person holding any office or appointment of profit or trust under the government of the United States be burgess.

Almost the only statutory qualification for any of the appointive officers requires that solicitors shall be "learned in the law."

Division of Authority - For the most part the corporate powers of boroughs are vested in the borough councils. To some extent their authority is shared, however, with the other elective officers, whose duties and powers we shall now consider. (a) The burgess has legislative, administrative and judicial functions. He presides over the organization meeting of council but may not vote unless his vote is required to effect the organization of council or to elect any officer elected at the organization meeting. In all cases where because of a tie vote the council is unable to enact an ordinance or fill a vacancy in its membership or in any other borough office, and the disagreement continues until the next regular meeting, it is the duty of the burgess to attend that meeting and cast the deciding vote. He may also veto any ordinance or resolution but his veto may be overridden by two-thirds of all the members elected to council or by a majority of council plus one when the number of councilmen is less than nine.

Council appoints the policemen and designates one of them as chief of police but the burgess has full control over the police force, directing the time during which, the place where, and the manner in which, the force shall perform their duties. He may suspend for cause and without pay any policeman until the succeeding regular meeting of the council, at which time the council may discharge or reinstate the policeman. This provision is the cause of much friction where the burgess and council are of different parties or factions and results in the demoralization of police work.

In addition the burgess has many of the powers of a justice of the peace, especially in connection with the enforcement of borough ordinances. It is his duty to preserve order in the borough, to enforce borough ordinances and regulations, to hear complaints, and to remove nuisances. He is charged with exacting faithful performance of the duties of appointive officers.

Much friction in the government of many boroughs could be obviated and responsibility for the proper conduct of borough affairs could be better secured by making the burgess a member of council with the same voice in legislation as any other member but with no veto power; also by providing that he be given sole authority to appoint and dismiss policemen in addition to the power he already has to suspend them and direct their work. His choice of policemen should, however, be limited to persons whose qualifications have been certified by the method to be described later.

The burgess might also be the president of council with council electing a president pro tem to take his place when he is unable to perform his duties. Under the present law the president of council discharges the duties of the burgess whenever the latter is absent or incapacitated, and during such time receives the same compensation that the burgess would receive. This is the only circumstance under which a member of council may now be legally paid for his official services. If the burgess were made a member of council he should be entitled to compensation as at present.

(b) The high constable has the authority of township constables; gives notice of the biennial elections of the borough by posting advertisements; and may act as a policeman, in which case it is unlawful for him to accept any compensation, in addition to his salary as a policeman, except public rewards and the legal mileage allowed to constables for traveling expenses. If he is not a policeman he may collect the costs and fees to which a constable is entitled.

It will be seen that the constable has no authority which materially reduces the responsibility of council for the management of borough affairs. On the other hand he has no powers important enough to warrant his retention as an elective officer. The few duties which he has should be exercised by an officer designated by council. In boroughs employing policemen these duties should be assigned to a policeman; in other boroughs they should be performed by the borough secretary.

(c) The elected assessors have the usual powers and duties of assessors in the other local subdivisions of the state. They are paid by the county and are essentially county employees although locally elected. Recommendations for the improvement of the assessment system have been made in Chapter IV above and will not be repeated here.

Dissatisfaction with the existing assessment system led to the adoption of the Act of April 26, 1929, P. L. 823, which provides that in counties having no county board of assessors the boroughs may appoint their own assessors who shall make the assessments upon which borough taxes are based. Such an assessment however is not used by the county, school district or poor district; it is therefore a duplication of the work of those units and an added cost. There is no reason for a borough having its own assessment if the assessment for general purposes is equitable. Steps should be taken to improve the general county assessment whereupon all boroughs should be required to use it as the basis for taxation.

(d) The tax collector performs the purely ministerial duty of collecting county, borough, school and poor district taxes within the borough. His compensation for collecting each tax is prescribed either by statute or by the legislative authorities of the unit for which he is acting. The maximum amount that he can be paid for collecting borough taxes is 5% of the amount collected. In addition he may retain out of the taxes collected actual and needful expenditures for printing, postage, books, blanks and forms.

The tax collector is thus the best paid officer in many boroughs and the cost of tax collections is one of the largest regular items of overhead expense. Dr. Nicholson has shown, in "Collection of Local Taxes in Pennsylvania" (Department of Revenue, Special Bulletin No. 1, 1932), that it costs Pennsylvania four times as much to collect its local taxes as it costs Ohio to collect almost the same amount. This is due almost entirely to the fact that Ohio centralizes the collection of its taxes. The same results could be obtained in Pennsylvania by having one main independent tax office in each county with deputies sent to different sections of the county from time to time. The employees of this office should be for the most part full-time salaried officers.

Electing the tax collector needlessly complicates and lengthens the ballot. In addition it frequently results in the choice of a collector who is unable to secure the bond which is required before he can undertake his duties and who is incompetent to keep the simple records which are required. As he has no duties which call for the exercise of discretion there is no reason why he should be elected and every reason why he should be appointed. This change should be made whether the collection of taxes is centralized in a county office or left to local collectors. In many boroughs the secretary could collect the taxes at considerable savings over the present system.

(e) The borough auditors or controller have the important function of acting as a check upon the other fiscal officers of the borough and school district and, in some cases, of the poor district as well. In view of the nature of their work it is quite proper that they should not be appointed by the officers whose administration they are supposed to control. On the other hand the elective process is poorly adapted to securing competent auditors. In some boroughs there is a tradition which results in the nonpolitical choice of qualified auditors and in the great majority of small boroughs the amounts spent are so small and the activities of the government so limited that even untrained persons are able to afford adequate protection against the misuse of public funds. Even in such boroughs irregularities are commonly overlooked by the auditors, many of whom only partially perform their own duties. In larger boroughs where the sums involved are greater and the activities of the borough more varied, the lack of training, incompetence, and occasional but rare dishonesty of the auditors have more serious results.

The annual statements of the auditors or controller are the only readily available source of information as to borough finances. In their customary form they are almost entirely useless, first, because of their lack of detail (or sometimes because of the excessive amount of unclassified detail) and second, because of their lack of uniformity in presenting such data as are given. Some which list all payments under the names of the recipients without explanation as to the purpose for which payment was made are probably sufficient to prevent dishonesty in small boroughs where all the recipients are personally known to most of the residents but are useless in larger places. In no case can they be used for comparisons of unit costs in different communities. Other statements give practically no detail at all. Following is a copy of all the data as to the receipts and expenditures of one borough of more than 10,000 population, as shown in the auditors' report filed with the Department of Internal Affairs.

Recipts (sic)

Balance last audit		$15,178.74
Received from taxes	$17,221.72	
Short term loan	5,000.00	
Other sourses (sic)	3,419.93	
		25,641.65
Total		$40,820.39

Vouchers redeemed from Jan. 10,1933 to Jan. 8, 1934, less sinking fund	$37,505.78	
Sinking fund	1,500.00	
Treasurer commission	750.11	
Amount of surcharge	598.49	
Balance on hand Jan. 8, 1934	466.01	$40,820.39

In the boroughs having auditors these officers make a single annual audit of the records of the various officials handling borough funds. In those few boroughs which have accepted the provision of the General Borough Act allowing them to have a single controller in place of the three auditors it is the duty of this officer to exercise more continuous control over borough finances. In addition to his annual audits he may supervise the accounts of all borough officers and require statements from them at any time. He must countersign all warrants but no warrant may be countersigned unless there is money in the treasury to pay it. It is his duty to see that no appropriation made by council is overdrawn, to suggest plans to council for the management and improvement of the borough finances, and to keep a complete set of books showing the affairs of the borough. For those boroughs which are large enough to warrant the employment of a controller this system provides a much better control of finances than is possible under the more common alternative plan. It suffers, however, from the same weakness, that a popular election is no guarantee that a competent controller will be chosen.

To secure strict conformity with the laws governing borough finances the annual audit should be made by full-time qualified employees of a central state office. Besides being better equipped than the average auditor to safeguard taxpayers against the misuse of public funds because of their better training and the smaller possibility that they would act in collusion with the officers whose accounts they audited, such officers would be able to render valuable assistance to the boroughs in recommending better and more uniform bookkeeping methods than are now in use. In some of the large boroughs where experience has shown that it is possible to elect qualified controllers it might be well to retain that officer to act as a day-to-day check upon the legality of the acts of the other elected officers. Whether or not there should be such an officer should be optional, however, with the voters of the borough for in many boroughs the incompetence or inactivity of the elected controllers has made the office a needless expense.

Administrative Systems - For the most part each borough council may establish the administrative system which it prefers. Many boroughs need almost no appointive officials simply because the boroughs perform no functions which necessitate their employment. On the other hand some of the larger boroughs require many employees and an extensive administrative system, particularly if they provide a variety of services for their inhabitants.

In every borough visited committees of council played a large part in the administration of borough affairs. With few exceptions the councilmanic committees are primarily concerned with administrative rather than legislative matters. There is ordinarily a separate committee to supervise each important borough activity or group of related activities. In most small boroughs and in many larger ones the committees actively supervise the various public works and services. In some places decisions are left largely to the chairman while in others actions are taken only by a majority of the committee. In other boroughs the committees are consulted only in respect to major problems, all routine matters being disposed of by an appointive officer who may have the title of secretary, manager, engineer, street commissioner, etc. This is the usual custom in boroughs employing a manager or full-time engineer. Even the decision as to major matters is left to such officers in a few boroughs. Local customs and regulations also govern the relations between the committees and council as a whole. Ordinarily committee decisions in administrative matters are approved by council.

Borough Managers - In small boroughs where the borough does little except maintain the streets and pay monthly for street lighting the system above described is fairly adequate if the chairman of the street committee is an able businessman and interested in his work or where a large degree of reliance is placed in the secretary or street commissioner. In larger boroughs, however, the committee system is too decentralized to be efficient. Recognizing this, some boroughs have availed themselves of the authority granted them in the General Borough Act to appoint a borough manager. Others have practically the same system but call their chief administrative officer the borough engineer, street commissioner, etc.

Such boroughs are with few exceptions the best managed and most active of all. Many of them are in the metropolitan areas surrounding Philadelphia and Pittsburgh where it is customary to perform a wide variety of services.

Budgeting and Purchasing - The same conditions in respect to budgeting and purchasing methods are found in boroughs as have been elsewhere described in this report as existing in townships, especially those of the first class. While some boroughs have excellent budgeting methods, the majority muddle along in a more or less hit-or-miss fashion depending largely upon the officers' general knowledge of past conditions, and in a few there appears to be no relation whatsoever between tax levies and plans for borough expenditures. In most of the units examined no great improvement in budgeting would be possible without first taking steps to improve greatly the fiscal records which are commonly very inadequate at present.

The chief defects observed in purchasing methods were that ordinarily no one officer is charged with the responsibility of securing information about current prices and that there is entirely too much day-to-day buying.

Personnel - No merit system for the selection of appointive borough employees is in operation yet many such employees and officers have had surprisingly long tenures. This is especially true in the case of the secretaries many of whom have held their offices for from ten to thirty years or even more. In the small boroughs the lack of a merit system is of little importance because of the nature of the work done. In the larger boroughs, however, the situation has some serious results particularly in demoralizing the police forces. To overcome this the head of the state police system should be authorized to prescribe minimum qualifications for local police officers. He should also have power to conduct qualifying examinations for applicants and to maintain a training school for successful candidates. Persons so chosen should be removed only for misconduct or inefficiency.

Borough Depositories - In the last few years the financial difficulties of many boroughs have been greatly intensified because of the loss of funds deposited in banks. Usually this has been most serious in the case of sinking funds. Permissive legislation has been passed authorizing boroughs to take steps to prevent the recurrence of such losses but many boroughs have not fully availed themselves of these powers. It would be desirable for the legislature to safeguard against such losses of sinking funds by providing that in the future boroughs might issue only serial bonds. It should also be mandatory that any bank in which there are deposits of borough funds in excess of the amount insured by the F. D. I. C. should furnish adequate collateral to safeguard the borough deposits.

Taxation - The borough council may levy and collect annually for general borough purposes a tax not exceeding 15 mills on the dollar on the valuation assessed for county purposes. Additional levies authorized for special purposes are as follows: for sinking fund, 5 mills; for playground maintenance, 2 mills; for the care of shade trees, .1 mill; for street lighting, 8 mills; for fire equipment, 3 mills; for municipal buildings, 2 mills; and for maintenance of public libraries, 2 mills. Except for the general levy most of these require the consent of the electors.

In addition to the foregoing levies the courts are authorized to order special levies under some circumstances. The most important of these is authorized in section 1302 of the General Borough Act, which provides that when the court of quarter sessions shall be shown that the debts due by any borough exceed the amount which the borough may collect in any one year by taxation, or when the proper officers refuse or neglect to levy a sufficient tax to pay the same, the court may direct a writ of mandamus to the proper borough officers to collect by special taxation an amount sufficient to pay the debt in one or more annual instalments.

The repeal of this provision is strongly urged by some persons who allege that its existence encourages borough officials to commit the borough to expenditures in excess of its ability to pay. It is further alleged that such commitments are often unjustifiable and are at times entered into by illegal collusion between borough officials and the concerns from which they purchase. This complaint was encountered only in parts of the state where it frequently happens that a large percentage of borough taxes are paid by a few coal companies. The fact that the tax burden falls very largely on non-resident and non-voting taxpayers is commonly believed to contribute in large part to the bad management of fiscal affairs there prevalent. In boroughs visited in connection with this survey in other parts of the state nothing was found to indicate that the powers granted the courts by section 1302 led to abuses. On the contrary it was the opinion of many borough officials interviewed that the provision was a reasonable and desirable safeguard to protect bona fide creditors.

Complete repeal of section 1302 would be too drastic. The courts must have power to provide for the payment of bona fide debts, especially in cases where judgments are rendered against boroughs without previous commitments having been made by borough officials, as in the case of an accident upon the highways. Section 1302, as it now stands, is also a desirable means of providing for the payment of due debts where for political considerations councils might choose to levy inadequate taxes well below the 15 mill general limit. It is also doubtful whether repeal of this section would materially improve the situation in those parts of the state where the power therein granted is abused. The basic difficulty in many boroughs in Lackawanna and a few of the other coal counties is that the elected officials are irresponsible to the wishes of the chief taxpayers. Coupled with this is an apparently high degree of extravagance and a lack of efficient businesslike administrative methods.

Liability of Officers - In view of this situation section 1302 might well be modified though it should not be repealed. To prevent borough officials from incurring current obligations in excess of the legal taxing capacity of the borough, the members of the council and the burgess voting for such expenditures should be made liable rather than the borough. Since the financial responsibility of some of these officers is slight, this provision should effectively meet conditions where undue influence is exerted upon members of council to purchase extravagantly. The court should retain its present authority to mandamus the proper borough officers to collect special taxes for the purpose of paying debts for which borough officers are not responsible and for paying due debts when the council fails to levy an adequate tax within the legal limits. The courts or a state administrative agency should be further empowered to relieve members of council from liability in cases where local conditions necessitate tax levies in excess of the general limits.

State Control of Borough Finances - In connection with this change in section 1302 another step should be taken to meet situations where locally elected officials are irresponsive to the wishes of the taxpayers. Centralized control of'such fiscal matters as budgeting, fixing tax rates and purchasing has in general little to recommend it because of the red tape involved and the lack of responsiveness to local opinion. Yet it would be desirable to establish a central agency authorized to exercise such control in cases where the irresponsibility, extravagance and inefficiency of locally elected officials are worse than its failings. This agency should be authorized to take full control of local fiscal affairs only when petitioned to do so by a majority in interest (but not in numbers) of the borough taxpayers.

Other Sources of Revenue - In addition to their power to tax, boroughs are given wide powers to make special assessments and have many of the miscellaneous sources of revenue available to other units of local government. The nature of these will be evident from the following tables.

Explanation of Tables - Tables 1 to 5 inclusive deal with the receipts of 191 typical boroughs. These represent from 15 to 39 per cent of all the boroughs in each of the six population groups shown (ranging from less than 500 inhabitants to more than 10,000 inhabitants) and are representative of the different geographic and economic areas in the state. Data are given for both 1930 and 1933.

Revenue receipts are classified as taxes, special assessments, licenses and permits, fines and forfeits, interest, gifts and grants, departmental earnings, water rents, other public utility earnings, and other miscellaneous revenue. Non-revenue receipts are those which do not change the relation between total assets and total liabilities of the boroughs. Included under this general classification are receipts from the sale of bonds, temporary loans, sales of borough property, refunds and payments for joint construction projects, and other miscellaneous non-revenue receipts.

Average Borough Receipts - Table I shows the average classified receipts of the boroughs in each of six population groups. No average figures are given for the entire group of 191 boroughs because the range of population is so great that such averages would be meaningless. Group averages rather than data for individual boroughs are used because they give a broader view of the situation.

In view of the insignificant financial operations of boroughs having less than 1,500 inhabitants, as shown by this table, a question is raised as to whether the separate existence of some of these boroughs is justified.

Per Capita Receipts - Table II lists borough receipts per capita, by classes of receipts and by population groups. Among the more significant facts demonstrated by this table are the following:

1. Per capita as well as actual borough taxes were materially less in small boroughs than in large boroughs.

2. Boroughs having from 2,500 to 5,000 inhabitants generally had higher per capita receipts from taxes, special assessments, gifts and grants, departmental earnings, water rents, bond sales and temporary loans than had larger boroughs, and much higher receipts from these sources than had smaller boroughs.

3. Special assessments are used to a very slight degree by small boroughs. Average per capita receipts from special assessments decreased from $.95 in 1930 to $.18 in 1933, indicating a notable decrease in public improvements in boroughs.

4. Average per capita receipts from licenses and permits were $.16 in 1930 and $.38 in 1933. Boroughs in the three larger population groups collected approximately 50 per cent more per capita than did those in the three smaller groups.

5. Fines and forfeits yielded an average of $.11 per capita in 1930 and $.04 per capita in 1933. There was no significant variation as between population groups.

6. Departmental earnings were almost non-existent in boroughs having less than 1,500 inhabitants but were of sufficient size in the larger boroughs so that their

TABLE I.   AVERAGE RECEIPTS OF 191 BOROUGHS, BY POPULATION GROUPS, 1930 AND 1933

Class of Receipts	Year	Average Receipts by Population Groups					
		10000 and up (a)	5000 to 10000 (b)	2500 to 5000 (c)	1500 to 2500 (d)	500 to 1500 (e)	Under 500 (f)
Revenue Receipts							
Taxes	1930	$155,406	$ 63,460	$32,757	$15,119	$ 6,387	$2,393
	1933	105,687	46,738	25,162	9,926	4,148	1,456
Special assessments	1930	16,220	5,342	4,869	1,693	109	21
	1933	2,076	833	1,317	481	9	11
Licenses and permits	1930	2,468	1,500	637	190	73	30
	1933	7,754	2,531	1,094	493	217	87
Fines and forfeits	1930	2,027	817	380	204	81	24
	1933	532	283	188	135	23	27
Interest	1930	1,454	96	294	52	21	30
	1933	1,031	185	142	43	15	22
Gifts and grants	1930	1,946	1,028	933	493	40	28
	1933	1,081	352	422	146	31	11
Departmental earnings	1930	4,318	1,780	1,453	176	10	1
	1933	2,663	1,751	1,287	91	6	3
Water rents	1930	15,391	5,539	4,733	1,704	502	-
	1933	13,387	5,086	4,553	2,698	477	-
Other public utilities	1930	7,375	7,835	3,702	2,499	141	-
	1933	5,660	18,549	3,549	909	107	-
Other revenue receipts	1930	1,265	644	185	67	78	15
	1933	1,209	235	179	43	77	9
Total revenue receipts	1930	207,872	88,041	49,942	22,197	7,442	2,542
	1933	141,080	76,544	37,893	14,965	5,110	1,626
Non-revenue receipts							
Bonds	1930	19,608	3,456	5,647	222	1,147	571
	1933	15,263	300	3,655	926	-	-
Temporary loans	1930	60,274	21,720	13,977	3,009	1,437	470
	1933	31,977	8,281	13,125	1,780	612	318
Sales	1930	357	176	207	40	15	38
	1933	222	53	31	177	15	7
Refunds and joint construction	1930	518	740	921	75	49	430
	1933	880	230	124	64	43	26
Other non-revenue receipts	1930	640	98	111	15	11	16
	1933	631	56	266	34	10	1
Total non-revenue receipts	1930	81,397	26,190	20,862	3,360	2,659	1,526
	1933	10,000	8,920	17,200	2,980	679	351
TOTAL RECEIPTS	1930	289,268	114,231	70,805	25,557	10,101	4,067
	1933	190,054	85,464	55,098	17,945	5,790	1,978

(a) Average for 19 boroughs.   (d) Average for 27 boroughs.
(b) Average for 20 boroughs.   (e) Average for 45 boroughs.
(c) Average for 45 boroughs.   (f) Average for 35 boroughs.

TABLE II. PER CAPITA RECEIPTS OF 191 BOROUGHS, BY CLASSES OF RECEIPTS
AND BY POPULATION GROUPS, 1930 AND 1933

Class of Receipts	Year	POPULATION GROUPS						
		191 boroughs	10000 and over	5000 to 10000	2500 to 5000	1500 to 2500	500 to 1500	Under 500
Revenue Receipts								
Taxes	1930	9.01	9.39	8.84	9.37	7.68	7.40	7.36
	1933	6.37	6.39	6.51	7.20	5.04	4.81	4.48
Special assessments	1930	.95	.98	.74	1.39	.86	.13	.07
	1933	.18	.13	.12	.38	.24	.01	.03
Licenses and permits	1930	.16	.15	.21	.18	.10	.08	.09
	1933	.38	.47	.35	.31	.25	.25	.27
Fines and forfeits	1930	.11	.12	.11	.11	.10	.09	.07
	1933	.04	.03	.04	.05	.07	.03	.08
Interest	1930	.06	.09	.01	.08	.03	.02	.09
	1933	.04	.06	.03	.04	.02	.02	.07
Gifts and grants	1930	.16	.12	.14	.27	.25	.05	.08
	1933	.07	.07	.05	.12	.07	.04	.03
Departmental earnings	1930	.26	.26	.25	.42	.09	.01	.00
	1933	.20	.16	.24	.37	.05	.01	.01
Water rents	1930	.95	.93	.77	1.35	.87	.58	.00
	1933	.91	.81	.71	1.30	1.37	.55	.00
Other public utilities	1930	.75	.45	1.09	1.06	1.27	.16	.00
	1933	.93	.34	2.58	1.01	.46	.12	.00
Other revenue receipts	1930	.07	.08	.09	.05	.03	.09	.04
	1933	.06	.07	.03	.05	.02	.09	.03
Total revenue receipts	1930	12.50	12.56	12.26	14.28	11.28	8.63	7.82
	1933	9.19	8.53	10.66	10.84	7.60	5.92	5.00
Non-revenue receipts								
Bonds	1930	1.08	1.19	.48	1.62	.11	1.33	1.76
	1933	.68	.92	.04	1.05	.47	.00	.00
Temporary loans	1930	3.30	3.64	3.02	4.00	1.53	1.67	1.45
	1933	2.02	1.93	1.15	3.75	.90	.71	.98
Sales	1930	.03	.02	.02	.06	.02	.02	.12
	1933	.02	.01	.01	.01	.09	.02	.02
Refunds and joint	1930	.12	.03	.10	.26	.04	.06	1.32
construction	1933	.04	.05	.03	.04	.03	.05	.08
Other non-revenue receipts	1930	.03	.04	.01	.03	.01	.01	.05
	1933	.04	.04	.01	.08	.02	.01	.00
Total non-revenue receipts	1930	4.55	4.92	3.65	5.97	1.71	3.08	4.69
	1933	2.79	2.96	1.24	4.92	1.51	.79	1.08
TOTAL RECEIPTS	1930	11.05	17.48	15.90	20.25	12.98	11.71	12.51
	1933	11.99	11.49	11.90	15.76	9.12	6.71	6.08

average per capita yield was $.26 in 1930 and $.20 in 1933.

7. None of the boroughs with less than 500 inhabitants collected any water rents. In the other groups of boroughs the range in 1930 was from $.58 to $1.35 per capita, with a general average of $.95. These figures changed little in 1933.

8. Other public utilities were less common than water works but where found yielded large sums.

9. Total revenue receipts of the 191 boroughs amounted to $12.50 per capita in 1930 and to $9.19 in 1933. The two smallest groups of boroughs were from 30 to 40 per cent or more below the average. On the other hand, the medium sized boroughs, with from 2,500 to 5,000 inhabitants, had considerably greater than average per capita receipts.

10. The average per capita receipts from bond sales were $1.08 in 1930 and $.68 in 1933.

11. The average per capita amount of temporary loans in 1930 was $3.30. In 1933 it was $2.02. The significance of these relatively large figures is greatly weakened by the fact that to a large extent they represent the repeated turnover of short term loans. Even so, temporary borrowing is a very substantial element in borough financing. This is especially so in the larger boroughs.

12. The total per capita receipts of the groups of boroughs with less than 2,500 inhabitants ranged in 1930 from $11.71 to $12.98; those of the larger groups from $15.90 to $20.25. In 1933 the range for the smaller boroughs was from $6.08 to $9.12 and for the larger boroughs, from $11.49 to $15.76.

Relative Importance of Different Classes of Receipts - Table III compares each class of receipts to total receipts. Non-revenue receipts, which were almost entirely debt items, were approximately one fourth of all receipts both years but were slightly less proportionately in 1933 than in 1930.

Although non-revenue receipts swell the income of boroughs temporarily, all expenses must eventually be met from revenue receipts. Table IV shows the relative importance of the different kinds of revenue receipts, approximately 70 per cent of which are taxes. The vast majority of taxes for borough purposes are levied upon property, with occupation and poll taxes making up the balance. The smallest boroughs are most dependent upon taxation (approximately 90 per cent) while the medium sized boroughs need taxes least (approximately 60 to 70 per cent).

Special assessments were 7.6 per cent of all revenue receipts in 1930 but only 2 per cent in 1933.

Water rents and the earnings of other public utility enterprises were, next to taxes, the most important source of borough revenue. For all 191 boroughs these were 13.6 per cent of total revenue receipts in 1930 and 20 per cent in 1933. However these figures give an incomplete idea of the true situation. Only 34 of the 191 boroughs had water works and only 9 had other utilities. In these boroughs utility earnings are generally about as large as tax receipts and in extreme cases as in Middletown the borough operates on the profit from its utilities without levying taxes.

Trends in Borough Financing - Table V shows that all but three items of receipts decreased from 1930 to 1933. Only licenses and permits showed a consistent increase in all groups of boroughs, this increase being due to repeal of the prohibition laws. The earnings of electric plants increased only because a new plant was put in operation. A percentage gain of 28.5 in miscellaneous non-revenue receipts is insignificant because it represented only a few hundred dollars. The small percentage of decrease in water rents is due to the 58.3 per cent increase in water rent collections in boroughs having from 1,500 to 2,500 inhabitants. Other groups showed big decreases in their revenue from this source.

TABLE III. RATIO OF EACH CLASS OF RECEIPTS TO TOTAL RECEIPTS OF 191 BOROUGHS, 1930 AND 1933, BY POPULATION GROUPS

Class of Receipts	Year	POPULATION GROUPS						
		191 boroughs	10000 and over	5000 to 10000	2500 to 5000	1500 to 2500	500 to 1500	Under 500
Revenue receipts								
Taxes	1930	52.8	53.7	55.6	46.2	59.2	63.2	58.8
	1933	53.2	55.6	54.7	45.7	55.3	71.6	73.6
Special assessments	1930	5.6	5.6	4.7	6.9	6.6	1.1	.5
	1933	1.5	1.1	1.0	2.4	2.7	.2	.5
Licenses and permits	1930	.9	.9	1.3	.9	.7	.7	.7
	1933	3.2	4.1	3.0	2.0	2.7	3.8	4.4
Fines and forfeits	1930	.7	.7	.7	.5	.8	.8	.6
	1933	.3	.3	.3	.3	.8	.4	1.4
Interest	1930	.4	.5	.1	.4	.2	.2	.7
	1933	.4	.5	.2	.3	.2	.3	1.1
Gifts and grants	1930	.9	.7	.9	1.3	1.9	.4	.7
	1933	.6	.6	.4	.8	.8	.5	.6
Departmental earnings	1930	1.5	1.5	1.6	2.1	.7	.1	.0
	1933	1.7	1.4	2.0	2.3	.5	.1	.1
Water rents	1930	5.6	5.3	4.8	6.7	6.7	5.0	.0
	1933	7.6	7.0	6.0	8.3	15.0	8.2	.0
Other public utilities	1930	4.4	2.5	6.9	5.2	9.8	1.4	.0
	1933	7.8	3.0	21.7	6.4	5.1	1.8	.0
Other revenue receipts	1930	.4	.4	.6	.3	.2	.8	.4
	1933	.5	.6	.3	.3	.2	1.3	.5
Total revenue receipts	1930	73.3	71.9	77.1	70.5	86.9	73.7	62.5
	1933	76.7	74.2	89.6	68.8	83.4	88.2	82.2
Non-revenue receipts								
Bonds	1930	6.3	6.8	3.0	8.0	.9	11.4	14.0
	1933	5.6	8.0	.4	6.6	5.2	.0	.0
Temporary loans	1930	19.3	20.8	19.0	19.7	11.8	14.2	11.6
	1933	16.8	16.8	9.7	23.8	9.9	10.6	16.1
Sales	1930	.2	.1	.2	.3	.2	.1	.9
	1933	.1	.1	.1	.1	1.0	.3	.3
Refunds & joint construction	1930	.7	.2	.6	1.2	.3	.5	10.6
	1933	.4	.5	.3	.2	.4	.7	1.3
Other non-revenue receipts	1930	.2	.2	.1	.2	.1	.1	.4
	1933	.3	.3	.1	.5	.2	.2	.0
Total non-revenue receipts	1930	26.7	28.1	22.9	29.5	13.1	26.3	37.5
	1933	23.3	25.8	10.4	31.2	16.6	11.7	17.8
Total receipts	1930	100.0	100.0	100.0	100.0	100.0	100.0	100.0
	1933	100.0	100.0	100.0	100.0	100.0	100.0	100.0

TABLE IV.   RATIO OF EACH CLASS OF REVENUE RECEIPTS TO TOTAL REVENUE RECEIPTS
OF 191 BOROUGHS, 1930 AND 1933 BY POPULATION GROUPS

Class of receipts	Year	POPULATION GROUPS						
		191 boroughs	10000 and over	5000 to 10000	2500 to 5000	1500 to 2500	500 to 1500	Under 500
Revenue receipts								
Taxes	1930	72.1	74.8	72.1	65.6	68.1	85.8	94.2
	1933	69.3	74.9	61.1	66.4	66.3	81.2	89.5
Special assessments	1930	7.6	7.8	6.1	9.7	7.6	1.5	.8
	1933	2.0	1.5	1.1	3.5	3.2	.2	.7
Licenses and permits	1930	1.3	1.2	1.7	1.3	.9	1.0	1.2
	1933	4.1	5.5	3.3	2.9	3.3	4.3	5.4
Fines and forfeits	1930	.9	1.0	.9	.8	.9	1.1	1.0
	1933	.5	.4	.4	.5	.9	.5	1.6
Interest	1930	.5	.7	.1	.6	.2	.3	1.2
	1933	.5	.7	.2	.4	.3	.3	1.4
Gifts and grants	1930	1.3	.9	1.2	1.9	2.2	.5	1.1
	1933	.8	.8	.5	1.1	1.0	.6	.7
Departmental earnings	1930	2.1	2.1	2.0	2.9	.8	.1	.0
	1933	2.2	1.9	2.3	3.4	.6	.1	.2
Water rents	1930	7.6	7.4	6.3	9.5	7.7	6.8	.0
	1933	9.9	9.5	6.6	12.0	18.0	9.3	.0
Other public utilities	1930	6.0	3.5	8.9	7.4	11.3	1.9	.0
	1933	10.1	4.0	24.2	9.4	6.1	2.1	.0
Other revenue receipts	1930	.6	.6	.7	.4	.3	1.0	.6
	1933	.6	.9	.3	.5	.3	1.5	.6
Total revenue receipts	1930	100.0	100.0	100.0	100.0	100.0	100.0	100.0
	1933	100.0	100.0	100.0	100.0	100.0	100.0	100.0

TABLE V. PER CENT OF CHANGE IN AMOUNT OF RECEIPTS OF 191 BOROUGHS, 1933 OVER 1930, BY CLASSES OF RECEIPTS AND POPULATION GROUPS. (All items are decreases unless preceded by a plus sign.)

Class of receipts	191 boroughs	10000 and up	5000 to 10000	2500 to 5000	1500 to 2500	500 to 1500	Under 500
Revenue Receipts							
Taxes	29.3	32.0	26.4	23.2	34.3	35.1	39.2
Special assessments	81.2	87.2	84.4	72.9	71.6	91.8	49.0
Licenses and permits	+137.6	+214.2	68.8	+71.7	+158.5	+199.5	+187.0
Fines and forfeits	63.6	73.7	65.3	50.6	34.0	71.5	10.5
Interest	30.0	29.1	+93.3	51.7	18.5	27.6	25.0
Gifts and grants	54.8	17.4	65.8	54.8	70.4	21.3	59.3
Departmental earnings	22.3	38.3	1.6	11.4	47.9	38.6	+213.3
Water rents	4.3	13.0	81.8	38.2	+58.3	5.1	.0
Other public utilities	+24.3	23.3	+136.8	41.1	63.6	24.0	.0
Other revenue	20.3	4.4	63.5	3.3	35.1	1.0	35.4
Total revenue receipts	26.4	32.1	13.1	24.2	32.6	31.3	36.0
Non-revenue receipts							
Bond sales	37.2	22.2	9.1	35.3	+316.7	100.0	100.0
Temporary loans	38.8	46.9	61.9	6.1	40.8	57.4	32.4
Sales	45.6	37.8	69.9	85.0	+346.8	.0	82.3
Refunds and joint construction	63.1	+70.0	68.9	86.5	14.8	12.5	93.9
Other non-revenue receipts	+28.5	1.4	43.1	+139.5	+130.0	14.1	95.3
Total non-revenue receipts	38.7	39.8	65.9	17.6	11.3	74.5	77.0
TOTAL RECEIPTS	29.7	34.3	25.2	22.2	29.8	42.7	51.4

Per Capita Borough Taxes - Table VI is based upon compilations of tax data prepared by the Bureau of Statistics of the Department of Internal Affairs and population figures of the 1930 census. The data presented deal with taxation in the boroughs of 10 representative counties in 1932. In those boroughs taxes ranged from $1.01 per capita to $55.04 per capita for borough purposes. A more detailed analysis follows:

Range	Number of Boroughs	Range	Number of Boroughs
Less than $2	11	$10 - $12	12
$2 - $4	30	12 - 15	7
4 - 6	39	15 - 20	3
6 - 8	31	20 - 50	3
8 - 10	18	More than $50	1

Fifteen of the 26 boroughs whose levies exceeded $10 per capita were in Delaware County.

The average tax rate for borough purposes was 15.5 mills, one borough levying no tax and another levying 40 mills. The frequency of various millages was as follows:

Millage	Number of Boroughs	Millage	Number of Boroughs
Under 5	6	20 - 25	8
5 - 10	33	25 - 30	6
10 - 15	46	30 - 40	3
15 - 20	52	40	1

Table VI also compares borough tax rates with those of the school districts, counties and poor districts taxing the same property. Except in 18 boroughs the school tax was greater than the borough tax, frequently being twice as high. In general, county and poor taxes were smaller than either school or borough levies. The borough millage ranged from 1.7 per cent to 63.8 per cent of the total millage levy for all local purposes, usually being about 30 per cent of the total as is shown below.

Per Cent of Total	Number of Boroughs	Per Cent of Total	Number of Boroughs
Less than 10	1	40 - 50	12
10 - 20	16	50 - 60	3
20 - 30	59	More than 60	2
30 - 40	61	Data incomplete	2

It is obvious that while economies in borough operation may be possible it is the school district which imposes the chief tax burden upon property.

Table VI also shows the total combined millage levied by all taxing authorities in these boroughs. These ranged between 16.75 and 99.5 mills, distributed as follows:

Total Millage	Number of Boroughs	Total Millage	Number of Boroughs
Less than 20	1	60 - 70	15
20 - 30	17	70 - 80	7
30 - 40	40	80 - 90	2
40 - 50	50	90 - 100	3
50 - 60	20		

Twenty-one of the 27 boroughs where the combined millaged exceeded 60 mills were in Forest, Lackawanna, Lycoming and Potter Counties. Most of the boroughs in Snyder and York Counties had less than average combined millages.

Tax Delinquency, 1932 - Table VII deals with tax delinquency in the same group of boroughs. Column 2 shows what proportion of the 1932 tax levy of each borough was delinquent at the end of that year. The range was as follows:

TABLE VI – TAX LEVIES IN BOROUGHS, 1932

Borough	Population	Taxable valuation for borough purposes	Borough taxes levied	Borough taxes per capita	Tax rate in mills per dollar of assessed valuation					Per cent borough millage is of total
					Borough	School	County	Poor	Total	
Centre County										
Bellefonte	4,804	2,052,769	34,897	7.26	17	21	8	4	50	34.0
Centre Hall	658	229,086	2,520	3.83	11	14	8	3	36	30.6
Howard	664	188,355	753	1.13	4	15	8	4	31	12.9
Milesburg	644	122,870	1,843	2.86	15	25	8	7	55	27.3
Millheim	659	295,230	3,248	4.93	11	15	8	3	37	29.7
Phillipsburg	3,600	1,555,280	34,216	9.50	22	31	8	4	65	33.8
Port Matilda	508	119,904	1,079	1.56	9	24	8	10	51	17.6
South Phillipsburg	480	56,225	2,137	4.45	38	35	8	10	91	41.8
Snow Shoe	520	104,850	1,887	3.63	18	41	8	4	71	25.4
State College	4,450	3,094,622	58,798	13.21	19	25	8	1	53	35.8
Unionville	304	87,210	872	2.87	10	23	8	6	47	21.3
Delaware county										
Aldan	2,269	2,727,579	38,186	16.82	14	21	4.25	2.5	41.75	33.5
Clifton Heights	5,057	2,950,350	44,255	8.75	15	25	4.25	2.5	46.75	32.1
Collingdale	7,857	3,934,395	55,082	7.01	14	31	4.25	2.5	51.75	27.1
Colwyn	2,064	1,369,225	20,538	9.95	15	22	4.25	2.5	43.75	34.3
Darby	9,899	5,176,250	75,056	7.58	14.5	29	4.25	2.5	50.25	28.9
E. Lansdowne	3,168	2,027,675	41,567	13.12	20.5	22.5	4.25	2.5	49.75	41.2
Eddystone	2,414	7,476,193	63,548	26.32	8.5	15	4.25	2.5	30.25	28.1
Folcroft	1,432	1,110,300	15,544	10.85	14	30	4.25	2.5	50.75	27.6
Glenolden	4,482	3,257,859	48,868	10.90	15	25	4.25	2.5	46.75	32.1
Lansdowne	9,542	10,640,286	167,585	17.56	15.75	23	4.25	2.5	45.5	34.6
Marcus Hook	4,867	5,104,902	56,154	11.54	11	17	4.25	2.5	34.75	31.7
Media	5,372	5,650,715	73,459	13.67	13	16	4.25	2.5	35.75	36.4
Millbourne	396	1,089,700	21,794	55.04	20	5	4.25	2.5	31.75	63.0
Norton	1,341	949,000	11,388	8.49	12	30	4.25	2.5	48.75	24.6
Norwood	3,878	2,921,650	37,981	9.79	13	23	4.25	2.5	42.75	30.4
Parkside	1,497	1,532,722	12,262	8.19	8	19	4.25	2.5	33.75	23.7
Prospect Park	4,623	3,953,295	59,299	12.83	15	21	4.25	2.5	42.75	35.1
Ridley Park	3,356	3,994,475	67,974	20.25	17	21	4.25	2.5	44.75	38.0
Rose Valley	303	436,319	4,303	14.20	10	0	4.25	2.5	16.75	59.7
Rutledge	789	736,810	11,052	14.01	15	21	4.25	2.5	42.75	35.1
Sharon Hill	3,825	4,830,068	45,886	12.00	9.5	18	4.25	2.5	34.25	27.7
Swarthmore	3,405	5,401,400	67,133	19.72	22.5	25	4.25	2.5	54.25	41.5
Trainer	1,648	1,557,845	12,463	7.56	8	15	4.25	2.5	29.75	26.9
Upland	2,500	1,099,961	14,849	5.94	13.5	29	4.25	2.5	49.25	27.4
Yeadon	5,430	7,480,488	127,168	2.34	17	19	4.25	2.5	42.75	39.8
Forest county										
Tionesta	670	228,250	5,478	8.18	24	33	10	7	74	32.4

TABLE VI - TAX LEVIES IN BOROUGHS, 1932

Borough	Population	Taxable valuation for borough purposes	Borough taxes levied	Borough taxes per capita	Tax rate in mills per dollar of assessed valuation					Per cent borough millage is of total
					Borough	School	County	Poor	Total	
Lackawanna county										
Archbald	9,587	3,724,453	100,560	10.49	27	40	6.75	6	79.75	33.9
Blakely	8,260	3,024,348	36,292	4.39	12	42	6.75	6	66.75	18.0
Clarks Green	694	586,110	5,275	7.60	9	20	6.75	.5	36.25	24.8
Clarks Summit	2,604	1,842,310	18,423	3.07	10	20	6.75	1	37.75	26.5
Dalton	1,072	912,681	11,409	10.64	12.5	19	6.75	1.5	39.75	31.4
Dickson City	12,395	5,100,974	86,717	7.00	17	39	6.75	6	68.75	24.7
Dunmore	22,627	13,940,679	195,169	8.63	14	25	6.75	4	49.75	28.1
Elmhurst	841	248,274	1,738	2.07	7	18	6.75	0	31.75	22.9
Glenburn	520	446,003	2,676	5.15	6	23	6.75	1	36.75	16.3
Gouldsboro	76	54,095	433	6.73	8	9	6.75	0	23.75	33.7
Jermyn	3,519	1,650,370	16,504	4.69	10	18	6.75	6	40.75	24.5
La Plume	332	219,637	1,428	4.30	6.5	15	6.75	.5	28.75	22.6
Mayfield	3,774	1,164,094	15,715	4.16	13.5	32	6.75	4.5	56.75	23.8
Moosic	4,557	1,402,008	21,030	4.61	15	32	6.75	4.5	58.25	25.8
Moscow	892	425,123	4,251	4.77	10	33	6.75	0	49.75	20.1
Old Forge	12,661	3,433,857	51,508	4.07	15	52	6.75	4	77.75	19.3
Olyphant	10,743	6,308,427	113,552	10.57	18	32	6.75	6	62.75	28.7
Taylor	10,428	6,101,775	76,272	7.31	12.5	37	6.75	4	60.25	20.7
Throop	8,027	5,064,731	131,683	1.64	26	36.5	6.75	6	75.25	34.6
Vandling	1,169	637,641	4,463	3.82	7	33	6.75	2	48.75	14.4
Winton	8,508	2,617,365	49,730	5.85	19	40	6.75	6	71.75	26.5
Lycoming county										
Du Boistown	1,049	196,650	2,950	2.81	15	35	8.5	5	63.5	23.6
Hughestown	1,868	511,725	9,723	5.21	19	35	8.5	3	65.5	29
Jersey Shore	5,781	1,472,950	45,661	7.90	31	52	8.5	8	99.5	31.2
Montgomery	1,903	716,571	17,964	9.44	25	28	8.5	10	71.5	35
Montoursville	2,710	934,273	15,883	5.86	17	32	8.5	3	60.5	28.1
Muncy	2,413	942,420	16,964	7.03	18	25	8.5	4	55.5	32.4
Picture Rocks	548	155,438	2,710	4.95	17	35	8.5	1	61.5	27.6
Salladasburg	227	48,770	481	2.12	11	22	8.5	3	44.5	24.7
South Williamsport	6,058	3,651,056	34,685	5.73	9.5	18.5	8.5	1	37.5	22.7
Potter County										
Austin	1,116	298,998	3,289	2.95	11	30	10	10	61	18
Coudersport	2,740	959,616	20,152	7.35	21	25	10	10	66	31.8
Galeton	2,200	419,284	16,771	7.62	40	35	10	10	95	42.1
Lewisville	514	165,112	4,628	8.99	28	35	10	10	83	33.7
Oswayo	133	27,849	278	2.09	10	30	10	10	60	16.7
Shinglehouse	1,380	244,730	7,097	3.14	29	35	10	10	84	34.5

TABLE VI - TAX LEVIES IN BOROUGHS, 1932

Borough	Population	Taxable valuation for borough purposes	Borough taxes levied	Borough taxes per capita	Tax rate in mills per dollar of assessed valuation					Per cent borough millage is of total
					Borough	School	County	Poor	Total	
Snyder county										
Beavertown	604	370,272	1,481	2.45	4	12	6	3	25	16
Freeburg	391	207,187	1,865	4.77	9	11	6	6	32	28.1
Middleburg	1,024	637,918	5,741	5.61	9	17	6	4	36	25
Selinsgrove	2,797	2,333,934	17,504	6.26	7.5	11	6	5	29.5	25.4
Shamokin Dam	738	516,356	2,582	3.24	5	10	6	2	23	21.7
Sullivan county										
Dushore	715	384,850	7,312	10.23	19	19	10	5	53	35.8
Eagles Mere	212	497,880	7,470	35.24	15	10	10	5	40	37.5
Forksville	105	40,205	201	1.91	5	16	10	5	36	13.9
Laporte	163	125,205	1,327	8.14	10.6	20	10	5	45.6	23.2
Westmoreland county										
Adamsburg	227	83,875	671	2.96	8	18	8.5	-	34.5	23.2
Arnold	10,575	4,515,450	81,278	7.69	18	30	8.5	-	56.5	21.9
Arona	457	153,210	460	1.01	3	18	8.5	-	29.5	10.2
Avonmore	1,240	529,107	7,937	6.59	15	18	8.5	-	41.5	36.1
Bolivar	783	304,130	4,866	6.21	16	25	8.5	-	49.5	32.3
Derry	3,046	1,278,295	16,618	5.50	13	20	8.5	-	41.5	31.3
Donegal	163					18	8.5	-	26.5	
East Vandergrift	2,441	436,510	7,857	3.22	18	35	8.6	-	61.5	29.3
Export	2,184	550,655	7,709	3.53	14	35	8.5	-	57.5	24.3
Hunkers	378	113,050	678	1.79	6	26	8.5	-	40.5	14.8
Hyde Park	736	297,885	5,362	7.29	18	25	8.5	-	51.5	25.0
Irwin	3,443	2,361,580	-35,424	10.29	15	25	8.5	-	48.5	30.9
Jeannette	15,126	8,446,555	143,591	9.49	17	23	8.5	-	48.5	35.1
Latrobe	10,644	6,612,485	99,187	9.32	15	24	8.5	-	47.5	35.8
Ligonier	1,978	1,157,695	16,208	8.52	14	18	8.5	-	40.5	34.6
Livermore	131	54,140	433	3.31	8	14	8.5	-	30.5	36.2
Madison	365	140,695	1,688	4.62	12	21	8.5	-	41.5	28.9
Manor	1,305	663,720	11,283	8.65	17	20	8.5	-	45.5	37.4
Mount Pleasant	5,869	2,812,948	45,007	7.67	16	24	8.5	-	48.5	33.0
New Alexandria	615	358,675	2,869	4.67	8	18	8.5	-	34.5	23.2
New Florence	796	325,410	2,603	3.27	8	28	8.5	-	44.5	18
New Salem	721	350,465	2,804	3.89	8	20	8.5	-	36.5	21.9
North Belle Vernon	3,072	1,325,300	21,205	6.90	16	33	8.5	-	57.5	27.8
North Irwin	1,064	490,725	4,907	4.61	10	23	8.5	-	41.5	41.5
Oklahoma	New Borough	308,510	3,085	-	10	(?)	8.5	-	(?)	(?)
Penn	926	480,158	10,083	10.89	21	16	8.5	-	45.5	46.2
Scottdale	6,714	4,341,440	78,146	11.64	18	22	8.5	-	48.5	37.1
Seward	742	249,720	3,746	5.05	15	-	8.5	-	23.5	63.8

TABLE VI – LEVIES IN BOROUGHS, 1932

Borough	Population	Taxable valuation for borough purposes	Borough taxes levied	Borough taxes per capita	Tax rate in mills per dollar of assessed valuation					Per cent borough millage is of total
					Borough	School	County	Poor	Total	
Westmoreland county (Con't).										
Smithton	709	330,845	4,301	6.07	13	16	8.5	–	37.5	34.7
South Greensburg	2,520	1,833,740	31,174	12.37	17	24	8.5	–	49.5	34.3
S. W. Greensburg	3,105	1,735,200	29,498	9.50	17	34	8.5	–	59.5	28.6
Suterville	918	238,470	3,577	3.90	15	30	8.5	–	53.5	28.0
Trafford	4,187*	2,170,235	31,469	7.52	14.5	30	8.5	–	53	27.4
Vandergrift	11,479	5,421,010	100,035	8.71	1	25	8.5	–	34.5	1.7
West Leechburg	1,044	569,010	6,259	6.00	11	22	8.5	–	41.5	26.5
West Newton	2,953	1,462,895	21,943	7.43	15	20	8.5	–	43.5	34.5
Youngstown	482	149,878	1,349	2.80	9	16	8.5	–	33.5	26.9
Youngwood	2,783	1,171,415	15,228	5.47	13	29	8.5	–	50.5	25.7
York county										
Cross Roads	153	60,347	603	3.74	10	15	8	–	33	30.3
Dallastown	2,849	794,366	22,242	7.81	28	24	8	–	60	46.7
Delta	762	317,865	6,357	8.34	20	12	8	–	40	50.0
Dillsburg	983	422,661	4,649	4.73	11	14	8	–	33	33.3
Dover	676	301,635	2,413	3.57	8	8	8	–	24	33.3
East Prospect	391	151,195	2,570	6.57	17	12	8	–	37	45.9
Fawn Grove	302	153,925	1,231	4.08	8	25	8	–	41	19.5
Felton	400	134,812	1,753	4.38	13	12	8	–	33	39.4
Franklintown	255	60,318	724	2.84	12	12	8	–	32	37.5
Glen Rock	1,309	526,905	6,323	4.83	12	20	8	–	40	30.0
Goldsboro	459	177,130	1,417	3.09	8	16	8	–	32	25.0
Hallam	771	373,407	3,361	4.36	9	14	8	–	31	29.0
Hanover	11,805	5,145,071	82,321	6.97	16	25	8	–	49	32.7
Jacobus	445	156,845	784	1.76	5	11	8	–	24	20.8
Jefferson	423	148,968	1,788	4.23	12	6	8	–	26	46.2
Lewisberry	226	68,850	413	1.83	6	10	8	–	24	25.0
Loganville	404	140,955	846	2.09	6	10	8	–	24	25.0
Manchester	940	348,885	3,489	3.71	10	16	8	–	34	29.4
Mount Wolf	999	370,601	5,930	5.94	16	25	8	–	49	32.7
New Freedom	1,125	454,483	6,817	6.06	15	23	8	–	46	32.6
New Salem	329	107,455	860	2.31	8	22	8	–	38	21.1
North York	2,416	913,024	15,065	6.24	16.5	23	8	–	47.5	34.7
Railroad	268	106,083	1,591	5.94	15	14	8	–	37	40.5
Red Lion	4,757	2,273,124	34,097	7.17	15	19	8	–	42	35.7
Seven Valleys	405	162,939	1,629	4.02	10	11.5	8	–	29.5	33.9
Shrewsbury	671	245,816	2,950	4.40	12	10	8	–	30	40.0
Spring Grove	1,236	580,645	6,387	5.17	11	13	8	–	32	34.4

* Includes 100 in Allegheny County.

TABLE VI - LEVIES IN BOROUGHS, 1932

Borough	Population	Taxable valuation for borough purposes	Borough taxes levied	Borough taxes per capita	Tax rate in mills per dollar of assessed valuation					Per cent borough millage is of total
					Borough	School	County	Poor	Total	
York county (Con't).										
Stewartstown	863	490,885	6,873	7.96	14	14	8	-	36	38.9
Wellsville	268	121,970	488	1.82	4	12	8	-	24	16.7
West York	5,381	2,343,880	35,158	6.53	15	25	8	-	48	31.3
Windsor	1,009	282,040	4,513	4.47	16	19	8	-	43	37.2
Winterstown	219	96,830	387	1.77	4	11	8	-	23	17.4
Wrightsville	2,247	719,274	10,430	4.64	14.5	16	8.8	-	38.5	37.7
Yoe	560	163,545	6,378	11.39	39	20	8	-	67	58.2
Yorkana	219	60,540	1,029	4.70	17	12	8	-	37	45.9
York Haven	770	320,540	4,808	6.24	15	18	8	-	41	36.6

TABLE VII. TAX DELINQUENCY, 1932

Borough	1932 taxes delinquent at end of year		Taxes for prior years collected in 1932	1932 taxes delinquent less back taxes collected 1932
	Amount	Per cent of tax levy		
Centre County				
Bellefonte	10,464	30.0	4,939	5,525
Centre Hall	353	14.0	-	353
Howard	119	15.8	-	119
Milesburg	783	42.5	373	410
Millheim	1,881	57.9	730	1,151
Philipsburg	25,708	75.1	11,541	14,167
Port Matilda	634	58.8	220	414
South Philipsburg	1,837	86.0	609	1,228
Snow Shoe	586	31.1	38	548
State College	20,594	35.0	11,744	8,850
Unionville	323	37.0	-	323
Delaware County				
Aldan	15,579	41.0	3,290	12,289
Clifton Heights	17,743	40.1	-	17,743
Collingdale	27,022	49.1	8,470	18,552
Colwyn	6,552	31.9	2,035	4,517
Darby	33,028	44.0	2,104	30,924
E. Lansdowne	13,080	31.5	-	13,080
Eddystone	8,483	13.3	858	7,625
Folcroft	7,796	50.2	3,709	4,087
Glenolden	16,242	33.2	3,955	12,287
Lansdowne	76,329	45.5	14,140	62,189
Marcus Hook	9,228	16.4	1,648	7,580
Media	21,927	29.8	11,096	10,831
Millbourne	2,640	12.1	53	2,587
Norton	6,307	55.4	647	5,660
Norwood	14,461	38.1	4,635	9,826
Parkside	5,414	44.2	4,884	530
Prospect Park	20,027	33.8	7,590	12,437
Ridley Park	22,884	33.7	650	22,234
Rose Valley	351	8.2	-	351
Rutledge	3,648	33.0	1,989	1,659
Sharon Hill	16,820	36.7	4,418	12,402
Swarthmore	15,586	23.2	54	15,532
Trainer	3,009	24.1	272	2,737
Upland	2,892	19.5	3,616	- 724
Yeadon	49,473	38.9	-	49,473
Forest County				
Tionesta	2,310	42.2	2,017	293
Lackawanna County				
Archbald	23,560	23.4	970	22,590
Blakely	17,360	47.8	15,920	1,440
Clarks Green	2,023	38.3	-	2,023
Clarks Summit	4,681	25.4	64	4,617
Dalton	4,485	39.3	3,332	1,153
Dickson City	30,984	35.7	2,152	28,832
Dunmore	52,674	27.0	44,350	8,324
Elmhurst	492	28.3	111	381
Glenburn	876	32.7	525	351
Gouldsboro	90	20.8	52	38
Jermyn	6,759	41.0	3,682	3,077
La Plume	810	56.7	1,150	- 340
Mayfield	9,973	63.5	4,820	5,153

TABLE VII. TAX DELINQUENCY, 1932

Borough	1932 taxes delinquent at end of year		Taxes for prior years collected in 1932	1932 taxes delinquent less back taxes collected 1932
	Amount	Per cent of tax levy		
Lackawanna County (Con't).				
Moosic	9,524	45.3	4,549	4,975
Moscow	1,508	35.5	1,298	210
Old Forge	27,729	53.8	2,597	25,132
Olyphant	18,743	16.5	4,474	14,269
Taylor	25,579	33.5	14,456	11,123
Throop	41,430	31.5	13,437	27,993
Vandling	1,321	29.6	773	548
Winton	17,098	34.4	13,171	3,927
Lycoming County				
Du Boistown	826	28.0	1,251	- 425
Hughestown	2,723	28.0	-	2,723
Jersey Shore	15,411	33.8	7,497	7,914
Montgomery	5,896	32.8	-	5,896
Montoursville	2,482	15.6	-	2,482
Muncy	4,620	27.2	93	4,527
Picture Rocks	1,013	37.4	158	855
Salladasburg	366	76.1	-	366
South Williamsport	12,699	36.6	719	11,980
Potter County				
Austin	2,017	61.3	397	1,620
Coudersport	3,036	15.0	1,634	1,402
Galeton	4,005	23.9	1,801	2,204
Lewisville	1,217	26.3	200	1,017
Oswayo	102	36.7	36	66
Shinglehouse	2,364	33.1	1,863	501
Snyder County				
Beavertown	509	34.4	-	509
Freeburg	715	38.3	-	715
Middleburg	2,534	44.1	-	2,534
Selinsgrove	6,261	35.8	2,510	3,751
Shamokin Dam	531	20.6	32	499
Sullivan County				
Dushore	2,308	31.6	720	1,588
Eagles Mere	943	12.6	-	943
Forksville	48	23.9	-	48
Laporte	78	5.9	-	78
Westmoreland County				
Adamsburg	198	29.5	166	32
Arnold	34,264	42.2	117	34,147
Arona	200	43.5	95	105
Avonmore	2,088	25.0	458	1,630
Bolivar	2,434	50.0	75	2,359
Derry	3,446	20.7	1,243	2,203
Donegal	-	----	-	-
East Vandergrift	3,492	44.4	-	3,492
Export	2,958	38.4	665	2,293
Hunkers	298	44.0	-	298
Hyde Park	1,451	27.1	231	1,221
Irwin	10,506	29.7	1,391	9,115
Jeannette	43,089	30.0	8,613	34,476
Latrobe	31,503	31.8	6,677	24,826
Ligonier	3,928	24.2	2,382	1,546
Livermore	94	21.7	17	77

TABLE VII. TAX DELINQUENCY, 1932

Borough	1932 taxes delinquent at end of year		Taxes for prior years collected in 1932	1932 taxes delinquent less back taxes collected 1932
	Amount	Per cent of tax levy		
Westmoreland County (Con't).				
Madison	407	24.1	-	407
Manor	3,391	21.2	776	2,615
Mount Pleasant	15,980	35.5	4,500	11,480
New Alexandria	509	17.7	222	287
New Florence	769	29.5	375	394
New Salem	887	31.6	244	643
North Belle Vernon	10,043	47.4	-	10,043
North Irwin	1,541	31.4	514	1,027
Oklahoma	975	31.6	-	975
Penn	4,370	43.3	100	4,270
Scottdale	30,421	38.9	1,852	28,569
Seward	2,051	54.8	517	1,534
Smithton	539	12.5	168	371
South Greensburg	8,602	27.6	-	8,602
S. W. Greensburg	6,606	27.0	1,923	4,683
Suterville	1,517	42.4	202	1,315
Trafford	12,504	39.7	1,359	12,504
Vandergrift	28,312	28.3	5,401	22,911
West Leechburg	626	10.0	266	360
West Newton	5,419	24.7	793	4,626
Youngstown	601	44.6	48	553
Youngwood	5,463	35.9	1,311	4,152
York County				
Cross Roads	44	7.8	-	44
Dallastown	3,976	17.8	933	3,043
Delta	1,480	23.3	-	1,480
Dillsburg	449	9.7	57	392
Dover	185	7.7	170	15
East Prospect	227	8.8	137	90
Fawn Grove	260	21.1	16	244
Felton	353	20.1	219	134
Franklintown	284	39.2	112	172
Glen Rock	891	14.1	-	891
Goldsboro	167	11.8	-	167
Hallam	1,677	49.9	991	686
Hanover	15,580	18.9	4,481	11,099
Jacobus	121	15.4	40	81
Jefferson	139	7.8	43	96
Lewisberry	164	39.7	-	164
Loganville	101	11.9	11	90
Manchester	762	21.8	440	322
Mount Wolf	1,257	21.1	580	677
New Freedom	1,642	24.1	200	1,442
New Salem	39	4.5	-	39
North York	2,725	18.1	1,008	1,717
Railroad	298	18.7	124	174
Red Lion	5,611	16.5	2,981	2,630
Seven Valleys	209	12.8	21	188
Shrewsbury	120	4.1	94	26
Spring Grove	708	11.1	619	89
Stewartstown	1,068	15.5	348	720
Wellsville	59	12.1	-	59
West York	5,826	16.6	502	5,324
Windsor	1,148	25.4	726	422
Winterstown	1	.3	-	1
Wrightsville	1,597	15.3	299	1,298
Yoe	654	10.3	66	588
Yorkana	200	19.4	101	99
York Haven	574	11.9	204	370

Per Cent Delinquent	Number of Boroughs	Per Cent Delinquent	Number of Boroughs
Less than 10	10	50 - 60	8
10 - 20	31	60 - 70	2
20 - 30	37	70 - 80	2
30 - 40	43	80 - 90	1
40 - 50	21		

Nearly half of these boroughs had one-third or more of their 1932 levy delinquent at the end of the year and one in twelve failed to collect half its levy. The situation was worse in Centre County and best in York County.

Column 3 shows the amount of taxes for prior years collected by these boroughs during 1932 and column 4 the amounts by which 1932 delinquencies exceeded such back tax collections. In only 3 of the 156 boroughs did back tax collections exceed 1932 delinquencies. For the 156 boroughs the 1932 delinquencies amounted to $1,184,594, against which $321,342 in back taxes, or 27.1 per cent of current delinquencies, were collected.

Relation between Total Millage and Delinquencies - Comparison of Tables VI and VII shows a direct relationship between the total millage levied and the percentage of delinquency. In boroughs where the total levy on property was less then 30 mills, the average borough tax delinquency was 23.9 per cent; where the total levy was from 30 to 40 mills, the average delinquency was 24.8 per cent; where the total levy was from 40 to 50 mills, the average delinquency was 30.1 per cent; and where the total levy was more than 50 mills, the average delinquency was 37.0 per cent, or more than half again as much as in boroughs with total levies of less than 30, or even 40, mills.

These figures clearly imply that the higher the property tax rate, the fewer the people who pay their taxes when due and the greater the burden upon those who do pay. Those who fail to pay their taxes when due lose their property or are subject to heavy penalties for late payment. Both those who pay promptly and those who do not suffer from high taxes and high percentages of delinquency. Neither group gains unless the legislature adopts the unwise policy of remitting penalties upon delinquent taxes.

This policy merely puts a premium upon failure to pay taxes promptly. Although it gives temporary relief to those who are unable to pay their taxes when due, this advantage is overweighed by the fact that it allows property owners who are able to pay to postpone payment without penalty. This results in a greater than necessary spread between current receipts and necessary current expenditures and thereby causes the governmental unit to borrow. This in turn increases still more the cost of government as well as the burden upon those who pay their taxes when due. In the long run it is therefore worse for those who are temporarily unable to pay their taxes when due.

Following are some of the possibilities which should be considered in regions having high total tax rates upon property.

1. Concerted action by county, borough, school and poor district authorities to reduce tax rates. Table VIII shows that in many boroughs the amount of delinquent taxes at the end of 1933 exceeded the levy for that year. Radically reduced current tax rates should permit the collection of substantial amounts of delinquent taxes and at the same time materially stimulate local business.

2. Some boroughs are, governmentally, living beyond their incomes. They should curtail their governmental services to what they can afford. The more general adoption of the plan of charging on a unit basis for such services as ash and garbage collection could be used to lessen the property tax while still furnishing these services to those who need them.

3. Some of the costs of local government are made mandatory by state law. In the case of boroughs such costs are not particularly burdensome and are usually justifiable, but the county, school and poor district governments which are also supported by the borough taxpayers have many mandatory expenses. Some of these costs are not justifiable, others are. Where they are, steps should be taken to perform county, poor and

school district functions more economically, by adopting for example, the changes in county organization recommended in Part I of this survey.

In addition, there is the possibility of the state paying a greater share of the cost of local functions made mandatory by state law, thereby reducing the tax rate on property. This would involve the imposition of new state taxes.

Tax Delinquencies, 1933 - Table VIII compares total borough tax delinquencies at the end of 1933 with the 1933 tax levies of 173 of the same boroughs whose receipts were shown in Tables 1 to 5. The percentage of the levy delinquent is summarized below:

Per Cent Delinquent	Number of Boroughs	Per Cent Delinquent	Number of Boroughs
Under 25	21	100 - 125	14
25 - 50	62	125 - 150	7
50 - 75	42	150 - 200	9
75 - 100	15	Over 200	3

High rates of delinquency were especially common in the boroughs in Lackawanna, Luzerne, Monroe, Beaver and Westmoreland Counties.

Borough Indebtedness - Table IX shows the net bonded debt, total debt and floating debt of 178 boroughs. These boroughs are listed in the order of their size as shown by the 1930 census. Sixty-four of the 178 had no bonded indebtedness or had sinking funds equal in amount to their outstanding bonds. Nearly all large boroughs had outstanding bonds but only 17 of the 65 boroughs with less than 1,000 population shown in this table had any bonded indebtedness in 1933. The range in per capita net bonded indebtedness was as follows:

Per Capita Debt	Number of Boroughs	Per Capita Debt	Number of Boroughs
Less than $5	13	$30 - $40	13
$ 5 - $10	16	$40 - $50	10
$10 - $20	30	$50 - $100	4
$20 - 30	27	Over $100	1

Wyomissing, with a per capita debt of $140.95, far exceeded any other shown.

In the 114 boroughs shown which had bonded debts the average ratio of debts to assessed valuation was 3.5% plus. The distribution was as follows:

Percentage of Assessed Valuation	Number of Boroughs	Percentage of Assessed Valuation	Number of Boroughs
2 or less	31	5 - 6	8
2 - 3	26	6 - 7	3
3 - 4	25	7 - 8	4
4 - 5	16	8 - 9	1

Temporary loans and judgments outstanding at the end of the year are included with net bonded indebtedness to give the totals shown in column 5. Eight boroughs accounted for approximately 55 per cent of all the temporary loans. These were Dunmore ($185,000), Mt. Carmel ($99,000), Norristown ($50,000), North East ($44,000), Narberth ($34,000), Quakertown ($33,000), North Wales ($32,000) and Blakely ($31,000).

Other debts of these boroughs included unpaid bills and outstanding warrants amounting to $420,403 in 58 boroughs. Of this total $260,158 was owed by 5 boroughs, namely, Dunmore, Mt. Carmel, Blakely, Arnold and Swoyersville. All of these had exceptionally high percentages of unpaid taxes, ranging from 99.6 to 212.6 of their 1933 tax levies. The first three, all in the anthracite region, were among the boroughs having most of the temporary loans. High tax rates, high rates of tax delinquency, inadequate budgeting and accounting methods and large amounts of unpaid bills occur together.

TABLE VIII. TAX DELINQUENCIES, 1933

Name of Borough	Borough Tax Levy - 1933 -	Total Taxes Outstand- ing Dec. 31, 1933	Per Cent of 1933 Tax Out- standing	Name of Borough	Borough Tax Levy - 1933 -	Total Taxes Outstand- ing Dec. 31, 1933	Per Cent of 1933 Tax Out- standing
Norristown	354,365	125,070	35.2	Union City	18,053	6,421	35.6
Dunmore	202,292	430,071	212.6	Royersford	36,667	13,475	36.7
Kingston	195,016	102,701	52.6	North East	36,317	18,326	50.5
Shamokin	142,237	172,296	121.1	Penbrook	24,658	8,410	30.1
Ambridge	172,206	76,111	44.2	Birdsboro	20,855	10,890	52.2
Pottstown	150,273	52,483	34.9	Jermyn	15,899	12,726	80.0
Mt.Carmel	98,019	97,636	99.6	Perkasie	27,580	14,314	51.9
N.Braddock	165,846	70,352	42.4	Freedom	28,607	13,813	48.3
Plymouth	112,204	97,558	86.9	E.Lansdowne	30,317	30,707	101.3
Jeannette	142,950	76,731	53.6	S.W.Greensburg	25,621	13,690	53.4
Steelton	148,166	48,166	32.5	Kennett Square	25,395	12,109	47.7
Berwick	82,552	56,040	67.8	N.Belle Vernon	21,133	35,611	168.5
Carlisle	91,434	44,154	48.2	Meyersdale	20,934	11,258	53.8
West Chester	84,153	23,298	27.6	Derry	16,670	12,639	75.8
Ellwood City	74,746	33,296	44.5	Hummelstown	14,562	4,651	31.9
Phoenixville	107,784	27,167	25.2	Lykens	14,634	8,288	56.6
Vandergrift	80,654	23,474	29.1	Spring City	16,320	7,118	43.6
Latrobe	98,729	53,388	54.0	Williamstown	12,026	6,535	54.3
Arnold	71,556	73,708	103.0	West Newton	18,220	8,577	47.1
New Brighton	69,876	52,376	75.5	Millersburg	16,547	7,778	47.0
Lansdowne	145,394	44,785	30.8	Youngwood	12,795	9,644	75.4
Minersville	56,537	23,065	40.8	Coudersport	21,112	9,358	44.3
Swoyersville	32,207	38,020	118.0	Mt. Joy	7,552	2,032	26.9
Milton	62,784	35,246	56.1	Hatboro	26,831	12,886	48.0
Lansdale	54,413	28,149	51.7	Oxford	28,701	12,505	43.5
Wilson	72,638	24,125	33.2	Clarks Summit	15,763	20,556	130.4
Blakely	38,966	41,457	106.4	W.Conshohocken	15,938	7,243	45.4
Collingdale	55,262	34,414	62.3	S. Greensburg	27,364	7,192	26.3
Scottdale	69,148	76,112	110.1	E.Vandergrift	7,716	3,101	40.2
Kulpmont	26,554	38,346	144.4	North Wales	22,579	8,962	39.7
E. Stroudsburg	52,578	30,589	58.2	Highspire	12,789	5,231	40.9
Midland	106,664	33,663	31.6	Parksburg	18,205	17,835	98.0
Stroudsburg	47,193	24,332	51.6	Watsontown	16,153	5,135	31.8
Mt.Pleasant	42,160	29,576	70.2	Galeton	13,836	14,586	105.4
Beaver	68,469	81,716	119.3	Export	7,469	7,350	98.4
Media	62,948	38,040	60.4	Fleetwood	20,094	8,545	42.5
Morrisville	47,425	22,390	47.2	Heidelberg	13,410	40,433	226.9
Clifton Heights	45,027	25,749	57.2	Whitaker	11,338	19,850	175.1
W. Reading	69,071	40,627	58.8	Colwyn	17,387	9,132	52.5
Quakertown	31,115	5,839	18.8	Conway	15,341	24,788	161.6
Jenkintown	82,068	26,743	32.6	Marion Heights	9,996	7,299	73.0
Narberth	80,270	27,663	34.5	Ligonier	13,466	8,699	64.6
Monaca	58,004	61,894	106.7	Baden	15,465	25,183	162.8
Doylestown	43,339	7,979	18.4	Rockledge	16,915	4,645	27.5
Moosic	21,525	22,507	104.6	Everson	9,462	18,498	195.5
Downingtown	46,731	25,499	54.5	Newtown	10,201	1,990	19.5
Shillington	42,927	18,438	43.0	Mohnton	14,580	12,996	89.1
Trafford	26,033	45,317	174.1	Bridgewater	14,288	12,082	84.6
Boyertown	2,938	3,849	131.0	South			
Souderton	26,859	16,824	62.6	Coatesville	20,626	1,368	6.6
				Sinking Spring	16,449	9,292	56.5

TABLE VIII.　TAX DELINQUENCIES, 1933

Name of Borough	Borough Tax Levy - 1933 -	Total Taxes Outstand- ing Dec. 31, 1933	Per Cent of 1933 Tax Out- standing	Name of Borough	Borough Tax Levy - 1933 -	Total Taxes Outstand- ing Dec. 31, 1933	Per Cent of 1933 Tax Out- standing
Albion	7,107	4,019	56.5	Delaware Water			
Paxtang	15,389	4,433	28.8	Gap	4,347	7,986	183.7
Malvern	13,734	4,643	33.8	Telford	2,866	2,794	97.5
West Grove	7,772	2,360	30.4	Trappe	5,325	2,792	52.4
Elizabethville	6,452	1,458	22.6	Schwenksville	4,701	3,833	81.5
Yardley	10,992	7,322	66.6	Langhorne Manor	4,609	500	10.8
Manor	9,905	5,915	59.7	Berrysburg	961	172	17.9
West Telford	11,281	5,777	51.2	Green Lane	1,473	359	24.4
Avonmore	6,350	3,335	52.5	Hunker	449	607	135.2
Vandling	4,350	3,744	86.1	Madison	1,094	768	70.2
				Homewood	980	861	89.9
Langhorne	5,624	2,228	39.6				
Royalton	2,260	1,008	44.6	Elverson	523	131	25.0
New Hope	6,750	3,884	57.5	Ivyland	2,334	403	17.3
Dalton	9,144	6,470	70.8	Dublin	2,344	215	9.2
North Irwin	3,910	2,872	73.5	Snydertown	1,086	319	29.4
Koppel	14,319	5,452	38.1	Glasgow	717	244	34.0
West Leechburg	5,105	361	17.1	Georgetown	408	169	41.4
Penn	9,063	4,394	48.5	Platea	1,578	2,231	141.4
Sutersville	3,592	4,966	138.3	Mill Village	1,741	772	44.3
Moscow	2,860	2,171	75.9	Adamsburg	433	74	17.1
				Hookstown	1,253	601	48.0
West Mayfield	11,209	12,493	111.5				
Red Hill	3,502	1,264	36.1	Elgin	398	583	146.5
New Florence	1,945	1,360	69.9	Donegal	415	95	22.9
S.Langhorne	6,831	2,813	41.2	Frankfort			
Bolivar	4,991	3,493	70.0	Springs	178	26	14.6
Waterford	5,249	2,312	44.0				
St.Lawrence	7,938	2,211	27.9				
Halifax	3,003	333	11.1				
Seward	2,449	1,410	57.6				
Riegelsville	4,121	402	7.3				
New Salem	1,399	2,116	151.3				
East Rochester	4,663	5,278	113.2				
Smithtown	4,320	1,353	31.3				
Herndon	1,621	510	31.5				
Trumbauersville	2,162	263	12.2				
Fallston	8,751	2,024	23.1				
Richlandtown	3,016	439	14.6				
Patterson Heights	7,765	8,102	104.3				
Gratz	2,171	638	29.4				
New Alexandria	2,848	884	31.0				
Eastvale	1,305	3,070	235.2				
Dauphin	1,986	532	26.8				
Beallsville	3,667	407	11.1				
Bally	3,465	587	16.9				
South Heights	5,241	6,296	120.1				
Youngstown	803	465	57.9				
Fairview	4,636	2,593	55.9				
Lawrenceville	2,911	954	32.8				
Arona	295	543	184.1				
Riverside	2,327	840	36.1				

TABLE IX. BOROUGH INDEBTEDNESS, 1933

Name of Borough	Assessed Valuation 1933	NET BONDED DEBT			Total Debt	Outstanding Warrants and Bills Payable
		Amount	Per Cent of 1933 valuation	Per Capita		
Norristown	24,311,770	922,393	3.8	25.73	972,393	------
Dunmore	12,425,159	589,602	4.7	26.06	775,026	101,302
Kingston	19,501,972	217,000	1.1	10.05	217,000	6,906
Shamokin	6,938,391	295,808	4.6	14.59	295,808	441
Ambridge	12,300,420	484,868	3.9	23.97	484,868	------
Pottstown	12,590,115	691,304	5.5	35.58	707,304	------
Mt. Carmel	4,170,900	243,412	5.8	13.55	342,436	59,965
Plymouth	6,960,369	76,500	1.1	4.62	76,500	283
Jeannette	8,408,815	393,691	4.7	26.03	393,691	------
Steelton	8,231,451	313,667	3.8	23.60	313,667	------
Berwick	8,689,695	151,652	1.7	11.98	151,652	305
Carlisle	11,429,205	637,294	5.6	50.59	637,294	------
West Chester	12,946,590	*499,000	3.9	40.49	*499,000	------
Elwood City	8,312,150	247,343	3.0	20.07	247,343	------
Phoenixville	7,983,974	151,384	2.0	12.58	151,384	12,764
Vandergrift	4,959,080	13,421	.3	1.17	13,421	------
Latrobe	6,581,135	286,265	4.3	26.89	303,765	------
Arnold	4,472,275	349,429	7.8	33.04	349,429	30,654
New Brighton	5,781,296	103,947	2.3	10.45	103,947	1,487
Lansdowne	9,692,961	360,617	3.7	37.79	389,879	------
Minersville	2,687,151	102,900	3.7	10.96	105,400	------
Swoyersville	2,683,917	157,514	5.9	17.25	158,514	30,103
Milton	3,579,770	78,000	2.2	8.17	78,000	5,278
Lansdale	4,534,415	201,723	4.4	24.07	201,723	18,403
Wilson	7,288,320	250,177	3.4	30.27	250,177	------
Blakely	3,247,169	99,532	3.1	12.05	130,923	38,134
Collingdale	3,678,895	247,133	6.7	31.45	247,133	------
Scottdale	4,321,735	193,380	4.5	28.80	193,380	6,177
Kulpmont	1,475,219	29,389	2.0	4.80	29,389	------
East Stroudsburg	4,381,505	171,380	3.9	28.10	192,705	65
Midland	7,876,725	336,268	4.3	55.98	336,268	------
Stroudsburg	4,794,370	126,742	2.6	21.26	126,742	------
Mt. Pleasant	2,810,215	91,072	3.2	15.52	91,072	4,193
Beaver	4,891,600	269,412	5.5	47.56	292,562	5,269
Media	5,722,580	220,659	3.9	41.08	224,659	------
Morrisville	3,500,000	104,691	3.0	19.50	104,691	------
West Reading	5,755,955	433,871	7.5	88.40	433,871	------
Quakertown	3,660,707	113,274	3.1	23.20	146,424	------
Jenkintown	6,312,550	223,015	3.5	46.49	233,015	------
Narberth	4,677,670	131,257	2.8	28.11	165,220	------
Monaca	3,866,918	111,118	2.9	23.94	111,118	10,535
Doylestown	4,333,925	144,000	3.3	31.46	152,500	------
Moosic	1,655,757	73,393	4.4	16.11	92,360	127
Downingtown	3,594,667	161,902	4.5	35.60	176,902	1,829
Northumberland	1,140,000	19,686	1.7	4.39	19,686	1,505
Trafford	2,169,415	48,472	2.2	11.58	50,472	3,106
Boyertown	4,052,690	158,418	3.9	40.18	158,418	------
Souderton	2,441,734	22,500	.9	5.83	22,500	------
Union City	1,504,435	32,215	2.1	8.50	40,578	2,939
Royersford	2,820,345	105,279	3.7	28.31	105,279	------

* - Figures for 1932.

TABLE IX. BOROUGH INDEBTEDNESS, 1933

Name of Borough	Assessed Valuation 1933	NET BONDED DEBT			Total Debt	Outstanding Warrants and Bills Payable
		Amount	Per Cent of 1933 Valuation	Per Capita		
North East	2,684,050	97,632	3.6	26.60	141,732	393
Penbrook	1,643,895	39,500	2.4	11.07	41,500	------
Jermyn	1,590,898	------		-----	11,650	------
Perkasie	2,758,022	------		-----	12,000	------
Freedom	1,787,940	47,910	2.7	14.85	60,410	------
East Lansdowne	2,021,115	129,026	6.4	40.73	135,526	6,501
Wyomissing	6,097,025	438,500	7.2	140.95	448,750	4,737
S.W.Greensburg	1,708,070	86,059	5.0	27.72	89,559	------
Kennett Square	2,539,480	98,968	3.9	32.02	99,968	------
N. Belle Vernon	1,320,360	65,497	5.0	21.32	65,497	9,350
Derry	1,282,275	43,492	3.4	14.28	54,723	39
Hummelstown	1,323,075	------	---	-----	-------	------
Lykens	975,533	------	---	-----	-------	------
Mt. Penn	3,386,195	108,383	3.2	35.92	108,383	------
Spring City	1,919,940	80,391	4.2	27.18	80,391	------
Williamstown	625,303	1,572	.3	.58	5,532	------
West Newton	1,457,590	19,000	1.3	6.43	19,000	------
Millersburg	1,504,270	16,780	1.1	5.77	16,780	------
Youngwood	1,182,915	1,700	.15	.61	1,700	1,510
Leetsdale	3,725,070	40,836	1.1	14.72	40,836	------
Coudersport	960,851	29,500	3.1	10.77	29,500	------
Mt. Joy	1,078,854	------	---	-----	-------	------
Hatboro	2,145,380	118,475	5.5	44.69	118,475	------
Oxford	2,609,215	30,500	1.2	11.70	30,500	------
Clarks Summit	1,576,300	24,774	1.6	9.51	24,774	1,554
West Conshohocken	1,275,027	15,188	1.2	5.89	15,188	3,194
South Greensburg	1,824,265	86,390	4.7	34.28	86,390	597
East Vandergrift	428,696	7,900	1.8	3.24	12,556	------
North Wales	1,658,517	50,270	3.0	21.00	82,520	------
Highspire	881,283	30,000	3.4	12.89	30,300	742
Parkesburg	1,137,854	73,497	6.5	32.12	78,802	1,908
Watsontown	769,190	------	---	-----	-------	------
Galeton	414,179	19,330	4.7	8.79	19,330	9,423
Export	533,500	13,650	2.6	6.25	13,650	12
Fleetwood	2,232,256	39,739	1.8	18.48	39,739	------
Whitaker	1,030,750	25,042	2.4	12.09	25,042	5,186
Colwyn	1,362,000	54,624	4.0	26.47	54,624	------
Marion Heights	384,350	------	---	-----	9,000	8,246
Ligonier	1,122,200	24,133	2.2	12.20	35,133	12,336
Baden	1,288,093	29,545	2.3	15.36	35,545	365
Newtown	1,700,005	------	---	-----	-------	------
Mohnton	1,205,088	5,137	.4	2.82	16,137	------
Bridgewater	793,800	21,500	2.7	12.00	26,000	------
South Coatesville	3,028,900	------	---	-----	-------	------
Sinking Spring	1,644,902	37,812	2.3	21.35	37,812	34
Albion	710,696	16,541	2.3	9.84	24,752	------
Paxtang	1,538,940	78,500	5.1	49.25	86,800	------
Girard	769,100	------	---	-----	-------	2,456
Malvern	1,373,350	26,395	1.9	17.02	26,395	------
Womelsdorf	1,069,475	9,000	.8	6.06	9,000	------

TABLE IX.   BOROUGH INDEBTEDNESS, 1933

Name of Borough	Assessed Valuation 1933	NET BONDED DEBT			Total Debt	Outstanding Warrants and Bills Payable
		Amount	Per Cent of 1933 Valuation	Per Capita		
West Grove	1,002,850	7,771	.8	5.65	7,771	------
Elizabethville	529,980	------	---	-----	6,300	------
Yardley	916,025	10,783	1.2	8.24	10,783	------
Manor	606,300	------	---	-----	------	108
West Telford	705,055	35,500	5.0	28.35	35,500	------
Avonmore	529,150	14,423	2.7	11.63	14,423	------
Vandling	543,752	------	---	-----	------	1,285
Langhorne	1,117,950	3,668	.3	3.20	3,668	------
Royalton	239,995	------	---	-----	------	1,125
New Hope	769,076	------	---	-----	------	------
Dalton	825,760	23,962	2.9	22.35	27,962	------
North Irwin	440,035	------	---	-----	------	------
West Leechburg	567,185	22,942	4.0	21.98	22,942	152
Penn	477,090	13,572	2.8	14.66	17,572	------
Sutersville	245,350	6,500	2.6	7.08	9,729	------
Moscow	476,610	------	---	-----	3,500	------
West Mayfield	1,120,899	20,275	1.8	23.14	20,275	720
Red Hill	500,286	------	---	-----	------	------
New Florence	319,305	------	---	-----	------	------
South Langhorne	620,420	651	.1	.83	9,901	------
Bolivar	310,155	9,199	3.0	11.75	9,199	------
Waterford	438,455	------	---	-----	------	------
Avondale	582,245	------	---	-----	------	------
St.Lawrence	587,970	50,040	8.5	65.84	50,765	2,100
Halifax	299,865	------	---	-----	3,000	------
Seward	244,900	------	---	-----	1,186	------
Riegelsville	515,186	------	---	-----	------	------
New Salem	349,810	------	---	-----	2,200	------
East Rochester	358,682	------	---	-----	------	10
Smithton	302,200	------	---	-----	2,475	14
Herndon	202,630	------	---	-----	------	------
Trumbauersville	243,495	------	---	-----	------	------
Millville	420,450	3,648	.9	5.48	4,448	------
Fallston	795,460	------	---	-----	------	30
Richlandtown	335,095	------	---	-----	------	------
Patterson Heights	597,317	4,394	.7	6.88	6,194	1,242
Atglen	404,780	1,394	.3	2.25	1,394	------
New Alexandria	355,985	------	---	-----	2,000	------
Eastvale	217,479	------	---	-----	50	534
Dauphin	223,330	------	---	-----	------	------
Modena	336,250	------	---	-----	------	------
Bally	314,025	25,000	8.0	43.18	32,000	1,477
South Heights	262,040	7,334	2.8	13.36	11,734	643
New Galilee	177,075	------	---	-----	------	------
Youngstown	136,015	------	---	-----	------	------
Mt. Pocono	537,454	------	---	-----	2,000	------
Fairview	299,112	1,198	.4	2.61	7,198	------
Lawrenceville	167,635	------	---	-----	------	45
Arona	147,675	------	---	-----	------	------
Turbotville	172,150	------	---	-----	------	------

TABLE IX. BOROUGH INDEBTEDNESS, 1933

| Name of Borough | Assessed Valuation 1933 | NET BONDED DEBT | | | Total Debt | Outstanding Warrants and Bills Payable |
		Amount	Per Cent of 1933 Valuation	Per Capita		
Riverside	161,925	------	---	------	------	------
Delaware Water Gap	543,385	17,300	3.2	39.05	17,300	450
New Britain	377,376	------	---	------	1,200	------
Telford	191,066	10,058	5.3	24.41	10,058	------
Trappe	600,145	17,614	2.9	42.96	30,114	------
Schwenksville	387,375	------	---	------	------	------
Langhorne Manor	542,210	------	---	------	2,000	------
Green Lane	245,515	------	---	------	1,200	------
Hunker	112,275	------	---	------	------	------
Madison	136,765	------	---	------	------	------
Homewood	98,000	------	---	------	------	------
Elverson	261,460	------	---	------	------	------
Uniontown	132,760	270	.2	.85	270	------
Ivyland	232,275	------	---	------	2,500	------
Dublin	234,399	------	---	------	3,000	------
Silverdale	175,265	------	---	------	------	------
Snydertown	72,370	------	---	------	------	------
Glasgow	55,765	------	---	------	------	------
Georgetown	102,290	------	---	------	------	------
Platea	119,274	5,300	4.4	21.29	5,300	------
Mill Village	94,110	------	---	------	5,000	------
Adamsburg	86,450	------	---	------	------	------
Hookstown	83,535	------	---	------	1,850	115
Elgin	66,380	------	---	------	190	------
Donegal	70,450	------	---	------	------	------
Livermore	48,150	------	---	------	------	------
Frankfort Springs	59,485	------	---	------	------	------
Gouldsboro	46,405	------	---	------	------	------
178 boroughs	420,430,912	13,982,453	#3.6		14,898,540	420,403

Based on assessed valuation of 114 boroughs having bonded indebtedness.

Borough Expenditures - Table X shows by population groups the average expenditures of 193 boroughs. With a few exceptions these are the same boroughs whose receipts were shown in Table I. Expenditures for the operation and maintenance of general governmental departments are classified under the headings, - general government, police protection, fire protection, inspection services, public health, sanitary services, street and bridge maintenance, street lighting, recreation and miscellaneous.

The chief items covered by the heading general government are the salaries and/or expenses of the burgess, secretary, controller or auditors, treasurer, tax collector, solicitor, and manager. A few related expenses are included in a few cases. Police protection includes some minor expenditures by a few boroughs for lockups and public pounds. Fire protection is self-explanatory. Inspection services include expenditures for building and plumbing inspection and for the inspection of weights and measures. Public health includes the cost of the board of health and health officers and, where found, expenditures for hospitals, food inspection, markets and comfort stations. Sanitary services cover the cost of sewage disposal, garbage and ash collection and street cleaning. Included with the cost of street and bridge maintenance is the salary and expenses of the borough engineer, whose duties are usually primarily connected with street building and maintenance. Street lighting and recreation are self-explanatory. The chief miscellaneous items are library appropriations, premiums on officers' bonds, compensation insurance and dues payable to the State Association of Boroughs.

A subtotal brings together expenditures for the operation and maintenance of the general departments and those for interest, the state tax on loans and the retirement of bonded debt. Expenditures for permanent improvements, repayment of temporary loans, non-governmental expenditures and public service enterprises are shown separately.

Mandatory Expenses - Few expenses of a mandatory nature are imposed upon boroughs. Each borough must have a minimum number of officers but there are few mandatory requirements as to their payment. Rather the law tends to prescribe maximum limits above which they may not be paid. There is some required expense for printing ordinances but an inactive borough needs few ordinances in the course of a year. Aside from the few clerical supplies that council must provide almost the only other expense which falls upon every borough is for the maintenance of streets and bridges. Each borough may exercise considerable option as to how well it will maintain its highways but each is responsible for their condition. However, many miles of roads within boroughs have been built wholly or partly at state expense and some of these are also maintained by the state. Another expenditure of a mandatory nature occurs occasionally when the State Department of Health orders the erection of a sewage disposal plant. The exercise of almost every other borough power requiring the expenditure of borough funds is optional, either with the borough officers or with the electorate or a group of property owners. In some cases, however, the exercise of a power that is optional makes certain expenditures mandatory. Thus, for instance, the adoption of a regulatory ordinance entails the cost of publication. Compensation insurance must be provided when employees are hired and when debts are incurred, they must, of course, be repaid.

Per Capita Expenditures - In Table XI are shown the average per capita expenditures, by population groups, of these 193 boroughs for each of the functions listed. The per capita figures for both 1930 and 1933 are based upon the populations of these boroughs in 1930. Among the more significant facts shown by this table are the following:

1. Per capita as well as actual expenditures of small boroughs are much less than those of large boroughs. In general per capita expenditures vary directly with the size of the boroughs. This is especially true of expenditures for police and fire protection, inspection services, public health and sanitary services.

2. A notable exception to the general rule is per capita expenditure for street lighting which is greater the smaller the borough.

3. The highest per capita expenditures occur not in the largest boroughs but in those with from 2,500 to 5,000 inhabitants. This is due to exceptionally high outlays for sanitary services, street and bridge maintenance, recreation, debt service, permanent improvements, and public service enterprises. A possible explanation is that

TABLE X.   EXPENDITURES OF 193 BOROUGHS, BY POPULATION GROUPS, 1930 AND 1933

Function	Year	Average Expenditures by Population Groups					
		1000 and up (a)	5000 to 10000 (b)	2500 to 5000 (c)	1500 to 2500 (d)	500 to 1500 (e)	Under 500 (f)
Operation and maintenance							
General government	1930	$14,841	$ 7,682	$ 3,629	$ 1,664	$ 794	$ 191
	1933	11,482	6,541	2,933	1,674	633	170
Police protection	1930	21,931	7,840	3,549	1,205	267	12
	1933	17,307	6,672	3,055	1,009	145	17
Fire protection	1930	17,764	6,277	2,649	1,282	530	73
	1933	14,130	5,092	2,677	961	537	131
Inspection services (g)	1930	445	199	40	14	-	-
	1933	278	120	10	10	-	-
Public health	1930	2,395	831	334	132	71	22
	1933	2,046	738	339	138	67	23
Sanitary services (h)	1930	13,067	5,036	2,843	897	151	9
	1933	9,149	4,675	2,234	595	69	9
Street and bridge maintenance	1930	31,839	20,295	10,355	5,943	2,011	770
	1933	20,618	8,745	6,838	2,902	1,323	555
Street lighting	1930	14,366	6,340	3,635	2,078	1,010	493
	1933	12,185	5,468	3,791	1,979	910	507
Recreation	1930	2,002	524	866	52	18	8
	1933	911	268	283	49	12	4
Miscellaneous	1930	4,047	1,709	757	379	159	51
	1933	3,740	1,009	812	363	201	68
Total operation and maintenance	1930	122,697	56,733	28,657	13,646	5,010	1,628
	1933	91,846	39,818	22,972	9,680	3,896	1,484
Bond and sinking funds	1930	17,072	7,302	5,750	1,600	764	173
	1933	20,521	9,113	4,966	1,525	498	129
Interest and tax on loans	1930	15,650	11,138	4,774	2,020	620	119
	1933	18,537	9,805	4,634	1,937	457	128
TOTAL operation, maintenance & debt service	1930	155,419	75,173	39,181	17,266	6,394	1,920
	1933	130,904	58,736	32,571	13,141	4,850	1,740
Permanent improvements	1930	51,245	8,827	12,898	4,585	1,338	1,155
	1933	14,023	3,001	4,310	1,198	385	23
Temporary loans	1930	74,941	23,001	13,708	2,323	1,552	617
	1933	31,239	7,802	14,087	2,221	726	305
Non-governmental expenditures	1930	412	756	130	141	45	4
	1933	630	247	145	29	15	2
Public service enterprises	1930	12,392	10,067	5,305	2,013	444	5
	1933	10,103	15,521	4,061	2,799	246	11
TOTAL EXPENDITURES	1930	294,409	117,824	71,222	26,328	9,773	3,702
	1933	186,899	85,307	55,175	19,387	6,220	2,081

(a) Average for 19 boroughs.   (d) Average for 26 boroughs.
(b) Average for 20 boroughs.   (e) Average for 48 boroughs.
(c) Average for 45 boroughs.   (f) Average for 35 boroughs.
(g) Includes building and plumbing inspection and weights and measures.
(h) Includes sewerage, garbage and ash collection and street cleaning.

TABLE XI. AVERAGE PER CAPITA EXPENDITURES OF 193 BOROUGHS, BY FUNCTIONS AND
BY POPULATION GROUPS, 1930 AND 1933

Function	Year	Population Groups						
		193 boroughs	10000 and up	5000 to 10000	2500 to 5000	1500 to 2500	500 to 1500	Under 500
Operation and maintenance								
General government	1930	.95	.90	1.07	1.04	.84	.92	.59
	1933	.78	.69	.91	.84	.85	.73	.52
Police protection	1930	1.08	1.33	1.09	1.02	.61	.31	.04
	1933	.88	1.05	.93	.87	.51	.17	.05
Fire protection	1930	.89	1.07	.87	.76	.65	.61	.23
	1933	.76	.85	.71	.77	.49	.62	.40
Inspection services (a)	1930	.02	.03	.03	.01	.01	.00	.00
	1933	.01	.02	.02	.00	.01	.00	.00
Public health	1930	.12	.14	.12	.10	.07	.08	.07
	1933	.11	.12	.10	.10	.07	.08	.07
Sanitary services (b)	1930	.71	.79	.70	.81	.46	.17	.03
	1933	.54	.55	.65	.64	.30	.08	.03
Street & bridge maintenance	1930	2.44	1.92	2.83	2.96	3.01	2.32	2.37
	1933	1.43	1.25	1.22	1.96	1.47	1.53	1.71
Street lighting	1930	.95	.87	.88	1.04	1.05	1.16	1.52
	1933	.87	.74	.76	1.09	1.00	1.05	1.56
Recreation	1930	.12	.12	.07	.25	.03	.02	.02
	1933	.05	.06	.04	.08	.02	.01	.01
Miscellaneous	1930	.23	.24	.24	.22	.19	.18	.16
	1933	.22	.23	.21	.23	.18	.23	.21
Total - operation	1930	7.51	7.42	7.90	8.20	6.92	5.78	5.01
and maintenance	1933	5.65	5.55	5.54	6.58	4.91	4.49	4.56
Bonds and sinking funds	1930	.93	1.03	1.02	1.65	.81	.88	.53
	1933	1.00	1.24	1.27	1.42	.77	.57	.40
Interest & tax on loans	1930	.94	.95	1.55	1.37	1.02	.72	.36
	1933	.96	1.12	1.37	1.37	.98	.53	.39
Total - operation and	1930	9.78	9.39	10.47	11.22	8.76	7.37	5.90
maintenance & debt service	1933	7.90	7.91	8.18	9.32	6.67	5.59	5.35
Permanent improvements	1930	2.61	3.10	1.23	3.69	2.33	1.54	3.55
	1933	.69	.85	.42	1.23	.61	.44	.07
Temporary loans	1930	3.49	4.53	3.20	3.92	1.18	1.79	1.89
	1933	1.86	1.89	1.09	4.03	1.13	.83	.94
Non-governmental	1930	.05	.02	.11	.04	.07	.06	.01
expenditures	1933	.04	.04	.03	.04	.01	.02	.01
Public service	1930	1.04	.75	1.40	1.52	1.02	.51	.02
enterprises	1933	1.07	.61	2.16	1.16	1.42	.28	.03
TOTAL EXPENDITURES	1930	17.27	17.79	16.40	20.39	13.36	11.29	11.38
	1933	11.96	11.30	11.88	15.80	9.84	7.17	6.40

(a) Includes building & plumbing inspection & weights and measures.
(b) Includes sewerage, garbage & ash collection & street cleaning.

these boroughs are too large to forego rendering the municipal services which small boroughs dispense with, yet not large enough to provide these services most economically.

4. The only important expenses of the average borough with less than 500 population are for street and bridge maintenance, street lighting, general government, fire protection and debt service. Inspection services took less than one-half cent per capita, recreation from 1 to 2 cents, sanitary services 3 cents, police protection from 4 to 5 cents, and public health 7 cents. Examination of the basic data from which this table was prepared discloses that few of the small boroughs spent anything at all for any of these functions except public health.

5. Average expenditures of 19 boroughs of more than 10,000 population (large enough to be cities of the third class if they chose) were $17.79 per capita in 1930 and $11.30 per capita in 1933. Their expenditures for general government were $.90 per cpaita in 1930 and $.69 per capita in 1933.

6. Computed upon the population of all the 193 boroughs upon which these tables are based, the per capita cost of operation and maintenance of public service enterprises was $1.04 in 1930 and $1.07 in 1933. They should be calculated, however, only upon the population of those boroughs which had such enterprises. Upon this basis they were $4.14 in 1930 and $3.76 in 1933.

7. Average per capita expenditures for public improvements were $2.61 in 1930 and $.69 in 1933. These are probably minimum figures in both cases because many borough financial reports make no distinction between expenditures for construction and for maintenance. It is likely therefore that considerable amounts shown in these tables for maintenance of streets and bridges and sanitary services were spent for new construction.

8. The large amounts shown for the repayment of temporary loans are misleading because there is no way to learn from the financial reports the extent to which these represent the repeated turnover of short term notes. Thus Norristown alone expended $338,000 for the repayment of temporary loans in 1933, and spent only $335,572 for all other purposes. No such sum was originally borrowed at any one time. In this instance the borough officials were taking advantage of the current financial situation and the excellent credit position of the borough to finance some major improvements by short term notes instead of by the sale of bonds. Because many boroughs resort to the practice of renewing short term notes at frequent intervals during the year, it is useless to attempt comparisons between actual or per capita expenditures for the payment of temporary loans and expenditures for other purposes. Nevertheless Tables II and XI indicate a marked reduction in the volume of such financing between 1930 and 1933.

Comparative Cost of Borough Functions - Table XII shows relative functional costs in percentages of total expenditures. The points of chief significance in this table are the following:

1. For the entire group, debt service was the greatest expense, taking 23.3 per cent of the total in 1930 and 29.5 per cent in 1933.

2. Of the operating expenses street and bridge maintenance was the most expensive item, taking 24.9 per cent of the total in 1930 and 17.9 per cent in 1933.

3. General government, police protection, fire protection and street lighting each took approximately 10 per cent of the total expenditures of the entire group both years, sanitary services about two thirds as much. All other operating costs combined were only about 5 per cent of the total.

4. As between different population groups there was pronounced variation in some items. These variations were greatest in the cases of police protection, fire protection, inspection services, sanitary services and debt service, which took relatively greater parts of the expenditures of the larger boroughs, and of street and bridge maintenance and street lighting, which took by far the most of the expenditures of the

TABLE XII. RATIO OF EXPENDITURES FOR SPECIFIED FUNCTIONS TO TOTAL EXPENDITURES
FOR OPERATION AND MAINTENANCE, INCLUDING DEBT SERVICE, OF 193 BOROUGHS,
BY POPULATION GROUPS, 1930 AND 1933

Function	Year	Population Groups						
		193 boroughs	1000 and over	5000 to 10000	2500 to 5000	1500 to 2500	500 to 1500	Under 500
Operation and maintenance								
General government	1930	9.8	9.5	10.2	9.3	9.6	12.4	9.9
	1933	9.7	8.8	11.1	9.0	12.7	13.0	9.8
Police protection	1930	11.0	14.1	10.4	9.1	7.0	4.2	.6
	1933	11.0	13.2	11.4	9.4	7.7	3.0	1.0
Fire protection	1930	9.1	11.4	8.4	6.8	7.4	8.3	3.8
	1933	9.5	10.8	8.7	8.2	7.3	11.1	7.5
Inspection services (a)	1930	.2	.3	.3	.1	.1	.0	.0
	1933	.1	.2	.2	.0	.1	.0	.0
Public health	1930	1.2	1.5	1.1	.9	.8	1.1	1.1
	1933	1.3	1.6	1.3	1.0	1.1	1.4	1.3
Sanitary services (b)	1930	7.2	8.4	6.7	7.3	5.2	2.4	.5
	1933	6.7	7.0	8.0	6.9	4.5	1.4	.5
Street & bridge maintenance	1930	24.9	20.5	27.0	26.4	34.4	31.5	40.1
	1933	17.9	15.8	14.9	21.0	22.1	27.3	31.9
Street lighting	1930	9.7	9.2	8.4	9.3	12.0	15.8	25.7
	1933	10.8	9.3	9.3	11.6	15.1	18.8	29.2
Recreation	1930	1.3	1.3	.7	2.2	.3	.3	.4
	1933	.6	.7	.4	.9	.4	.3	.2
Miscellaneous	1930	2.3	2.6	2.3	1.9	2.2	2.5	2.7
	1933	2.8	2.9	2.6	2.5	2.8	4.1	3.9
Bonds and sinking funds	1930	11.6	11.0	9.7	14.7	9.3	12.0	9.0
	1933	15.0	15.7	15.5	15.2	11.6	10.3	7.4
Interest & tax on loans	1930	11.7	10.1	14.8	12.2	11.7	9.7	6.2
	1933	14.5	14.2	16.7	14.2	14.7	9.4	7.3
Total - operation and maintenance & debt service	1930	100.0	100.0	100.0	100.0	100.0	100.0	100.0
	1933	100.0	100.0	100.0	100.0	100.0	100.0	100.0

(a) Includes building and plumbing inspection and weights and measures.
(b) Includes sewerage, garbage and ash collection and street cleaning.

very small boroughs. The relative cost of street lighting in the smallest boroughs was nearly three times the average.

5. In general, the larger the borough the more varied are the services it renders. Many of the small boroughs perform only the functions of second class townships while the larger boroughs frequently render all the services usually performed by cities. Thus, for example, McEwensville, with 210 inhabitants, spent in 1933 $135 for general government, $5 for public health, $115 for maintenance of streets and bridges, and $362 for street lighting, whereas Chambersburg, with 13,788 inhabitants, maintained police and fire departments, a comfort station, parks and playgrounds, sanitary sewers, a refuse incinerator, public market, municipal ambulance, library, water department and electric light department, in addition to the minimum services performed by all boroughs.

Comparison of 1930 and 1933 Expenditures - Table XIII shows the percentage of increase or decrease from 1930 to 1933 in the amount expended for each important borough function. All items in the table are decreases unless preceded by a plus sign. Taking the 193 boroughs as a single group only three items show any increase and these are relatively small. Payments for bond retirement increased 6.1 per cent and interest payments 14.8 per cent. Both of these were due to substantial increases on the part of the largest boroughs only. Average payments for public service enterprises increased slightly but by no means uniformly.

The cost of borough operation and maintenance decreased 24.8 per cent. This decrease was fairly uniform in the five larger population groups, but the boroughs with less than 500 inhabitants were able to curtail their already small expenditures only about one third as much. The biggest percentage decreases for the entire group were in expenditures for recreation, inspection services and street and bridge maintenance. Only the last was large enough originally to make the reduction of material importance. Substantial reductions were also made in expenditures for general government, police and fire protection. Permanent improvements were reduced 70.8 per cent. In some cases the trend was not uniform in the boroughs of different sizes.

Compensation of Principal Borough Officers - Table XIV shows, by population groups, the average compensation of the principal borough officers. The table is based upon the Pennsylvania Economic Council's analysis of the financial statements of 193 boroughs. Because of the incomparability of these statements the data are necessarily incomplete. After eliminating all boroughs from which comparable data were unobtainable 181 boroughs remained for which data were secured for one or all of the six officers shown in these tables.

Almost all of the boroughs with more than 10,000 inhabitants pay their burgesses regular salaries. As the boroughs become smaller, however, the percentage doing so decreases until only about one of every seven boroughs with less than 500 inhabitants pays the burgess a salary. The auditors or controller and the tax collector are paid in all boroughs, and the secretary also with a very few exceptions. Many of the small boroughs pay a solicitor only when his services are required and even more pay a small retainer to cover routine advice and pay additional sums for extra services. The treasurer is not uncommonly unpaid even in large boroughs, if he is an official of the borough depository.

CONCLUSIONS

A. Reduction in Number of Boroughs.

1. FINDING. The ease with which boroughs may be created has led to the creation of many uneconomically small boroughs.

2. FINDING. The chief reasons leading to the incorporation of new boroughs have been desire for a greater degree of local control in municipal matters, desire to escape payment of greater amounts of township taxes than were returned to the territory comprising the new borough, and desire for better or more numerous governmental services than the township offered.

TABLE XIII. PER CENT OF CHANGE IN AMOUNT OF EXPENDITURES OF 193 BOROUGHS, 1933 OVER 1930, BY FUNCTIONS AND BY POPULATION GROUPS. (All items are decreases unless preceded by a plus sign.)

Function	193 boroughs	10000 and up	5000 to 10000	2500 to 5000	1500 to 2500	500 to 1500	Under 500
				Population Groups			
Operation and maintenance							
General government	18.4	22.6	14.9	19.2	+0.6	20.3	10.8
Police protection	18.5	21.1	14.9	20.1	16.3	45.7	+43.0
Fire protection	15.1	20.5	18.9	+1.1	25.1	+1.3	+78.0
Inspection services (a)	42.4	37.5	39.7	74.1	28.0	3.0	0.0
Public health	9.8	14.6	11.2	+1.3	+4.5	3.1	+6.9
Sanitary services (b)	23.8	30.0	7.2	21.4	33.7	54.8	1.6
Street and bridge maintenance	41.1	35.2	56.9	34.0	51.2	34.2	27.9
Street lighting	8.6	15.2	13.8	+4.3	4.8	9.8	+2.9
Recreation	58.7	54.5	50.8	67.3	6.7	31.7	50.5
Miscellaneous	3.1	7.6	11.7	+7.3	4.3	+26.6	+32.0
Total - operation & maintenance	24.8	25.1	29.8	19.8	29.1	22.2	8.9
Bonds and sinking funds	+6.1	+20.2	+24.8	13.6	4.7	34.9	25.6
Interest and tax on loans	+14.8	+18.4	12.0	2.5	4.1	26.3	7.5
Total - operation and maintenance and debt service	18.2	15.8	21.9	16.9	23.9	24.1	9.4
Permanent improvements	70.8	72.6	66.0	66.6	73.0	71.3	98.0
Temporary loans	44.1	58.3	66.1	+2.8	4.4	53.2	50.6
Non-governmental expenditures	28.3	52.9	67.4	+11.6	79.8	67.6	59.4
Public-service enterprises	+2.8	18.5	+54.2	23.4	+39.0	44.6	+121.8
TOTAL EXPENDITURES	30.7	36.5	27.6	22.5	26.4	36.5	43.8

(a) Includes building and plumbing inspection and weights and measures.
(b) Includes sewerage, garbage and ash collection and street cleaning.

TABLE XIV.   AVERAGE COMPENSATION PAID PRINCIPAL BOROUGH OFFICERS

| | Burgess | | | Secretary | | |
Population group	Number of boroughs	Average Compensation 1930	1933	Number of boroughs	Average Compensation 1930	1933
Over 10,000	16	$1,265	$1,032	15	$1,922	$1,747
5,000 to 10,000	10	492	468	15	1,378	1,392
2,500 to 5,000	21	214	194	39	520	597
1,500 to 2,500	15	105	98	23	237	226
500 to 1,500	12	60	52	41	114	108
Less than 500	5	43	53	23	34	36

| | Treasurer | | | Solicitor | | |
Population group	Number of boroughs	Average Compensation 1930	1933	Number of boroughs	Average Compensation 1930	1933
Over 10,000	8	$1,811	$ 967	17	$1,716	$1,272
5,000 to 10,000	11	296	324	16	677	654
2,500 to 5,000	26	211	172	30	317	258
1,500 to 2,500	15	109	123	19	163	186
500 to 1,500	22	53	45	33	144	97
Less than 500	19	30	23	12	66	28

| | Auditors or Controller | | | Tax Collector | | |
Population group	Number of boroughs	Average Compensation 1930	1933	Number of boroughs	Average Compensation 1930	1933
Over 10,000	12	$ 610	$ 566	15	$3,214	$2,431
5,000 to 10,000	13	217	209	16	1,864	1,345
2,500 to 5,000	34	70	71	42	960	709
1,500 to 2,500	20	54	68	22	376	359
500 to 1,500	39	22	20	39	228	154
Less than 500	22	11	11	28	81	53

RECOMMENDATION. The courts of quarter sessions should be directed to refuse charters to new boroughs unless there is an especially strong reason for their creation, such as that the district desires to provide municipal services which the township is either unauthorized or unsuited to supply.

3. FINDING. Many boroughs form natural political units with adjacent boroughs or other thickly populated regions.

4. FINDING. The waste caused by the existence of too many small boroughs is not shown in the direct expenditures of the boroughs themselves, for these are low in amount and per capita, but in the additional costs which their existence imposes upon adjacent townships, the school system, and the county, state and national governments.
RECOMMENDATION. That the Bureau of Municipalities survey conditions in one or more populous counties and propose consolidations, pointing out the economies which would result, and that these proposals be submitted to a vote of the people in the districts concerned.

5. FINDING. The present procedure for annulling borough charters is cumbersome. Boroughs whose charters are annulled may revert to the townships from which they were formed or may become new townships. Those which have given up their charters in recent years have formed new townships so that the number of governmental units was not decreased by their action.
RECOMMENDATION. It should be easier to have borough charters annulled and there should be no option to create new townships. All abandoned boroughs should be consolidated with some existing unit or units of local government.

6. FINDING. Boroughs are usually coterminous with borough school districts.
RECOMMENDATION. That this wholly illogical system be abandoned where small boroughs are permitted to exist.

7. FINDING. Boroughs may exercise most of the powers of third class cities and have a more flexible system of government.

B. Internal Structure and Procedure

8. FINDING. The present division of authority between the burgess and council frequently causes friction and makes the government of boroughs less efficient and responsible to the will of the people.
RECOMMENDATION. That the burgess be made a member of council with a vote but no veto power and that he be given sole authority to appoint and dismiss policemen.

9. FINDING. The high constable has no duties which warrant his retention as an elected officer.
RECOMMENDATION. That the office of high constable be abolished and his duties be vested in an appointed officer to be designated by the borough council.

10. FINDING. With a properly administered county-wide assessment system there would be no need for boroughs to make separate assessments for borough purposes.
RECOMMENDATION. That after the county assessment system has been improved the act authorizing separate assessments for borough purposes be repealed.

11. FINDING. There is no reason why assessors and tax collectors should be elected or for there being such officers in each borough.
RECOMMENDATION. That their duties be transferred to appointed county officers and that these elective positions be abolished.

12. FINDING. It is proper that auditors or controllers should be independent of the authorities whose accounts they audit but the elective system does not assure the selection of suitable persons. Consequently their work is often poorly done.
RECOMMENDATION. That the auditors of local accounts should be full-time qualified accountants employed by and working under the supervision of a state department.

13. FINDING. Complete lack of uniformity, and often of system, in accounting methods

makes comparisons almost impossible and affords taxpayers inadequate information concerning borough fiscal affairs.

RECOMMENDATION. That the Bureau of Municipalities in the Department of Internal Affairs, or some other state agency, be given full authority and adequate appropriations to devise and require conformance with uniform systems of accounts; and that to facilitate this, boroughs be classified.

14. FINDING. Decentralized supervision of administrative affairs in many boroughs results in some inefficiency in such matters as financial planning, purchasing, etc.

RECOMMENDATION. Large and active boroughs should adopt the manager system to secure the advantages which it has demonstrated where it is in use. The system is particularly adapted to boroughs of more than 5,000 population and to smaller boroughs which operate utilities.

15. FINDING. Political considerations govern the selection of appointive borough officers, especially in the police departments where the effects are demoralizing.

RECOMMENDATION. That the head of the state police force be authorized to prescribe minimum qualifications for local police officers, to conduct qualifying examinations for candidates, and to maintain a training school for successful applicants.

16. FINDING. Losses in closed banks have caused major financial difficulties in many boroughs, the largest losses usually being of sinking funds.

RECOMMENDATIONS. (a) That boroughs be authorized to issue only serial bonds in the future, to obviate the necessity for maintaining large sinking funds; and

(b) That boroughs whose deposits in any depository exceed the amount insured by the F. D. I. C. require their depositories to furnish adequate collateral to safeguard their deposits.

17. FINDING. The statutory provision that the courts may impose extra tax levies to pay the debts of boroughs encourages extravagant commitments where members of council are irresponsible, although the provision is essentially desirable.

RECOMMENDATIONS. (a) That the courts should retain their power to impose special taxes to pay debts for which borough officers are not responsible (such as judgments in connection with highway accidents) and to pay due debts when council fails to levy an adequate tax within its legal limits.

(b) That the burgess and members of council voting for expenditures in excess of the legal resources of the borough from taxes and other sources be made liable for the excess expenditures, but that the courts or a state administrative officer be authorized to relieve local officers from liability in cases where local conditions necessitate tax levies in excess of the general limits.

(c) That a state administrative officer be authorized to exercise control over borough budgeting, tax rates and purchasing, but only when petitioned to do so by a majority in interest, though not in numbers, of the borough taxpayers.

18. FINDING. The higher the combined tax rate for all local purposes, the fewer are the people who pay their taxes when due and the greater is the burden upon those who do pay. Remitting penalties upon delinquent taxes increases their burden still more.

RECOMMENDATION. That county, borough, school and poor district authorities take concerted action to reduce taxes by instituting all possible economies; that to lessen the property tax charges be made for special services to individuals whenever possible; and that the state pay a larger share of local expenses made mandatory by state law.

CHAPTER XXV

THE THIRD CLASS CITY

Purpose and Scope of Study:- It is the purpose of this part of the Survey to de-scribe the system of government created for third class cities and the actual day-to-day operation of this system. To do this, it has been necessary to examine not only the Constitution and statutes of the State in so far as they pertain to this type of govern-ment, but also to study the official acts, records and reports of a number of such ci-ties. The method adopted was to ascertain first the powers and functions delegated to such cities together with the statutory organization of offices, boards and commissions and then to examine the actual administrative policies and practices adopted by the dif-ferent cities and the results obtained thereby. Special emphasis has been placed on the problem of raising revenue, including bonded debts; also on expenditures, both total and functional. In comparing the revenue and expenditure figures of different cities, the figures for both 1930 and 1933 are presented in order to show to what extent the cities have been able to re-adjust themselves to present economic conditions. Certain admini-strative practices have also been considered. Throughout the report the aim has been to present a clear and accurate description of existing local conditions but not to set forth the supposed mistakes or irregularities of individual officials.

Number of Cities Studied and Sources of Materials Used:- At the present time there are forty-four third class cities with populations ranging from 7,152 in Corry to 115,967 in Erie according to the United States Census of 1930. Twenty-eight of these cities con-tain less than 30,000 people each and these are the only cities covered by this study. One of these, New Kensington, only became a city in 1934 and no attempt was made to in-clude it with the others studied.

In order to obtain first hand information, the following eighteen cities were visi-ted during the summer of 1934:- Beaver Falls, Butler, Carbondale, Clairton, Coatesville, Connellsville, Farrell, Franklin, Greensburg, Meadville, Monongahela, Nanticoke, Oil City, Sharon, Sunbury, Titusville, Uniontown and Washington. During these visits offic-ial records and reports were obtained and interviews were had with numerous officers and their assistants. Many council meetings also were attended and the manner of conducting city business was observed. As a rule, much of the material desired was received from the city clerk, controller or treasurer, although valuable information concerning poli-cies was obtained from mayors and other members of council. While nine of the cities included in the survey were not visited, nevertheless information was secured concerning them by means of personal correspondence and an examination and study of their official reports and other documents.

A large part of the statistical data presented in reference to the levy and collec-tion of taxes was taken from the official figures compiled by the Bureau of Statistics, Department of Internal Affairs in Harrisburg, and much valuable material was likewise obtained from the Bureau of Municipal Affairs in the same department. Several of the tables dealing with revenue and expenditures are based on the reports of directors of accounts and finance as analyzed by the expert accountants of the Pennsylvania Economic Council.

In 1927 the Legislature divided cities into four classes as follows:-
"Those containing a population of one million or over shall con-stitute the first class. Those containing a population of five hundred thousand and under one million shall constitute the second class. Those containing a population of one hundred and thirty-five thousand and under five hundred thousand shall constitute the second class A. Those contain-ing a population under one hundred and thirty-five thousand shall consti-tute the third class." (P. L. March 9, 1927, 18)

As the law now stands, Philadelphia is the only city of the first class, Pittsburgh

* This is a condensed summary of a larger collection of data and conclusions which could not be included because of limitations of space.

the only city of the second class, and Scranton the only city of the second class A. This classification enables the Legislature to enact separate laws for the government of each of the three largest cities in the State in accordance with local needs. At the present time, there are forty-four cities of the third class, each with a population of less than one hundred and thirty-five thousand and while newly created cities must have at least 10,000 inhabitants when created, there are four cities of smaller size now in existence. Table I shows the county in which each of the cities is located, the date of the present charter and population in 1920 and 1930, together with the percentage of increase or decrease during the ten year period.

From 1920 to 1930 the population in fifteen of these cities remained about the same. DuBois and Farrell show decreases of 15.2 and 7.9 per cent respectively, but the population increased as much as from 15.0 to 45.6 per cent in twelve cities. In some of these cases, however, the increase was in part due to the annexation of contiguous territory. In 1910, there were only twenty-six third-class cities and they contained 11.5 per cent of the State's population whereas in 1930 there were forty-two such cities, or an increase of 61.5 per cent and they contained 16.0 per cent of the people residing in the State. Since 1930 two additional cities have been chartered. At the present time, there are forty-nine boroughs, fifteen first-class townships and nine second-class townships with populations in excess of the number required for a third-class city, but several of these units are spread over considerable territory and are not, for that reason, well suited to become cities.

Of the eighteen cities visited, eleven might readily absorb adjoining boroughs or townships or parts of such units. In some cases as many as five or six thousand people reside in the territory surrounding or adjoining a city and frequently it is impossible from appearances to determine where the city ends and the other unit begins. Without question mergers or consolidations in these instances would result in reducing the total number of officials employed. It would also reduce the cost of police and fire protection, the number and cost of schools, and at the same time, permit such combined units to carry on certain functions together which, separately, one unit is not in a position to do.

Laws Governing Cities of the Third Class:- At the present time, all cities of the third class are governed principally by the Act of June 23, 1931 (P.L. 932), officially known as "The Third-Class City Law" and referred to in this study simply as the "Code." This enactment brings together in one body of laws some two hundred and fifty separate statutes and amendments dealing with various matters affecting cities of the third class. However, the code is based primarily on three acts of Assembly. The first of these was the Wallace Act of 1874 (P.L. 230) which, in addition to classifying all cities, provided in considerable detail for the incorporation, government and regulation of third class cities. For the next fifteen years, this act, with some amendments thereto, remained substantially the law governing such cities. In 1889 a new statute was passed which re-enacted the law of 1874 with some new features (1889 P.L. 277). In 1913 the Clark Act made further changes, the most important of which was the abolition of the bi-cameral system of councils and the substitution of the commission form of government. (1913 P.L. 568)

Powers and Functions of Third-Class Cities:- The broad fundamental powers and functions of a city are:- (1) To provide for the administration of justice and the maintenance of law and order; (2) to provide police protection; (3) to provide fire protection; (4) to exercise the right of eminent domain, and to provide for city planning and zoning; (5) to open, widen, extend, straighten or vacate any highway, street, avenue, alley or lane; (6) to locate, build and maintain bridges or viaducts; (7) to construct or cause to be constructed sewers of all kinds; (8) to provide for the collection and removal of garbage, ashes and other waste and refuse material; (9) to acquire land for parks, recreation centres, swimming pools, etc. and maintain and operate the same; (10) to purchase or erect and maintain all proper works, machinery, buildings, cisterns, reservoirs and pipes necessary to furnish water for the city and for anyone therein who desires such service; (11) to assess personal and real property; (12) to levy and collect taxes, license fees, fines, etc.; and (13) to issue and sell bonds and provide for the payment of the same.

TABLE I

POPULATION 1920-30

IN PENNSYLVANIA CITIES OF THE THIRD CLASS

City	County	Date of Present City Charter	Population 1930	Population 1920	Change in Percent
Allentown	Lehigh	Sept. 23, 1874	92,563	73,502	25.9
Altoona	Blair	Aug. 2, 1884	82,054	60,331	36.0
Beaver Falls†	Beaver	Dec. 3, 1928	17,147	15,445*	11.0
Bethlehem	Lehigh and Northampton	July 17, 1917	57,892	50,358	15.0
Bradford	McKean	Jan. 14, 1879	19,306	15,525	24.4
Butler	Butler	Mar. 19, 1917	23,568	23,778	-0.9
Carbondale	Lackawanna	June 21, 1886	20,061	18,640	7.6
Chester	Delaware	Jan. 17, 1890	59,164	58,030	2.0
Clairton‡	Allegheny	Sept. 14, 1921	15,291	10,777*	41.9
Coatesville	Chester	Apr. 27, 1915	14,582	14,515	0.5
Connellsville	Fayette	May 12, 1911	13,290	13,804	-3.7
Corry	Erie	Oct. 1, 1896	7,152	7,228	-1.1
DuBois	Clearfield	Dec. 28, 1914	11,595	13,681	-15.2
Duquesne	Allegheny	Oct. 24, 1917	21,396	19,011	12.5
Easton	Northampton	Jan. 12, 1887	34,468	33,813	1.9
Erie	Erie	Mar. 27, 1878	115,967	93,372	24.2
Farrell	Mercer	Oct. 15, 1930	14,359	15,586*	-7.9
Franklin	Venango	Jan. 14, 1909	10,254	9,970	2.8
Greensburg	Westmoreland	Jan. 3, 1928	16,508	15,033*	9.8
Harrisburg	Dauphin	Aug. 25, 1874	80,339	75,917	5.8
Hazleton	Luzerne	Dec. 4, 1891	36,765	32,277	13.9
Johnstown	Cambria	Dec. 18, 1889	66,993	67,327	-0.5
Lancaster	Lancaster	May 27, 1924	59,949	53,150	12.8
Lebanon	Lebanon	Nov. 25, 1885	25,561	24,643	3.7
Lock Haven	Clinton	Aug. 22, 1913	9,668	8,557	13.0
McKeesport	Allegheny	Jan. 15, 1891	54,632	46,781	16.8
Meadville	Crawford	Aug. 13, 1891	16,698	14,568	14.6
Monessen	Westmoreland	Jan. 18, 1921	20,268	18,179*	11.5
Monongahela	Washington	Sept. 30, 1912	8,675	8,688	-0.15
Nanticoke	Luzerne	Sept. 11, 1925	26,043	22,614*	15.2
New Castle	Lawrence	May 17, 1875	48,674	44,938	8.3
New Kensington§	Westmoreland	Nov. 10, 1933	23,002*	15,803*	45.6
Oil City	Venango	Feb. 17, 1881	22,075	21,274	3.8
Pittston	Luzerne	Dec. 10, 1894	18,246	18,497	-1.4
Pottsville	Schuylkill	Mar. 22, 1911	24,300	21,876	11.1
Reading	Berks	July 8, 1874	111,171	107,784	3.1
Sharon	Mercer	Dec. 17, 1918	25,908	21,747*	19.1
Sunbury	Northumberland	Dec. 29, 1920	15,626	15,721*	-0.6
Titusville	Crawford	Dec. 24, 1877	8,055	8,432	-4.5
Uniontown	Fayette	Dec. 19, 1913	19,544	15,692	24.5
Washington	Washington	Apr. 3, 1923	24,545	21,480*	14.3
Wilkes-Barre	Luzerne	Sept. 22, 1898	86,626	73,833	17.3
Williamsport	Lycoming	Feb. 4, 1876	45,729	36,198	26.3
York	York	Jan. 11, 1887	55,254	47,512	16.3

(*) Borough
† Beaver Falls and College Hill boroughs consolidated in 1930. Combined population in 1920: 15,445.
‡ Clairton, North Clairton and Wilson boroughs consolidated in 1922. Combined population in 1920: 10,777.
§ The population figures for both 1920 and 1930 cover the boroughs of New Kensington and Parnassus, which together in 1934 became the City of New Kensington.

ORGANIZATION AND ADMINISTRATION

Governmental Organization by Offices:- The elected officers of a third-class city are a mayor, four councilmen, a treasurer and a controller. They are all elected at large, at the same time, and for a term of four years. Council fixes the salary of its own members and of the mayor. The salary of a councilman, however, may not be less than $250 nor more than $4500; that of the mayor not less than $500 nor more than $5250. The law does not require that the mayor or a member of council possess any special qualification, training or experience. The treasurer and controller, however, are required to be "competent accountants."

In addition to the elected officers above mentioned there are four important appointed officials whose positions are created by statute. These are, clerk, solicitor, assessor, and engineer. All four are "elected" by council for a term of four years and their compensation is fixed by ordinance. There are no qualifications prescribed for the city clerk. The engineer, however, must be a "registered engineer" and the solicitor must "be learned in the law and qualified to practice in the Supreme Court of the Commonwealth."

The Mayor:- In addition to his functions as a member and president of council, the mayor has certain other important duties to perform. He is the chief executive of the city and as such he must see that the ordinances of the city and the laws of the Commonwealth relating to the government of the city are enforced. It is the mayor's duty to supervise the conduct of all city officers and to hear and investigate all complaints lodged against them for non-performance or improper performance of their respective duties. The mayor has the right at any time to call on any official of the city or head of department for any information which he may require concerning the affairs of any officer or department. The mayor also possesses a limited amount of judicial and semi-judicial authority.

The Director of Accounts and Finance:- The law imposes certain specific duties on the councilman who is placed in charge of the Department of Accounts and Finance. He is made responsible to a large extent for the proper handling of the public funds and the records pertaining thereto. All warrants drawn against the city must be countersigned by him. He has supervision and control of the accounts of all of the departments and must keep a record of accounts under appropriate titles to show separately and distinctly all of the assets and property vested in the city and all receipts and expenditures of each department. It is his duty to prepare annually certain statements and reports, the contents of which will be discussed later.

The City Treasurer:- It is the duty of the treasurer to demand and receive all moneys payable to the city from whatever source and to deposit the same in such banks or financial depositories as council may direct. It is also his duty to honor and pay all warrants drawn against the city which are countersigned by the director of accounts and finance and the city controller. The treasurer in a third class city in addition to having the statutory and customary powers and duties of a treasurer has numerous functions connected with the collection of city, school and poor taxes.

The City Controller:- One of the principal duties of the controller is to examine, audit and settle all accounts in which the city is concerned, either as debtor or creditor. He must also examine and audit the accounts of all officers, bureaus, and departments which receive and disburse public moneys, or which are charged with the management, control or custody of city funds. If as the result of any examination or audit he discovers any shortage, default, irregularity, or mismanagement, it is his duty to report the same to council. The controller must countersign all warrants for the payment of moneys out of the city treasury if and when he is satisfied of the legality of such payment.

The City Solicitor:- The solicitor has charge of all legal matters pertaining to the city. He prepares all legal papers, such as contracts, mortgages, deeds, bonds, etc.

He must commence and prosecute all suits necessary for the protection and best interest of the city and defend any suit brought against the city. When called upon, he must furnish to the council or the mayor, or the head of any department, a written opinion on any law matter touching their functions.

The City Engineer:- The engineer has control of public works construction and maintenance, of the city. He is responsible for the drawing of all plans and specifications and the making of estimates for all city engineering work contemplated or undertaken. He is required to perform such duties as council shall prescribe with reference to the construction, maintenance and repair of all streets, roads, pavements, sewers, bridges, etc.

Organization Chart:- The organization of the city government by departments, commissions, boards and principal offices is shown by the following chart. The third-class city code specifically creates three civil service boards and a city planning commission. Council may create a board of health, a recreation board, a shade tree commission and a water and lighting commission.

The City Council:- Council is vested with broad legislative, executive and administrative powers in conducting the affairs of third-class cities. It possesses the sole legislative power of the city and may pass any ordinance or resolution necessary to the exercise of any of the numerous powers or functions authorized by law. The law requires that there be at least one stated meeting of council in each month and such other meetings as council by ordinance may determine. The mayor as president of council may call special meetings and on the demand of two councilmen must call such meeting upon twenty-four hour notice. Three members constitute a quorum. The mayor does not possess any veto power.

Council meets regularly once a week in only four of the eighteen cities visited, twice a month in eleven, and only once a month in the other three. In many cities council requires the attendance of the solicitor at all council meetings in order that it may have legal advice on all matters coming before it. Sometimes the controller or treasurer is asked to attend in order to present financial data and not infrequently the assessor and engineer are invited or attend voluntarily. All meetings are open to the public as the law requires but in a majority of cities attendance by the public is slight except on very special occasions. In several cities a local newspaper reporter attends every meeting and sees that the public is informed of all important actions.

Distribution of Functions Among the Departments:- The law requires that all the executive and administrative powers, authority and duties shall be distributed among five separate departments as follows:-

            1 - Department of Public Affairs.
            2 - Department of Accounts and Finance.
            3 - Department of Public Safety.
            4 - Department of Streets and Public Improvements.
            5 - Department of Parks and Public Property.

It is left to council to determine the powers and duties to be performed by each department and it may make such rules and regulations as are deemed necessary or proper for the efficient and economical administration of such departments.

The law places the mayor at the head of the Department of Public Affairs but council at its biennial organization meeting determines by majority vote how the remaining four departments are to be distributed among the four councilmen. This distribution may be altered or changed at any time by council if it appears that the public service would be benefited thereby. Too frequently a department is not organized along the lines of efficient and economical operation but rather to satisfy the personal preferences or prejudices of the individual councilman who directs it. Sometimes when council is divided politically or factionally, the minority members, regardless of their qualifications, training or experience, are assigned to the two departments with the smallest expenditures.

ORGANIZATION CHART - CITIES OF THE THIRD CLASS

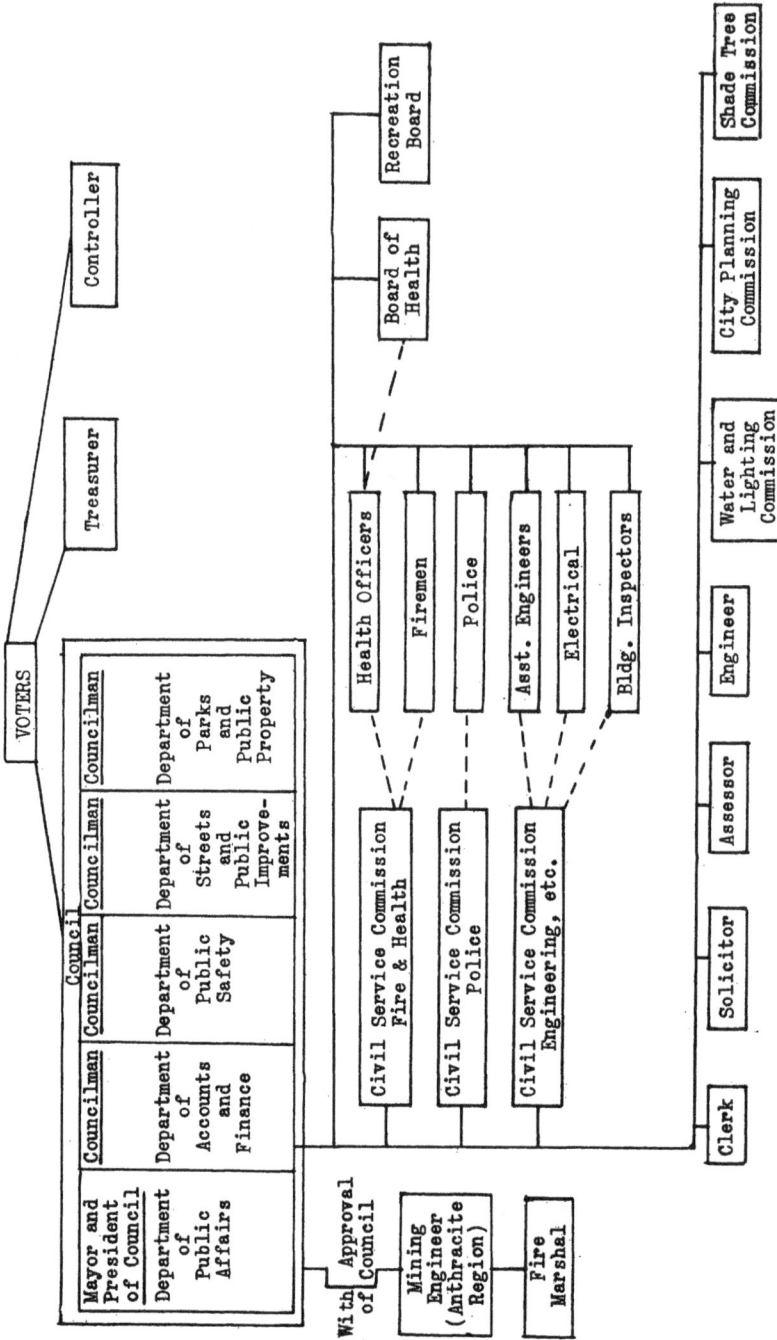

VOTERS

Controller

Treasurer

Council

Mayor and President of Council

Councilman	Councilman	Councilman	Councilman
Department of Public Affairs	Department of Accounts and Finance	Department of Public Safety	Department of Streets and Public Improvements	Department of Parks and Public Property

With Approval of Council

Mining Engineer (Anthracite Region)

Fire Marshal

Civil Service Commission Fire & Health

Civil Service Commission Police

Civil Service Commission Engineering, etc.

Health Officers

Firemen

Police

Asst. Engineers

Electrical

Bldg. Inspectors

Board of Health

Recreation Board

Clerk

Solicitor

Assessor

Engineer

Water and Lighting Commission

City Planning Commission

Shade Tree Commission

Supervision and Control of Departments:- One of the distinctive features of the commission form of government as it exists in Pennsylvania is the extreme division of administrative power and responsibility among the directors of the five major departments. While it is highly desirable that departmental responsibility be definitely placed, it is not desirable that separate and independent branches of government be created, not subject to some strong central authority. Although council has sole legislative authority and may make regulations concerning the work of all departments, nevertheless, each councilman as director of a particular department has almost complete control over its affairs. The law provides that the mayor shall supervise the conduct of all offices but in practice this means little and very few mayors attempt to do more than manage the department specifically entrusted to their care. Since the treasurer and controller are elected to office and are charged by law with the performance of certain definite duties, it would be extremely difficult for any mayor to dictate to them how they should carry out their respective function. While the lack of central control over all administrative matters is a weakness of the commission plan of government most of the cities visited conduct their affairs fairly well without such control.

Another defect in the commission plan is that both legislative policy and administrative detail are placed in the same hands. This is contrary to the generally accepted principles of government. While the officials who decide broad policies should be elected by the people whom they represent it does not follow that the same men are best equipped to carry out all the details of administration. These latter are dependent on expert ability, training and experience and it is too much to hope that men with these qualities can be secured by means of a popular election. This is in fact one of the principal arguments advanced on behalf of the city manager form of government. While quite a few boroughs in Pennsylvania have appointed managers, none of the cities studied have done so. However, several of the smaller ones have permitted the clerk or in some cases the engineer to act somewhat in the capacity of a city manager.

The adoption of the manager plan would concentrate all administrative reponsibility in a single trained executive and where conditions permitted, would take politics out of the details of administration. While no city should be compelled to adopt such a plan, every city should be given the opportunity to decide for itself whether such plan should be given a trial. Certainly some of the larger cities might find it a more efficient and economical method than the one in use. Usually the principle of proportional representation in the election of councilmen is recommended in connection with the adoption of the city manager plan and while not absolutely necessary it is favorable to the success of the manager form.

Annual Budget:- It is generally conceded that neither private nor public business can be successfully conducted without some form of annual financial planning. The law governing third-class cities expressly provides for an annual "budget" in the following manner:

"Annual Budget - The director of accounts and finance shall, at the first stated meeting in December in each year, present to council a detailed statement of the estimated receipts, expenditures, and liabilities of every kind, for the ensuing year, with the balance of unexpended appropriations, and all other information of value as a basis for fixing the levy and tax rate for the next fiscal year.

"The several departments of the city government shall, before the first stated meeting in December, present to council an estimate of the probable receipts and expenditures and of the amount required by each of said departments for the public service during the ensuing year as a basis for making the annual appropriations thereto." (1931 P.L. 932, Sec. 1809)

The law further provides that after these estimates are submitted to council it shall proceed to make the annual appropriations and fix the tax rate at such figure as (with all sources of revenue) will fully cover the aggregate amount of estimated expenditures. Outside of being called a budget in the title of the section there is little in the wording of the law to show that a budget in the real sense of the word is intended at all. No one official is charged with the responsibility of preparing it. The director of accounts and finance is required to submit estimates of expected revenue and expenditures as a _basis_ on which to fix the tax levy but the second paragraph of the section requires that each department also furnish _council_ with like information.

In view of these meager and indefinite requirements, it is not surprising to find that there is very little real "budget making" in many cities. Information concerning the preparation of the budget was secured from fourteen of the cities visited and in only four of them was a budget prepared by the director of accounts and finance from data submitted to him by the heads of the various departments. In three cities each department furnished council with a tentative budget and in three others the city clerk assembled the necessary information and presented it to council in a form resembling a budget. In two additional cities the city clerk and director of accounts and finance prepared estimates for council and in one city the clerk and engineer followed the same method.

While council in all cases must finally approve and adopt a budget there are several cities where the usual process amounts to no more than changing a few figures in the present budget. In many cases, council first estimates the probable revenue that the city will receive during the year and then proceeds to appropriate the sum to the different departments. In only a few of the cities studied is there any attempt to ascertain _first_ the actual needs of the departments, bureaus, etc. and then proceed to fix a tax rate which, based on the assessed valuation of property with allowance for delinquencies, exemptions, etc., will produce enough money together with other revenue to cover all expenditures.

Every city should be required to prepare an annual budget based on a standard classification of all items and it would be desirable if all cities with the same general governmental structure would adopt a standard form. A public hearing after proper notice should be provided and the budget as adopted should be published. These are well-established rules of good practice. (See A. E. Buck "Budgeting For Small Cities") Some one official should be charged with the duty of preparing the budget for council and ordinarily it should be the executive or administrative head of the government. This presents a difficult problem in third-class cities since under the commission form there is no single directing head so far as administration is concerned. No matter how carefully a budget has been prepared its usefulness depends almost entirely on the way it is carried out and this in turn depends in no small measure on centralized administrative control which is lacking under the commission plan.

Purchase of Supplies:- The law provides that all work and materials required by any city where the cost exceeds five hundred dollars shall be furnished and performed under a written contract. Such contract must be awarded to the lowest responsible bidder after advertising three times, (each publication on a different day) in not more than two newspapers. Bids are not to be opened until at least ten days after the first advertisement. The above requirements, however, do not apply to contracts for the repair of streets, sewers or other public improvements, to street cleaning, collecting of rubbish, ashes and garbage, or to contracts made with the Commonwealth where the city agrees to pay a portion of the cost of any public improvement. Where a new street is constructed or an old one rebuilt on a new base or a sewer is constructed or grading done, such work must be done under written contract given to the lowest responsible bidder.

Each city may by ordinance establish a purchasing department which shall have supervision over the purchase and distribution of all supplies required by the different departments. The law provides that if a purchasing department is established it shall be attached to the department of accounts and finance or such other department as council may determine and be operated in accordance with the rules and regulations prescribed by council. While such department is authorized in the hope that it will assist council in the elimination of waste and extravagance in the purchase and distribution of supplies, only one of the eighteen cities visited has such a department at the present time. Most of the cities purchase on a "hand to mouth" basis with the head of each of the major departments deciding when and what to buy. In two or three cities a committee of one or two councilmen acts as a purchasing agent while in a few others, all purchases are made through the city clerk or director of accounts and finance on requisitions submitted by the head of each department.

The fact that all contracts must pass through the hands of the director of accounts and finance or that the city clerk orders supplies on the request of department heads, does not amount to centralized purchasing in any sense of the word. Goods are purchased in the amount and at the time required without any attempt being made to combine the needs of all departments or to anticipate future requirements. Supplies are bought piece-meal or in small quantities when purchases in larger volume and on specifications would result in substantial saving. The lack of centralized purchasing and long range planning is due in a large degree, to the desire of each councilman to control as far as possible all expenditures made for his department. The interest of the city as a whole is frequently not considered.

Except possibly in the smaller cities there should be a central purchasing agent responsible for the buying of all supplies and materials and where possible the city and school authorities should cooperate in this matter.

All of the cities secure bids for contracts in excess of $500, and the old practice of deliberately evading the law by making two or more separate agreements each for an amount of less than $500 is followed only in a few cities. In fact, in eight of the eighteen cities visited it is now the general practice to advertise for bids on contracts involving considerably less than $500. In fifteen cities advantage is taken of discounts for prompt payment on contracts whenever possible.

Selection of Personnel:- The law creates three civil service boards for each city as follows:-

"-----(a) A board for the examination of applicants for appointment to any position in the police department; (b) a board for the examination of applicants for appointment to any salaried position in the fire department (except volunteer departments), or as sanitary policemen or inspectors of the health department; (c) a board for the examination of applicants for appointment to any position in the engineering or electrical departments, or to the position of building inspectors." (1931 P. L. 932, Sec. 4402)

Each board consists of three citizens elected by council for a term of four years. One member of the first and second boards must be an educator and one a physician. One person may be appointed to two or more boards. The members of these boards receive no salary or other compensation for their services. Each board may appoint a secretary and define his duties but his compensation is fixed by council.

The boards are authorized to prepare and adopt such rules and regulations covering the selection and appointment of applicants for positions as are adapted to securing the best service for the public. There does not appear to be any good reason for three separate boards and only a few of the twenty-eight cities covered in this Survey have more than one board and until very recently a few did not have even one. In several cities, the city clerk acts as secretary to the board and in others he is the third member of the board.

In many of the smaller cities the board exists in name only or acts in a perfunctory manner. Several cities have only four or five policemen, no paid firemen, only one health officer and maybe a building inspector; the problem of selecting employees is therefore not a serious one. Even some of the larger cities have volunteer fire departments and practically all civil service appointments are in connection with the police department. Clerks, assistants, secretaries, etc. employed in the general offices are not subject to civil service regulations.

According to some city officials, there is as much politics involved in the selection of employees under civil service rules as there was before the law was passed. One city treasurer declared that when two of the councilmen changed their party after election, the health officer, fire chief and police chief had to support the ticket of that party in the next election in order to hold their civil service positions. In another city it was stated that the last examination prepared by the civil service board was passed by every candidate and that it was so simple and easy that any sixth grade student could have passed. A councilman in charge of the Department of Streets and Improvements in one city stated that two civil service appointees in the engineering bureau were in his opinion entirely incompetent for the work they were doing but that it was impossible to have them removed. Councilmen in other cities claim that they are unable to effect economies because of the difficulty in reducing the number of civil service employees.

A civil service commission if properly conducted is desirable in the larger cities where many persons are employed annually. In the smaller places where there is no need for a civil service system provisions of the present law are generally ignored.

Pensions:- The law provides for the creation of three pension funds, one for police, one for firemen and one for other employees. The police fund is mandatory, the other two optional. To maintain the police fund the city is required to deduct from a policeman's pay an equal and proportionate monthly sum which shall not exceed three per cent of his year's salary. As to the city's contribution, the law provides as follows:-

"There shall be paid annually to the organization or association, constituting and having in charge the distribution of police pension funds in every city, a sum of money not exceeding one per centum of all city taxes collected by the city other than taxes levied to pay interest on or extinguish the debt of the city or any part thereof." (1931 P. L. 932, Sec. 4305)

While the creation of the fund is mandatory and the payment of money into it mandatory, still since the amount of such payment is to be a sum "not exceeding one per centum---" the deposit of even one cent annually would fulfill all legal requirements. As the law now stands a city must create such a fund, must make a monthly charge against a policeman's salary, but need not contribute anything to the fund. This may explain in part why the policemen are not enthusiastic about the law and why only four of the twenty-one cities for which such information was received have established a fund, in spite of the fact that the State law requires that it be done.

Only two or three of the cities examined have pension funds for firemen to which the cities themselves contribute. In nearly all cities, however, there exists one or more firemen's relief associations which receive from the State a share of the tax on foreign insurance companies. Most firemen are required to pay monthly or annual dues into the relief fund of such associations. Only one of the cities studied has a pension fund for employees other than firemen and policemen.

Every city should make some provision for retiring its superannuated employees and especially its policemen and firemen. Provision should also be made to assist the widows and orphans of those who have lost their lives in the course of duty. The law governing these matters should be mandatory and there should be a practicable method of compelling compliance with the requirements of the law. Both the city and the employee should share in building up a fund sufficient to provide such protection.

Accounting Methods - The law does not prescribe any particular system of keeping accounts in third class cities. The director of accounts and finance has supervision over the accounts of all departments and must keep a record of such accounts under appropriate titles to show separately and distinctly: (1) All assets and property vested in the city, (2) all trusts in care of the city, (3) debts owing by the city, and (4) all the receipts and expenditures of the various departments. The law provides that the accounts of the city treasurer "shall be kept in such manner as to clearly exhibit all the items of receipts and expenditures and the sources from whence the moneys are received and the objects for which the same are disbursed." The treasurer is required to "keep separate and distinct accounts of the receipts and expenditures of the city, the sinking fund and the water and lighting department, respectively, and also of every special fund which may come into his hands."

An examination of the books and records in many cities discloses the fact that there is little uniformity in accounting methods. Bookkeeping practices adopted years ago are still in use although they no longer meet present day needs. When a new treasurer or director of accounts and finance takes office, he usually follows the system already in use either through lack of knowledge or unwillingness to take the trouble of making a change. In one city, the Bureau of Municipalities at Harrisburg aided the local official in establishing a modern bookkeeping system but after one year the system was discarded because the city clerk, who did most of the bookkeeping work, could not follow it readily. In many cities the single entry method of bookkeeping is still used; in some of the larger places examined the double entry system has been recently adopted. In a few cities both systems are in use - the single entry for the general account and double entry for sinking fund or other special accounts. Some cities do not create separate funds when modern accounting principles would seem to require them - while a few others have unnecessarily complicated their accounts by establishing special or separate funds for nearly every function performed. In these latter places transfers between funds are of frequent occurrence and it is often difficult to ascertain the city's exact financial condition.

The law requires the treasurer to be a "competent accountant" but only a few of them are so qualified and as long as they are elected rather than appointed there is little reason to believe that they will possess the training or experience necessary for the work performed. In some cities the man elected is not an accountant and finds it necessary to appoint an assistant who knows how to handle the work properly. The director of accounts and finance who has control over all records, accounts, etc. is one of the councilmen and is not by statute required to possess any special qualification, training or experience for the important duties imposed upon him by law.

While the books and records in most of the cities are kept in much better shape than those usually found in boroughs and townships, and are satisfactory in that they are accurate and contain full details as a rule, still there is room for improvement in many places. Due to the lack of uniformity in the accounting methods used by the different cities, it is very difficult to make comparisons of the receipts and expenditures of cities of the same class or size. A State agency should be given authority to formulate a uniform accounting system applicable to the needs of third class cities and the use of such system should not only be mandatory but subject to the supervision of State officials. Many local officers today admit the need of better accounting methods and express a desire to learn how the cost of various functions compares as between their own and other cities. The accrual basis of accounting should be adopted so that income or receipts will be shown for the period in which they were actually received and expenditures for the period in which the liability was incurred and not the period in which the money was finally paid. While the legal requirement of a uniform, modern system of keeping accounts is highly desirable it is not as important as the selection of properly qualified men to perform the work of keeping the necessary books, records, accounts, etc.

Auditing:- The law makes ample provision for the auditing of all public accounts. The city controller is authorized to examine, audit and settle all accounts in which the city is concerned and all accounts of bureaus, officers and departments which collect and disburse public moneys. He also audits or secures an audit by a certified public accountant of all funds handled by associations or organizations of municipal employees for the benefit of such employees. If he discovers any default, irregularity or mismanagement, it is his duty to report the same to council for its consideration. Once a year, or oftener if council directs, the controller submits a report of the audits he has made and prepares a summary showing the fiscal condition of the city.

The controller like the treasurer is elected by the people and is by law required to be a "competent accountant." In some cases he is so qualified but in many others, he is not an accountant at all, although he may have had a little experience in bookkeeping. Some controllers were found who could not explain important items in their annual financial summary for 1933 and several admitted that the city clerk or some other officer helped them prepare the statement. The amount of work involved in most cities is not sufficient to justify the full time services of a competent man and for that reason, council should be permitted to employ an expert accountant on a part time basis. Such a man could be found in one of the local banks or reputable business concerns in the city.

Reporting:- The director of accounts and finance is required by law to make an annual report to council in March of each year of the public accounts of the city for the preceding fiscal year. This report must contain the following: (1) Sources from which all revenue and funds are derived; (2) a statement in detail of the several appropriations made by council, the amount drawn on each appropriation and the balance outstanding to the debit or credit of such appropriation at the close of the fiscal year; (3) a statement of the present funded and floating indebtedness of the city, and the date and maturity of the funded debt; (4) a schedule of assets and the character and value thereof; and (5) the amount of taxable property located in the city. The report must be published once in not more than two local newspapers and council may also have it published in pamphlet form. In most cities the report is published in pamphlet form together with other data concerning the city and its governmental organization and functions. The reports are not always prepared on time and in some cases they are not published until September or October.

There is more similarity in the reports of third class cities than in the reports of other local units of government. This is due in part to the fact that the law specifies that the report contain certain items of information as previously shown. However, many city reports do not include all the information required by law and there is considerable difference in the way in which certain facts are presented. Sources of revenue are generally shown in detail although some cities include all taxes, personal, occupational, and property in one sum and do not indicate what part of such amount represents delinquent taxes. One report shows "Miscellaneous receipts of $611,760." This was the proceeds of a bond issue and should have been so listed.

It is in the treatment of expenditures that the greatest difference of reporting is found. The reports for about half of the cities examined fail to include a detailed statement of the appropriations made by council. Several of the reports show the total amount appropriated for each of the five major departments but the detailed statement required by law is omitted. The 1933 reports for the Cities of Pottsville and Uniontown are typical in this respect. Reports of other cities frequently show by departments and in some cases by bureaus not only the amount appropriated for each principal item of expense but transfers to and from particular appropriations, the amount actually spent and charged against an appropriation and the balance unexpended at the end of the year. While many cities do not show the amount of money appropriated for various purposes they do give the amount actually spent for salaries, labor, supplies, improvements, etc. Sometimes this information is shown for each department or bureau separately but in a few cities it is all presented together so that it impossible to determine from the report the cost of particular services or the salary received by individual officers.

The report of one city for 1930 showed "Payroll" $129,254.02 and "Current Expenditures" $61,439.64, with no further breakdown. Frequently expenditures in amounts varying from $5,000 to $40,000 are grouped together so that it is impossible to learn the real facts without an examination of the books and records of the city. Sometimes a report will show the figures in detail for two or three items and then show a miscellaneous item two or three times as large as any of those given. A few cities go to the other extreme and show all details even the cost of a typewriter ribbon. From some reports, it is impossible to determine the yearly cost of a particular item or function because the figures shown include payments made on obligations incurred the year before. Some cities have many special accounts and frequently these accounts are not included in the published report with the result that the reader does not obtain a full and complete picture of the financial conditions.

In regard to debts, assets and property valuation hardly any two cities report the data in the same manner. Nearly all of the cities show the amount of funded debt and either the date of issue or maturity, and a few show both. In some reports not only is each bond issue shown separately but the purpose for which the money was borrowed is shown as well as the rate of interest paid. Only a few of the reports are prepared in a way to indicate clearly which of the bonds are of the special assessment type. Most reports give a summary statement concerning the sinking fund, but not full details.

The law also requires a statement of the city's floating indebtedness. Many cities do not include this item in their annual reports. Sometimes temporary loans, notes and judgements against the city are shown but not unpaid bills or even outstanding warrants. Many cities do not report their assets and some that do, fail to indicate their character as the law requires. Only two or three of the reports examined showed the total of all outstanding taxes including those liened or returned to the county treasurer for collection. Some of the cities which do show their assets place an absurd value on streets and other physical property in order to make it appear that assets are in excess of liabilities. A majority of the reports do not disclose the assessed valuation of taxable property as the law requires and several of those which do, give the valuation for the succeeding year, and not the year covered by the report.

The method of bookkeeping in most cities is much better than the reports would seem to indicate. The fault lies mainly with the officer who prepares the annual report and not with the information with which he has to work. There is not only a need for a more detailed report but also for uniformity in all reports. A State agency should prepare standard reporting forms suitable to the needs of third class cities and to the system of bookkeeping required of such cities. This agency should also be given sufficient power to compel local officers to submit annual reports in proper form within a certain time after the close of the fiscal year. At the present time, State officials experience great difficulty in obtaining desired information from local units of government even where the law provides that such information must be furnished to the officials on request.

Relations of Cities to Other Units of Government - It is a common practice for cities to enter into contracts with the State or with the county in which they are located for the purpose of joint construction of highways and bridges. Some cities have also made contracts with adjoining boroughs or townships for the construction of sewers, sewage disposal plants, etc. In these cases, the boroughs or townships usually pay a part of the original cost of construction and agree to pay a part of the annual operating expenses. Several of the cities studied furnish fire protection to such local units and are paid an agreed annual charge, while others furnish such service only when called upon and are usually paid on an hourly basis. In many places the firemen assist outside fire departments without making any charge whatsoever.

There are numerous opportunities for boroughs and townships to save money by co-operating with nearby cities. This is especially true in regard to police and fire protection, sewers, parks and playgrounds. The cities would also receive a benefit from such joint undertakings since the total cost involved would be borne by a larger number of people. The Bureau of Municipal Affairs in Harrisburg has advised with the officers of several cities and adjoining boroughs and townships in connection with the purchase and operation of parks and playgrounds. This bureau or some other State agency should take the lead in suggesting the possibility of such joint action and especially in regard to creating larger areas of police and fire protection. In many of the smaller units these are the principal items of expense but in spite of this, they have poorly paid and poorly equipped police and fire departments which as a rule cannot furnish adequate service. A consolidation of the facilities and personnel of many of these units with those of larger units would benefit all parties concerned.

REVENUE AND NON-REVENUE RECEIPTS.

Sources of Revenue:- The chief sources of city revenue are:

1 - Taxes levied on real estate, occupations and persons.
2 - Special assessments against real estate for streets, sidewalks, sewers, shade trees, etc.
3 - Licenses and fees required in connection with the carrying on of certain trades, occupations, businesses and professions.
4 - Fines and forfeits imposed for the violation of city ordinances or the laws of the Commonwealth.
5 - Interest received on bank deposits or assets in the sinking fund.
6 - Gifts and grants from the State or other units of government or from individuals.
7 - Earnings of departments, water-works or other municipal enterprises.

In addition to such revenue, a city frequently obtains money from non-revenue sources such as the sale of bonds, temporary loans, refunds and the sale of property.

Of all the above the principal source of income is the tax on real estate, occupations and persons. Before the city can secure funds from property taxes, three distinct steps must be taken. First, the value of the property must be determined by the assessor; second, the city council must decide upon the tax rate in mills for each $100 valuation of real estate; and third, the taxes as levied must be collected. Each of these problems will be presented separately.

Assessment of Property:- The law requires the city assessor to make or cause to be made, during the year of the triennial assessment for county purposes, a full, just, equal, and impartial assessment of all property, taxable for county purposes, and all matters and things within the city which are subject by law to taxation for city purposes. The assessor is also required to make an accurate and complete list of, and place a just valuation on, all property within the city which is exempt from taxation. He does not include in his assessment the value of certain personal property such as money at interest, mortgages, stocks and bonds, on which the State levies a four mill tax for county purposes. It is the further duty of the assessor to prepare and return annually a list of all inhabitants, male and female, over twenty-one years of age. In arriving at the value of any piece of property, the assessor or his assistants are required by law to value "----- the property at such sum as the same would, in his judgment, bring at a fair public sale thereof."

In addition to the assessment of real estate made by the city assessor a second separate and distinct assessment of the same property is made for county purposes. This assessment is made by county assessors elected in each ward of the city. These assessors are required by law to assess the property "according to the actual value thereof, and at such rates and prices for which the same would separately bona fide sell." This has been construed by the courts as meaning the full market value at a public sale after due notice of such sale is given.

Comparison of Assessments made by City and County Assessors:- With substantially the same basis of value to work on it is interesting to see how the assessments made by the city assessor and those made on the same property at the same time by the county assessors compare. In order to make such a comparison, assessment figures for the years 1933 and 1934 were secured from the city assessors or other city officials and like information from county commissioners concerning the county assessments. Every effort was made to exclude from these figures all assessments on persons, occupations, money at interest, etc. and include only assessments of taxable real estate. Table II presents these figures for twenty-seven third class cities with a 1930 population of less than 30,000. The total assessed valuations of taxable real estate as found in each city by the city and county assessors are shown for each year separately. The percentage which the county valuation forms of the city valuation is also shown as well as the percent of increase or decrease in each total valuation in 1934 as compared with each total valuation in 1933.

In 1933, there were only five cities where the county valuations exceeded the city valuations and in three of these the difference between the two assessments was less than one per cent. In the Cities of Lebanon and Franklin, the county values were 110.4 and 104.1 per cent respectively of the city values. In 1934, the county valuations exceeded those of the city in Nanticoke, Lebanon and Franklin only. In twelve cities the county assessments for 1933 were less than 90 per cent of the city assessments. In the Cities of Washington, Uniontown, Bradford, and Monongahela, the value placed on all real estate by the county was less than 60 per cent of the value placed on the same property by the city assessors. In Monongahela, which is in the same county as the City of Washington, the county valuation was only 37.1 per cent of the city's valuation. The table shows that in 1934 there were still twelve cities where the county values were placed at less than 90 per cent of the city values, but only three cases where they fell below 60 per cent of such values. In the Cities of Carbondale, Clairton, Farrell, Lock Haven, Monessen and Sunbury, the difference between the two assessments for 1933 varied only slightly more than two per cent.

TABLE II

A COMPARISON OF THE ASSESSED VALUATION PLACED BY THE CITY AND COUNTY ASSESSORS FOR THE YEARS 1933 AND 1934 ON REAL ESTATE LOCATED IN CERTAIN THIRD CLASS CITIES

City	Population 1930	Assessed Valuation of Taxable Real Estate for 1933 — Assessment made by City Assessor	Assessment made by County Assessor	County Assessed Valuation as a percentage of City Assessed Valuation	Assessed Valuation of Taxable Real Estate for 1934 — Assessment made by City Assessor	Assessment made by County Assessor	County Assessed Valuation as a percentage of City Assessed Valuation	Per Cent of Increase or Decrease in City Valuation for 1934 over 1933	Per Cent of Increase or Decrease in County Valuation of 1934 over 1933
Nanticoke	26,043	$13,915,693	$13,213,586	95.0	$12,689,219	$12,996,065	102.4	− 8.8	− 1.6
Sharon	25,908	18,580,850	17,890,150	96.3	18,294,775	16,529,055	90.3	− 1.5	− 7.6
Lebanon	25,561	24,534,530	27,084,929	110.4	24,413,756	26,810,670	109.8	− .5	− 1.0
Washington	24,545	26,098,130	14,612,305	56.0	23,405,755	14,278,305	61.0	−10.3	− 2.3
Pottsville	24,300	16,634,256	14,381,934	86.5	16,578,086	14,215,878	98.8	− .3	− 1.2
Butler	23,568	33,342,520	31,036,630	93.1	27,247,120	28,089,155	96.1	−12.3	− 9.5
Oil City	22,075	15,569,755	13,634,190	87.6	15,379,230	13,478,850	87.6	− 1.2	− 1.1
Duquesne	21,396	25,249,550	21,235,600	84.1	25,164,850	21,539,250	85.6	+ .3	+ 1.4
Monessen	20,268	15,390,705	15,406,050	100.1	15,238,875	13,191,650	86.6	− 1.0	−14.4
Carbondale	20,061	10,809,622	10,614,449	98.2	10,859,814	10,516,657	96.8	+ .5	− .9
Uniontown	19,544	20,790,700	11,947,070	57.5	20,737,875	11,787,155	56.8	− .3	− 1.3
Bradford	19,306	11,870,355	6,963,963	58.7	11,967,740	6,976,332	58.3	+ .8	+ .2
Pittston	18,246	10,999,211	9,205,659	83.7	10,870,510	9,159,203	84.3	− 1.2	− .5
Beaver Falls	17,147	13,268,095	12,250,506	92.3	12,664,860	11,639,672	91.9	− 4.5	− 5.0
Meadville	16,698	10,368,915	9,793,475	94.5	10,380,610	9,736,160	93.8	+ .1	− .6
Greensburg	16,508	14,786,850	14,154,350	95.7	14,611,780	14,011,750	95.9	− 1.2	− 1.0
Sunbury	15,626	6,125,290	5,987,445	97.7	6,089,000	5,910,190	97.1	+ .6	− 1.3
Clairton	15,291	26,163,950	26,200,500	100.2	25,951,870	25,510,750	98.3	− .8	− 2.6
Coatesville	14,582	14,774,010	11,666,795	79.0	13,348,840	11,480,976	86.0	− 9.6	− 1.6
Farrell	14,359	10,536,891	10,596,085	100.6	10,131,270	10,073,529	99.4	− 3.8	− 4.9
Connellsville	13,290	7,652,335	6,156,825	85.5	7,474,590	6,110,700	81.8	− 2.3	− .7
Du Bois	11,595	4,528,104	3,203,082	70.7	4,505,551	3,123,985	69.3	− .5	− 2.5
Franklin	10,254	6,798,080	7,079,990	104.1	6,768,350	7,055,940	104.2	− .4	− .3
Lock Haven	9,668	4,902,610	4,788,055	97.7	4,882,235	4,737,170	97.0	− .4	− 1.1
Monongahela	8,675	11,961,968	4,441,220	37.1	9,535,681	4,377,900	45.9	−20.3	− 1.4
Titusville	8,055	5,903,580	5,346,865	90.6	5,870,143	5,289,733	90.1	− .6	− 1.1
Corry	7,152	5,831,335	3,780,439	64.8	4,394,250	3,367,812	76.6	−24.6	−10.9

In view of the fact that there has been considerable demand made in the last two years to reduce the assessed valuation of real estate, it is interesting to see what changes were made in the assessment rolls for 1934 as compared with the rolls of 1933. In considering the figures here presented the reader should bear in mind that some new buildings have been erected, or improvements made to old ones, in 1933, but not enough to materially affect the results shown. Table II shows only three instances where the 1934 city assessments exceeded those of 1933 and the only two cases where the county figures for 1934 exceeded those of 1933, and in all cases the increase was less than two per cent. On the other hand, in twenty-four out of the twenty-seven cities, the city assessors made reductions in valuations ranging from .3 to 24.6 per cent. In the cities of Washington and Butler, the decrease was more than ten per cent and in Mononga-hela and Corry, more than twenty per cent. The average decrease for the twenty-four cities was 4.4 per cent. In twenty-five of the cities, the county assessors made re-ductions ranging from .3 to 14.4 per cent but only in the cities of Monessen and Corry did the decrease amount to more than 10 per cent. The average county reduction for the twenty-five cities was 3.0 per cent.

Interviews with city and county officials (not including assessors) in many sec-tions of the State, disclosed no one who seriously contended that a duplication of assessments was necessary or desirable. In many cities the county assessors merely copy the assessments made by the city assessor. It has been estimated that the elimin-ation of the present dual system would save the forty-four third class cities at least $250,000 annually.

In 1931, the Legislature created in all third class counties a "Board for the Assessment and Revision of Taxes." (1931 P. L. 348) This board is composed of three members appointed by the county commissioners for a term of four years. The board is authorized to divide the county into convenient districts and appoint subordinate as-sessors for such districts. While the law does not abolish the office of city assessor in the third class cities located in third class counties, it does make provision whereby such cities (there are eight of them) may by ordinance accept and adopt the law and thereby do away with the present duplication of assessments.

The Legislature should establish a system by which all property within a county would be assessed by a single agency. This could be done by extending the provision of the third class county law to cover counties of the fourth, fifth and sixth classes. An even more simple and economical method of assessment should be provided for the less thickly populated counties of the seventh and eighth classes.

The Levy of Taxes:- After the value of all taxable property has been determined for a given year, it is the duty of council to fix the tax rate in mills and authorize the preparation of the tax duplicate. The Legislature has set certain limits beyond which council may not go in levying taxes. The purposes for which taxes may be levied and the limitation in mills for each purpose are as follows:

1 - A tax for general revenue purposes not exceeding fifteen mills on the dollar on all persons and property taxable for county purposes.
2 - An additional tax not exceeding ten mills on the dollar on all such persons and property, for the payment of interest on bonded indebtedness, for the payment of loans to support the government and to make necessary improvements in the city.
3 - A poll tax for general revenue purposes not exceeding one dollar annually on all inhabitants above the age of twenty-one years.
4 - A tax not exceeding ten mills on the dollar for the purpose of maintaining and supporting a Bureau of Charities and Welfare, provided that the care of the poor has not been transferred to the county.
5 - A tax not exceeding two mills on the dollar for the purpose of maintaining and operating parks, playgrounds, swimming pools, recreation centers, etc.
6 - A tax not exceeding two mills on the dollar for the purpose of establishing a public library if such project is approved by the voters.
7 - A tax not exceeding one-tenth of one mill on the dollar for the purpose of de-fraying the cost and expenses of caring for shade trees.

In all, council may levy a total of 29.1 mills (or 39.1 mills in those cities having a separate poor district). This does not include the one dollar poll tax or any of the special assessments which may be imposed and which will be discussed later. During the last three years the total millage in the twenty-seven cities studied has ranged from 6.5 per cent in Monongahela to 20.75 per cent in Pittston.

Total Amount of Taxes Levied:- In view of the fact that a city may reduce its taxable valuation and at the same time increase its millage or vice versa, the only accurate and reliable way to determine an increase or decrease in the tax burden is to ascertain the total amount of taxes levied in any year and compare such figures for different years. This has been done for twenty-seven cities and the results are shown in Table III. The years covered are 1927 and 1930 to 1933 inclusive. The year 1927 was used because it is the first year for which reliable information could be obtained and for the additional reason that it covers a period when conditions were nearer to normal than any subsequent year. Information for 1934 is not yet available. The figures showing the amount of taxes levied for the years 1927, 1930, 1931 and 1932 were obtained from the Bureau of Statistics, Department of Internal Affairs, Harrisburg. Those for the year 1933 were obtained directly from the local officials concerned.

Because of the fact that a taxpayer living in a third class city pays not only a city tax, but also a school tax, county tax and maybe a poor tax, data on such taxes are also included in Table III, in order to give a complete picture of the tax burden imposed in any one year on an individual taxpayer. The table shows the total amount of taxes levied for each separate kind of tax, for each city and for each year separately. It also shows what percentage each of the different taxes form of the total of all taxes levied, as well as the amount levied per capita by each of the tax levying authorities. The populations shown for 1930 are the Federal Census figures and all others are estimates only, worked out by interpolation. Four year averages are presented for each item of information listed in the table. It will be observed that data on poor taxes are not shown for each city. This is due to the fact that in the counties in which these cities are located no separate poor tax is levied, but appropriations are made from county funds to take care of poor relief.

Amount of Taxes Levied for City Purposes:- It should be interesting to a city taxpayer to know what percentages of his tax dollar go to pay for city government, for the support of schools and for county and poor purposes respectively. In the twenty-seven cities studied, the average amount of taxes levied for the last four years for city, school, county and poor purposes combined was $16,877,662 annually. Of this amount, $5,312,342 or 31.5 per cent of the total was levied annually for the operation and maintenance of city government and the various functions performed by it. For the support of the schools, taxes were annually levied in the amount of $7.761,057 or 46.0 per cent of the total levy, while $3,804,263 or 22.5 per cent was levied for county and poor purposes combined. Almost one-half of every dollar levied was for the support of the schools and their administration.

The figures for 1933 show that in the City of Pittston the city tax levy was 41.4 per cent of the total tax levy whereas in Sunbury, taxes for city government amounted to only 21.1 per cent of the total. When the percentages for 1933 are compared with those of 1927, it is seen that in fourteen cities the 1933 figures are higher, while in thirteen cities they are lower. In sixteen of the cities there was only about a three point difference between the figures of 1927 and 1933. In the Cities of Washington, Oil City, Duquesne, Carbondale, Pittston and DuBois, the variation for the two years is more pronounced. In seven cities the trend for the last three years has been down while in eight cities it has been up.

City Taxes Levied Per Capita:- Table III also shows the amount of taxes levied per person for city purposes for each of the cities for the five years previously referred to. In fourteen of these cities, the city tax per person was less in 1933 than it was in 1927. In the other thirteen cities, the levy was higher in 1933 but in all of these cities except Sunbury and DuBois, the levy for 1933 was less than that for 1932. In some cities a substantial reduction was made between 1930 and 1933 as for instance; Nanticoke from $11.30 to $8.74; Sharon from $12.13 to $8.63; Duquesne from $15.44 to $8.56; Uniontown from $16.42 to $12.87; Bradford from $12.23 to $9.49; and Clairton from $15.74 to $11.45.

TABLE III
TAXES LEVIED

City	Population	Year	CITY			SCHOOL			COUNTY			POOR			TOTAL		
			Total City Taxes Levied	City taxes levied as a percentage of all taxes levied	City Taxes Levied Per Capita	Total School Taxes Levied	School taxes levied as a percentage of all taxes levied	School Taxes Levied Per Capita	Total County Taxes Levied	County taxes levied as a percentage of all taxes levied	County Taxes Levied Per Capita	Total Poor Taxes Levied	Poor taxes levied as a percentage of all taxes levied	Poor Taxes Levied Per Capita	Total of all Taxes Levied	Total Per Cent	Total of all Taxes Levied Per Capita
Nanticoke	25,015	1927	$309,688	34.1	$12.38	$429,897	47.3	$17.19	$122,283	13.5	$4.89	$46,463	5.1	$1.86	$908,331	100	$36.31
	26,043	1930	294,394	27.9	11.30	542,631	51.4	20.84	160,457	15.2	6.16	57,243	5.4	2.20	1,054,725	100	40.50
	26,386	1931	253,931	29.6	9.62	431,254	50.3	16.34	118,607	13.8	4.50	53,728	6.3	2.04	857,520	100	32.50
	26,729	1932	262,693	30.2	9.83	419,998	48.3	15.71	113,412	13.0	4.24	73,439	8.4	2.75	869,542	100	32.53
	27,072	1933	236,566	29.6	8.74	377,828	49.2	13.96	122,179	15.3	4.51	63,587	7.9	2.35	800,160	100	29.56
Average	26,558	1930-33	261,896	29.3	9.87	442,928	49.3	16.71	128,664	14.3	4.85	61,999	7.0	2.33	895,486	100	33.77
Sharon	24,659	1927	278,516	30.9	11.29	409,271	45.4	16.60	213,870	23.7	8.67				901,657	100	36.57
	25,908	1930	314,338	29.4	12.13	517,159	48.3	19.96	239,253	22.3	9.23				1,070,745	100	41.33
	26,324	1931	313,881	30.1	11.92	498,902	47.8	18.95	231,055	22.1	8.78	84,609	8.4	3.16	1,043,838	100	39.65
	26,740	1932	293,625	29.3	10.98	460,851	45.9	17.23	164,642	16.4	6.16				1,003,727	100	37.54
	27,156	1933	234,295	28.8	8.63	347,533	42.7	12.80	231,958	28.5	8.54				813,786	100	29.97
Average	25,532	1930-33	299,034	29.4	10.91	456,111	46.1	17.23	216,727	22.3	8.17	84,609	8.4	3.16	988,024	100	37.12
Lebanon	25,287	1927	230,720	36.3	9.12	306,094	48.1	12.10	88,431	13.9	3.50	10,954	1.7	.43	636,199	100	25.16
	25,561	1930	245,630	39.4	9.61	256,243	41.1	10.02	108,316	17.4	4.24	13,963	2.2	.55	624,152	100	24.42
	25,653	1931	248,186	38.9	9.67	258,396	40.5	10.07	107,147	16.8	4.18	24,881	3.9	.97	638,560	100	24.89
	25,745	1932	250,331	39.7	9.72	258,396	40.9	10.04	105,664	16.6	4.10	16,816	2.7	.65	631,207	100	24.52
	25,837	1933	244,500	39.0	9.46	258,481	41.2	10.00	104,011	16.6	4.03	19,739	3.1	.76	626,731	100	24.26
Average	25,699	1930-33	247,149	39.2	9.61	257,879	40.9	10.03	106,285	16.8	4.13	18,850	2.9	.73	630,162	100	24.52
Washington	23,629	1927	262,683	38.2	11.12	287,099	41.7	12.15	138,285	20.1	5.85				688,067	100	29.12
	24,545	1930	297,989	36.0	12.14	362,898	43.9	14.79	166,575	20.1	6.79				827,462	100	33.71
	24,852	1931	299,504	34.3	12.05	402,925	46.2	16.21	169,545	19.4	6.82				871,974	100	35.09
	25,159	1932	301,343	34.3	11.98	412,327	46.9	16.39	166,126	18.9	6.60				879,796	100	34.97
	25,466	1933	301,223	33.8	11.83	426,671	47.8	16.75	163,826	18.4	6.43				891,719	100	35.02
Average	25,006	1930-33	300,015	34.6	12.00	401,205	46.2	16.03	166,518	19.2	6.66				867,738	100	34.69
Pottsville	23,570	1927	251,225	31.3	10.66	386,841	48.1	16.41	165,384	20.6	7.02				803,450	100	34.09
	24,300	1930	257,615	32.3	10.60	373,152	46.8	15.36	167,384	21.0	6.89				798,151	100	32.85
	24,542	1931	257,621	32.0	10.50	376,540	46.8	15.34	170,521	21.2	6.95				804,682	100	32.79
	24,784	1932	226,654	28.7	9.15	386,584	48.9	15.60	176,690	22.4	7.13				789,928	100	31.87
	25,026	1933	224,640	30.8	8.98	383,602	52.7	15.33	120,254	16.5	4.81				728,496	100	29.11
Average	24,663	1930-33	241,682	30.9	9.80	379,970	48.8	15.40	158,712	20.2	6.44				780,314	100	31.65
Butler	23,631	1927	322,765	36.3	13.66	320,950	36.1	13.58	245,474	27.6	10.39				889,189	100	37.63
	23,568	1930	277,991	31.5	11.80	304,078	34.4	12.90	301,670	34.1	12.80				883,739	100	37.50
	23,547	1931	364,364	37.3	15.47	305,258	31.2	12.96	308,239	31.5	13.09				977,861	100	41.53
	23,526	1932	324,934	35.1	13.81	300,809	32.5	12.79	300,500	32.4	12.77				926,243	100	39.37
	23,505	1933	306,712	35.9	13.05	266,957	31.2	11.36	281,653	32.9	11.98				855,322	100	36.39
Average	23,537	1930-33	318,500	35.0	13.53	294,276	32.8	12.50	298,016	32.7	12.66				910,791	100	38.70

TABLE III (cont.)
TAXES LEVIED (cont.)

City	Population	Year	CITY Total City Taxes Levied	City taxes levied as a percentage of total taxes levied	City Taxes Levied Per Capita	SCHOOL Total School Taxes Levied	School taxes levied as a percentage of total taxes levied	School Taxes Levied Per Capita	COUNTY Total County Taxes Levied	County taxes levied as a percentage of total taxes levied	County Taxes Levied Per Capita	POOR Total Poor Taxes Levied	Poor taxes levied as a percentage of total taxes levied	Poor Taxes Levied Per Capita	TOTAL Total of all Taxes Levied	Total Per Cent	Total of all Taxes Levied Per Capita
Oil City	21,834	1937	$225,000	29.2	$10.81	$360,170	46.8	$16.50	$170,967	22.2	$7.83	$14,107	1.8	$.65	$770,244	100	$35.23
	22,075	1930	225,536	38.4	12.98	349,959	40.9	15.85	205,824	24.0	9.32	14,662	1.7	.66	855,981	100	38.77
	22,155	1931	291,565	34.8	13.16	385,919	40.1	15.16	195,586	23.3	8.88	14,718	1.8	.66	837,888	100	37.82
	22,235	1932	290,039	37.1	13.04	327,962	41.9	14.75	149,995	19.2	6.75	14,774	1.9	.66	782,770	100	35.20
	22,315	1933	290,039	37.8	13.00	322,988	42.1	14.47	109,902	14.3	4.93	44,130	5.8	1.98	767,059	100	34.37
Average	22,195	1930-33	289,320	35.7	13.03	334,207	41.2	15.05	165,314	20.2	7.45	22,071	2.8	.99	810,912	100	36.54
Duquesne	20,684	1927	245,841	32.7	11.93	317,597	41.9	15.35	165,099	21.8	7.98	25,942	3.6	1.30	758,479	100	36.57
	21,396	1930	330,384	34.5	15.44	387,868	42.0	18.13	205,811	21.5	9.61	34,662	3.6	1.62	958,725	100	44.81
	21,635	1931	254,386	30.4	11.76	355,729	42.5	16.44	195,126	23.3	9.02	31,934	3.8	1.48	837,175	100	38.70
	21,874	1932	215,689	26.6	9.87	335,736	41.4	15.35	195,490	24.1	8.94	63,815	7.9	2.92	810,930	100	37.07
	22,113	1933	189,365	25.4	8.56	297,429	39.9	13.45	194,813	26.1	8.81	63,706	8.5	2.88	745,313	100	33.70
Average	21,755	1930-33	247,506	29.2	11.40	344,191	41.0	15.84	197,810	23.7	8.09	49,529	5.9	2.22	838,035	100	38.57
Monessen	19,642	1927	179,619	26.4	9.14	387,064	57.0	19.71	112,579	16.6	5.73				679,262	100	34.58
	20,268	1930	242,877	30.0	11.98	425,142	52.6	20.97	140,692	17.4	6.94				808,711	100	39.90
	20,477	1931	243,758	29.9	11.90	428,528	52.6	20.93	141,780	17.4	6.92				814,061	100	39.75
	20,686	1932	219,525	27.9	10.63	430,155	54.5	20.79	138,998	17.6	6.72				788,978	100	38.14
	20,895	1933	203,184	26.3	9.72	427,853	55.4	20.48	141,205	18.3	6.76				772,242	100	36.96
Average	20,582	1930-33	227,410	28.5	11.05	427,920	53.7	20.79	140,669	17.6	6.83				795,998	100	38.68
Carbondale	19,634	1927	150,824	22.1	7.68	391,826	57.4	19.96	96,175	14.1	4.90	43,367	6.4	2.21	682,192	100	34.75
	20,061	1930	159,915	25.8	7.97	309,356	50.0	15.42	103,377	16.7	5.15	45,497	7.4	2.27	618,145	100	30.81
	20,203	1931	130,395	27.2	8.93	329,492	49.6	16.31	95,726	14.4	4.74	58,234	8.8	2.88	663,847	100	32.86
	20,345	1932	178,572	27.1	8.78	327,370	50.7	16.09	80,449	12.5	3.95	58,984	9.1	2.90	645,375	100	31.72
	20,487	1933	168,000	27.1	8.20	327,099	52.8	15.97	70,521	11.4	3.44	54,072	8.7	2.64	619,692	100	30.25
Average	20,274	1930-33	171,720	27.0	8.47	323,329	50.8	15.94	87,518	13.7	4.32	54,197	8.5	2.67	636,765	100	31.41
Uniontown	18,387	1927	303,615	35.9	16.51	385,824	45.7	20.98	143,578	17.0	7.81	11,804	1.4	.64	844,821	100	45.95
	19,544	1930	320,867	32.8	16.42	432,525	44.2	22.13	205,757	21.0	10.53	19,475	2.0	.99	978,624	100	50.07
	19,929	1931	314,915	33.3	15.80	424,924	44.9	21.32	187,143	19.8	9.39	19,141	2.0	.96	946,123	100	47.47
	20,314	1932	289,870	33.1	14.27	382,927	43.7	18.85	184,028	21.0	9.06	18,959	2.2	.93	875,784	100	43.11
	20,699	1933	266,496	32.3	12.87	360,000	43.7	17.39	178,787	21.7	8.64	18,924	2.3	.91	824,207	100	39.82
Average	20,122	1930-33	298,037	32.9	14.81	400,094	44.2	19.88	188,929	20.8	9.40	19,125	2.1	.94	906,185	100	45.08
Bradford	18,171	1927	165,746	32.2	8.57	242,126	50.1	13.92	60,891	12.6	3.35	24,312	5.0	1.34	493,075	100	26.58
	19,306	1930	236,014	33.6	12.23	317,458	45.2	16.44	122,659	17.5	6.35	25,659	3.7	1.33	701,790	100	36.35
	19,684	1931	240,897	34.9	12.24	323,764	46.5	16.45	105,675	15.2	5.37	26,181	3.8	1.33	696,517	100	35.38
	20,062	1932	241,082	36.2	12.01	291,509	43.8	14.53	107,413	16.1	5.35	26,065	3.9	1.30	666,019	100	33.20
	20,440	1933	193,933	35.3	9.49	255,846	46.6	12.52	71,568	13.	3.50	27,855	5.1	1.36	549,202	100	26.87
Average	19,873	1930-33	227,969	35.0	11.49	297,144	45.5	14.98	101,828	15.4	5.14	26,440	4.1	1.33	653,382	100	32.95

TABLE III (Cont'd.)
TAXES LEVIED

City	Population	Year	CITY Total City Taxes Levied	CITY City taxes levied as a percentage of all taxes levied	CITY City Taxes Levied Per Capita	SCHOOL Total School Taxes Levied	SCHOOL School taxes levied as a percentage of all taxes levied	SCHOOL School Taxes Levied Per Capita	COUNTY Total County Taxes Levied	COUNTY County taxes levied as a percentage of all taxes levied	COUNTY County Taxes Levied Per Capita	POOR Total Poor Taxes Levied	POOR Poor taxes levied as a percentage of all taxes levied	POOR Poor Taxes Levied Per Capita	TOTAL Total of all Taxes Levied	TOTAL Total Per Cent	TOTAL Total of all Taxes Levied Per Capita
Pittston	18,322	1927	$214,472	35.7	$11.71	$245,764	40.9	$13.41	$77,035	12.8	$4.20	$63,269	10.5	$3.45	$600,540	100	$32.78
	18,246	1930	192,986	28.2	10.58	354,643	51.8	19.43	99,791	14.6	5.47	37,525	5.5	2.06	684,945	100	37.54
	18,221	1931	181,700	33.4	9.97	289,109	43.9	13.12	83,548	15.3	4.58	40,364	7.4	2.22	544,721	100	29.90
	18,196	1932	246,442	40.6	13.54	239,109	39.4	13.14	81,419	13.4	4.47	39,742	6.6	2.18	606,712	100	33.34
	18,171	1933	240,102	41.4	13.21	217,479	37.5	11.97	83,501	14.4	4.60-	39,421	6.8	2.17	580,503	100	31.95
Average	18,209	1930-33	215,308	35.6	11.82	262,585	43.5	14.41	87,065	14.4	4.78	39,263	6.5	2.16	604,220	100	33.18
Beaver Falls	16,573	1927	218,878	30.6	13.21	335,640	46.9	20.25	135,974	19.0	8.20	25,109	3.5	1.51	715,596	100	43.17
	17,147	1930	231,881	33.0	13.52	343,699	48.9	20.04	108,213	15.4	6.31	19,629	2.8	1.14	703,432	100	41.02
	17,337	1931	240,855	32.4	13.89	362,019	48.7	20.88	114,523	15.4	6.60	25,977	3.5	1.50	743,374	100	42.88
	17,527	1932	218,554	31.1	12.18	338,490	48.5	19.03	114,617	16.7	6.54	26,018	3.8	1.48	687,679	100	39.24
	17,717	1933	185,753	29.3	10.48	308,019	48.6	17.39	114,285	18.0	6.45	25,858	4.1	1.46	633,915	100	35.78
Average	17,432	1930-33	218,013	31.4	12.51	336,807	48.6	19.33	112,910	16.3	6.47	24,371	3.5	1.39	692,100	100	39.73
Meadville	16,059	1927	171,261	32.9	10.66	241,005	46.3	15.01	108,524	20.8	6.76				520,790	100	32.43
	16,698	1930	203,668	33.4	12.20	274,605	45.0	16.45	131,389	21.6	7.87				609,662	100	36.51
	16,911	1931	194,474	31.5	11.50	283,534	46.0	16.77	138,771	22.5	8.20				616,779	100	36.47
	17,124	1932	197,719	34.2	11.55	254,851	44.1	14.88	125,878	21.8	7.35				578,448	100	33.78
	17,337	1933	182,453	35.5	10.52	215,115	41.9	12.41	116,066	22.6	6.69				513,634	100	29.63
Average	17,018	1930-33	194,579	33.6	11.44	257,026	44.2	15.12	128,026	22.1	7.52				579,630	100	34.09
Greensburg	16,069	1927	220,065	31.0	13.69	351,886	49.6	21.90	137,646	19.4	8.57				709,587	100	44.16
	16,508	1930	223,546	31.0	13.54	344,898	47.8	20.89	152,906	21.2	9.26				721,280	100	43.69
	16,656	1931	235,032	31.8	14.11	351,689	47.5	21.11	153,499	20.7	9.22				740,160	100	44.44
	16,804	1932	234,015	33.2	13.99	321,196	45.5	19.11	150,493	21.3	8.96				705,704	100	42.00
	16,952	1933	283,570	34.2	13.78	292,114	42.7	17.23	158,210	23.1	9.33				683,894	100	40.34
Average	16,790	1930-33	231,541	32.5	13.84	318,442	45.8	19.58	153,777	21.5	9.19				712,759	100	42.61
Sunbury	15,651	1927	88,182	20.3	5.31	260,261	63.6	16.63	46,784	11.4	2.99	19,196	4.7	1.23	409,423	100	26.16
	15,626	1930	92,478	21.2	5.92	275,029	63.0	17.60	53,537	12.3	3.42	15,356	3.5	.98	436,400	100	27.93
	15,617	1931	87,616	20.5	5.86	271,597	60.9	17.39	58,804	13.2	3.77	24,394	5.5	1.56	446,311	100	28.58
	15,608	1932	87,757	20.2	5.62	252,853	58.2	16.20	61,080	14.1	3.91	32,500	7.5	2.08	434,190	100	27.82
	15,599	1933	88,935	21.1	5.70	254,740	60.4	16.33	45,533	10.8	2.92	32,490	7.7	2.08	421,698	100	27.03
Average	15,613	1930-33	90,172	20.7	5.77	263,555	60.6	16.88	54,739	12.6	3.50	26,185	6.0	1.67	434,650	100	27.84
Clairton	13,984	1927	197,811	25.3	14.20	385,226	49.3	27.65	170,114	21.8	12.21	27,483	3.5	1.97	780,634	100	56.02
	15,291	1930	240,657	25.1	15.74	445,598	46.5	29.14	232,802	24.3	15.22	40,146	4.2	2.63	959,203	100	62.73
	15,742	1931	242,102	25.3	15.38	449,081	47.0	28.53	225,520	23.6	14.33	38,461	4.2	2.44	955,164	100	60.68
	16,193	1932	243,088	24.4	15.01	446,896	44.4	27.60	228,275	23.6	14.10	78,580	7.9	4.85	996,789	100	61.56
	16,644	1933	190,655	21.8	11.45	381,265	43.6	22.91	223,879	25.6	13.45	78,601	9.0	4.72	874,400	100	52.54
Average	15,968	1930-33	229,126	24.1	14.39	430,710	45.4	27.04	227,619	24.1	14.27	58,934	6.3	3.66	946,389	100	59.37

TABLE III (Cont'd.)
TAXES LEVIED

City	Population	Year	Total City Taxes Levied	City taxes levied as a per-cent-age of all taxes levied	City Taxes Levied Per Capita	Total School Taxes Levied	School taxes levied as a per-cent-age of all taxes levied	School Taxes Levied Per Capita	Total County Taxes Levied	County taxes levied as a per-cent-age of all taxes levied	County Taxes Levied Per Capita	Total Poor Taxes Levied	Poor taxes levied as a per-cent-age of all taxes levied	Poor Taxes Levied Per Capita	Total of all Taxes Levied	Total Per Cent	Total of all Taxes Levied Per Capita
Coatesville	14,564	1927	$131,370	32.8	$9.02	$211,712	52.9	$14.54	$57,302	14.3	$3.93				$400,384	100	$27.49
	14,582	1930	180,882	38.2	12.40	228,980	48.4	15.70	63,224	13.4	4.34				473,026	100	32.44
	14,589	1931	184,037	39.1	12.61	223,687	47.5	15.33	63,330	13.4	4.34				470,994	100	32.28
	14,596	1932	169,733	38.3	11.63	211,881	47.8	14.51	61,440	13.9	4.21				443,004	100	30.35
	14,603	1933	147,740	35.8	10.12	212,672	51.5	14.56	52,574	12.7	3.60				412,986	100	28.28
Average	14,598	1930-33	170,583	37.8	11.69	219,278	48.8	15.02	60,142	13.3	4.12				450,003	100	30.83
Farrell	14,725	1927	116,538	24.5	7.91	240,732	50.6	16.35	118,824	25.0	8.07				476,094	100	32.33
	14,359	1930	147,594	28.2	10.28	246,008	47.1	17.13	128,889	24.7	8.98				522,491	100	36.39
	14,236	1931	155,832	28.3	10.99	260,384	47.3	18.36	134,077	24.4	9.45	$48,806	8.8	$3.46	550,293	100	38.66
	14,113	1932	159,260	28.8	11.28	254,514	46.3	18.03	92,744	16.7	6.57				555,324	100	39.35
	13,990	1933	141,070	28.0	10.09	231,719	46.0	16.56	131,456	26.1	9.40				504,245	100	36.04
Average	14,175	1930-33	150,939	28.3	10.65	248,156	46.5	17.52	121,792	22.9	8.60				533,088	100	37.61
Connellsville	13,447	1927	158,454	32.0	11.78	245,537	49.5	18.26	84,690	17.1	6.30	6,988	1.4	.52	495,669	100	36.86
	13,290	1930	168,894	32.0	12.71	233,048	45.1	17.54	104,241	20.2	7.84	10,446	2.0	.79	516,629	100	38.87
	13,239	1931	171,272	34.0	12.94	234,064	46.5	17.68	87,973	17.5	6.64	9,837	2.0	.74	503,146	100	38.00
	13,188	1932	167,667	34.4	12.71	222,298	45.6	16.86	87,917	18.0	6.66	9,952	2.0	.75	487,834	100	36.99
	13,137	1933	166,559	35.1	12.70	211,484	44.5	16.10	87,303	18.4	6.65	9,914	2.1	.75	475,560	100	36.20
Average	13,214	1930-33	168,673	34.0	12.76	225,224	45.4	17.04	91,859	18.5	6.94	10,037	2.0	.75	495,792	100	37.51
Du Bois	12,218	1927	87,143	24.4	7.13	211,049	59.1	17.27	38,058	10.7	3.11	20,714	5.8	1.70	356,964	100	29.22
	11,595	1930	110,762	29.3	9.55	209,549	55.4	18.07	35,381	9.4	3.05	22,430	5.9	1.93	378,122	100	32.61
	11,386	1931	106,991	29.5	9.40	198,792	54.7	17.46	35,408	9.7	3.11	22,076	6.1	1.94	363,267	100	31.90
	11,177	1932	106,603	30.0	9.54	180,138	50.7	16.12	35,254	9.9	3.15	33,240	9.4	2.97	355,235	100	31.78
	10,968	1933	106,781	30.0	9.74	181,504	51.	16.55	34,344	9.7	3.13	32,987	9.3	3.01	355,616	100	32.42
Average	11,282	1930-33	107,784	29.7	9.55	192,496	52.9	17.05	35,097	9.6	3.11	27,683	7.6	2.46	363,060	100	32.17
Franklin	10,166	1927	93,718	27.9	9.22	159,943	47.7	15.73	74,025	22.1	7.28	7,715	2.3	.76	335,401	100	32.99
	10,254	1930	100,025	27.4	9.75	175,183	48.0	17.08	82,228	22.5	8.02	7,874	2.2	.77	365,310	100	35.63
	10,282	1931	99,513	27.5	9.68	174,537	48.3	16.98	79,604	22.0	7.74	7,780	2.2	.76	361,434	100	35.15
	10,310	1932	99,455	29.5	9.65	161,828	48.0	15.69	68,011	20.2	6.60	7,766	2.3	.75	337,060	100	32.69
	10,338	1933	84,881	29.6	8.21	133,471	46.5	12.91	45,507	15.9	4.40	23,145	8.1	2.24	287,004	100	27.76
Average	10,296	1930-33	95,969	28.5	9.32	161,255	47.7	15.66	68,838	20.1	6.69	11,641	3.7	1.13	337,702	100	32.80
Lock Haven	9,334	1927	67,023	30.2	7.18	109,932	49.6	11.78	44,766	20.2	4.80				221,721	100	23.76
	9,668	1930	68,995	26.6	7.11	126,911	49.1	13.13	36,623	14.2	3.79	26,014	10.0	2.69	258,243	100	26.71
	9,779	1931	69,665	28.3	7.12	124,265	50.5	12.71	21,272	8.6	2.18	30,729	12.5	3.14	245,931	100	25.15
	9,890	1932	59,273	25.7	5.99	119,043	51.5	12.04	22,163	9.6	2.24	30,524	13.2	3.09	231,003	100	23.36
	10,001	1933	59,437	26.8	5.94	104,764	47.3	10.48	26,856	12.	2.68	30,597	13.8	3.06	221,654	100	22.16
Average	9,835	1930-33	64,268	26.8	6.54	118,746	49.6	12.09	26,729	11.1	2.72	29,466	12.3	2.97	289,208	100	24.34

TABLE III (Concluded)
TAXES LEVIED

City	Population	Year	CITY			SCHOOL			COUNTY			POOR			TOTAL		
			Total City Taxes Levied	City taxes levied as a per-cent-age of total of all taxes levied	City Taxes Levied Per Capita	Total School Taxes Levied	School taxes levied as a per-cent-age of total of all taxes levied	School Taxes Levied Per Capita	Total County Taxes Levied	County taxes levied as a per-cent-age of total of all taxes levied	County Taxes Levied Per Capita	Total Poor Taxes Levied	Poor taxes levied as a per-cent-age of total of all taxes levied	Poor Taxes Levied Per Capita	Total of all Taxes Levied	Total Per Cent	Total of all Taxes Levied Per Capita
Monongahela	8,681	1927	$ 90,624	30.8	$10.44	$162,243	55.1	$18.70	$ 41,758	14.2	$4.81				$294,625	100	$33.96
	8,675	1930	91,530	30.2	10.55	163,270	53.9	18.82	48,060	15.9	5.54				302,860	100	34.91
	8,674	1931	93,637	31.4	10.80	156,724	52.5	18.07	47,937	16.1	5.53				298,298	100	34.39
	8,673	1932	87,297	31.4	10.07	144,485	52.0	16.66	46,215	16.6	5.33				277,997	100	32.05
	8,672	1933	81,131	31.4	9.36	132,027	51.0	15.22	45,510	17.6	5.25				258,668	100	29.83
Average	8,674	1930-33	86,399	31.1	10.19	149,127	52.3	17.19	46,931	16.5	5.41				284,456	100	32.79
Titusville	8,166	1927	96,508	33.8	11.82	114,435	40.1	14.01	74,715	26.2	9.15				285,653	100	34.98
	8,055	1930	100,198	33.8	12.44	113,392	38.2	14.08	82,954	28.0	10.30				296,544	100	36.81
	8,017	1931	85,650	30.7	10.68	113,183	40.6	14.12	80,188	28.7	10.00				279,021	100	34.80
	7,979	1932	90,685	33.0	11.37	111,544	40.6	13.98	72,469	26.4	9.08				274,698	100	34.43
	7,941	1933	84,320	34.1	10.62	96,417	39.0	12.14	66,406	26.9	8.36				247,143	100	31.12
Average	7,998	1930-33	90,213	32.9	11.27	108,634	39.6	13.58	75,504	27.5	9.43				274,352	100	34.29
Corry	7,172	1927	90,215	38.4	12.58	113,873	48.5	15.88	30,923	13.2	4.31				235,011	100	32.77
	7,152	1930	79,532	35.0	11.12	116,633	51.4	16.30	30,916	13.6	4.32				227,081	100	31.75
	7,144	1931	78,858	35.0	11.04	113,714	50.7	15.92	31,909	14.2	4.47				224,481	100	31.42
	7,136	1932	78,682	34.1	11.03	113,291	49.	15.88	39,027	16.9	5.47				231,000	100	32.37
	7,128	1933	69,276	34.1	9.72	95,411	46.9	13.39	38,594	19.0	5.41				203,281	100	28.52
Average	7,140	1930-33	76,587	34.5	10.72	109,762	49.5	15.37	35,112	15.9	4.91				221,461	100	31.01

Total Taxes Levied Per Capita:- The $16,877,662 annually levied for the last four years in the twenty-seven cities averaged together, amounts to a yearly outlay of $35.61 per person. The amount of taxes levied per capita differs so materially between the various cities that a clear understanding of the matter can only be had by an examination of the figures for each city separately. Cities with a high per capita levy in 1933 are: Clairton - $52.54; Greensburg - $40.34; Uniontown - $39.82; Monessen - $36.96; Butler - $36.39; Connellsville - $36.20; and Farrell - $36.04.

One of the most significant things shown by Table III is that in twenty out of the twenty-seven cities, there has been a reduction in the total per capita levies for 1933 as compared with the levies in 1927. Furthermore, in fourteen of the cities there has been a reduction in the amount of taxes levied per person in each of the last three years as compared with the corresponding levies made in 1930, the year in which for a majority of the cities, taxes reached their peak figure. Taking the four year average of all taxes levied in the twenty-seven cities, the total per capita levy of $35.61 is divided among the different taxes as follows:- City taxes amounted to $11.21 per person; School taxes to $16.37 per person and County and Poor taxes to $8.03 per person.

The Collection of Taxes:- The third and last step in the process of taxation is the collection. The law makes the treasurer of a third class city the collector of all city, school and poor taxes assessed or levied in the city.

Each of the tax levying authorities fixes the treasurer's compensation or commission for tax collection. In the case of city and poor taxes the law requires, however, that in no event shall the compensation be less than one-fourth of one per centum of all taxes collected prior to the penalty period nor less than five per centum on all taxes collected after a penalty has been incurred. His compensation for the collection of school taxes is fixed by the board of school directors. All necessary printing and stationery supplies are furnished by the respective taxing units. The collector may employ as many deputies, assistants or clerks as he may require but their compensation must be paid by him personally.

As the result of a long, hard fight to consolidate the collection of taxes in fewer hands, and to abolish fees and commissions in the payment for such services, the Legislature in 1933 amended the law concerning the collection of taxes in third class cities. Since this law only applies to a city when a new city treasurer is elected the laws previously discussed are still in force in those cities where a treasurer was not elected in 1933 or 1934. The Act of 1933 continues the city treasurer as the collector of all city, school and poor taxes levied and collected in the city, but leaves the collection of county taxes in the city to the appointees of the county commissioners. It is to be regretted that the Legislature did not consolidate all tax collection work in the hands of one official, the city treasurer.

The Act named (1933 P. L. 91) did however abolish the fee or commission basis of payment in so far as city, school and poor taxes are concerned. It divided cities into different groups on the basis of population and established salaries for each group as follows:

Population of 25,000 and under - $1,600 -- $3,500
Between 25,000 and 50,000- $3,500 -- $5,000
Between 50,000 and 55,000- $6,000
Over 55,000   - $7,500

It will be observed that a minimum and maximum salary is not provided for the last two population groups. McKeesport is the only city (1930 census) with a population between 50,000 and 55,000, but there are eleven cities with popluation ranging from 55,254 (York) to 115,967 (Erie), and six of these have a population in excess of 80,000. By this arrangement the collectors in the Cities of York and Erie received the same salary in spite of the great difference in population and amount of taxes collected. It is difficult to see any logic or justice in this classification. The Legislature should re-classify the cities for salary purposes and provide minimum and maximum salaries for all population groups. If the collection of all local taxes is not transferred to a single office in each county in the State then the present salaries provided for the last two population groups should be substantially reduced. Under the new law the city and school district pay in equal proportion all of the expenses incurred in the collection of taxes.

This law has not been in effect long enough or in a sufficient number of cities to determine whether its adoption will result in the saving of much money. Whether there will be any saving at all, will depend on the joint action of the officials authorized to fix the salaries and determine the number of clerks and assistants necessary to perform the work involved. In one city already operating under the new plan of payment, the collector's salary has been fixed at $5,000 annually which is the maximum allowed by law for cities with a population of between 25,000 and 50,000 and the city in question has less than 27,000 people. With nearly the minimum population of the group to which it belongs this city starts the first collector off with the maximum salary for the group. In this office there are two full time employees and one who works about six months of the year. Before the salary law went into effect there was only one full time clerk and a part-time assistant. In two or three smaller cities salaries have been fixed at or near the $3,500 maximum permitted by the Legislature.

Extent of Tax Delinquency:- Tax delinquency has been a problem for many years but in the last three years it has assumed major importance. With hundreds of thousands of people out of work and thousands of homes, offices, stores and factories unoccupied, taxpayers have experienced considerable difficulty in meeting their tax obligations in full and many have failed to pay anything at all in the last year or two. In order to show the extent of tax delinquency in twenty-seven cities during the past four years, Table IV is presented. All data for the years 1930, 1931 and 1932 were obtained from the Department of Internal Affairs, Harrisburg, while all figures for 1933 were obtained from city officials direct or from their published reports. Nothing in this table pertains to school, county or poor taxes.

An examination of column three shows that in all but four of the cities the percentage of tax delinquency increased in each of the last three years. In only four cities was the amount of unpaid taxes less than 20.0 per cent of the total taxes levied. Even in Clairton, the city with the lowest delinquency in 1933, 16.8 per cent of the money due was not paid during the year and in this city a very large portion of all taxes are paid by large corporations. The most serious situation is in Pittston where 51.2 per cent of 1933 taxes remained uncollected at the end of the year. Other cities with very large delinquencies in 1933 are Connellsville, 49 per cent; Uniontown, 48 per cent; and Beaver Falls with 42.2 per cent. There are eleven additional cities where the delinquent taxes amount to from 30.0 to 40.0 per cent of the total taxes levied.

The city authorities were not suddenly confronted in 1933 with this problem of tax collection because in almost every case the percentage of unpaid taxes had been increasing year by year for the last three or four years. This very definite trend is shown in the following ten cities:-

City	Per Cent of Current Taxes Remaining Unpaid at the End of Each Year.			
	1930	1931	1932	1933
Nanticoke	4.4	16.6	18.5	25.5
Lebanon	13.7	14.4	18.2	22.1
Washington	15.1	19.7	28.7	32.0
Butler	3.9	18.5	24.5	30.7
Duquesne	7.3	11.8	26.9	33.7
Monessen	8.6	13.9	24.8	30.7
Beaver Falls	16.0	24.5	35.7	42.2
Greensburg	10.0	14.6	27.3	36.0
DuBois	19.4	23.7	32.5	38.0
Franklin	10.7	13.8	19.6	24.0

TABLE IV
DELINQUENT TAXES

City	Year	Total City Taxes Levied	Current taxes delinquent at end of Tax Year	Per Cent of the Total Taxes levied remaining delinquent at the end of Tax Year	Taxes Collected in Current Year from levies of Previous Years	Amount by which Current delinquent taxes exceeded taxes collected from levies of previous years (Col. 2-Col. 4)	Total amount of all Taxes Outstanding December 31,1933	Total Outstanding Taxes as a percentage of the 1933 levy
		$						
Nanticoke	1930	294,394	$12,926	4.4	$ 16,930			
	1931	253,931	42,138	16.6	19,735	$22,403		
	1932	262,693	48,609	18.5	2,503	46,106		
	1933	236,566	60,213	25.5	25,501	34,712	$75,000	31.7
Sharon	1930	314,333	51,309	16.3	40,126	11,183		
	1931	313,881	46,741	14.9	42,862	3,879		
	1932	293,625	102,613	34.9	29,665	72,948		
	1933	234,295	85,868	36.6	19,358	66,510	168,881	72.1
Lebanon	1930	245,630	33,673	13.7	33,659	14		
	1931	248,136	35,712	14.4	36,858			
	1932	250,331	45,559	18.2	37,934	7,625		
	1933	244,500	53,972	22.1	35,702	18,270	163,995	67.1
Pottsville	1930	257,615	26,694	10.4	29,297			
	1931	257,621	29,988	11.6	22,086	7,902		
	1932	226,654	41,308	18.2	23,602	17,706		
	1933	224,640	46,797	20.8	18,249	28,548	91,737	40.8
Washington	1930	297,989	44,931	15.1	2,500	42,431		
	1931	299,504	59,129	19.7	21,771	37,358		
	1932	301,343	86,582	28.7	33,913	52,669		
	1933	301,223	96,345	32.0	41,531	54,814	195,187	64.8
Butler	1930	277,991	10,915	3.9		10,915		
	1931	364,364	67,299	18.5	19,929	47,370		
	1932	324,934	79,632	24.5	40,771	38,861		
	1933	306,712	94,265	30.7	36,093	58,172	168,041	54.8
Oil City	1930	285,536	42,421	14.9	21,947	20,474		
	1931	291,665	58,467	20.0	25,450	33,017		
	1932	290,039	61,478	21.2	68,714			
	1933	290,039	80,761	27.8	39,843	40,918	135,692	46.8
Duquesne	1930	330,384	24,161	7.3	20,105	4,056		
	1931	254,386	30,116	11.8	12,026	18,090		
	1932	215,889	58,023	26.9	10,559	47,464		
	1933	189,365	63,832	33.7	11,469	52,363	166,839	88.1
Monessen	1930	242,877	21,004	8.6	18,095	2,909		
	1931	243,753	33,773	13.9	14,954	18,819		
	1932	219,825	54,472	24.8	10,693	43,779		
	1933	203,184	62,456	30.7	18,774	43,682	132,938	65.4

TABLE IV (Cont'd)
DELINQUENT TAXES

City	Year	Total City Taxes Levied	Current taxes delinquent at end of Tax Year	Per Cent of the Total Taxes levied remaining delinquent at the end of Tax Year	Taxes Collected in Current Year from levies of Previous Years	Amount by which Current delinquent taxes exceeded taxes collected from levies of previous years (Col. 2- Col. 4)	Total amount of all Taxes Outstanding December 31,1933	Total Outstanding Taxes as a percentage of the 1933 levy
		$						
	1930	159,915	$40,549	25.4	$ 20,193	$20,356		
	1931	180,395	64,059	35.5	14,165	49,894		
Carbondale	1932	178,572	64,712	36.2	32,953	31,759		
	* 1933							
	1930	320,867	65,167	20.3	48,068	17,099		
	1931	314,915	83,973	26.7	46,905	37,068	$	
Uniontown	1932	289,870	102,262	35.3	41,377	60,885	250,235	
	1933	266,496	127,958	48.0	36,519	91,439		93.9
	1930	236,014	36,073	15.3	13,148	22,925		
	1931	240,897	71,912	29.9	32,347	39,565		
Bradford	1932	241,032	68,053	28.2	28,297	39,756		
	1933	193,933	61,488	31.7	49,695	11,793	146,190	75.4
	1930	192,986	28,643	14.8		28,643		
	1931	181,700	27,356	15.1		27,356		
Pittston	1932	246,442	95,161	38.6	21,055	74,106		
	1933	240,102	122,934	51.2	33,312	89,622	338,078	140.8
	1930	231,891	37,006	16.0	31,280	5,726		
	1931	240,855	58,964	24.5	25,138	33,826		
Beaver Falls	1932	213,554	76,183	35.7	23,714	52,469		
	1933	185,753	78,303	42.2	20,472	57,831	206,341	111.1
	1930	203,668	13,410	6.6	7,447	5,963		
	1931	194,474	6,824	3.5	11,330			
Meadville	1932	197,719	32,395	16.4	20,665	11,730		
	1933	182,453	47,311	25.9	17,634	29,677	72,615	39.8
	1930	223,546	22,413	10.0	22,167	246		
	1931	235,032	34,272	14.6	19,696	14,576		
Greensburg	1932	234,015	63,926	27.3	26,902	37,024		
	1933	233,570	84,071	36.0	29,526	54,545	132,283	56.6
	1930	92,478	13,951	15.1	13,115	836		
	1931	91,516	12,789	14.0	11,140	1,649		
Sunbury	1932	87,757	17,783	20.3	11,847	5,936		
	1933	88,935	17,627	19.8	7,246	10,381	37,425	42.1
	1930	240,657	18,955	7.9	8,244	10,711		
	1931	242,102	26,730	11.0	8,981	17,749		
Clairton	1932	243,088	37,405	15.4	8,891	28,514		
	1933	190,655	32,063	16.8	11,705	20,358	92,161	48.3

* The figures for 1933 were not obtainable.

TABLE IV (Concluded)
DELINQUENT TAXES

City	Year	Total City Taxes Levied	Current taxes delinquent at end of Tax Year	Per Cent of the Total Taxes levied remaining delinquent at the end of Tax Year	Taxes Collected in Current Year from levies of Previous Years	Amount by which Current delinquent taxes exceeded taxes collected from levies of previous years (Col. 2- Col. 4)	Total amount of all Taxes Outstanding December 31, 1933	Total Outstanding Taxes as a percentage of the 1933 levy
		$						
Coatesville	1930	180,822	$ 8,158	4.5	$ 2,026	$ 6,132		
	1931	184,037	14,470	7.9	2,204	12,266		
	1932	169,733	34,330	20.2	9,729	24,601		
	1933	147,740	31,150	21.1	19,301	11,849	$ 61,188	41.4
Farrell	1930	147,594	12,100	8.2	8,604	3,496		
	1931	155,832	25,758	16.5	6,477	19,281		
	1932	159,260	37,454	23.5		37,454		
	1933	141,070	50,280	35.6	20,305	29,975	93,252	66.1
Connellsville	1930	168,894	23,008	13.6	24,504			
	1931	171,272	54,766	32.0	25,634	29,132		
	1932	167,667	67,146	40.0	18,079	49,067		
	1933	166,859	81,842	49.0	779	81,063	196,282	117.6
Du Bois	1930	110,762	21,464	19.4	14,953	6,511		
	1931	106,991	25,403	23.7	17,449	7,954		
	1932	106,603	34,629	32.5	15,141	19,488		
	1933	106,781	40,540	38.0	18,631	21,909	70,818	66.3
Franklin	1930	100,025	10,721	10.7	7,015	3,706		
	1931	99,513	13,706	13.8	7,223	6,483		
	1932	99,455	19,504	19.6	10,322	9,182		
	1933	84,881	20,369	24.0	9,861	10,508	36,216	42.7
Lock Haven	1930	68,695	6,820	9.9	7,231			
	1931	69,665	7,878	11.3	5,972	1,906		
	1932	59,273	7,891	13.3	7,235	656		
	1933	59,437	10,328	17.4	4,014	6,314	14,634	24.6
Monongahela	1930	91,530	2,235	2.4	371	1,864		
	1931	93,637	6,977	7.5	1,389	5,588		
	1932	87,297	25,646	29.4	2,114	23,532		
	1933	81,131	31,114	38.4	5,808	25,306	22,246	27.4
Titusville	1930	100,198	2,607	2.6	3,816			
	1931	85,650	1,761	2.1	3,300			
	1932	90,685	12,186	13.4	8,908	3,278		
	1933	84,320	14,523	17.2	11,950	2,573	25,304	30.0
Corry	1930	79,532	10,639	13.4	8,219	2,420		
	1931	78,858	15,439	19.6	6,894	8,545		
	1932	78,682	15,440	19.6	6,902	8,538		
	1933	69,276	23,816	34.4	9,450	14,366	54,539	78.7

Not only have tax delinquencies increased each year in most cities but the cities have not been able to collect in one year unpaid taxes from previous years in an amount equal to their current delinquencies. This means that the total amount of unpaid taxes is increasing yearly. The extent to which this is true in each city can be seen from an examination of columns four and five in Table IV. The yearly increase in the total amount of outstanding taxes has been especially large in the Cities of Butler, Duquesne, Monessen, Uniontown, Pittston, Beaver Falls, Greensburg and Connellsville.

A great deal of difficulty is encountered when an attempt is made to ascertain the total amount of all outstanding taxes for any city. In some cities the officials do not have the information at hand and in a few cases only an audit of all city books for several years would show the true figure. Column five of Table IV shows the total a- mount of unpaid taxes for each city at the end of 1933 based on information taken from official reports and supplemented by data submitted by city controllers or treasurers. These figures should be considered as showing the minimum amounts only. An examination of them shows that in many cities a very considerable amount of money is outstanding. A few cities could operate on their present scale for over a year without levying any taxes if they could collect all back taxes due them. In fifteen of the twenty-six cities unpaid taxes are outstanding in amount equal to more than 50.0 per cent of the 1933 levy. In only one city (Lock Haven) is the amount less than 25.0 per cent while in Pittston, Connellsville and Beaver Falls it is in excess of 100.0 per cent.

In spite of the extent to which tax delinquency has been permitted to go, there seems very little to be done about it at the present time. The attitude of the Legis- lature, as well as that of most city and county officials, is that until there is more economic recovery it is useless, in fact cruel, to attempt to enforce the tax collection laws rigorously and thereby force thousands of people to lose their homes or even their personal property. The only immediate relief seems to lie in the direction of further economies in the performance of the many functions or even the elimination of some ser- vices for the time being.

A Comparison of Receipts for the Years 1930 and 1933 in Twenty-Three Cities:- In order to show the present trend in receipts from all sources, figures have been secured for twenty-three third class cities. Although as previously pointed out the chief source of income is from taxes, there are other sources of revenue from which 5.0 to 20.0 per cent of the total receipts are obtained. Such sources are special assessments, licenses and permits, fines and forfeits, earnings of departments, gifts and grants, interest, water rent in some cases and other miscellaneous items. In addition, there are the non-revenue receipts already described, i.e. the sale of bonds, temporary loans, refunds and joint constructions and the sale of real and personal property. In Table V, a comparison is made of the amount of money received by twenty-three cities from each of the important sources of revenue for the years 1930 and 1933. The total revenue re- ceipts (less water rent) for each city are shown in column 6, the total non-revenue re- ceipts in column 12, and the total of all receipts in column 13. Revenue receipts per capita are shown in column 7 and the per cent of decrease in revenue receipts in 1933 as compared with 1930 is shown in column 8.

The most significant thing shown by this table is the appalling decrease in re- venue receipts since 1930. For the year 1933, the four Cities of Sharon, Duquesne, Meadville and Corry received less than one-half as much money as they received in 1930. In eleven additional cities 1933 revenue receipts were from 25.0 to 50.0 per cent less than like receipts for 1930. For the City of Sunbury, the decrease was only 16.8 per cent and this was the smallest drop in any of the twenty-three cities. An examination of the table will show that the large falling off in total income from revenue occurred in the items of taxes and special assessments, and particularly the latter. In 1933 the twenty-three cities together collected taxes in the amount of $3,283,072 as compared with $4,385,185 in 1930. This was a decrease of 25.1 per cent. In 1930, they received on account of special assessments the sum of $978,781 while in 1933 they only received $81,310 from this source - a decrease of 91.7 per cent.

A majority of the cities received more from licenses and permits in 1933 than they did in 1930. For the twenty-three cities together this item of revenue increased 92.2 per cent. This increase was due mainly to the issue of licenses and permits for the

TABLE V RECEIPTS

City	Year	Taxes	Special Assessments	Licenses and Permits	Other Revenue Receipts	Water Rents	Revenue Receipts Less Water Rents			Sale of Bonds	Temporary Loans	Other Non-Revenue Receipts	Total Non-Revenue Receipts	Grand Total of Receipts
							Amount	Per Capita	Per Cent Decrease					
Nanticoke	1930	$244,899	$94,485	$2,982	$4,838	$0	$347,204	$13.33		$100,000	$60,000	$3,537	$163,537	$510,741
	1933	201,855	11,838	22,301	1,278	0	237,272	8.76	31.6	0	0	12,024	12,024	249,296
Sharon	1930	307,935	65,013	2,213	60,957	0	436,118	16.83		50,000	56,023	12,846	118,869	554,987
	1933	181,104	9,285	6,286	6,538	0	203,313	7.49	53.4	90,000	45,025	354	135,379	388,692
Lebanon	1930	245,666	49,672	6,532	34,061	88,862	335,931	13.14		0	65,000	4,380	69,380	494,173
	1933	226,231	484	10,994	10,598	60,664	248,307	9.61	26.1	0	18,500	9,851	28,351	337,322
Washington	1930	278,688	30,575	6,593	19,284	0	335,140	13.65		0	0	294	294	335,434
	1933	240,392	4,406	9,515	17,508	0	271,921	10.68	18.9	0	8,500	1,710	10,210	282,131
Pottsville	1930	257,420	2,454	4,693	6,606	0	271,173	11.16		47,039	16,000	90,733	153,772	424,945
	1933	195,422	924	14,460	1,969	0	212,775	8.50	21.5	30,000	52,500	3,435	85,935	298,710
Oil City	1930	261,979	84,453	19,815	21,283	115,697	387,530	17.56		711	30,000	751	31,462	534,689
	1933	246,295	8,739	16,752	18,520	101,440	290,406	13.01	25.1	0	0	1,057	1,057	392,903
Duquesne	1930	326,250	31,797	5,091	23,104	72,878	386,242	18.05		600,000	0	11,851	611,851	1,070,971
	1933	141,976	3,668	6,832	3,025	30,418	155,501	7.03	59.7	0	117,854	916	118,770	304,689
Monessen	1930	240,384	20,556	2,475	11,793	0	275,208	13.58		0	35,000	1,248	36,248	311,456
	1933	158,698	8	8,523	1,080	0	168,309	8.05	38.8	0	60,000	101	60,101	228,410
Bradford	1930	210,991	54,587	15,457	11,493	98,389	292,528	15.15		148,949	2,231	11,270	162,450	563,367
	1933	195,562	15,336	17,339	7,415	95,365	235,652	11.53	19.4	65,000	27,000	2,949	94,949	425,966
Pittston	1930	208,398	4,016	993	3,092	0	216,499	11.87		0	0	65	65	216,564
	1933	145,260	78	10,732	1,589	0	157,659	8.68	27.2	0	0	7	7	157,666
Beaver Falls	1930	221,828	1,709	549	9,424	0	233,510	13.62		0	82,000	0	82,000	315,510
	1933	127,709	0	6,496	4,198	0	138,403	7.81	40.7	0	0	0	0	138,403
Meadville	1930	195,693	352,509	1,341	11,723	86,659	561,266	33.61		0	22,300	197	22,497	670,422
	1933	152,605	910	5,090	9,049	65,025	167,654	9.67	70.1	0	39,000	177	39,177	271,856
Greensburg	1930	218,408	11,799	705	16,702	0	247,614	15.00		668,290	0	9,540	677,824	925,438
	1933	177,093	1,625	6,770	5,497	0	190,985	11.27	22.9	0	0	52	52	191,097

TABLE V RECEIPTS (Cont'd)

City	Year	REVENUE RECEIPTS					Revenue Receipts Less Water Rents			NON-REVENUE RECEIPTS				Grand Total of Receipts
		Taxes	Special Assessments	Licenses and Permits	Other Revenue Receipts	Water Rents	Amount	Per Capita	Per Cent Decrease	Sale of Bonds	Temporary Loans	Other Non-Revenue Receipts	Total Non-Revenue Receipts	
Sunbury	1930	$92,819	$2,258	$12,092	$4,475	0	$111,644	$7.14		$20,000	$5,000	$26,792	$51,792	$163,436
	1933	78,342	517	12,394	1,648	0	92,901	5.96	16.8	0	8,000	1,336	9,336	102,237
Clairton	1930	231,896	40,040	0	18,233	0	290,169	18.98		0	0	0	0	290,169
	1933	169,127	4,776	5,748	14,383	0	194,084	11.66	33.1	0	0	1,750	1,750	195,784
Coatesville	1930	174,690	12,439	3,195	5,789	72,261	196,113	13.45		30,147	22,962	13,236	66,345	334,719
	1933	135,457	1,048	5,270	5,730	52,106	147,505	10.10	24.8	17,106	7,500	242	24,848	224,459
Farrell	1930	142,886	13,119	698	8,670	0	165,373	11.52		0	45,000	130	45,130	210,503
	1933	109,905	2,350	5,216	3,388	0	120,854	8.64	26.9	0	0	589	589	121,443
Du Bois	1930	101,785	58,620	1,190	22,784	46,900	184,379	15.90		0	55,000	5,980	60,980	292,259
	1933	82,460	2,048	4,910	5,447	41,939	94,865	8.65	48.3	0	0	445	445	137,249
Franklin	1930	96,672	3,604	201	9,228	81,833	109,705	10.70		40,000	0	16,000	56,000	247,538
	1933	74,028	2,069	3,018	5,815	71,645	84,930	8.22	22.6	0	0	22	22	156,597
Lock Haven	1930	67,631	5,317	1,382	30,276	55,727	104,606	10.82		10,000	12,528	58	22,586	182,919
	1933	51,085	5,947	3,321	3,152	49,077	63,505	6.35	39.3	0	0	2,125	2,125	114,707
Monongahela	1930	89,382	51	7,300	6,587	0	103,320	11.91		0	3,258	5,548	8,806	112,126
	1933	56,509	0	6,249	3,596	0	66,354	7.65	35.8	0	0	0	0	66,354
Titusville	1930	92,941	20,998	4,939	8,563	33,799	127,441	15.82		46,000	0	879	46,879	208,119
	1933	81,759	5,225	5,277	5,285	30,467	97,546	12.28	23.5	0	0	185	185	128,198
Corry	1930	75,944	18,710	1,337	43,870	0	139,861	19.56		0	0	448	448	140,309
	1933	54,198	29	2,149	3,197	0	59,573	8.36	57.4	0	0	190	190	59,763

sale of beer during the last half of 1933. The increase in this form of revenue is most noticeable in the Cities of Nanticoke, Pottsville, Pittston, Monessen, Sharon, Greensburg, Farrell and a few other places where there is a large foreign population and where heavy industries are located. At the same time however there was a substantial reduction in the amount received from fines and forfeits. For the twenty-three cities the decrease amounted to 60.1 per cent.

There are also noteworthy features of the non-revenue receipts. In twelve of the twenty-three cities for which information was secured for both 1930 and 1933 some money was obtained in 1930 from the sale of bonds. In the Cities of Duquesne and Greensburg bonds were sold in excess of half a million dollars. In eight of these twelve cities, no bonds were issued in 1933. In eleven additional cities there were no bonds sold in either 1930 or 1933. Temporary loans were obtained in 1930 by fifteen of the twenty-three cities and eight of these found it necessary to secure loans again in 1933. These loans ranged in amounts from a few thousand to over one hundred thousand dollars and usually were for the purpose of meeting payrolls falling due prior to the annual collection of taxes. Only six of the twenty-three cities were able to get along in both years without such loans.

With such substantial falling off in revenue, many cities are having considerable difficulty in supplying usual or even necessary services. In some instances, they are only able to get along by floating temporary loans or by issuing bonds against uncollected taxes. If the cost of operating the schools (which usually amounts to from 40.0 to 60.0 per cent of the total local tax burden) could be shifted in whole or in part from real estate there is reason to believe that the other taxing districts would be able to collect enough taxes from real estate and other sources to perform their respective functions. If the burden on real estate is not lessened, then some new source of income for local governments must be found or many services now demanded by the people must be abandoned.

Bonded Indebtedness:- One of the causes of the present financial embarrassment of many cities is the amount of outstanding bonded indebtedness on which payments of principal and interest are regularly required. In past years, when the future looked bright, many cities borrowed large sums of money to build streets, public buildings, water works, sewers, playgrounds, fire houses, etc. At the time such loans were secured, the burden as spread over twenty or thirty years did not look heavy. Today, however, with revenues sharply reduced, and yearly more difficult to collect, some cities find that previously acquired fixed charges make it all but impossible to adjust income to expenditures.

In addition to the provisions contained in the Constitution relating to the creation of debts by any municipality, the third class city code specifically deals with debt limitations. The law provides that a city council may borrow money on the credit of the city and pledge the revenue thereof for the payment of the same if the amount involved, based on the assessed valuation of the taxable property located in the city, is kept within certain limitations, as follows:-

1 - Not exceeding two per centum of such value if the loan is by councilmanic action solely.
2 - Not exceeding seven per centum of such value if the loan is with the consent of a majority of the voters.
3 - Not exceeding ten per centum of such value if the loan is approved by three-fifths of the electors voting on such proposed indebtedness. (1931 P. L. 932, Sec. 2702)

Council is also authorized to issue bonds for the purpose of funding any existing indebtedness. Such bonds if issued must be payable in not less than one year or more than twenty years. At one time, bonds to renew or extend a debt could be issued to run for an additional thirty years but due to the abuse of such provisions by certain municipalities the time limit was lowered to twenty years. Much of the present bonded debt in some cities is in the form of funding bonds extending debts which should have been paid within the time limits of the original indebtedness. Many bonds now maturing must be refunded because the financial condition of many cities is such that the money necessary to redeem them is not available.

State Supervision in Connection with the Issue of Municipal Bonds:- Whenever any county, third class city, borough, township, poor district, or school district of the second, third and fourth class, wishes to incur a debt or increase its existing indebtedness by the issuance of bonds or other evidence of indebtedness (except temporary loans for less than one year in anticipation of revenue) it is necessary to secure the approval of the State Department of Internal Affairs. (1927 P. L. 91) This department must be furnished with a copy of the ordinance and other proceedings had by the municipality in question, together with a statement showing the assessed valuation of taxable property, the total amount of existing indebtedness and the several amounts claimed as permitted deductions in ascertaining the real indebtedness existing at the time. These deductions are:- (1) All delinquent taxes; (2) solvent debts due and collectable; (3) revenue due within the year; (4) cash in the treasury not marked for any particular purpose; and (5) cash securities in the sinking fund.

If upon examination of all the record, the department finds that the proposed debt is not within the limitations imposed by the Constitution or that the proceedings are not in accordance with the law governing such matters, then the department must disapprove the proposed debt and notify the municipality of that fact. At the present time about 50 per cent of the applications are rejected when first presented because of improper procedure or the failure to submit the necessary data in the proper form. If, however, full corrections are made, the department issues its certificate of approval and the bonds may be sold, otherwise not. It should be noted that this department has no discretion in regard to the questions whether the money should be borrowed - or for what purpose it should be used - or as to the amount involved - its sole authority has to do with the regularity of the proceedings and the compliance with the Constitution and laws of the Commonwealth.

The department also has the authority to investigate the sinking funds of the various municipalities covered by this act, as well as all records, books and papers relating to such funds. If it finds that proper funds have not been created, or if created, have not been properly managed, or if taxes have not been levied and collected as required by law to pay the interest and principal on existing bonds, then it becomes its duty to order the municipality to create such fund, or provide sufficient revenue for an existing fund. While the law granting the department this authority was enacted in 1927, the department policy so far has been to act mainly in an advisory capacity. Some investigations of sinking funds have been made and irregularities corrected, but the department has not been aggressive along such lines. Now that economic conditions have caused many sinking funds to get out of balance and in need of investigation and supervision, the department has not been provided by the Legislature with sufficient money or personnel to undertake this important work and therefore practically nothing is being done in the matter.

The Amount of Overlapping Bonded Debt in Certain Third Class Cities:- Since every separate taxing authority within a political subdivision of the State is considered as a municipality or quasi-municipality for the purpose of incurring debt, it follows that each may incur separately a debt not exceeding the Constitutional limitations. A city for instance may borrow up to seven per cent of the valuation of taxable property and the school district within the city may do the same thing - and if there is a separate poor district within the city it may do likewise and, of course, the county in which the city is located may also borrow to the same limit. If all four taxing units borrowed to their capacity, it would mean that debts could be incurred in an amount equal to 28 per cent of the value of taxable property within the city. If the third class city code is constitutional in regard to the ten per cent debt limit previously referred to, then the total debt of all taxing districts in a city might equal 31 per cent of the valuation.

An attempt was made to obtain full and accurate information as to net bonded indebtedness of twenty-seven cities. The results are shown in Table VI. The figures were secured either from the reports of the director of accounts and finance, from personal correspondence, or from records compiled by the Department of Internal Affairs. Only city bonds outstanding December 31, 1933 are included. School bonds are included as of the end of the 1932-1933 fiscal year. In the case of county bonds information was not obtainable down to the end of 1933 for every county and it was necessary to take the

TABLE VI

NET BONDED INDEBTEDNESS AT THE END OF 1933
IN CERTAIN THIRD CLASS CITIES

City	Population 1933	City Bonded Debt (1)	School Bonded Debt	County Bonded Debt	Total Bonded Debt	Assessed Valuation of Taxable Real Estate for 1934 (3)	Total Bonded Debt as a percentage of Real Estate Valuation	Per Capita Total Bonded Debt
Nanticoke	27,072	220,669	411,951	174,007*	806,627	12,689,219	6.4	$29.80
Sharon	27,156	670,528	1,078,674	770,820*	2,520,022	18,294,775	13.8	92.80
Lebanon	25,837	528,200	375,000	200,101*	1,103,301	24,413,756	4.5	42.70
Washington	25,466	767,141	1,259,000	431,420	2,457,561	23,405,756	10.5	96.50
Pottsville	25,026	557,650	920,841	241,572*	1,720,063	16,578,086	10.4	68.73
Butler	23,505	1,084,142	305,729	776,040*	2,165,911	29,247,120	7.4	92.15
Oil City	22,315	733,936	720,000	564,688	2,018,624	15,379,230	13.1	90.46
Duquesne	22,113	1,370,833	613,291	1,110,131*	3,094,255	25,164,850	12.3	139.93
Monessen	20,895	746,696	908,037	311,460*	1,966,193	15,238,875	12.9	94.10
Carbondale	20,487	422,008	539,875	90,826*	1,052,709	10,859,814	9.7	51.38
Uniontown	20,699	298,585	928,500 (2)	458,499*	1,685,584	20,737,875	8.1	81.43
Bradford	20,440	487,805	337,428	123,374*	968,607	11,967,740	8.1	47.39
Pittston	18,171	209,542	216,982	121,804*	548,328	10,870,510	5.0	30.18
Beaver Falls	17,717	420,100	796,446	619,500	1,836,046	12,664,860	14.5	103.63
Meadville	17,337	537,148	250,225	419,814*	1,207,187	10,380,610	11.6	69.63
Greensburg	16,952	816,026	962,174	286,291*	2,064,491	14,611,780	14.1	121.78
Clairton	16,644	553,444	974,742	1,295,153*	2,823,339	25,951,870	10.9	169.63
Sunbury	15,599	89,769	346,627	80,645	517,041	6,089,000	8.5	33.15
Coatesville	14,603	1,066,906	244,730	43,328*	1,354,964	13,348,840	10.2	92.79
Farrell	13,990	442,331	481,984	444,414*	1,368,729	10,131,270	13.5	97.84
Connellsville	13,137	408,740	102,587	239,074*	750,401	7,474,590	10.0	57.12
Du Bois	10,968	294,808	106,497	215,695*	617,000	4,505,551	13.7	56.25
Franklin	10,338	231,591	18,000	295,177	544,768	6,768,350	8.0	52.70
Lock Haven	10,001	224,524	262,854	None	487,378	4,882,235	10.0	48.73
Monongahela	8,672	229,990	365,225	134,310	729,525	9,535,681	7.7	84.12
Titusville	7,941	267,329	55,368	225,594*	548,291	5,870,143	9.3	69.05
Corry	7,128	65,300	138,428	123,300	327,028	4,394,250	7.4	45.88

* Net Bonded Indebtedness at the end of 1932.
(1) Includes Special Assessment Bonds.
(2) As of July 1, 1933.
(3) Assessment made by City Assessor.

bonded indebtedness at the close of 1932 for those cities marked with an asterisk. In regard to these county bonds an explanation is necessary. When a county borrows money and issues bonds therefor, the debt created is the debt of the entire county and it is not in any way divided or apportioned as between the cities, boroughs or townships of the county. When, however, a tax is levied by the county to redeem these bonds or for general purposes, the tax rate in mills is the same for all units within the county so that each city, borough or township pays its share of the county bonded debt as represented by its proportion of the total assessed valuation of all taxable property within the county.

If there are any outstanding bonds sold by separate poor districts within these cities, such bonds have not been included in this table. The assessed valuations used in Table VI are those made by the city assessors for the year 1934 and are on the whole higher than the valuations placed on the same property by the county assessors. Table VI not only shows the amount of bonded debt (less sinking fund requirements) of each of the taxing units in the city, but the ratio of such debt to the assessed valuation of the taxable real estate located within the city. The highest ratio of bonded debt to property values is found in the City of Beaver Falls where the total of such debts amounts to 14.5 per cent of the taxable property. The average bonded debt for all twenty-seven cities amounts to 10.4 per cent of the average assessed valuation of real estate.

When the per capita burden of indebtedness is considered, one finds wide variations between the different cities. In Nanticoke, for instance, total bonded debts amount to only $29.80 for each person, whereas in Clairton, such debts amount to $169.63 per capita. In Duquesne, Greensburg and Beaver Falls, the per capita debt is also in excess of $100.00. In nine additional cities, there is a debt burden of over $75.00 per person, while in only seven of the twenty-seven cities does it amount to less than $50.00. The average per capita bonded debt for all of the twenty-seven cities studied is $78.77. The figures presented here are significant in that they show the extent to which the various taxing districts are either unwilling or unable to finance their affairs out of current revenue.

The data thus far presented on indebtedness have dealt with the bonded debt incurred by each of the taxing units within a city. The officials of the city government are not responsible for the incurring of debts by the school board or board of county commissioners. The facts concerning these units have been included so that the reader may see the total burden of all overlapping bonded indebtedness placed on the same real estate by the separate action of the different taxing authorities.

The Amount of City Contracted Debt Outstanding December 31, 1933:- Table VII shows the total amount of outstanding debt incurred exclusively by the city in the operation and maintenance of its governmental and allied functions.

This table shows that at the end of 1933 several of the cities owed substantial sums of money in addition to their outstanding bonds. Considering all the cities it will be seen that the per capita total city debt ranges from $6.97 in Sunbury to $73.06 in Coatesville.

Need for Some Centralized Control Over the Creation of Bonded Debt:- With several separate tax levying authorities each permitted to impose fixed debts on the same taxpayer and with laws which allow such debts to amount to from 28 to 31 per cent of the assessed valuations of taxable real estate, it is highly desirable to provide some form of check or control over the actions of such authorities. In Indiana, the State Tax Commission is empowered to veto local bond issues upon the appeal of a fixed number of local taxpayers. Pennsylvania has no State Tax Commission or like body but it would be possible to create such a body with some control over the question of assessments, tax levies and the issuance of bonds. Such a commission could deal with the question of bond issue only, but since there is such a close relation between assessments, tax rates and bond issues, all three problems should be considered together. Since there

TABLE VII
TOTAL CITY INDEBTEDNESS AS OF DECEMBER 31, 1933

City	Regular Bonds	Special Assessment Bonds	Judgments, Temporary Loans, etc.	Total of All Bonds Loans, etc.	Total of All Bonds, Loans, etc. Per Capita	Outstanding Warrants and Unpaid Current Bills Dec. 31, 1933
Nanticoke	$ 220,669	$	$ 171,242	$ 391,911	$ 14.48	$ 90,486
Sharon	670,528		27,925	698,453	25.72	
Lebanon	528,200 ①			528,200	20.44	6,981
Washington	654,141	113,000	11,322	778,463	30.57	31,541
Pottsville	557,650		27,500	585,150	23.38	
Butler	1,084,142			1,084,142	46.12	
Oil City	666,504 ②	67,432		733,936	32.89	
Duquesne	1,370,833 ③		50,000	1,420,833	64.25	
Monessen	746,696		61,585	808,281	38.68	27,728
Carbondale	422,008			422,008	20.60	
Uniontown	298,585		12,400	310,985	15.02	25,497
Bradford	430,305	57,500		487,805	23.87	24,904
Pittston	209,542		172,342	381,884	21.02	38,422
Beaver Falls	420,100			420,100	23.71	26,210
Meadville	192,148	345,000		537,148	30.98	
Greensburg	816,026			816,026	48.14	
Sunbury	89,769		19,000	108,769	6.97	
Clairton	471,444	82,000		553,444	33.25	5,582
Coatesville	1,066,906 ④			1,066,906	73.06	
Farrell	442,331			442,331	31.62	
Connellsville	372,240	36,500		408,740	31.11	
Du Bois	259,500 ⑤	35,308		294,808	26.88	
Franklin	231,591 ⑥			231,591	22.40	
Lock Haven	224,524 ⑦			224,524	22.45	1,724
Monongahela	229,990			229,990	26.52	15,658
Titusville	195,329	72,000		267,329	33.66	
Corry	65,300			65,300	9.16	14,634

① Includes $64,800 sewer and paving bonds.
② Includes $154,000 water works bonds.
③ May include some assessment bonds.
④ Includes $327,700 water works bonds.
⑤ Includes $141,000 water works bonds.
⑥ Includes $101,257 water works bonds.
⑦ Includes $152,000 water works bonds.

are over 5,000 separate taxing districts in this State, it would be impossible for any one board or commission to pass annually on such matters as they pertained to each individual district, but this difficulty could be overcome by the creation of regional or even county boards under the control or supervision of the State Commission. A majority of the cases could possibly be disposed of by the local boards with the right of appeal to the State Commission which would also act originally in certain exceptional types of cases.

The New York "Commission for the Revision of the Tax Laws" has recommended that each political subdivision be permitted to tax and bond itself within certain fixed limits but not in excess of such limits without the consent of a state board or commission. This would mean that the state board would not have to pass annually on the tax levies or bond issues of each taxing unit but only on those going beyond certain limits to be fixed by law. If the separate taxing districts within a single political unit cannot agree as to how to stay within the fixed limits, then the matter would be turned over to the state commission for final decision. Such a procedure has the advantage of leaving the responsibility in the first instance entirely with the local authorities, the State interfering only when such authorities fail to agree. (See also Wm. Anderson "The Reorganization of Local Government in Minnesota.)

Pennsylvania should adopt some such plan and should materially reduce the present debt limits which permit within a single governmental unit a total overlapping bonded debt of between 28 and 31 per cent of the assessed value of taxable property. It should be recalled however that at present no political unit can be bonded in excess of 8.0 per cent of its valuation, (2.0 per cent for each taxing district) without a vote of the people. In units without a separate poor district, the above limit would be only 6.0 per cent. While this limit appears reasonable, the trouble is that in the past the political organization of the local unit has not had much difficulty in getting a sufficient number of people to vote in favor of large bond issues, and if the people approve, total debts may be incurred in excess of 28 per cent of the value of their property.

During the last ten years the local taxing authorities have increased their indebtedness on the assumption of constantly increasing population and rising property values. They have generally insisted that the amount of money borrowed and the use to which it was put were matters of only local concern and that the State should not have any control or supervision over their action. Today things are different and the same officials are now demanding that the State relieve them of some of the burden by appropriating money for poor relief, roads and schools. If the State is called upon to pay a part of the cost of such services it is not only right but necessary that the State exercise some control or supervision over local expenditures and especially those involving long term obligations.

<u>EXPENDITURES.</u>

   <u>Total Operating Cost of Certain Cities Compared</u> - The extent to which cities have been able to reduce their total operating costs since 1930 is of paramount interest and importance. Column 2 of Table VIII, shows the per cent of increase or decrease from 1930 to 1933 in the cost of all general departments, exclusive of debt service, permanent improvements, temporary loans and public service enterprises. In a period of three years, twenty-three cities taken together, reduced their general expenses by 25.4 per cent. The average per capita cost for general services in 1933 was $6.85. Costs per person were highest in Bradford ($9.34) and lowest in Sunbury ($5.32).

   <u>Cost of Debt Service</u>:- Many cities find it difficult to reduce their expenditures materially because such a large part of every tax dollar must be set aside to meet fixed charges. The most important of these charges are redemption of bonds, and interest and State Tax on bonds. The cost of debt service for twenty-two cities averaged 30.6 per cent of the total operating cost for 1933. The average cost of debt service per capita was $3.01 and the range was from $0.33 in Sunbury to $7.61 in Bradford. (See Table IX)

   <u>A Comparison of the Salaries paid to Principal Officers in 1930 and 1934</u>:- From

TABLE VIII

TOTAL OPERATING COST OF TWENTY-THREE THIRD CLASS CITIES, - 1930 AND 1933; PER CENT OF CHANGE IN SUCH COSTS 1933 OVER 1930; AND PER CAPITA OPERATING COST, - 1933

City	Population	Year	Total Cost of Operating All Departments Exclusive of Debt Service for 1930 & 1933	Per Cent of Increase or Decrease in the Cost of Operating All Departments Exclusive of Debt Service for 1933 over 1930	Per Capita Cost of Operating All Departments Exclusive of Debt Service for 1933	Total Cost of Operating All Departments Including Debt Service for 1933 & 1930	Per Cent of Increase or Decrease in the Cost of Operating All Departments Including Debt Service for 1933 over 1930	Per Capita Cost of Operating All Departments Including Debt Service for 1933
			1	2	3	4	5	6
Nanticoke	26,043 / 27,072	1930 / 1933	$230,701 / 152,826	-33.8	$8.86 / 5.65	$275,750 / 219,248	-20.5	$10.59 / 8.10
Sharon	25,908 / 27,156	1930 / 1933	250,207 / 175,001	-30.1	9.66 / 6.44	348,956 / 218,647	-37.3	13.47 / 8.05
Lebanon	25,561 / 25,837	1930 / 1933	217,336 / 183,482	-15.6	8.50 / 7.10	342,840 / 268,332	-21.7	13.41 / 10.39
Washington	24,545 / 25,466	1930 / 1933	232,387 / 176,764	-23.9	9.47 / 6.94	343,300 / 291,058	-15.2	13.99 / 11.43
Pottsville	24,300 / 25,026	1930 / 1933	185,273 / 164,119	-11.4	7.62 / 6.56	288,622 / 210,992	-26.9	11.88 / 8.43
Oil City	22,075 / 22,315	1930 / 1933	196,505 / 184,336	-6.2	8.90 / 8.26	278,726 / 241,552	-13.3	12.63 / 10.82
Duquesne	21,396 / 22,113	1930 / 1933	314,799 / 119,300	-62.1	14.71 / 5.40	407,350 / 237,707	-41.6	19.04 / 10.75
Monessen	20,268 / 20,895	1930 / 1933	165,600 / 113,544	-31.4	8.17 / 5.43	230,164 / 177,493	-22.9	11.36 / 8.49
Bradford	19,306 / 20,440	1930 / 1933	217,932 / 191,000	-12.4	11.29 / 9.34	345,675 / 346,464	+ 0.2	17.91 / 16.95
Pittston	18,246 / 18,171	1930 / 1933	150,811 / 140,749	- 6.7	8.27 / 7.75	212,694 / 183,859	-13.6	11.66 / 10.12

TABLE VIII (Cont'd.)

			1	2	3	4	5	6
Beaver Falls	17,147 17,717	1930 1933	$155,834 104,486	-33.0	$ 9.09 5.90	$204,054 164,994	-19.1	$11.90 9.31
Meadville	16,698 17,337	1930 1933	213,156 132,055	-38.0	12.77 7.62	238,466 160,717	-32.6	14.28 9.27
Greensburg	16,508 16,952	1930 1933	174,932 140,464	-19.7	10.60 8.29	242,621 187,005	-22.9	14.70 11.03
Sunbury	15,626 15,599	1930 1933	95,718 83,001	-13.3	6.13 5.32	109,752 88,181	-19.7	7.02 5.65
Clairton	15,291 16,644	1930 1933	211,141 134,577	-36.3	13.81 8.09	261,371 202,458	-22.5	17.09 12.16
Coatesville	14,582 14,603	1930 1933	142,327 96,582	-32.1	9.76 6.61	226,581 168,763	-25.5	15.54 11.56
Farrell	14,359 13,990	1930 1933	111,543 88,223	-20.9	7.77 6.31	120,995 118,014	- 2.5	8.43 8.44
DuBois	11,595 10,968	1930 1933	84,664 79,416	- 6.2	7.30 7.24	112,725 104,918	- 6.9	9.72 9.57
Franklin	10,254 10,338	1930 1933	77,237 63,150	-18.2	7.53 6.11	99,615 97,892	- 1.7	9.71 9.47
Lock Haven	9,668 10,001	1930 1933	67,734 75,784	+11.9	7.01 7.58	94,544 89,761	- 5.1	9.78 8.98
Monongahela	8,675 8,672	1930 1933	63,202 46,789	-26.0	7.29 5.40	104,925 58,406	-44.3	12.10 6.74
Titusville	8,055 7,941	1930 1933	70,998 61,395	-13.5	8.81 7.73	148,256 99,186	-33.1	18.41 12.49
Corry	7,152 7,128	1930 1933	65,791 50,764	-22.8	9.20 7.12	81,154 61,205	-24.6	11.35 8.59
Average of 23 cities	17,495	1930 1933	$160,688 $119,904	-25.4	$ 6.85	$222,571 $173,776	-21.9	$ 9.93

TABLE IX

COMPARATIVE COST OF DEBT SERVICE IN TWENTY-TWO THIRD CLASS CITIES, - 1930 AND 1933; PER CENT OF CHANGE, 1933 OVER 1930; PER CAPITA COST, - 1933; AND RATIO OF DEBT COST TO TOTAL OPERATING COST INCLUDING DEBT SERVICE, 1933

City	Population	Year	Cost of Debt Service for 1930 & 1933 ①	Per Cent of Increase or Decrease in Cost of Debt Service for 1933 over 1930	Per Capita Cost of Debt Service for 1933	Total Cost of Operating All Departments ② Including Debt Service for 1933	Cost of Debt Service as a Percentage of Total Operating Cost for 1933
Nanticoke	26,043 27,072	1930 1933	$ 45,049 66,422	+47.4	$2.45	$219,248	30.3
Sharon	25,908 27,156	1930 1933	98,749 43,646	-55.8	1.61	218,647	20.0
Lebanon	25,561 25,837	1930 1933	125,504 84,850	-32.4	3.28	268,332	31.6
Washington	24,545 25,466	1930 1933	110,913 114,294	+ 3.0	4.49	291,058	39.3
Pottsville	24,300 25,026	1930 1933	103,349 46,873	-54.6	1.87	210,992	22.2
Oil City	22,075 22,315	1930 1933	82,221 57,216	-30.4	2.56	241,552	23.7
Duquesne	21,396 22,113	1930 1933	92,551 118,403	+27.9	5.35	237,707	49.8
Monessen	20,268 20,895	1930 1933	64,564 63,949	- 1.0	3.06	177,493	36.0
Bradford	19,306 20,440	1930 1933	127,743 155,464	+21.7	7.61	346,464	44.9
Pittston	18,246 18,171	1930 1933	61,883 43,110	-30.3	2.37	183,859	23.5
Beaver Falls	17,147 17,717	1930 1933	48,220 60,508	+25.5	3.42	164,994	36.7
Meadville	16,698 17,337	1930 1933	25,310 28,662	+13.2	1.65	160,717	17.8
Greensburg	16,508 16,952	1930 1933	67,689 46,541	-31.2	2.75	187,005	24.9

TABLE IX (Cont'd.)

COMPARATIVE COST OF DEBT SERVICE IN TWENTY-TWO THIRD CLASS CITIES, - 1930 AND 1933; PER CENT OF CHANGE, 1933 OVER 1930; PER CAPITA COST, - 1933; AND RATIO OF DEBT COST TO TOTAL OPERATING COST INCLUDING DEBT SERVICE, 1933

City	Population	Year	Cost of Debt Service for 1930 & 1933 ①	Per Cent of Increase or Decrease in Cost of Debt Service for 1933 over 1930	Per Capita Cost of Debt Service for 1933	Total Cost of Operating All Departments ② Including Debt Service for 1933	Cost of Debt Service as a Percentage of Total Operating Cost for 1933
Sunbury	15,626 / 15,599	1930 / 1933	$ 14,034 / 5,180	-63.1	$ .33	$ 88,181	5.9
Coatesville	14,582 / 14,603	1930 / 1933	84,254 / 72,181	-14.3	4.94	168,763	42.8
Farrell	14,359 / 13,990	1930 / 1933	9,452 / 29,791	+215.2*	2.13	118,014	25.2
Du Bois	11,595 / 10,968	1930 / 1933	28,061 / 25,502	- 9.1	2.33	104,918	24.3
Franklin	10,254 / 10,338	1930 / 1933	22,378 / 34,742	+55.3	3.36	97,892	35.5
Lock Haven	9,668 / 10,001	1930 / 1933	26,810 / 13,977	-47.9	1.40	89,761	15.6
Monongahela	8,675 / 8,672	1930 / 1933	41,723 / 11,617	-72.2	1.34	58,406	19.9
Titusville	8,055 / 7,941	1930 / 1933	77,258 / 37,791	-51.1	4.76	99,186	38.1
Corry	7,152 / 7,128	1930 / 1933	15,363 / 10,441	-32.0	1.46	61,205	17.1
Average of 22 Cities	17,533	1930 / 1933	62,412 / 52,780	-15.4	3.01	172,472	30.6

* Farrell was a borough in 1930.
① Includes payment of bonds, sinking fund charges, interest and State Tax on bonds.
② Does not include permanent improvements, temporary loans, public service enterprises or non-governmental expenses.

1930 to 1934 the total salaries paid to the principal officers in five cities rose by increases ranging from 0.5 to 13.3 per cent and in twenty-two cities decreased from 0.1 per cent to 42.4 per cent. The Cities of Lebanon, Pottsville, Carbondale, Uniontown, Pittston and Meadville continue to pay the same or even higher salaries than they paid in 1930 in spite of the financial difficulties which some of them are facing. (See Table X) In five cities the mayor is paid the lowest salary possible under the code ($500), and in several other cities the sum paid is only slightly higher. In six cities members of council receive $300 which is only $50 in excess of the minimum fixed by statute, and in five additional cities they receive less than $500 annually. In half the cities studied, the salary of the clerk, who is a full time employee, has been reduced since 1930. The compensation of the solicitors, the assessor and the engineer has been very substantially reduced in several cities.

The Cost of Operating General Administrative Offices:- A very large part of the general administrative work involved in conducting the affairs of a city is distributed among the following offices:- Mayor and Council, Clerk, Controller, Treasurer, Assessor and Solicitor. While the financial reports of most cities are not prepared in such a way as to make possible a comparison of the costs incurred by each of these offices separately, it is possible to compare costs for the offices as a group. Included under the heading "Cost of Operating General Offices" are such items as:- Salaries of officials, assistants, clerks and stenographers; printing; advertising; postage; telephone; insurance; premiums on bonds; and office supplies and equipment. Many cities were able to reduce administrative expenses during the last three years - the average reduction amounting to 14.1 per cent. Office expenses for all the cities averaged 8.5 per cent of the total operating cost for 1933. The offices now being considered are not as a rule overmanned. Several of the cities employ one full-time secretary or stenographer whose time is distributed between the different offices according to current need. Nearly every city recognizes the economy in having as few full-time employees as possible and engaging extra part-time help when necessary (See Table XI).

The Cost of Police Protection:- One of the primary functions of local government is to protect persons and property and maintain law and order. The laws governing third class cities make police protection a mandatory function but each city is permitted to decide for itself what such protection shall consist of and how much money shall be spent in securing it. The principal item of police cost is salaries. Many of the cities have a captain or lieutenant or both, and several of the larger ones have three desk sergeants - each working on an eight hour shift in order to keep the police stations open continually. Ordinary patrolmen are paid from $100 to $150 per month while captains and lieutenants receive from $5 to $10 more. Only five cities have a policeman per 1000 population while four of them have only one policeman for every 2000 people. The police staff in several cities is inadequate and in many cases the police department lacks essential equipment. Twenty of the twenty-two cities reduced the cost of police protection between 1930 and 1933 - the average reduction for the twenty-two cities was 29.0 per cent. The average per capita cost was $1.17. (See Table XII.)

The Cost of Fire Protection:- While fire protection is not a mandatory function, it is nevertheless one of great importance. In several places it is furnished by volunteer organizations. In a majority of these cases, however, a number of drivers are employed on a full-time basis and in some instances the city contributes a certain amount of money annually to assist each volunteer fire company. Nine cities have a regular full-time force and as a rule such firemen are paid on about the same basis as patrolmen. In several places a skeleton full-time personnel is maintained and part-time firemen are paid for each hour of actual service. One of the principal costs of fire protection is the annual charge made by water companies for supplying fire hydrants and water. The Cities of Lebanon, Oil City, Duquesne, Bradford, Meadville, Coatesville, DuBois, Franklin, Lock Haven, Titusville operate their own water works and own the fire hydrants. The average reduction in the cost of fire protection between 1930 and 1933 was 9.5 per cent. The average per capita cost was $1.21. (See Table XIII)

The Cost of Public Health:- Included in the cost of public health are such items as food inspection, health officers, hospitals, comfort stations, sewage disposal, collection and disposal of garbage, rubbish and ashes, and street cleaning. Eight of the twenty-seven cities included in this study operate sewage disposal plants while

TABLE X

SALARIES OF THE PRINCIPAL OFFICIALS OF TWENTY-SEVEN THIRD CLASS CITIES, – 1930 AND 1934

City	Year	Mayor	Councilmen (Four) ②	Controller	Treasurer ③	Clerk	Solicitor	Assessor	Engineer	Chief of Police	Fire Chief	Health Officer	Supt. of Streets	Miscellaneous ⑤	Total Salaries ⑥	Per Cent Increase or Decrease	Salary Cost Per Capita
Nanticoke	1930	$1200	$4400	$1100	$1000	$1800	$2000	$2100	$1800	$2391	$1000	$1800	$2000	$3060	$24,651		$ .95
	1934	1200	4500	1125	2500	1500	1500	1200	1200	1800	1000	1620	1800	1560	20,005	-18.8	.77
Sharon	1930	1200	3000	850	100	2100	2400	2400	4000	2400	2400	1800	0	1800	24,350		.94
	1934	1200	3000	850	100	1740	1320	1680	2700	1740	1680	1500	0	240	17,650	-27.5	.68
Lebanon	1930	1500	4800	750	50	1800	1200	1500	2700	2100	100	600	2400	3960	23,410		.92
	1934	1500	4800	1200	50	1800	1200	1500	1944	2100	100	600	2400	3690	22,884	- 2.5	.89
Pottsville	1930	1800	4800	1200	500	1500	2000	1200	2400	1980	1980	1620	1920	950	23,350		.96
	1934	1800	4800	1200	500	1500	2000	1200	2400	1980	1980	1620	0	0	20,480	-12.3	.84
Washington	1930	1200	5250	1125	1200	3000	2400	1800	3200	2340	2340	1500	0	4200	28,355		1.16
	1934	2500	4500	1125	100	2000	1800	1500	3000	2250	2250	1800	0	1800	24,525	-13.5	1.00
Butler	1930	2000	6400	1600	100	2100	1800	2000	3500	2097	2190	1875	0	2000	27,562		1.17
	1934	2000	6400	1600	100	1728	1000	1600	2800	1632	1752	1548	1872	1600	25,532	- 7.4	1.08
Oil City	1930	1200	3600	1800	1975	2400	1800	2100	3300	2100		2400	0	0	20,700		.94
	1934	1200	3600	900	2000	2160	1800	1800	2820	1920	1980	2160	1920	1200	23,460	+13.3	1.06
Duquesne	1930	1200	4500	1125	2400	3300	3120	1800	4044	3300	3300	2160	0	4200	32,049		1.50
	1934	1200	4500	1125	100	2100	1800	1200	3234	2100	2100	4.80 per day	2100	560	22,019	-31.3	1.03
Monessen	1930	2400	3900	1200	100	3000	1500	1500	4200	2580	2220	3000	0	0	25,500		1.26
	1934	1600	3456	864	3120	2250	900	1200		1980		2250	0	0	14,700	-42.4	.73
Carbondale	1930	1200	3000	750	700	1000	2400	2000	2400	2100	500	1200	0	2300	18,850		.94
	1934	1200	3000	750	700	1000	2400	2000	2700	1980	500	1200	0	2100	18,880	- 0.1	.94
Uniontown	1930	2400	3000	750	100	2280	1500	1200	2580	2100	1980	2100	0	1620	21,610		1.10
	1934	2400	3000	750	100	2400	1500		2160	1890	1890	1620	1890	2112	21,612	+ 0.5	1.11
Bradford	1930	800	2000	600	100	2700	2700	1800	4000	2400	2400	2520	0	2400	24,320		1.26
	1934	720	1800	540	100	2430	2430	2160	3600	2160	2160	2160	0	3474	23,634	- 2.8	1.22
Pittston	1930	1500	4000	1000	1452	2000	1650	1200	1650	1867	500	1800	0	800	17,967		.98
	1934	1500	4000	1000	1452	2000	1650	1200	1650	2100	500	1800	1800	250	19,450	+ 8.3	1.07
Beaver Falls	1930	1200	4500	1500	100	1860	2000	1800	3000	2100			0	0	15,860		.92
	1934	900	2880	720	1625	1500	1200	1200	1800	1710	1570		0	0	13,480	-15.0	.79

TABLE X (Cont'd.)

SALARIES OF THE PRINCIPAL OFFICIALS OF TWENTY-SEVEN THIRD CLASS CITIES. - 1930 AND 1934

City	Year	Mayor	Council-men (Four) ②	Control-ler	Treas-urer ③	Clerk	Solici-tor	Assessor	Engineer	Chief of Police	Fire Chief	Health Officer	Supt. of Streets	Miscellan-eous ⑤	Total Salaries ⑥	Per Cent Increase or Decrease	Salary Cost Capita
Meadville	1930	$720	$1800	$1200	$100	$2100	$900	$2100	$3600	$2520	$900	$1500	$1920	$6000	$25,260		$1.51
	1934	720	1800	1200	100	2400	900	2100	3600	2520	900	1600	1920	2180	21,840	-13.5	1.31
Greensburg	1930	1200	3000	750	100	2400	1200		3000	2160	0	1300	0	2220	17,230		1.04
	1934	1200	3000	750	85	2280	1539	2565	2565	1846	0	900	0	2052	18,697	+ 8.5	1.13
Sunbury	1930	1200	3000	750	100	1200	1500	900	1500	1680	420	720	0	0	12,870		.82
	1934	1200	3000	750	1500	1200	900	600	900	1566	480	720	0	0	11,316	-12.1	.72
Clairton	1930	800	2400	800	300	3000	2100	1500	3600	2640	2400	1800	2340	2700	26,080		1.71
	1934	800	2400	800	300	2640	1200		3168	2280	2100	1080	1920	1800	20,188	-22.6	1.32
Coatesville	1930	1800	2400	600	400	1620	1200	750	3000	1620		1200	1800	1620	17,610		1.21
	1934	1620	2160	540	360	1620	1500	675	2700	1458	150	600	1200	960	15,183	-13.8	1.04
Farrell ①	1930																
	1934	500	1800	600	60	1680	1080	1380	1680	1800	1620	1380	0	0	13,520	--	.94
Connellsville	1930	500	1800	450	100	2100	1600	900	2400	1920	2100	1500	1950	0	17,220		1.30
	1934	500	1600	450	100	1920	1200	900	1800	1920	1920	1320	0	0	13,730	-20.3	1.03
Du Bois	1930	1200	1200	600	300	2100	1000	1000	2700	1800	0	1500	0	0	13,100		1.13
	1934	1200	1200	600	1500	2280	1000	1000	2700	1800	0	300	0	0	12,080	- 7.8	1.04
Franklin	1930	500	1200	1600	500	800	800	700	2100	1920	480	900	1800	2100	14,900		1.45
	1934	500	1200	1600	500	800	800	1200	2100	1920	480	900	1800	1800	15,100	+ 1.3	1.47
Lock Haven	1930	600	1200	300	1500	600	400	500	1500	1320	150	900	1000	1080	9,050		.83
	1934	600	1200	300	1500	600	400	500	1380	1320	150	540	0	250	7,240	-20.0	.75
Monongahela	1930	500	1200	300	600	1600	800	600	2400		120	1800	0	0	9,520		1.10
	1934	500	1200	300	540	480 ④ 1	500	500	2100		120	1200	0	0	6,421	-32.6	.74
Titusville	1930	500	1200	1500	2300	600	800	$7 per day	1500	1860		2100	1500	600	12,160		1.51
	1934	500	1200	1500	2250	600	800	max.600 $ 7 per day	1500	1674	594	300	1406	1674	11,746	- 3.4	1.46
Corry	1930	500	2000	250	500	480	1000	500	1500	1750	300	0	1500	0	9,780		1.37
	1934	600	1200	300	600	480	720	$4.50 per day	1200	1620	216	0	1458	0	7,794	-20.3	1.09

① Farrell was a Borough in 1930 and the salary figures are not available.
② The total salary paid to all four councilmen.
③ Some treasurers receive a commission for collecting taxes in addition to salary shown here.
④ In 1934 one man served as clerk, assessor and engineer for a total salary of $2601.
⑤ This item includes salaries paid to milk and food inspectors, sealer of weights and measures, municipal electrician, plumbing and building inspector and secretary to the board of health, but does not include salaries paid to clerks, stenographers, assistants, etc.
⑥ Total salaries do not include any salary or commission paid to city treasurer.

TABLE XI

COMPARATIVE COST OF GENERAL OFFICES IN TWENTY-ONE THIRD CLASS CITIES, - 1930 AND 1933; PER CENT OF CHANGE, 1933 OVER 1930; PER CAPITA COST, - 1933; AND RATIO OF COST OF GENERAL OFFICES TO TOTAL OPERATING COST INCLUDING DEBT SERVICE, 1933

City	Population	Year	Cost of Operating General① Offices for 1930 & 1933	Per Cent of Increase or Decrease in Cost of Operating General Offices for 1933 over 1930	Per Capita Cost of Operating General Offices for 1933	Total Cost of Operating All Departments② Including Debt Service for 1933	Cost of Operating General Offices as a percentage of Total Operating Cost for 1933
Nanticoke	26,043 / 27,072	1930 / 1933	$29,222 / 20,528	-29.8	$ .76	$ 219,248	9.4
Sharon	25,908 / 27,156	1930 / 1933	21,953 / 19,499	-11.2	.72	218,647	8.9
Lebanon	25,561 / 25,837	1930 / 1933	26,810 / 23,507	-12.3	.91	268,332	8.8
Washington	24,545 / 25,466	1930 / 1933	28,342 / 24,015	-15.3	.94	291,058	8.3
Pottsville	24,300 / 25,026	1930 / 1933	20,661 / 19,345	- 6.4	.77	210,992	9.2
Oil City	22,075 / 22,315	1930 / 1933	22,543 / 24,790	+10.0	1.11	241,552	10.3
Monessen	20,268 / 20,895	1930 / 1933	20,585 / 13,075	-36.5	.63	177,493	7.4
Bradford	19,306 / 20,440	1930 / 1933	22,426 / 17,117	-23.7	.84	346,464	4.9
Pittston	18,246 / 18,171	1930 / 1933	19,240 / 19,591	+ 1.8	1.08	183,859	10.7
Beaver Falls	17,147 / 17,717	1930 / 1933	20,068 / 12,723	-36.6	.72	164,994	7.7
Meadville	16,698 / 17,337	1930 / 1933	14,511 / 13,332	- 8.1	.77	160,717	8.3

TABLE XI (Cont'd.)

COMPARATIVE COST OF GENERAL OFFICES IN TWENTY-ONE THIRD CLASS CITIES, - 1930 AND 1933; PER CENT OF CHANGE, 1933 OVER 1930; PER CAPITA COST, - 1933; AND RATIO OF COST OF GENERAL OFFICES TO TOTAL OPERATING COST INCLUDING DEBT SERVICE, 1933

City	Population	Year	Cost of Operating General(1) Offices for 1930 & 1933	Per Cent of Increase or Decrease in Cost of Operating General Offices for 1933 over 1930	Per Capita Cost of Operating General Offices for 1933	Total Cost of Operating All Departments(2) Including Debt Service for 1933	Cost of Operating General Offices as a percentage of Total Operating Cost for 1933
Greensburg	16,508 16,952	1930 1933	$18,384 16,479	-10.4	$ .97	$ 187,005	8.8
Sunbury	15,626 15,599	1930 1933	13,418 11,277	-16.0	.72	88,181	12.8
Coatesville	14,582 14,603	1930 1933	11,617 10,501	- 9.6	.72	168,763	6.2
Farrell	14,359 13,990	1930 1933	8,153 9,212	+13.0	.66	118,014	7.8
DuBois	11,595 10,968	1930 1933	11,981 11,266	- 6.0	1.03	104,918	10.7
Franklin	10,254 10,388	1930 1933	7,558 7,151	- 5.4	.69	97,892	7.3
Lock Haven	9,668 10,001	1930 1933	7,677 7,903	+ 2.9	.79	89,761	8.8
Monongahela	8,675 8,672	1930 1933	11,357 5,995	-47.2	.69	58,406	10.3
Titusville	8,055 7,941	1930 1933	9,426 8,290	-12.1	1.04	99,186	8.4
Corry	7,152 7,128	1930 1933	7,369 6,295	-14.6	.88	61,205	10.3
Average of 21 cities	17,315	1930 1933	$16,728 $14,375	-14.1	$ .83	$169,366	8.5

(1) Includes mayor and council, clerk, controller, treasurer, assessor and solicitor.
(2) Does not include permanent improvements, temporary loans, public service enterprises or non-governmental expenses.

TABLE XII

COMPARATIVE COST OF POLICE PROTECTION IN TWENTY-TWO THIRD CLASS CITIES, - 1930 AND 1933; PER CENT OF CHANGE, 1933 OVER 1930; PER CAPITA COST, - 1933; AND RATIO OF POLICE COST TO TOTAL OPERATING COST INCLUDING DEBT SERVICE, 1933

City	Population	Year	Cost of Police Protection for 1930 and 1933	Per Cent of Increase or Decrease in Cost of Police Protection for 1933 over 1930	Per Capita Cost of Police Protection for 1933	Total Cost of Operating All Departments ① Including Debt Service for 1933	Cost of Police Protection as a percentage of Total Operating Cost for 1933
Nanticoke	26,043 / 27,072	1930 / 1933	$57,132 / 17,392	-69.6	$ .64	$ 219,248	7.9
Sharon	25,908 / 27,156	1930 / 1933	50,396 / 32,122	-36.3	1.18	218,647	14.7
Lebanon	25,561 / 25,837	1930 / 1933	41,720 / 37,074	-11.1	1.43	268,332	13.8
Washington	24,545 / 25,466	1930 / 1933	36,176 / 30,692	-15.2	1.21	291,058	10.5
Pottsville	24,300 / 25,026	1930 / 1933	28,531 / 28,546	+ .0005	1.14	210,992	13.5
Oil City	22,075 / 22,315	1930 / 1933	30,081 / 28,107	- 6.6	1.26	241,552	11.6
Duquesne	21,396 / 22,113	1930 / 1933	76,077 / 39,525	-48.0	1.79	237,707	16.6
Monessen	20,268 / 20,895	1930 / 1933	31,112 / 20,391	-34.5	.98	177,493	11.5
Bradford	19,306 / 20,440	1930 / 1933	28,654 / 23,689	-17.3	1.16	346,464	6.8
Pittston	18,246 / 18,171	1930 / 1933	34,622 / 28,596	-17.4	1.57	183,859	15.6
Beaver Falls	17,147 / 17,717	1930 / 1933	28,228 / 17,271	-38.8	.97	164,994	10.5
Meadville	16,698 / 17,337	1930 / 1933	22,746 / 18,557	-18.4	1.07	160,717	11.6

TABLE XII (Cont'd.)

COMPARATIVE COST OF POLICE PROTECTION IN TWENTY-TWO THIRD CLASS CITIES, - 1930 AND 1933; PER CENT OF CHANGE, 1933 OVER 1930; PER CAPITA COST, - 1933; AND RATIO OF POLICE COST TO TOTAL OPERATING COST INCLUDING DEBT SERVICE, 1933

City	Population	Year	Cost of Police Protection for 1930 and 1933	Per Cent of Increase or Decrease in Cost of Police Protection for 1933 over 1930	Per Capita Cost of Police Protection for 1933	Total Cost of Operating All Departments① Including Debt Service for 1933	Cost of Police Protection as a percentage of Total Operating Cost for 1933
Greensburg	16,508 16,952	1930 1933	$36,668 28,320	-22.8	$1.67	$ 187,005	15.1
Sunbury	15,626 15,599	1930 1933	11,192 11,012	- 1.6	.71	88,181	12.5
Coatesville	14,582 14,603	1930 1933	27,154 19,082	-29.7	1.31	168,763	11.3
Farrell	14,359 13,990	1930 1933	25,393 16,087	-36.6	1.15	118,014	13.6
DuBois	11,595 10,968	1930 1933	8,788 8,820	+ .4	.80	104,918	8.4
Franklin	10,254 10,338	1930 1933	15,096 12,919	-14.4	1.25	97,892	13.2
Lock Haven	9,668 10,001	1930 1933	11,119 8,805	-20.8	.88	89,761	9.8
Monongahela	8,675 8,672	1930 1933	9,225 5,113	-44.6	.59	58,406	8.8
Titusville	8,055 7,941	1930 1933	14,180 11,602	-18.2	1.46	99,186	11.7
Corry	7,152 7,128	1930 1933	9,150 6,072	-33.6	.85	61,205	9.9
Average of 22 cities	17,533	1930 1933	28,792 20,445	-29.0	1.17	172,472	11.9

① Does not include permanent improvements, temporary loans, public service enterprises or non-governmental expenses.

TABLE XIII

COMPARATIVE COST OF FIRE PROTECTION IN TWENTY-TWO THIRD CLASS CITIES, - 1930 AND 1933; PER CENT OF CHANGE, 1933 OVER 1930; PER CAPITA COST, - 1933; AND RATIO OF FIRE COST TO TOTAL OPERATING COST INCLUDING DEBT SERVICE, 1933

City	Population	Year	Cost of Fire Protection for 1930 and 1933	Per Cent of Increase or Decrease in the Cost of Fire Protection for 1933 over 1930	Per Capita Cost of Fire Protection for 1933	Total Cost of Operating All Departments① Including Debt Service for 1933	Cost of Fire Protection as a percentage of Total Operating Cost for 1933
Nanticoke	26,043 27,072	1930 1933	$26,435 24,871	- 5.9	$ .92	$ 219,248	11.3
Sharon	25,908 27,156	1930 1933	43,330 33,026	-23.8	1.22	218,647	15.1
Lebanon	25,561 25,837	1930 1933	21,238 23,663	+11.4	.92	268,332	8.8
Washington	24,545 25,466	1930 1933	29,019 27,256	- 6.1	1.07	291,058	9.4
Pottsville	24,300 25,026	1930 1933	30,355 28,219	- 7.0	1.13	210,992	13.4
Oil City	22,075 22,315	1930 1933	40,305 38,328	- 4.9	1.72	241,552	15.9
Duquesne	21,396 22,118	1930 1933	49,019 32,348	-34.0	1.46	237,707	13.6
Monessen	20,268 20,895	1930 1933	18,562 18,236	- 1.8	.87	177,493	10.3
Bradford	19,306 20,440	1930 1933	39,910 38,192	- 4.3	1.87	346,464	11.0
Pittston	18,246 18,171	1930 1933	17,422 22,619	+29.8	1.24	183,859	12.8
Beaver Falls	17,147 17,717	1930 1933	27,544 24,996	- 9.3	1.41	164,994	15.2
Meadville	16,698 17,337	1930 1933	19,973 19,274	- 3.5	1.11	160,717	12.0

TABLE XIII (Cont'd.)

COMPARATIVE COST OF FIRE PROTECTION IN TWENTY-TWO THIRD CLASS CITIES, - 1930 AND 1933; PER CENT OF CHANGE, 1933 OVER 1930; PER CAPITA COST, - 1933; AND RATIO OF FIRE COST TO TOTAL OPERATING COST INCLUDING DEBT SERVICE, 1933

City	Population	Year	Cost of Fire Protection for 1930 and 1933	Per Cent of Increase or Decrease in the Cost of Fire Protection for 1933 over 1930	Per Capita Cost of Fire Protection for 1933	Total Cost of Operating All Departments① Including Debt Service for 1933	Cost of Fire Protection as a percentage of Total Operating Cost for 1933
Greensburg	16,508 16,952	1930 1933	$26,897 21,630	-19.6	$1.28	$187,005	11.6
Sunbury	15,626 15,599	1930 1933	26,547 28,342	+ 6.8	1.82	88,181	32.1
Coatesville	14,582 14,603	1930 1933	11,882 7,118	-40.1	.49	168,763	4.2
Farrell	14,359 13,990	1930 1933	19,286 15,840	-17.9	1.13	118,014	13.4
DuBois	11,595 10,968	1930 1933	11,613 7,949	-31.6	.72	104,918	7.6
Franklin	10,254 10,338	1930 1933	13,353 9,775	-26.8	.95	97,892	10.0
Lock Haven	9,668 10,001	1930 1933	8,600 14,162	+64.7	1.42	89,761	15.8
Monongahela	8,675 8,672	1930 1933	6,315 5,390	-14.6	.62	58,406	9.2
Titusville	8,055 7,941	1930 1933	11,678 11,243	- 3.7	1.42	99,186	11.3
Corry	7,152 7,128	1930 1933	15,392 13,054	-15.2	1.83	61,205	21.3
Average of 22 Cities	17,533	1930 1933	$23,394 $21,160	- 9.5	$1.21	$172,472	12.3

① Does not include permanent improvements, temporary loans, public service enterprises or non-governmental expenses.

the others drain their sewers into rivers or creeks. There are three different methods
employed in the collection of garbage: First, in eleven cities the individual citizen
makes his own arrangement by private contract; second, in five cities the collection is
handled by city employees; third, in the remaining eleven, the city makes a contract
with a private concern for collecting all garbage.

In a few places, private contractors pay concessions to the city for the right to
collect garbage and sometimes a city receives as much as six or seven thousand dollars
a year for such concession. Expenditures for the purpose of protecting the public
health have decreased 39.3 per cent since 1930. In many cases these savings have resulted from curtailing the activities of the street cleaning department but in others
they were due to more favorable contracts for the collection and disposal of garbage.
In a few cities savings were effected by the actual discontinuance of strictly public
health services such as the physical examination of employees in hotels, restaurants,
drug stores and other establishments handling food stuffs; inspectors of food and milk
depots; etc. Such services should be broadened and extended in many of the cities.
The average per capita cost for these services was only $0.65 in 1933. It was as low as
$0.09 in Sunbury and as high as $2.16 in Meadville. (See Table XIV.)

The Cost of Maintaining Streets and Bridges:- It is a customary function of local
units of government to provide and maintain streets and bridges within their respective
territories. The cost of repairing these streets is an important item of expense in
most cities. The cost of new streets is not included in Table XV but is considered
later under permanent improvements. However, since from 75 to 90 per cent of the cost
of the engineering department is due to street work, the overhead cost of such department is included. Twenty-six of the cities studied have an average of 47 miles of
streets, lanes and alleys open for public use of which 21 miles or 44.6 per cent is of
a hard surface type such as concrete, asphalt, stone or brick. Between 1930 and 1933
twenty-two cities reduced street maintenance costs by an average of 37.6 per cent. The
average per capita cost was $1.25 for 1933. In some cases reduced expenditures were
due to lower cost of materials and to the payment of lower wages but in many cases they
were due to the fact that the cities postponed street repairs for a later day. If this
policy is continued for another year or two, many cities will be required to build new
streets as some of them were almost beyond repair in the summer of 1934.

The Cost of Street Lighting:- All of the cities studied light their streets by
electricity and the cost involved is an important item for most of them. Meadville and
Titusville operate their own electric plants. Several of the larger cities maintain
"white-ways" in the principal business sections and this materially increases the cost.
In the last two years lighting costs have been reduced in many cities. One or more of
three methods have been used to bring about these savings: First, by reducing the rates
charged; second, by eliminating certain lights entirely or using every other one after
ten or eleven P.M.; and third, by decreasing the candle power of the lights used. Some
city officials feel that their cities are so inadequately policed that full street
lighting is necessary during the entire night. Street lighting costs in 1933 averaged
$18,189 for each city. This is only 7.3 per cent less than was spent in 1930 and is
the smallest decrease in the cost of any of the items of major importance. The average
per capita cost for 1933 was $1.04. (See Table XVI.)

Functional Distribution of Total Operating Costs including Debt Service:- Table
XVII gives a composite picture of how a dollar spent for general operating purposes is
distributed between the different functions of government. It is noteworthy that in
all the cities except Sunbury, Lock Haven and Corry, debt service is by far the largest
item of expense. If this item is not included it is found that the largest single item
for 1933 in each city is as follows:- Police Protection:- Lebanon, Farrell and Franklin; Fire Protection - Oil City, Bradford, Beaver Falls, Sunbury and Corry; Public
Health (including sewers, garbage collection, street cleaning, etc.) - Meadville;
Maintenance of Streets - Nanticoke, Washington, Pottsville, Pittston, Greensburg,
Coatesville, DuBois, Lock Haven, Monongahela and Titusville - ten of the twenty-one
cities; Street Lighting - Sharon and Monessen. If that portion of the debt service
traceable to the building of new streets was added to the cost of street maintenance,
the total of such costs would show streets as the most important item of expense.

TABLE XIV

COMPARATIVE COST OF PUBLIC HEALTH IN TWENTY-ONE THIRD CLASS CITIES, - 1930 AND 1933; PER CENT OF CHANGE, 1933 OVER 1930; PER CAPITA COST, - 1933; AND RATIO OF HEALTH COST TO TOTAL OPERATING COST INCLUDING DEBT SERVICE, 1933

City	Population	Year	Cost of Public Health① for 1930 and 1933	Per Cent of Increase or Decrease in Cost of Public Health for 1933 over 1930	Per Capita Cost of Public Health for 1933	Total Cost of Operating All Departments② Including Debt Service for 1933	Public Health as a percentage of Total Operating Cost for 1933
Nanticoke	26,043 / 27,072	1930 / 1933	$7,622 / 2,593	-66.0	$.10	$ 219,248	1.2
Sharon	25,908 / 27,156	1930 / 1933	42,439 / 17,998	-57.6	.66	218,647	8.2
Lebanon	25,561 / 25,837	1930 / 1933	34,579 / 28,937	-16.3	1.12	268,332	10.8
Washington	24,545 / 25,466	1930 / 1933	32,844 / 17,804	-45.8	.70	291,058	6.2
Pottsville	24,300 / 25,026	1930 / 1933	12,899 / 14,806	+14.8	.59	210,992	7.0
Oil City	22,075 / 22,315	1930 / 1933	17,818 / 12,520	-29.7	.56	241,552	5.2
Monessen	20,268 / 20,895	1930 / 1933	38,946 / 19,662	-49.5	.94	177,493	11.1
Bradford	19,306 / 20,440	1930 / 1933	45,280 / 13,484	-70.2	1.54	346,464	9.1
Pittston	18,246 / 18,171	1930 / 1933	7,874 / 6,963	-11.6	.38	183,859	3.8
Beaver Falls	17,147 / 17,717	1930 / 1933	15,519 / 9,578	-38.3	.54	164,994	5.8
Meadville	16,698 / 17,337	1930 / 1933	59,339 / 37,498	-36.8	2.16	160,717	23.3
Greensburg	16,508 / 16,952	1930 / 1933	12,501 / 5,079	-59.4	.30	187,005	2.7

TABLE XIV (Cont'd.)

COMPARATIVE COST OF PUBLIC HEALTH IN TWENTY-ONE THIRD CLASS CITIES, - 1930 AND 1933; PER CENT OF CHANGE, 1933 OVER 1930; PER CAPITA COST, - 1933; AND RATIO OF HEALTH COST TO TOTAL OPERATING COST INCLUDING DEBT SERVICE, 1933

City	Population	Year	Cost of Public Health[1] for 1930 and 1933	Per Cent of Increase or Decrease in Cost of Public Health for 1933 over 1930	Per Capita Cost of Public Health for 1933	Total Cost of Operating All Departments[2] Including Debt Service for 1933	Public Health as a Percentage of Total Operating Cost for 1933
Sunbury	15,626 / 15,599	1930 / 1933	$5,705 / 1,364	-76.1	$ .09	$ 88,181	1.6
Coatesville	14,582 / 14,603	1930 / 1933	17,661 / 15,107	-14.5	1.03	168,763	9.0
Farrell	14,359 / 13,990	1930 / 1933	12,349 / 10,972	-11.2	.78	118,014	9.3
Du Bois	11,595 / 10,968	1930 / 1933	11,006 / 12,036	+ 9.4	1.10	104,918	11.5
Franklin	10,254 / 10,338	1930 / 1933	3,892 / 3,492	-10.3	.34	97,892	3.6
Lock Haven	9,668 / 10,001	1930 / 1933	2,491 / 2,041	-18.1	.20	89,761	2.3
Monongahela	8,675 / 8,672	1930 / 1933	4,963 / 2,388	-51.9	.28	58,406	4.1
Titusville	8,055 / 7,941	1930 / 1933	3,435 / 2,799	-18.5	.35	99,186	2.8
Corry	7,152 / 7,128	1930 / 1933	2,897 / 894	-69.1	.13	61,205	1.5
Average of 21 cities	17,315	1930 / 1933	$18,669 / $11,334	-39.3	$0.65	$169,366	6.7

[1] Includes costs for sewerage, garbage and ash collection and street cleaning.
[2] Does not include permanent improvements, temporary loans, public service enterprises or non-governmental expenses.

TABLE XV

COMPARATIVE COST OF STREET MAINTENANCE IN TWENTY-TWO THIRD CLASS CITIES, - 1930 AND 1933; PER CENT OF CHANGE, 1933 OVER 1930; PER CAPITA COST, - 1933; AND RATIO OF MAINTENANCE COST TO TOTAL OPERATING COST INCLUDING DEBT SERVICE, 1933

City	Population	Year	Cost of Maintenance of Streets ① for 1930 & 1933	Per Cent of Increase or Decrease in Cost of Maintenance of Streets for 1933 over 1930	Per Capita Cost of Maintenance of Streets for 1933	Total Cost of Operating All Departments ② Including Debt Service for 1933	Street Maintenance as a percentage of Total Operating Cost for 1933
Nanticoke	26,043 27,072	1930 1933	$77,517 55,227	-28.8	$2.04	$219,248	25.2
Sharon	25,908 27,156	1930 1933	29,383 24,355	-17.1	.90	218,647	11.1
Lebanon	25,561 25,837	1930 1933	44,816 23,288	-48.0	.90	268,332	8.7
Washington	24,545 25,466	1930 1933	45,474 32,261	-29.1	1.27	291,058	11.1
Pottsville	24,300 25,026	1930 1933	52,165 37,790	-27.6	1.51	210,992	17.9
Oil City	22,075 22,315	1930 1933	41,451 33,263	-19.8	1.49	241,552	13.8
Duquesne	21,396 22,113	1930 1933	70,482 13,986	-80.2	.63	237,707	5.9
Monessen	20,268 20,895	1930 1933	27,127 8,060	-70.3	.39	177,493	4.5
Bradford	19,306 20,440	1930 1933	28,530 25,402	-11.0	1.24	346,464	7.3
Pittston	18,246 18,171	1930 1933	41,250 36,044	-12.6	1.98	183,859	19.6
Beaver Falls	17,147 17,717	1930 1933	30,869 15,754	-49.0	.89	164,994	9.6
Meadville	16,698 17,337	1930 1933	68,275 18,123	-73.5	1.05	160,717	11.3

TABLE XV (Cont'd.)

COMPARATIVE COST OF STREET MAINTENANCE IN TWENTY-TWO THIRD CLASS CITIES, - 1930 AND 1933; PER CENT OF CHANGE, 1933 OVER 1930; PER CAPITA COST, - 1933; AND RATIO OF MAINTENANCE COST TO TOTAL OPERATING COST INCLUDING DEBT SERVICE, 1933

City	Population	Year	Cost of Maintenance of Streets① for 1930 & 1933	Per Cent of Increase or Decrease in Cost of Maintenance of Streets for 1933 over 1930	Per Capita Cost of Maintenance of Streets for 1933	Total Cost of Operating All Departments② Including Debt Service for 1933	Street Maintenance as a percentage of Total Operating Cost for 1933
Greensburg	16,508 16,952	1930 1933	$43,365 32,892	-24.2	$1.94	$187,005	17.6
Sunbury	15,626 15,599	1930 1933	19,004 11,651	-38.7	.75	88,181	13.2
Coatesville	14,582 14,603	1930 1933	45,017 22,470	-50.1	1.54	168,763	13.3
Farrell	14,359 13,990	1930 1933	22,329 13,036	-41.6	.93	118,014	11.1
DuBois	11,595 10,968	1930 1933	12,633 13,088	+ 3.6	1.19	104,918	12.5
Franklin	10,254 10,338	1930 1933	12,629 10,818	-14.3	1.05	97,892	11.1
Lock Haven	9,668 10,001	1930 1933	21,830 23,503	+ 7.7	2.35	89,761	26.2
Monongahela	8,675 8,672	1930 1933	11,093 11,221	+ 1.2	1.29	58,406	19.2
Titusville	8,055 7,941	1930 1933	13,935 12,013	-13.8	1.51	99,186	12.1
Corry	7,152 7,128	1930 1933	15,331 8,658	-43.5	1.21	61,205	14.2
Average of 22 cities	17,533	1930 1933	$35,204 $21,950	-37.6	$1.25	$172,472	12.7

① Includes bridges.
② Does not include permanent improvements, temporary loans, public service enterprises or non-governmental expenses.

TABLE XVI

COMPARATIVE COST OF STREET LIGHTING IN TWENTY-TWO THIRD CLASS CITIES, - 1930 AND 1933; PER CENT OF CHANGE, 1933 OVER 1930; PER CAPITA COST, - 1933; AND RATIO OF LIGHTING COST TO TOTAL OPERATING COST INCLUDING DEBT SERVICE, 1933

City	Population	Year	Cost of Street Lighting for 1930 & 1933	Per Cent of Increase or Decrease in Cost of Street Lighting for 1933 over 1930	Per Capita Cost of Street Lighting for 1933	Total Cost of Operating All Departments ① Including Debt Service for 1933	Street Lighting as a percentage of Total Operating Cost for 1933
Nanticoke	26,043 / 27,072	1930 / 1933	$11,915 / 20,000	+67.9	$ .74	$ 219,248	9.1
Sharon	25,908 / 27,156	1930 / 1933	39,076 / 34,331	-12.1	1.26	218,647	15.7
Lebanon	25,561 / 25,837	1930 / 1933	28,975 / 27,468	-5.2	1.06	268,332	10.2
Washington	24,545 / 25,466	1930 / 1933	44,648 / 30,301	-32.1	1.19	291,058	10.4
Pottsville	24,300 / 25,026	1930 / 1933	34,285 / 28,421	-17.1	1.14	210,992	13.5
Oil City	22,075 / 22,315	1930 / 1933	22,072 / 21,714	- 1.6	.97	241,552	9.0
Duquesne	21,396 / 22,113	1930 / 1933	22,253 / 10,574	-52.5	.48	237,707	4.5
Monessen	20,268 / 20,895	1930 / 1933	14,883 / 21,783	+46.4	1.04	177,493	12.3
Bradford	19,306 / 20,440	1930 / 1933	24,063 / 27,142	+12.8	1.33	346,464	7.8
Pittston	18,246 / 18,171	1930 / 1933	15,685 / 15,493	- 1.2	.85	183,859	8.4
Beaver Falls	17,147 / 17,717	1930 / 1933	25,304 / 17,908	-29.2	1.01	164,994	10.9
Meadville	16,698 / 17,337	1930 / 1933	15,653 / 13,100	-16.3	.76	160,717	8.2

TABLE XVI (Cont'd.)

COMPARATIVE COST OF STREET LIGHTING IN TWENTY-TWO THIRD CLASS CITIES, - 1930 AND 1933; PER CENT OF CHANGE, 1933 OVER 1930; PER CAPITA COST, - 1933; AND RATIO OF LIGHTING COST TO TOTAL OPERATING COST INCLUDING DEBT SERVICE, 1933

City	Population	Year	Cost of Street Lighting for 1930 & 1933	Per Cent of Increase or Decrease in Cost of Street Lighting for 1933 over 1930	Per Capita Cost of Street Lighting for 1933	Total Cost of Operating All Departments① Including Debt Service for 1933	Street Lighting as a percentage of Total Operating Cost for 1933
Greensburg	16,508 16,952	1930 1933	$22,863 26,222	+14.7	$1.55	$187,005	14.0
Sunbury	15,626 15,599	1930 1933	14,477 13,541	- 6.5	.87	88,181	15.4
Coatesville	14,582 14,608	1930 1933	15,820 14,986	- 5.3	1.03	168,763	8.9
Farrell	14,359 13,990	1930 1933	12,950 12,017	- 7.2	.86	118,014	10.2
DuBois	11,595 10,968	1930 1933	11,460 11,619	+ 1.4	1.06	104,918	11.1
Franklin	10,254 10,338	1930 1933	12,629 12,682	+ .4	1.23	97,892	13.0
Lock Haven	9,668 10,001	1930 1933	11,429 10,952	- 4.2	1.10	89,761	12.2
Monongahela	8,675 8,672	1930 1933	10,721 10,777	+ .5	1.24	58,406	18.5
Titusville	8,055 7,941	1930 1933	8,573 7,087	-17.3	.89	99,186	7.2
Corry	7,152 7,128	1930 1933	12,083 12,044	- .3	1.69	61,205	19.7
Average of 22 cities	17,533	1930 1933	$19,628 18,189	- 7.3	$1.04	$172,472	10.5

① Does not include permanent improvements, temporary loans, public service enterprises or non-governmental expenses.

TABLE XVII

RATIO OF EXPENDITURES FOR SPECIFIED FUNCTIONS

TO TOTAL OPERATING EXPENDITURES INCLUDING DEBT SERVICE

IN TWENTY-ONE THIRD CLASS CITIES, - 1933

City	Debt Service	General Offices	Police Protection	Fire Protection	Public Health	Street Maintenance	Street Lighting	Other Purposes
	%	%	%	%	%	%	%	%
Nanticoke	30.3	9.4	7.9	11.3	1.2	25.2	9.1	5.6
Sharon	20.0	8.9	14.7	15.1	8.2	11.1	15.7	6.3
Lebanon	31.6	8.8	13.8	8.8	10.8	8.7	10.2	7.3
Washington	39.3	8.3	10.5	9.4	6.2	11.1	10.4	4.8
Pottsville	22.2	9.2	13.5	13.4	7.0	17.9	13.5	3.3
Oil City	23.7	10.3	11.6	15.9	5.2	13.8	9.0	10.5
Monessen	36.0	7.4	11.5	10.3	11.1	4.5	12.3	6.9
Bradford	44.9	4.9	6.8	11.0	9.1	7.3	7.8	8.2
Pittston	23.5	10.7	15.6	12.3	3.8	19.6	8.4	6.1
Beaver Falls	36.7	7.7	10.5	15.2	5.8	9.6	10.9	3.6
Meadville	17.8	8.3	11.6	12.0	23.3	11.3	8.2	7.5
Greensburg	24.9	8.8	15.1	11.6	2.7	17.6	14.0	5.3
Sunbury	5.9	12.8	12.5	32.1	1.6	13.2	15.4	6.5
Coatesville	42.8	6.2	11.3	4.2	9.0	13.3	8.9	4.3
Farrell	25.2	7.8	13.6	13.4	9.3	11.1	10.2	9.4
DuBois	24.3	10.7	8.4	7.6	11.5	12.5	11.1	13.9
Franklin	35.5	7.3	13.2	10.0	3.6	11.1	13.0	6.3
Lock Haven	15.6	8.8	9.8	15.8	2.3	26.2	12.2	9.3
Monongahela	19.9	10.3	8.8	9.2	4.1	19.2	18.5	10.0
Titusville	38.1	8.4	11.7	11.3	2.8	12.1	7.2	8.4
Corry	17.1	10.3	9.9	21.3	1.5	14.2	19.7	6.0

Expenditures for Permanent Improvements:- In 1930 nearly every one of the cities studied made substantial expenditures for permanent improvements. Twenty-one of them show capital outlays for that year in excess of $50,000 and ten range from $100,000 to over $400,000. Together twenty-five cities spent $2,818,750 or an average of $112,750 each. In 1933, twenty-three of these cities spent only $267,453 or an average of $11,628. In 1930, this money was spent for new streets, sewers, parks, buildings, fire trucks, etc. It is an interesting fact that $2,363,711 or 83.9 per cent of the $2,818,750 total improvement cost went for street construction. In 1933 only $102,956 or 38.5 per cent of the $267,453 spent in twenty-three of the cities for improvements was for new streets. Nearly every city has stopped making new improvements of any importance except in a few cases where the Federal Government has made generous gifts of Federal funds.

CONCLUSIONS.

FINDING: The data collected disclose a number of thickly populated boroughs or townships adjacent to third class cities and for all practical purposes forming integral parts of such cities. Their governmental needs are similar to those of the cities.
RECOMMENDATION: The Legislature should authorize a comprehensive inquiry to set forth the extent of this similarity of needs wherever it occurs throughout the state and to draft statutory methods of merger or consolidation which will aid in bringing about a satisfactory union between areas with similar needs. Where actual mergers are not possible, plans for creating regional fire, police, sewer and school districts for such combined areas should be drafted.

FINDING: There are forty-four third class cities ranging in population from 7,152 to 115,967 - twenty-eight of which have a population of less than 27,000. Although they are all governed by the same laws, these laws were enacted primarily to fit the needs of cities with a population of 50,000 and upwards.
RECOMMENDATION: Either these cities should be reclassified with appropriate legislation for their government or the present requirement of a minimum population of 10,000 should be materially increased.

FINDING: Under the commission form of government, all legislative authority resides in council while the supervision and control of all administrative matters are placed in the hands of the councilmen individually. In thus combining legislative and executive functions the generally accepted division of governmental authority is violated and strong, effective control over problems of administration is made impossible.
RECOMMENDATION: Responsibility for the administrative side of the city's business should be placed in the hands of one man, either the mayor or a city manager. Opportunity should be given to the people of each city to decide for themselves whether or not they want a city manager form of government.
RECOMMENDATION: The treasurer and controller should be appointed not elected.

FINDING: The law requires that all executive and administrative functions be placed in five specifically designated departments.
RECOMMENDATION: Whether the business of a city is conducted by a mayor or a manager every city should have the right to determine the number of departments and the distribution of functions between such departments.

FINDING: Nearly all of the cities studied purchase materials and supplies on a "hand to mouth" basis. Many of them could save money by buying on specifications and in larger quantities.
RECOMMENDATION: In the larger cities, there should be a central agent whose duty it is to make all purchases required by the city, and where possible such agent should cooperate with the school authorities in regard to the purchase of their materials and supplies.

FINDING: While the accounting methods in use in many of the third class cities are superior to those usually found in other types of local government, there is still much room for improvement.
RECOMMENDATION: A State agency should be given the authority, in consultation

with local officials, to devise standard systems of accounting applicable to the needs of each type of local government.

FINDING: Financial reports published by many cities do not contain the information required by law and there is a considerable lack of uniformity in the methods of preparing such reports. Uniformity in reporting is urgently needed.

RECOMMENDATION: The state agency recommended above should also be authorized to prepare standard forms for reporting financial data suitable to the method of accounting required of each type of government. Such agency should be clothed with ample authority to compel the adoption and use of the prescribed accounting systems and reporting forms.

FINDING: Although budgetary control of receipts and expenditures is essential for the efficient operation of any business such control is almost entirely lacking in a majority of the cities studied.

RECOMMENDATION: The law should not only require a budget, as it now does, but should provide for:- (a) public budget hearings; (b) publication of budget before final adoption; (c) review of budget by the county court on the application of at least 20 per cent of the citizens owning taxable real estate. The state agency recommended to supervise accounts and reports should prepare standard budget forms to be used by each type of government. ·

FINDING: The present system, whereby the city and county each make a separate assessment of real estate, unnecessarily increases the cost of government and should be abolished.

RECOMMENDATION: There should be but one triennial assessment and it should be made by a county board of assessors similar to that created in 1931 for third class counties. Until such boards are established in the county, the only assessment of property in third class cities should be made by the county assessor.

FINDING: The city treasurer is the collector of city, school and poor taxes levied in third class cities.

RECOMMENDATION: Until such time as the collection of all local taxes within a . county is placed in one central office, the city treasurer should also collect county taxes levied in cities.

FINDING: A recent law (not yet in operation in all cities) provides that the collector shall be paid a salary ranging from $1,600 in the smaller cities to $7,500 in those with a population of 55,000 or over.

RECOMMENDATION: The salary in these latter cities is excessive, bears no relation to either the amount of money collected or the number of people living in such cities and should be reduced.

FINDING: Tax delinquency has increased rapidly in the last three years and in 1933 fifteen of the cities studied failed to collect from 30 to 51 per cent of their current taxes. At the end of 1933 several cities had taxes outstanding in amounts equal to their 1933 tax duplicates. Unless either the burden on real estate (which accounts for about 80 per cent of the taxes levied) is materially reduced by effecting still greater economies or other forms of taxation are developed there is little that can be done now to reduce the amount of unpaid taxes. While there may be instances where undue leniency has been shown it is generally admitted by those who are in possession of the facts that great hardship and injustice would result if the more severe tax collection laws were rigidly enforced at this time.

FINDING: The State is in real need of a modern tax collection code which would centralize the collection of all local taxes in a county office, compensate all collectors on a salary basis and bring together and make uniform all the hodgepodge of tax laws now in existence.

FINDING: The total bonded debt (city, school and city's share of county) exclusive of special assessment bonds, in twenty-seven cities at the end of 1933 ranged from 4.5 per cent to 14.5 per cent of the assessed valuation of taxable real estate. While

the total bonded debt in each city is well within the liberal limits fixed by law the
debt burden on the taxpayer is indicated by the fact that the payment of principal
and interest on city bonds alone in twenty-two of these cities in 1933 accounted for
30.6 per cent of total operating costs.
        RECOMMENDATION: The existing debt limits should be reduced and a State agency
created with authority to act jointly with local agencies in approving the issue of
bonds by the separate authorities within a given municipality.

        FINDING: As measured by 1930 expenditures, a majority of the cities made sub-
stantial reductions in general operating cost in 1933. The average reduction for all
cities studied was 25.4 per cent.

        FINDING: Salaries of principal officials (particularly those appointed) have been
materially reduced in many cities and in a few cases offices have been abolished and
the work combined with that of another office. In several cities, however, little or
no reduction has been made in the salaries paid in 1930 and in some of these cases the
cities involved have been and still are in serious financial difficulty. As a rule
offices are not overmanned but some engineering departments are still retaining assistant engineers and other technical employees when new construction work is almost at a
complete standstill.
        RECOMMENDATION: Salaries of all officers and employees not in line with present
day price levels should be increased or decreased as the case may be, and whenever
possible, only part-time assistants employed.

        FINDING: Police and fire protection expenses have been reduced 29.0 and 9.5 per
cent respectively in the last three years - chiefly as a result of the lower salaries
paid.

        FINDING: Expenditures for the purpose of protecting the public health have de-
creased 39.3 per cent. In many cases these savings have resulted from curtailing the
activities of the street cleaning department but in some instances they were due to
more favorable contracts for the collection and disposal of garbage. In a few cities
they were due to the discontinuance of strictly public health services such as the
physical examination of employees in hotels, restaurants, drug stores and other establishments handling food stuffs, inspection of food and milk depots, etc.
        RECOMMENDATION: In many cities the public health services should be broadened
and strengthened.

        FINDING: The cost of maintaining streets has been reduced to such an extent (37.6
per cent - average) during the last two or three years that many of them are now
hardly safe for travel and will likely have to be rebuilt instead of merely being
repaired.
        RECOMMENDATION: It is highly desirable that existing streets, buildings and
equipment be kept in repair so that early replacements will not be necessary. It is
not economical to permit valuable property to deteriorate in order to cut down present
costs, although some cities are doing this instead of saving by other means.

        FINDING: The average amount spent by twenty-five cities for permanent improvements
decreased from $112,750 in 1930 to $11,628 (for twenty-three cities) in 1933. New
construction work of all kinds has practically stopped.
        RECOMMENDATION: Until the present debt burden has been greatly reduced, it would
be advisable for most cities to continue to defer making permanent improvements which
would necessitate the borrowing of money.

        FINDING: While the amount spent annually for street lighting has been reduced
recently in some cities, the average reduction for all the cities amounted to only 7.3
per cent. The decrease for the three years in street lighting costs was smaller than
for any other important item, and many cities have as yet failed to secure reductions
of any importance. In a few places, particularly Monongahela, Corry, Sharon and
Sunbury, the amount spent for lighting is out of all proportion to other important expenditures.

FINDING: Frequently the cost per capita of different functions varies greatly even between cities with practically the same population.

FINDING: Under the commission form of government, the head of each major department is responsible to council for expenditures made in his department but in many cities the money appropriated for a particular department is spent with little or no regard as to what other departments are doing.

RECOMMENDATION: Closer cooperation between the heads of the various departments is essential if systematic saving is to be attained. In the last analysis, whether expenditures are wisely and economically made depends to a very large extent on the training, experience and honesty of the men elected to council and their subordinates. The time to control expenditures is when the annual budget is prepared and if members of council would invite the principal taxpayers and business leaders of the city to participate (in advisory capacity) in the preparation of the budget, many economies might be effected.

INDEX OF TABLES AND CHARTS

Part I   The County

Chapter III   **Page**

Chapter III		
Table I	County Personnel - All Offices - 1933	44
Table II	Population per County Employee and Salary Costs, by County Offices, 1933	45
Table III	Receipts of County Fee Offices, 1933	47
Table IV	Ratio of Cost of Personal Services to Total Current County Expenditures for 1933	50
Chapter IV		
Table I	Personal Property Valuation in 12 Counties, 1932	63
Table II	Range in Volume and Cost of Assessments	65
Table III	Range in Cost of Assessments within Six Counties, 1933	66
Chapter V		
Table I	Population and Registered Votes in Election Districts, 1933	72
Table II	Total Number of Registered Voters; Average Number per Election District; and Average Cost per Registered Voter, in Specified Counties, 1933	78
Table III	Cost of Election in Election Districts - 1933 Election	80
Table IV	High Cost Election Districts, 1933 Election	82
Table V	Total Cost of Elections for 1933, in 12 Counties	84
Chapter VI		
Table I	Number of Justices by Counties - 1933	90
Table II	Cost of Justices of the Peace and Constables Paid by Counties	92
Chapter VIII		
Table I	Total Revenue of County per capita	127
Table II	Property, Poll and Occupational Taxes	128
Table III	County Liquor Revenue	128
Table IV	Gasoline Tax	129
Table V	Revenue from Court Costs and Fines	129
Table VI	Total Departmental Earnings	130
Table VII	Institutional Earnings	130
Table VIII	Total Expenditures per capita	131
Table IX	Total Administrative Expense of County per capita	131
Table X	Total Judicial Expense of County per capita	132
Table XI	Law Enforcement Offices	132
Table XII	Clerical Offices	133
Table XIII	Per Capita Expenditures for Charities and Corrections, 1930 and 1933	133
Table XIV	Total Highway Cost	134
Table XV	County Buildings and Court House Maintenance Costs	134
Table XVI	Financial and Administrative Costs	135
Table XVII	Percentage of Expenditures of Departments of Total Expenditure of County for 1933	135
Table XVIII	Total per capita Debt Service	137
Table XIX	Total per capita Interest Paid	137
Table XX	Total Net Bonded Debt per capita	137

Part II   The Second Class Township

Chapter XIII		
Table I	Summary of Assessments and Taxes for Township Purposes by Counties, 1932	199
Table II	Second Class Township Road Receipts, 1932	203
Table III	Second Class Township Expenditures by Counties 1932	205
Chapter XIV		
Table I	Highway Mileage by Governmental Divisions	212
Table II	Losch Act Townships - Analysis by Type of Road - 1933	218
Table III	Second Class Township Bridges 1932	220
Table IV	Second Class Township Purchases of Materials and Equipment by Items	222

Page

Table V   Assessed Value per Mile and Expenditures per Capita-1932   226

Part III   The First Class Township

Chapter XVI
  Table I   Diversity of Conditions   252
Chapter XVIII
  Table I   Comparison of Highway and Total Expenditures, 1933,
                  for 50 Townships   270
Chapter XX
  Table I   Comparison of Appropriations, Receipts and Expenditures   294
Chapter XXII
  Table I   Distribution of Tax Levy and Millage   317
  Table II   Tax Delinquency   327
  Table III   Summary of Receipts - 1933, for 50 Townships, Total
                  Population 378, 702.   330
  Table IV   Summary of Receipts - 1933, for 50 Townships, by
                  Population Groups   331
Chapter XXIII
  Table I   Functional Costs   336
  Table II   Summary of Functional Costs - 1933   348
  Table IIIA   Summary of Functional Costs - 1933 - For 50 Townships,
                  by Population Groups   350
  Table IIIB   Summary of Functional Costs - 1933 - For 50 Townships,
                  by Population Groups   351
  Table IIIC   Summary of Functional Costs - 1933 - For 50 Townships,
                  by Population Groups   352
  Table IIID   Summary of Functional Costs - 1933 - For 50 Townships,
                  by Population Groups   353
  Table IV   Analysis of Current Operations   354

Part IV   The Borough

Chapter XXIV
  Table I   Average Receipts of 191 Boroughs, by Population Groups,
                  1930 and 1933   370
  Table II   Per capita Receipts of 191 Boroughs, By Classes of
                  Receipts and by Population Groups, 1930 and 1933   371
  Table III   Ratio of each Class of Receipts to Total Receipts of
                  191 Boroughs, 1930 and 1933, by Population Groups   373
  Table IV   Ratio of each Class of Revenue Receipts to Total
                  Revenue Receipts of 191 Boroughs, 1930 and 1933 by
                  Population Groups   374
  Table V   Per Cent of Change in Amount of Receipts of 191 Boroughs,
                  1933 over 1930, by Classes of Receipts and Population
                  Groups   375
  Table VI   Tax Levies in Boroughs, 1932   377
  Table VII   Tax Delinquency, 1932   382
  Table VIII   Tax Delinquencies, 1933   387
  Table IX   Borough Indebtedness, 1933   389
  Table X   Expenditures of 193 Boroughs, by Population Groups,
                  1930 and 1933   394
  Table XI   Average per capita Expenditures of 193 Boroughs, by
                  Functions and by Population Groups, 1930 and 1933   395
  Table XII   Ratio of Expenditures for Specified Functions to Total
                  Expenditures for Operation and Maintenance, Including
                  Debt Service, 193 Boroughs, by Population Groups,
                  1930 and 1933   397
  Table XIII   Per Cent of Change in Amount of Expenditures of 193
                  Boroughs, 1933 over 1930, by Functions and by Popula-
                  tion Groups   399
  Table XIV   Average Compensation Paid Principal Borough Officers   400

Part V   The Third Class City

Chapter XXV   Page
    Table I   Population   )20-30 in Pennsylvania Cities of the
                   Third Class   405
    Table II   A Comparison of the Assessed Valuation Placed by the City
                   and County Assessors for the Years 1933 and 1934 on
                   Real Estate Located in Certain Third Class Cities   417
    Table III   Taxes Levied   420
    Table IV   Delinquent Taxes   427
    Table V   Receipts   431
    Table VI   Net Bonded Indebtedness at the End of 1933 in Certain
                   Third Class Cities   435
    Table VII   Total City Indebtedness as of December 31, 1933   437
    Table VIII   Total Operating Cost of Twenty-Three Third Class Cities -
                   1930 and 1933   439
    Table IX   Comparative Cost of Debt Service in Twenty-Two Third
                   Class Cities   441
    Table X   Salaries of the Principal Officials of Twenty-Seven
                   Third Class Cities   444
    Table XI   Comparative Cost of General Offices in Twenty-One
                   Third Class Cities   446
    Table XII   Comparative Cost of Police Protection in Twenty-Two
                   Third Class Cities   448
    Table XIII   Comparative Cost of Fire Protection in Twenty-Two Third
                   Class Cities   450
    Table XIV   Comparative Cost of Public Health in Twenty-One Third
                   Class Cities   453
    Table XV   Comparative Cost of Street Maintenance in Twenty-One
                   Third Class Cities   455
    Table XVI   Comparative Cost of Street Lighting in Twenty-Two Third
                   Class Cities   457
    Table XVII   Ratio of Expenditures for Specified Functions to Total
                   Operating Expenditures Including Debt Service in Twenty-
                   one Third Class Cities   459

INDEX OF CHARTS

Part I   The County

Chapter II   Page
   Chart I   Pennsylvania County Officers   14

Chapter III
   Chart II   Montgomery County   40
   Chart III   Forest County   41
   Chart IV   Proposed Form of County Organization   42

Part II   The Second Class Township

Chapter X
   Chart I   Pennsylvania Township and County Lines, 1932   145A
   Chart II   Fifty-Seven Townships in Chester County   145B
   Chart III   Township per capita Taxable Valuation 1932   146A
   Chart IV   Density of Township Population per Square Mile, 1930   147A

Chapter XI
   Chart V   Second Class Township Administration   155

Part III   The First Class Township

Chapter XVII
   Chart I   Governmental Organization, Lower Merion Township   261
   Chart II   Governmental Organization, Baldwin Township   262
   Chart III   Governmental Organization, West Lebanon Township   263

Part IV   The Borough

Chapter XXIV
   Chart I   Organization of Typical Small Borough not Divided into
                   Wards   362
   Chart II   Organization of Large Borough Divided into Wards   363

Part V   The Third Class City

Chapter XXV
   Chart I   Organization Chart - Cities of the Third Class   408

GENERAL INDEX

BIBLIOGRAPHICAL NOTE	Page
Local government in other states, reports on	10
Local government, Pennsylvania, special studies in	9-10

BOROUGH

Administration	
Budgeting and purchasing	367
Depositories	367-8
Managers	367
Personnel	367
State control of finances	368-9
Systems in general	366-7
Taxation	368
Liability of officers	368
Corporate powers	
Appropriation of public funds	361
Elastic clause in Borough Act	361
General powers	360
Miscellaneous	361
Offices, power to create	360
Prohibitory power	361
Regulatory power	361
Services, power to provide	360-1
Expenditures	
Average per capita, by function and population groups-1930 and 1933--Table	395
Comparison--1930 and 1933	398
Comparison by functions-1930 and 1933--Table	399
Functions, comparative cost	396-98
Table	397
In general	393
Mandatory expenses	393
Officers, compensation of	398
Table	400
Per capita expenditures	393-6
By population groups, 1930 and 1933--Table	394
Form of government, relative advantages of	360
Indebtedness	386
1933-Table	389-92(incl.)
Laws governing	357-9
Charters, annulment of	359
Creation	358
Limits, changes of	358-9
Multiplicity of governmental units, wastefulness caused by	359-60
Organization	
Division of authority	
Assessors (elected)	365
Auditors	365-66
Burgess	364
Controller	365-66
High constable	364-65
Tax collector	365
In general	361
Large borough-chart	363
Officers, qualification of	361-64
Small borough,-chart	362

BOROUGH   Page

    Receipts
        Average borough receipts   369
        Average receipt by population groups-1930 and 1933-Table   370
        Classes, relative importance of   372
        Financing, trends in   372
        Per capita receipts   369-72
            By classes and population groups-1930 and 1933-Table   371
        Percent of change-1933 over 1930-by classes and popula-
            tion groups-Table   375
        Ratio of each class of revenue receipts to total
            revenue receipts-1930 and 1933--Table   374
        Ratio of each class to total receipts-1930 and
            1933--Table   373
        Sources of   368-69
        Study-number of units studied and sources of material   357
        Study-purpose and scope   357
        Tax delinquency, 1932-   376-85
            Table   382-84(incl.)
        Tax delinquency, 1933   386
            Table   387-88
        Tax delinquency and total millage, relation between   385-386
        Tax levies-1932-Table   377-81(incl.)
        Taxes, per capita   376

CONTENTS, TABLE OF   III

CITY, THIRD CLASS
    Administration
        Accounting methods   412-13
        Auditing   413
        Budget (annual)   409-10
        Departments, supervision and control of   409
        Pensions   412
        Personnel, selection of   411-12
        Relations with other governmental units   415
        Reporting   413-15(incl.)
        Supplies, purchase of   410-11

    Expenditures
        Administrative offices(general)   443
            Comparative Table   446-7
        Debt service, cost of   438
            Comparative Table   441-42
        Fire protection   443
            Comparative Table   450-51
        Offices, salaries of-comparison 1930 and 1934   438-43
            Table   444-5
        Operating costs
            Functional distribution   452
                Table   459
            Total compared in certain cities   438
               Table   439-40
        Permanent improvements   460
        Police protection   443
            Comparative Table   448-9
        Public health   443-52
            Comparative Table   453-54
        Street Lighting   452
            Comparative Table   457-8
        Street maintenance-Comparative Table   455-6
        Streets and bridges-maintenance   452

CITY, THIRD CLASS   Page
    Laws governing   404
    Organization
        Chart   408
        City Council   407
        Departments-distribution of functions   407
        By offices
            Controller   406
            Director of accounts and finance   406
            Engineer   407
            Mayor   406
            Solicitor   406-7
            Treasurer   406

    Population-1920--1930-Table   405
    Powers and functions   404
    Receipts
        All sources-comparison 1930 and 1933   430-33
            Table   431-2
        Non-revenue
            Bonded indebtedness
                Centralized control, need for   436-38
                In general   433
                Outstanding at end of 1933-Table   435
                Overlapping in certain cities   434-36
                State supervision   434
            Indebtedness-total outstanding at end of 1933   436
                Table   437
            Sources of   416
        Revenue
            Assessment of property   416
                Comparision between city and county assessments   416-18
                Table   417
            Sources of   416
            Tax delinquency, extent of   426-30
                Table   427-9(incl.))
            Taxes
                Amount levied for city purposes   419
                Collection of   425-6
                Levy of   418-9
                Per capita levy for city purposes   419
                Per capita total levy   425
                Total amount levied   420-4(incl.)
            Taxes levied-Table showing date   419
    Study-Purpose and Scope   403
    Study-Number of cities covered and sources of material   403-4

COUNTY
    Accounting systems   22-3
    Assessments
        Cost of   64-6(incl.)
            Range in cost-table   66
            Range in volume and cost-Table   65-66
        Statutory provisions
            Assessing authorities   54
            Compensation   54
            Inter-triennial assessment   56
            Optional reassessments   56
            Revision and appeals   56
            Taxables   53-4
            Tax duplicates   57

COUNTY

	Page
Assessments	
Statutory provisions	
Third class counties-board of assessment and	
revision of taxes	57
Triennial assessment	54-6 (incl.)
Issuing of precepts	55
Returns of assessors	56
Valuation of occupations	55
Valuation of personal property	55-6
Valuation of real estate	55
System in operation	
Fourth to eighth class counties-	
Occupational assessment	62-3
Personal property tax	63-4
Personal property valuation-comparative table	63
Personnel	61-2
Real estate-relationship between assessed and	
true valuation	64
Records	62
Third class counties	
Application of system	58-61
Personnel and salaries	57
Auditors	16-19-20-26
Function of	26
Audits, independent	20
Budget, the	18
Chief clerk	25
Classification	
Origin and development	11-12
Present classification	11-12
Clerk of court's office, functions of	27
Clerk of Orphans' Court, functions of	27-8
Commissioners	
Origin	11
Powers	15-39-42 (incl.)
Constitutional provisions as to personnel	35-43
Controller	16-18-20
Controller's office, functions of	25-6
Coroner, functions of	27
Courts	
As administrative agencies	23-4
Powers of	23-4,39-41(incl.)
District attorney's office, functions of	27
Election Administration	
Cost	
Total cost of elections	82-4(incl.)
Table	84
Costs in election districts	79
Table	80-2(incl.)
Table showing high cost	82
Districts	
Constitutional provisions	70
Number per local subdivision	71
Population and registered voters per district	71-2
Population and registered votes - Table	72
Statutory provisions	70-1
Registration, actual operation	
In boroughs and townships	75-6
Personal and periodic registration	74
Personal and permanent registration	74-5
Suggestions for improvements	76-7

COUNTY

Election Administration | Page
Registration, cost of | 77-9(incl.)
    Table | 78-79
Registration, process of
    In boroughs and townships | 73-4
    Personal and periodic registration | 72-3
    Personal and permanent registration | 73
Engineer | 25
Expenditures
    Administrative costs-Table | 131
    Administrative offices, relative cost | 34-7(incl.)
        Comparative table of 11 items | 135
        Table showing relation to total expenditures | 135
    Charities and corrections | 133
        Table showing per capita expenditure | 133
    County buildings, maintenance of-Table | 134
    Court expenses-Table | 132
    Debt Service
        Cost, table showing total per capita | 137
        Interest, table showing total per capita | 137
    Highways | 133-4
        Table showing total cost | 134
    Joint responsibility | 15
    Law Enforcement | 132-3
        Clerical offices-Table | 133
        Law enforcement offices-Table | 132
    Mandatory expenditures | 15
    Mileage | 21-2
    Net bonded indebtedness, per capita-Table | 138
    Total, per capita-Table | 131
Fees, system of handling | 16
Finance
    Accounting systems and financial records | 22-23
    Auditors | 16-19-20
    Auditors' annual reports | 20-1
    Budget, the | 18
    Collection of taxes | 17
    Commissioners-power of appropriation | 15
    Controller | 16-20
    Controller's annual reports | 20-21
    Ex-officio boards | 17
    Fee officers | 16
    Finance and personnel | 22
    Independent Audits | 20
    Joint responsibility in expenditures | 15
    Mandatory expenditures | 15
    Modern methods, lack of | 17-23(incl.)
    Purchasing of supplies | 18-19
    Responsibility, lack of | 15-17(incl.)
    Summary | 17
    System of handling fees | 16
    Treasurer | 16
Government, history and development | 11-12
Jury commissioners, functions of | 28
Justices of the Peace, Aldermen and Constables
    Constitutional and statutory provisions-election | 87-88
    Cost to County
        Cost in 12 counties-Table | 92
        Payment by county | 91-2
    Court costs
        Analysis | 100-101
        Assessment | 101

COUNTY

Justices of the Peace, Aldermen and Constables   Page
    Court costs
        Computation and collection
            In general   101-2
            Adams county   109
            Centre county   109
            Chester county   108
            Erie county   107-8
            Lackawanna county   102-3
            Monroe county   109-10
            Montgomery county   107
            Northumberland county   108-9
            Union county   110-11
            Westmoreland county   103-6 (incl.)
        Statutory regulation   88-101
    Fees and compensation   88-90(incl.)
    General data
        Adams county   100
        Centre county   97-100(incl.)
        Erie county   94-97(incl.)
        Forest county   100
        Lackawanna county   92-4(incl.)
        Monroe county   100
        Montgomery county   94
        Northumberland county   77
        Union county   100
        Westmoreland County   94
    Justices
        Number by counties-Table   90
        Number in counties   90
    Statutory provisions-court costs   88-101
Mileage   21-22
Officers
    Fee officers   16
    Functions of   24-8(incl.)
Offices
    Elective   12-13
    Structural arrangement of-chart   14
Origin of   11
Pension system   42-43
Personnel
    Appointed officers and employees   39-42(incl.)
    Appointment, chart showing operation of
        Forest County   41
        Montgomery County   40
        Tioga County   39
    Constitutional provision   35-43(incl.)
    Cost of personnel, ratio to total expenditures-Table 49-50
    County commissioners, appointment by   39-42(incl.)
    County employees, actual turnover of   49
    Departmental heads, necessity of   50
    Elected officers, differences among counties in
        respect to   36-7
    Employee and salary data for 12 offices in 12
        counties-Table   45-6
    Employees-number,population per employee and salary
        data-Table   44
    Head of department, appointment by   39-42(incl.)
    Judges of district court, appointment by   39-41(incl.)
    Number of employees   37-8
    Organization, charts showing proposed forms of
        With county Manager   42
        Without county manager   41

COUNTY

    Personnel   Page

	Page
Personnel	
Pension System	42-3
Receipts of county fee offices-Table	47
Salaries of officers and employees	37-9(incl.)
Salary board in operation	47-9(incl.)
Solicitors	25-49
Statutory provisions	36-43(incl.)
Personnel, finance and	22
Prothonotary's office, functions of	27
Receipts of county fee offices-Table	47
Recorder of deeds office, functions of	27
Records	
Board of county commissioner's office	116-17
Clerk of quarter sessions and oyer and terminer	119-20
Clerk of orphan's court	121-22
Controller's office	117-18
Prothonotary's office	122-25(incl.)
Recorder of deeds office	120
Register of wills office	121
Treasurer's office	118-9
Records, financial	22-3
Register of wills office	27-8
Reports	
Auditors' annual	20-1
Controllers' annual	20-1
Revenues	
Court costs and fines-Table	129
Departmental earnings-Table	130
Institutional earnings-Table	130-1
Liquor revenue-Table	128
Per capita revenue-Table	127-28
Taxes	
Gasoline-Table	129
Property, poll and occupational-Table	128
Salaries	
Data for 12 offices in 12 counties-Table	45-6
Fees	38-9
Of employees-Table showing data	44
Of officers and employees	37-39(incl.)
Regulation by head of department or county court	38
Salary board	37-47-49(incl.)
Statutory regulation	37
Sheriff's office, functions of	26-7
Solicitors	25-49
State forest in counties	149
Statutory provisions as to personnel	36-43(incl.)
Supplies, purchasing of	18-19
Taxes, collection of	17
Treasurer	16
Treasurer's office, functions of-	26

INDEX OF CHARTS   468

INDEX OF TABLES   465-7(incl.)

MAJOR CONCLUSIONS, SUMMARY

	Page
Constitutional Changes-Summary	9
Fiscal Problems-	
Accounting, recording and reporting	5-6
Assessment	3-4
Auditing	6

MAJOR CONCLUSIONS, SUMMARY   Page

    Fiscal Problems-
        Budgeting   4-5
        Debt   5
        Tax Collection   4
        Tax levy   4
    Internal Organization
        Changes recommended
            Appointive offices, increase of   1-2
            Elective offices, reduction of   1-2
            Functional organization   2
            Responsibility   2
    Local Units-Relation to State and to Each Other
        State Supervision   7-8
        State Survey   8
    Local Units-Special Recommendations
        Borough   8-9
        City, third class   9
        County   8
        Township, first class   8
        Township, second class   8
    Particular Functions, Special Recommendations
        Elections   6
        Local highways   7
        Local Justice, administration of   6
        Welfare   7
    Personnel
        Merit System, the   3
        Training, guidance and supervision   2-3
    Recommendations-Spirit and Scope   1

PREFACE

    Acknowledgements   II
    Questions presented   I
    Survey-general method   I

TOWNSHIP, FIRST CLASS

    Accounting procedure
        Accounting system, purpose of   277
        Accounts, types of   278-9
        Books, types of   281-2
        Expenditure accounting   280-2(incl.)
        Functional Costs   282-3
        Practices, Keeping of accounts   278
        Reporting, uniform,
            Legislation   283-4
            Reporting form   284
            Reports   284-5
        Revenue accounting   279-80
    Administration
        Agreements, inter-municipal   273
        Fire protection   272
        General, Control of   267-8
        Health   271
        Highway   268-71(incl.)
            Expenditures in relation to total expenditures,
                Table showing   270
            Road Force   269
            Road Supervisor   268-9

TOWNSHIP, FIRST CLASS

	Page
Administration	
Highway	
Secretary or manager	269
Ward Commissioner	268
Miscellaneous Services	272-3
Police	271-2
Sewer	271
Zoning	272
Administrative responsibility	
Board of Commissioners	265-6
Auditing Procedure	
Auditing authority, the	305-6
Controller	305-6
Court appointed auditor	305-6
Elective auditors	305-6
Auditing forms	309-11(incl.)
Personnel--under elective system	306-7
Practices	
Statutory provisions	307
Under elective system	307-8
Statements or reports	
Publication and filing of	308-9
Statutory provisions	308-9
Budgetary procedure	
Budget	
Preparation and enactment by commissioners	290
Preparation by finance committee	291
Preparation by other groups	291
Preparation by secretary	291
Budget system, elements of	289
Budgetary Control	300-3(incl.)
Adherance to budget plan	301-3(incl.)
Method of maintaining	300-1
Table illustrating	302
Budget Ordinance	298-300(incl.)
Publication	298-300(incl.)
Estimates	
Adjustment to available data	297-8
Collection and preparation of	289-90
Expenditure-comparision of appropriations,receipts and expenditures,-Table	294-5
Revenue	291-3(incl.)
Comparative table	292
Over-appropriation	
In relation to expenditures	296-7
In relation to receipts	296
Under-appropriation	297
Classification	
Historical summary	243-4
Purpose	243
Statutory provisions	244-5
Commissioners, powers	265-6
Committees, (standing)	266
Corporate Powers	
Appropriation for specific purposes	247
Blanket authority	247-8
Mandatory and optional	245
Miscellaneous	247
Offices, power to create	245-6
Police powers, regulatory and/or	246-7
Primary powers	246

TOWNSHIP, FIRST CLASS

Page

Corporate Powers
    Public Utilities and other enterprises,-erection,
        maintenance, establishment and purchase   247
    Regulatory and/or police powers   246-7
Diversity
    Comparative table   252-3
    Nature and extent
        Future trend, reflected by tendency to cluster
          around metropolitan centers   249-50
        Population, density   248
        Population, range of   248
        Tax rate, variations in   248-9
Expenditures
    Highway in relation to total,-Table   270
Functional Costs
    Current operations
        Analysis   349-56
        Table   354-5
    Expenditures-mandatory   335-49
    Expenditures of individual townships   335
    Functions-relative costs   335
    Summary-Table   348
    Summary-by population groups-Table   350-3(incl.)
    Table   336-47(incl.)
    Variation with population changes   349
Internal Organization
    Appointive offices
        Auditor, court appointed   257
        Board of health   257
        Boards,-park, recreation and adjustment   257-8
        Manager   258-9
        Miscellaneous   258-9
        Secretary   257
        Solicitor and engineer   257
    Chart
        Baldwin Twp.   262
        Lower Merion Twp.   261
        West Lebanon Twp.   263
    Diversity   255-6
    Elective offices
        Auditors   256
        Commissioners   256
        Controller   256
        Miscellaneous   257
        Treasurer-tax collector   256
    Summary   259-60
Legislative responsibility
    Board of Commissioners   **265-6**
Organization   **266-7**
Receipts
    Summary   329
    Table Showing Summary   330
    Table showing summary by population groups   331-2
Revenues
    Tax delinquency   326-29
        Table   327-28
        Table showing growth-1930--1933   326
    Tax exonerations   325-6
    Tax levy and millage-table showing distribution   317-23(incl.)

TOWNSHIP, FIRST CLASS

    Revenues   Page
        Taxes, collection of
            County duplicate   325
            Installment payments   325
            Procedure   324-5
            Statutory provisions   324-5
        Taxes, levy of   315-16-24
            Statutory provisions   315-16
    Statutory Provisions
        Administrative powers   265-6
        Legislative powers   265
    Wards, division into   267

TOWNSHIP, SECOND CLASS

    Administration
        Corporate powers-illustration of excercise   153-4
        Corporate powers of supervisors   153
            Statutory provisions   153
        Relations between townships   162-3
        Relations between township and county   163
        Requirements of modern conditions   163
        Township officers
            Assessors   159
            Auditors   159
            Bond requirements   154
            Compensation   156-7
            Elective and appointive   154-5
                Chart   155
            Engineer   161-2
            In general   154
            Number of   156
            Police   162
            Road districts and road masters   161
            Road master and superintendent, duties of   161
            Secretary, duties of   160-1
            Secretary-treasurer   160
                Compensation of   160
            Solicitor   162
            Supervisors,
                Compensation of   157
                Corporate powers of   153
                Duties of   157-8
                Financial authority of   158-9
                As road masters   161
            Tax collector   159
                Compensation of   159
            Treasurer, duties of   160
    Annual Report
        Errors by township officers   169-70
        To State Highway Department   167-239-40
    Assessments and Levies
        1932--Table showing summary by counties   199-202(incl.)
    Audit Procedure
        Audit of
            Contracts and purchases   181
            Disbursements   180-1
            Receipts   180
            Surety Bonds   181

TOWNSHIP, SECOND CLASS
                                                                          Page
Budget
    Appeals by taxpayers   170-1
    Procedure   170
Classification-based on density of population   150
Consolidations
    Grounds for   237
    Re-organization of township government   237-8
    Of townships   236
Debt
    Bonded debt
        By counties, 1932   234
        Relation to other local governments   234
    Causes of   233
    Defaults   235-6
    Per mile   233-4
    Total debt   235
    Townships having none   233
    Unfunded debt   234-5
Diversity
    Economic conditions   146
    Population
        Classification based on density   150
        Density   147
        Largest townships in reference to   149-50
        Range   147-8
        Small townships in state forests   148-9
        Trends and distribution   147
    State forests
        Small townships in   148-9
    State forest in townships   149
    Types   145-6
Expenditures
    By counties-Table   205-10(incl.)
    Earth roads, maintenance of   189-90
    Miscellaneous road expenditures   191
    Road expenditures, classification of   188-9
    Roads, permanent improvement of   190
    Road masters' wages   190-1
    1932--by townships   187-8
Fiscal Problems
    Accounting and reporting   167
    Annual report
        Errors by township officers   169-70
        To State Highway Department   167-239-40
    Annual settlement-by collector   178
    Assessments
        Percent of true value-differences among townships   173-4
        Should they be related to income value   176
    Audit procedure
        Audit of contracts and purchases   181
        Audit of disbursements   180-1
        Audit or receipts   180
        Audit of surety bonds   181
    Auditor--is qualified accountant required   179-80
    Budget appeals by tax payers   170-1
    Budget procedure   170
    Expenditures
        Earth roads, maintenance of   189-90
        Road expenditures, classification of   188-9
        Road expenditures, miscellaneous   191

TOWNSHIP, SECOND CLASS

		Page
Fiscal Problems		
Expenditures		
Roads, permanent improvement of		190
Road Masters' wages		190-1
1932--by townships		187-8
Fiscal Year		'171
Levies-mandamus by court in, special cases		170
Mandatory expenditures		171-2
Real property tax		
Relatively heavier in rural areas		175
Should there be a limit		176-7
Receipts		
Cash road tax		192
County aid		194
Applications for		194
Forest road aid		193-4
In General		191
Other sources		194-5
State reward		192-3
Township roads		
Alternatives to increasing revenue for		196-7
Proposed state sources of revenue for		195-6
Records, minutes and		167-9(incl.)
Tax collection,-what constitutes good procedure		177-8
Tax collector's commission		178-9
Tax delinquency		
Causes of		175-6
1932--by townships		174
Tax levies		
Comparison based on 1930-trend from 1926 to 1932		185-6
Comparison with other local units-1932		185-6
Per capita by townships-1932		186-7
Procedure in collecting		177
Tax procedure		170
Tax rate, differences in		172-3
Tax Studies on various aspects		175
Mandatory Expenditures		171-2
Origins and Legal Basis		
Constitution of 1873		140
Constitutional revision, reasons for		140-2(incl.)
Constitutional Status undesirable		141-2
Debt limitation and retirement		141
Special legislation		141
Voting machines		141
Creation of townships		142
Early township planning		139
Origin in Pennsylvania		139-40
Second Class township code, subjects excluded from		142
Statutory basis-early		142
Losch Act		143
Present		143
Receipts		
Cash road tax		192
County aid		194
Forest road aid		193-4
In general		191
Other sources		194-5
Road receipts 1932-Table		203-4
State reward		192-3
Roads and Bridges		
Bids-competitive not required		223-4

TOWNSHIP, SECOND CLASS

     Roads and Bridges   Page

	Page
Roads and Bridges	
Contracts, interest in	
Engineers and architects-prohibited	224
Supervisors and road masters-prohibited	224
Materials and equipment, purchases	
Approval of Highway Department	219
By items-Table	222
Over $200	223-39
Parkinson Act	
Allocation	
How estimated	225-8
Where 1934 allocation exceeded 1932 maintenance cost	229
Effect on township tax levies	225
Maintenance costs-where 1934 allocation exceeded by	228-9
Mileage basis for maintenance allocation	228
Revision of mileage basis	229
Public Highway program	230
Public roads,	
Mileage in Pennsylvania	211-13(incl.)
Mileage in Pennsylvania-Table	212-3
Records	224
Supervisors may close temporarily	224
Public works-contracts over $500	223
Roads contracts, with private corporations	224-5
Roads, vacation of	216
State assistance,-conditions of	230-1
Township bridges, 1932	217-19
Table	220-1
Township highway system-1923 and 1932 compared	225
Township road machinery	223
Township roads	
Assessed value per mile and expenditures per capita-Table	226-7
Classification	216
Differences in total mileage	214
In "Fair" condition	215
In Losch Act townships	217
Analysis-Table	218
Mainly unimproved	214
Miles in counties-average per township	216
Range of mileage	215-16
Readjustment of mileage under State control	215
System divided into districts;road masters' wages	216-17
Workmen's compensation	224
State Financial Control	
Annual audit, publication	239
Annual report, filing with Highway Department	167-239-40
Indebtedness, control over	
Amount of issue	238
Bonds, methods of selling	239
Interest rates	239
State supervision	238-9
Tax levies for indebtedness	239
Term of Bond	239
Materials and equipment, purchases over $200	223-39
Tax Levy, maximum	238
Survey, methods employed	139
Tax Delinquency	
Causes of	175-6
1932-by townships	174

TOWNSHIP, SECOND CLASS

    Tax Levies   Page
        Comparison based on 1930-trend from 1926 to 1932   185-6
        Comparison with other local units-1932   185-6
        Levies, assessments and-1932-Table showing
            Summary by counties   199-202(incl.)
        Per capita by townships-1932   186-7

www.ingramcontent.com/pod-product-compliance
Lightning Source LLC
Chambersburg PA
CBHW081340190326
41458CB00018B/6056